Graduate Texts in Physics

Graduate Texts in Physics

Graduate Texts in Physics publishes core learning/teaching material for graduate-and advanced-level undergraduate courses on topics of current and emerging fields within physics, both pure and applied. These textbooks serve students at the MS- or PhD-level and their instructors as comprehensive sources of principles, definitions, derivations, experiments and applications (as relevant) for their mastery and teaching, respectively. International in scope and relevance, the textbooks correspond to course syllabi sufficiently to serve as required reading. Their didactic style, comprehensiveness and coverage offundamental material also make them suitable as introductions or references for scientists entering, or requiring timely knowledge of, a research field.

More information about this series at http://www.springer.com/series/8431

Edward Morse

Nuclear Fusion

 Springer

Edward Morse
Department of Nuclear Engineering
University of California, Berkeley
Berkeley, CA, USA

ISSN 1868-4513   ISSN 1868-4521   (electronic)
Graduate Texts in Physics
ISBN 978-3-030-07462-3   ISBN 978-3-319-98171-0   (eBook)
https://doi.org/10.1007/978-3-319-98171-0

This Springer imprint is published by the registered company Springer Nature Switzerland AG
The registered company address is: Gewerbestrasse 11, 6330 Cham, Switzerland

To my aunt, Jane R. Morse

Preface

This book is the result of teaching three fusion-related courses at Berkeley over the past 40 years. The topics in fusion and plasma physics are constantly evolving, however, and it is my hope that the subject matter ages well. I have attempted to give a historical perspective where possible, with the realization that some of the experiments and technology described here might only be in the historical archive as well someday. There are some concepts that are only briefly described, as they are only currently being pursued at a low level. These include a good many alternative magnetic confinement concepts. I personally worked on alternative magnetic confinement concepts for almost 20 years and have fond thoughts about spheromaks in particular. But I wanted to write about current concepts and projects where students can become involved. I showed the table of contents of this book to Ken Fowler, my colleague and former department chair, who came to Berkeley after running the magnetic fusion program at Livermore for decades before. He chided me by repeating two sentences from Edward Teller's 1981 book on fusion: "It has only 16 chapters. I cannot exclude the possibility that had we written 20 chapters, the additional four might have contained the final solution." This book has only twelve chapters, so perhaps even more has been left out.

The level of this text covers a wide range, with some material being immediately within the grasp of undergraduates in their junior or senior year with adequate knowledge of electromagnetic theory in vector form. (I usually ask them "do you know what the upside-down triangles stand for?") Some of the concepts of quantum mechanics appear in discussions on the fusion reactions, radiation losses, and superconductivity, but are not central to the applications to the fusion-relevant material. Engineering students have typically seen the Navier-Stokes equation, and the discussion of magnetohydrodynamics might follow more easily for them than others. Special functions such as the hypergeometric functions appear in places, but only as solutions to an equation and never with any knowledge of these required. (Other than perhaps finding the function's name in *Mathematica* or *Matlab*.) I would encourage instructors to give exercises around things like calculating MHD equilibria using Whittaker functions using these code packages, as calculating one's own tokamak equilibrium gives a strong feeling of accomplishment. Of the three courses mentioned earlier, two are graduate level, and it is my hope that this book gives enough depth to get started in fusion-related research. There are many references included for each chapter, and instructors should stress that students refer to original sources whenever possible. (This is easy to do these days with most journals being available online: DOI numbers and URLs are given where possible.) The book uses mostly SI units, but other units are unavoidable in plasma physics. Inertial confinement concepts usually use cgs units, for example, and temperatures T are universally quoted as what would technically be called kT, i.e. temperatures are in energy units such as eV or keV. The popular *NRL Plasma Formulary*, available online or with free hardcopies by request, is a good companion to this book, but instructors should caution their students on the disparate units used there in some places (eV for temperature, cgs Gaussian units for almost everything else).

I must acknowledge many people's assistance in preparing this manuscript. Firstly, there is my editor at Springer, Denise Penrose, who has encouraged me in this project at every turn. Tom Dolan at the University of Illinois also has contributed by reviewing every chapter. I have used his 1982 textbook in my own courses for decades and have a great respect for his comments and enthusiasm. Ralph Moir at Livermore reviewed some of the early chapters and has been particularly outspoken on instilling the right kind of guarded optimism to the younger generation regarding the subject matter. Charles Cerjan at Livermore did a very thorough review of the chapter on inertial confinement, and Peter Seidl also provided input, especially on the heavy ion program. Lisa Zemelman has done a thorough edit of the entire book for grammatical and syntax errors. My students have used draft sections of the book as reading material for the fusion courses at Berkeley and deserve credit for detecting errors. I am grateful to Springer for giving me the opportunity to publish this work and for providing pre-production assistance.

Berkeley, CA, USA   Edward Morse
June 2018

Contents

About the Textbook

The pursuit of nuclear fusion as an energy source requires a broad knowledge of several disciplines. These include plasma physics, atomic physics, electromagnetics, materials science, computational modeling, superconducting magnet technology, accelerators, lasers, and health physics. "Nuclear Fusion" distills and combines these disparate subjects to create a concise and coherent foundation to both fusion science and technology. It examines all aspects of physics and technology underlying the major magnetic and inertial confinement approaches to developing nuclear fusion energy. It further chronicles latest developments in the field and reflects the multi-faceted nature of fusion research, preparing advanced undergraduate and graduate students in physics and engineering to launch into successful and diverse fusion-related research.

"Nuclear Fusion" reflects Dr. Morse's research in both magnetic and inertial confinement fusion, working with the world's top laboratories, and embodies his extensive 40-year career in teaching three courses in fusion plasma physics and fusion technology at University of California, Berkeley.

About the Author

Dr. Edward Morse is Professor of Nuclear Engineering at the University of California, Berkeley, where for over 35 years he has taught the department's three senior undergraduate and graduate courses on fusion, plasma physics, and fusion technology. He has authored over 140 publications in the areas of plasma physics, mathematics, fusion technology, lasers, microwave sources, neutron imaging, plasma diagnostics, and homeland security applications. For several years he operated the largest fusion neutron source in the USA. Frequently consulted by the media to explain the underlying science and technology of nuclear energy policy and events, Dr. Morse is also a consultant and expert witness in applications of fusion neutrons to oil exploration.

Acronyms

ADCM Asymmetric Deflection Compensation Magnets. 253
ALARA "as low as reasonably achievable" (radiation safety). 490
Alcator "Alto Campo Toro" (tokamak at MIT). 11
Argus early Livermore laser fusion experiment. 15
ARIES Advanced Reactor Innovation and Evaluation Study. 469

Baseball mirror machine at Livermore. 9

CBET crossed-beam energy transfer. 349, 365
CESM Co-Extracted Electron Suppression Magnets. 253
CFR Code of Federal Regulations (US). 490
COL combined operating license. 489
CPP continuous phase plate. 350
Cyclops early Livermore laser fusion experiment. 15

DBTT ductile-to-brittle transition temperature. 430
DCA direct configuration accounting. 93
DKDP deuterated potassium dihydrogen phosphate. 399
DOE Department of Energy (US). xiv, 17, 492
Doublet tokamak at General Atomics, San diego. 11

ECA export credit agency. 508
ECH electron cyclotron heating. 257
ECH&CD electron cyclotron heating and current drive. 316, 319, 322, 323
EG extraction grid. 253
EIS environmental impact statement. 489
ELM Edge localized mode. xi, 170, 171
EM electromagnetic. 362
EOS equation of state. 385
EPA Environmental Protection Agency (US) (US). 489, 508
ERATO finite element MHD stability code. 145
EURATOM European Atomic Energy Community. 13

FOA final optical assembly. 350

Gandalf code to model heat transients in superconducting filaments. 456
GEKKO laser in Japan. 18
GSI Gesellschaft für Schwerionenforschung (heavy ion research center). 404

HCX High Current Experiment. 405
HHFW high harmonic fast wave. 331
HNB heating neutral beam. 251

IBT same as CBET. 365
ICF inertial confinement fusion. 347
ICH&CD ion cyclotron heating and current drive. 295, 300
ICRF ion cyclotron resonance (or range of) frequency. 257
ICRP International Commission on Radiological Protection. 490, 509
IDC interest during construction. 484
IRR internal rate of return. 485
ISS04 International Stellarator Database. 232
ITER International Thermonuclear Experimental Reactor. 13, 483

Janus early Livermore laser fusion experiment. 15
JET Joint European Torus. 12

KDP potassium dihydrogen phosphate. 16, 399

LASNEX Livermor code for laser fusion dynamics simulation. 94
LBNL Lawrence Berkeley National Laboratory. 405
LHH&CD lower hybrid heating and current drive. 310
LLE Laboratory for Laser Energetics. 16, 18
LMJ Laser Mégajoule, in France. 18
LPI laser-plasma interactions. 347
LTE local thermodynamic equilibrium. 91

MAMuG multi-aperture, multi-grid. 251
MAST Mega Ampere Spherical Tokamak. 331
MD major disruption. 416
MHD magnetohtydrodynamics. x, 111
MIG magnetron injection gun. 328

NDCX Neutralized Drift Compression Experiment. 405
NIF National Ignition Facility. 18
NLTE non-local thermodynamic equilibrium. 93
Nova laser fusion experiment at Livermore. 16
NPP nuclear power plant. 482
NPV net present value. 484
NRC Nuclear Regulatory Commission (US). 489, 510
NSTX National Spherical Tokamak Experimen. 331
NTLF National Tritium Labeling Facility. 491, 510

NTM neoclassical tearing modes. 169
NUREG Nuclear Regulatory Commission Regulation. 494

ODS oxide dispersion strengthening. 438
OECD Organisation for Economic Co-operation. 1, 488
Omega laser at LLE. 18

PAG preferential absorption glide. 435
PEPC resulting plasma electrode Pockels cell. 402
PIES Princeton iterative equilibrium solver. 220, 221
PKA primary knock-on atom. 423
PLT Princeton Large Torus. 12

QEOS "quotidian" equation of state. 386, 396
Q factor same as RBE. 490

RAFM reduced activation ferritic/ martensitic steels. 438
RBE relative biological effectiveness (multiplication factor for dose relative to X-
   ray exposure). 490
REBCO rare earth barium copper oxide. 453
Rem radiation-equivalent-man. 490
RF radio-frequency. 510
RTNS Rotating Target Neutron Source. 491

SBS stimulated Brillouin scattering. xii, 348, 364
Scylla theta-pinch at Los Alamos. 9
Shiva early Livermore laser fusion experiment. 15
SIPA stress-induced preferential absorption. 435
SIPN stress-induced preferential nucleation of loops. 435
SLACCAD SLAC code for accelerator design. 252
SRS stimulated Raman scattering. xii, 15, 348, 362

T-3 early Russian tokamak. 10
TBR tritium breeding ratio. 468
TE transverse electric. 312
TEM transverse electric and magnetic. 310
TF (1) toroidal field (magnetic confinement), (2) Thomas-Fermi model (inertial
   confinement). 390
TFR Tokamak Fontenay-aux-Roses. 11
TFTR Tokamak Fusion Test Reactor. 12, 239
tKD transmission Kikuchi diffraction. 441
TM transverse magnetic. 312
TMT thermomechanical treatment. 438
TTF Frascati Turbulent Tokamak. 12

VDE vertical displacement event. 417

WKB Wentzel-Kramers-Brillouin. 263, 288

WPPS Washington Public Power Supply System. 481

ZETA Zero Energy Thermonuclear Assembly. 7

Chapter 1
Fundamental Concepts

1.1 The Promise of Fusion Energy

A key feature of the twentieth century had been an insatiable demand for energy. The quest for energy has created economic busts and booms, pollution, health problems, and even war. A good fraction of the energy now used is from unsustainable sources. The energy demand worldwide is expected to increase greatly in the coming decades, especially from the developing nations, i.e. the non-OECD sector (see Fig. 1.1). While more difficult to predict, the later decades of the twenty-first century will likely show an even greater demand for energy, which correlates with the standard of living. How this increased demand for energy is handled is still uncertain. Clearly the development of an energy source that is safe, clean, sustainable, and non-intrusive would be welcomed by most people.

Nuclear fusion, the process by which certain light atomic nuclei react with each other to release energy, is a bright candidate for serving mankind's energy needs. The fuel supplies are abundant: deuterium, for example, represents an inventory of $1.38 \cdot 10^{43}$ atoms in seawater. If these deuterium atoms were all consumed in nuclear fusion reactions (using the "catalyzed-DD" process, as will be outlined later), about 1.6×10^{31} J would be released. At current energy consumption levels ($6.3 \cdot 10^{20}$ J per year, according to an International Energy Agency analysis for 2018) this deuterium supply would last for 25 billion years: longer than estimates of how long it will be until the sun becomes a red giant and subsumes the earth [29].

And now to the safety question. The reaction chain given above consumes its own radioactive by-product (tritium), but it also produces neutrons. These neutrons will generally leave the area where the fusion reactions are happening and deposit their energy in a "blanket" where the kinetic energy of the neutrons will be converted to thermal energy. The structural materials composing this blanket may interact with these neutrons, however, producing radioactive materials in the blanket. In normal operation, the blanket materials will, upon de-commissioning of the reactor, be set aside to "cool off" for a while and then eventually buried or stored as low-level

© Springer Nature Switzerland AG 2018
E. Morse, *Nuclear Fusion*, Graduate Texts in Physics,
https://doi.org/10.1007/978-3-319-98171-0_1

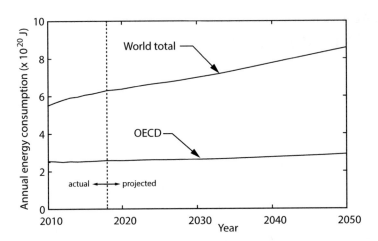

Fig. 1.1 Energy use worldwide, with projections for the next three decades. Source: US Energy Information Administration

waste. The radioactivity levels are much lower than the radioactivity associated with the fission power cycle. What about a fire? Some radioactivity might be released in this case, but studies have shown that with intelligent choices for the structural materials, little radioactivity would be released, and not enough to give a bystander outside the perimeter fence for the reactor a life-threatening dose. As a general rule-of-thumb, about 10^5 times less radiological risk is present in a fusion reactor as opposed to a fission reactor [6]. Also, nuclear runaway is not possible in a fusion reactor, as the amount of fuel in the reactor at any one time is only enough for about 100 s of operation (for the magnetic concepts) or a few second's worth (inertial fusion).

An additional safety-related question is the risk posed by fusion on nuclear weapon proliferation. This can be approached from several vantage points. First, there is the question of direct diversion of fusion fuel such as deuterium and tritium (and, for that matter, lithium) from a fusion reactor to either a state-supported weapon program or a terrorist organization. But since no "pure fusion" weapons exist, nor do most experts believe that they are possible, emphasis is placed on the security of fission weapon relevant material such as ^{233}U, ^{235}U, ^{237}Np, and ^{239}Pu and not on the light elements involved with fusion. Furthermore, deuterium and lithium are naturally occurring and available through ordinary commercial channels, and tritium can be produced in fission nuclear reactors. From another perspective, could weapon-usable material be bred in a fusion reactor? The answer is yes, but with difficulty. Reference [16] shows some possible ways to do this. The key point is that the design of a fusion reactor makes it far less desirable for weapon-grade material production than, say, a fission research-type reactor in the 10–100 MW power range.

What about intrusiveness? Fusion reactors are a concentrated source of energy, not unlike fission power from this perspective. The footprint of a fusion reactor facility is much smaller than, say, wind farms or solar panel arrays. Siting the fusion power plants near populated areas has far less risk than with fission plants, and so siting issues are more about public opinion than public health issues or emergency scenarios.

One question is always on people's minds when it comes to talking about fusion. Why is it taking so long? The answer is that early enthusiasm in the public media created a picture that fusion power was going to be available in just a few years after serious experimentation started in the 1950s. No one at that time knew the complexity of the physics that would unfold, nor the enormous technological challenges which would come up during the active phase of experimental development in the 1960s and 1970s. And four decades later, we do not have a working reactor, but we have a clearer plan of what it will take and what it will cost. The "crash program" mentality of the seventies has given way to the "cathedral builder" mentality, referring to the cathedrals of the middle ages being built over many generations. Hopefully this text will be useful to those willing to pick up the trowel.

In this age when every product carries some sort of warning, a warning should be given for this book. There is no certain outcome for the future of fusion power, as it is still in the research phase. The generation of a successful "grid-ready" fusion device has not happened yet. Furthermore, the end result might not be economical even if it all works out. (For a well-written criticism of fusion by an insider in the field, Lidsky's 1983 paper entitled "The Trouble with Fusion" [22] is a must-read.) Fusion has been "oversold" from time to time, to governments and to private investors. The purpose here is not to sell a prospectus, but to describe a great technological challenge, even if, as critics might say, "... it always will be."

1.2 History of Fusion Research

By 1920 the possibility of reactions among the light nuclei was proposed byEddington [13]. The idea that nuclear reactions could take place between two positively charged nuclei at energies far lower than that required by classical physics to allow the nuclei to touch was postulated with the development of quantum tunnelling theory by Friedrich Hund in 1926 and was turned into a quantitative theory by George Gamow in 1928 [15]. During the next decade, an American chemist, Harold Urey, discovered deuterium in 1931 while working at Johns Hopkins University, and the discovery of the neutron was made in 1932 by James Chadwick at the Cavendish Laboratory in Cambridge University. The other fusion-relevant nuclei soon followed: An Australian-born physicist, Mark Oliphant, discovered ^{3}He and tritium in 1932, also at the Cavendish Laboratory. In that same year, Oliphant and coworkers observed nuclear reactions among the light nuclei. Hans Bethe wrote a review article in 1936 which elucidated the mass-energy relations between the light nuclei [3].

1.2.1 Magnetic Confinement Experiments

The first fusion experiment was performed by Arthur Kantrowitz and Paul Jacobs in 1938 at the Langley Field Laboratory of National Advisory Committee for Aeronautics (NACA), the predecessor of NASA [19]. This experiment used an aluminum torus about four feet in diameter and with an eighteen-inch bore. It was wound with copper cables which were connected to a power supply for the wind tunnel at Langley, which could output several hundred kilowatts. The power injection for the plasma was provided by a 150-W radio transmitter. Diagnostics were fairly simple-minded: Dental X-ray films were placed near the torus on the theory that bremsstrahlung from the plasma would be detectable at their "predicted" temperature of one kilo-electron volt. This experimental work was shut down after less than a year, primarily because it was fairly off-mission for an aeronautical laboratory.

World War II caused most efforts in experimental plasma physics to stop. However, the Manhattan Project introduced many physicists to the idea of using nuclear fusion to enhance the yield of nuclear weapons, led by Edward Teller. The new bomb was called the "Super," but its design was not very advanced by the end of the war, and the director of the laboratory at Los Alamos, J. Robert Oppenheimer, did not want to pursue this option after the war ended with the detonation of two fission-based nuclear weapons in Japan. But talk of potential fusion power reactors was frequently part of the discussion at that time. After the war, however, many new projects emerged based upon some ideas that had been formulated before the war. One of these was the "pinch effect."

The possibility of using the pinch effect to produce fusion plasma conditions was proposed by Bennett in 1934 [2, 10], and early experiments along these lines started after WWII [17]. The pinch effect uses the self-generated magnetic field of a current-carrying plasma to provide magnetic confinement. Figure 1.2 illustrates the basic concept. A Briton, James Tuck, was sent to Los Alamos during the war to be a part of the Manhattan Project, and returned home in 1946. There he found that the pinch idea resonated with other physicists at the Clarendon Laboratory at Oxford University, notably George Thomson and Moses Blackman [9]. In fact, Thomson and Blackman filed a patent for a pinch-type device in 1946. By 1947, both the linear pinch (called the z-pinch) and a toroidal version of the same current-carrying plasma were built at Oxford, as Ph. D. thesis projects for A. A. Ware and S. W. Cousins. While Tuck received money to build a version of the pinch as a neutron source rather than a reactor, he was given an offer to return to Los Alamos, which he accepted, and the machine in England was never built.

The year 1951 was a pivotal year in fusion research. Most notably in the press was a story about a fusion project in Argentina during the regime of Juan Peron. Peron had funded a German scientist named Ronald Richter to work on a fusion scheme on a remote island laboratory. On March 25 of this year, Peron announced that there had been a "complete success" at achieving nuclear fusion. The secrecy surrounding the project allowed for the spurious (if not fraudulent) report, but it

Fig. 1.2 The Z-pinch
concept

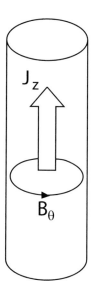

did excite the international media to ask what the American and British laboratories
were doing about developing nuclear fusion. One organization was fairly outspoken
about the need for a fusion research program was Princeton University, through
a collaboration of Lyman Spitzer, Jr. and John A. Wheeler. Wheeler had visited
Los Alamos in 1950 under the aegis of Project Matterhorn, Princeton's arm of the
Manhattan Project. The research on thermonuclear weapons (Teller's "Super") had
picked up again following the first Soviet nuclear test in 1949, and some of the
research work for Los Alamos was conducted on the Princeton campus. Some of
the plasma concepts emerging from the weapons work had some applicability to the
idea of a controlled thermonuclear reactor. Spitzer contemplated how to confine
the plasma for a fusion reactor and decided that a magnetic field encircling an
evacuated toroidal-type structure would be the best solution. However, Spitzer was
aware of a conundrum, first introduced by Enrico Fermi, that shows that the single-
particle orbits of electrons and ions drift in opposite directions in a toroidal magnetic
field, causing a charge separation and ultimately an electric field that would cause
both species to be carried out radially (see Sects. 3.4.4 and 3.4.5 for details).
Spitzer planned to remedy this effect by twisting the torus into a figure-eight shape.
Spitzer also noted the need for tritium in order to get the $100\times$ higher reactivity in
deuterium-tritium plasmas over purely deuterium plasmas. Given the lack of natural
tritium due to its short (12.3 year) half-life, he envisioned surrounding the vacuum
vessel with a blanket containing lithium. The neutrons produced in the plasma
would then produce tritium upon interaction with the lithium outside the reactor
vacuum chamber. He also did some calculations that showed the advantage of
running currents through the plasma as a source of internal ohmic heating. Spitzer,
Wheeler, and another colleague coined the term "stellarator" for the proposed
device, implying a connection to the production of energy in stars by thermonuclear

reactions. By July 1951, the Atomic Energy Commission gave Spitzer $50,000 to work on the Stellarator concept.

The Los Alamos program had not fielded a fusion experiment thus far, but Tuck and co-workers, who were very busy getting cross section data in support of the thermonuclear weapons program, still wanted to develop an experiment around the toroidal pinch. They convinced the director of the laboratory, Norris Bradbury, to give them $50,000 to build such a device. Tuck and another Los Alamos weapon scientist Stanislaus Ulam called the device the "Perhapsatron," a sort of irreverent counter to the lofty name "Stellarator." Construction started in 1952. They quickly learned of the presence of magnetohydrodynamic (MHD) instabilities in their device, as predicted theoretically by Martin Kruskal and Martin Schwarzschild at Princeton.

Another magnetic confinement scheme emerged in 1952, this time from the University of California, which had acquired a former navy training camp in Livermore, California. While the laboratory was primarily devoted to thermonuclear weapons research and development, the first director of the laboratory, Herbert York, was very enthused about the possibility of fusion as an energy source and was looking around at various approaches. York knew that any scheme to be built and studied at Livermore had to be sufficiently different from the Stellarator at Princeton and the pinch program at Los Alamos to get traction in Washington. He proposed a linear magnetic field with end loss being prevented by radiofrequency cavities at both ends. A recent Ph. D. graduate from Stanford University, Richard F. Post, was hired to work on the project, but quickly came to the conclusion that the "radiofrequency dam" idea was not going to be successful, and turned to the idea of pinching down the magnetic fields at the end of the tube, creating increased magnetic fields there to excite the "magnetic mirror effect" (as explained in Sect. 3.4.6). The resultant configuration is called the "mirror machine." The radiofrequency heating equipment was used instead to raise the portion of the particle energy in the directions perpendicular to the magnetic field, which in turn increases the population trapped in the mirror.

Starting in 1951, two Soviet scientists, Andrei Sakharov and Igor Tamm, designed what would come to be known as the tokamak. The name came from the Russian words "toroidalnya kamera ee magnetnaya katushka." This device was a type of toroidal pinch, like the Los Alamos approach, but with a large stabilizing toroidal magnetic field. Figure 1.3 shows the general features of early tokamak designs.

The criterion for stability of the toroidal pinch configuration was worked out by Vitaly Shafranov in the mid-1950s [14] and independently by Kruskal at Princeton [21]. The tokamak concept was not greeted with enthusiasm at first by Lev Artsimovich, the head of the Laboratory of Measuring Instruments of the USSR Academy of Sciences, which later became known as the Kurchatov Institute. But when the experimental measurements showed substantially better confinement parameters than other fusion schemes, Artsimovich became an ardent supporter of the concept.

Fig. 1.3 Schematic of early tokamak design

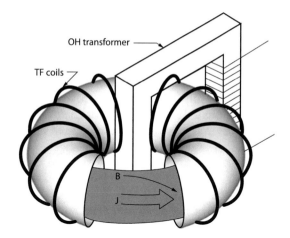

Fig. 1.4 Photograph of the ZETA machine circa 1958. From [31]. Used with permission, Springer Nature

It would be many years before scientists in the West would appreciate the importance of the tokamak as a potential path for a working fusion reactor. Part of this was due to secrecy. The fusion programs in the West and in the Soviet Union would not be fully disclosed to all parties until 1958, when an agreement was reached by all parties concerned to de-classify most facets of fusion research. The first open meeting on fusion was held in Geneva, Switzerland in that year, and it featured talks by several Soviet scientists on experimental results with their machines.

A pinch device was also being prepared in England at Harwell. This machine was called ZETA, for Zero Energy Thermonuclear Assembly. Construction work was begun in 1954, and the first shots on the machine were in 1956 [4, 5, 30]. Figure 1.4 shows a photograph of the machine. The machine was built on a very large scale for that time and had a 1.0 m minor radius and a 3.0 m major radius. The plasma current was (eventually) 500 kA, but the toroidal field was only around 0.11 T, much less than the lower limit for stability described by Kruskal and Shafranov. It was found that the plasma went through a violent phase of instability during startup, but then went into a "relaxed" state after some of the drive power was stopped. A feature of these early experiments was the presence of a strong neutron signal. At first these neutrons were thought to be from thermonuclear reactions, but later it was shown that there was a directionality to this neutron signal, and the average neutron energy was higher in the direction that the current was flowing toroidally [12]. This meant that these neutrons were not of thermonuclear origin, but rather produced by either deuterons accelerated by electric fields due to instabilities or by runaway electrons. However, the early exuberance felt by the ZETA team upon seeing these neutrons caused them to notify the press, which responded with wild enthusiasm. Once it was discovered that these neutrons were not coming from fusion, the news statement was retracted, causing much disappointment in the public, and contributing to the decision to shut the machine down in the mid-1960s. Neutrons had also been observed on the Los Alamos Perhapsatron and later pinch-type devices, and these also could be traced to a non-thermonuclear origin [11].

The fusion program fell into a slump in the early 1960s in the US and in the UK, despite fairly healthy budgets. The US fusion budget was running at $24 million to $32 million between 1960 and 1964 [9]. But there were many contenders on the playing field vying for resources. The Princeton group had produced a new stellarator, the Model C, and it started operating in 1961. However, confinement was poor, and many in the plasma community believed that the characteristic heat leakage rate from a plasma was not that given by classical collision theory but by "Bohm diffusion," an empirical scaling law for diffusion of heat and particles form the plasma which held that confinement time τ would scale like the magnetic field B divided by plasma temperature T rather than the classical theory which held that $\tau \propto B^2\sqrt{T}$. This discouraging scaling seemed omnipresent in all fusion devices in the West. Furthermore, a theorist, Marshall Rosenbluth, had postulated that all plasma configurations would be unstable if the magnetic field appeared as a concave surface against the plasma. This ruled out stability in all of the pinches and mirrors at Los Alamos and Livermore. Remaining in question would be two machines at Livermore: one used a levitating internal current-carrying ring (the "Levitron"), and the other used a circulating relativistic electron ring (the "Astron"). Both of these were phased out by 1973.

The mirror program at Livermore was rescued by a discovery made in the Soviet Union by M. S. Ioffe and communicated to the Americans at a conference in Salzburg, Austria in September 1961. This was the discovery that placing some bars horizontally around the plasma, which imposes a multipole magnetic field, would reduce the bulk instabilities and raise the confinement times by a factor of 35. These "Ioffe bars" essentially created a geometry such that the magnetic lines curved away from the plasma and created a "magnetic well" with the minimum magnetic field at the center of the plasma. By 1964, Livermore had put Ioffe bars on two experimental devices, called "Alice" and "Toy Top." Both showed substantial reduction in the bulk instability modes and improved confinement, but only by a

Fig. 1.6 Schematic for
θ-pinch

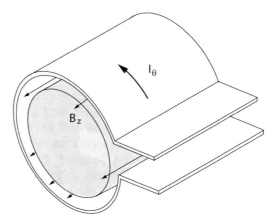

factor of ten or so. What Richard Post and his coworkers discovered was that a
new type of instability—a "microinstability" arose which presented itself a copious
generation of radiofrequency energy near the fundamental and harmonics of the ion
cyclotron frequency, caused by the lack of particles with velocity vectors within
some defined angle with the magnetic field: the "loss cone." Nevertheless, a new
design for an experiment emerged (based on a design from Culham Laboratory in
England) which melded the mirror magnetic coils and the Ioffe bars together so
that the magnet became a single coil shaped like the seam of a baseball. Thus the
experiment was called "Baseball." A picture of a magnet of this type is shown as
Fig. 1.5.

At Los Alamos, a different type of pinch was being explored in the mid-1960s,
based on earlier small experiments. This was the theta-pinch(θ-pinch) concept. In
the θ-pinch, an external coil surrounds the cylindrical plasma and is driven by a
sharply rising current in the circumferential direction around the plasma, as shown
in Fig. 1.6. The first such machine was called Scylla and was disclosed at the

Geneva Conference in 1958. Like the other pinches and ZETA, it also produced neutrons, and they also turned out to be of non-thermonuclear origin. However, four generations of Scylla were eventually built and tested, with the last one, Scylla IV, reaching temperatures of 8 keV. However, confinement was poor due to the open geometry. By 1966 Los Alamos proposed a toroidal version of Scylla that would have closed magnetic field lines, called "Scyllac" with the ending "c" for "closed." But the Los Alamos team, then headed by Fred Ribe, knew that the curvature required for a toroidal layout would induce instability. So a decision was made to add a sinusoidal ripple to the pinch coils, effectively giving it some similarity to the stellarator. In order to observe the physics of this scheme in a very large diameter, only partial sectors of the pinch coils were built. Instabilities were observed with complete disruption of the plasma after a few microseconds. A program to institute feedback control was initiated in the late 1960s.

1968 was a seminal year in the history of nuclear fusion research. A conference was held in Novosibirsk in August of that year, and Lev Artsimovich gave a talk at the conference relating the performance of the T-3 and TM-3 tokamaks. Artsimovich claimed that electron temperatures up to 1.0 keV had been observed, with ion temperatures as high as 500 eV. The confinement time gave a Lawson parameter (see Chap. 4) $n\tau = 10^{18}\,\mathrm{m}^{-3}\,\mathrm{s}$, an order of magnitude higher than anything seen in the West. This exceeded the Bohm confinement estimate given earlier by a factor of fifty. Some American plasma physicists were skeptical of these results. The Soviets had not used laser Thomson scattering to measure the temperature, but rather had measured the shift in the magnetic fields from the vacuum configuration as the plasma was formed as an indicator of the pressure in the plasma and then inferred the temperature by dividing by the measured electron density, since $p = nT$. Thomson scattering is considered the "gold standard" for measuring temperature, since there are no competing spurious processes that can interfere with the measurement. In order to test the validity of these claims, a team of British scientists offered to bring a complete Thomson scattering system to the Soviet Union and measure the electron temperature on the T-3 tokamak themselves. They did so in 1969, and confirmed Artsimovich's claims. An article based on these measurements was published in *Nature* in the same year [27]. This publication marks the most important communication in the history of magnetically confined plasma research. Figure 1.7 shows the actual experimental data from this famous measurement.

Just before the laser Thomson measurement on T-3, a decision was made in the US to convert Princeton's Model C Stellarator into a tokamak, which would later be named the Symmetric Tokamak, or ST. The ST tokamak was to confirm the Soviet results. It became operational in May 1970, and quickly reproduced the T-3 result. A team at Oak Ridge National Laboratory had also been planning a T-3-sized machine called Ormak. This machine was re-designed to become Ormak II, with a larger major radius than T-3 or ST after the Princeton result came in on the ST tokamak. Ormak II became the first machine to have pellet injection for fueling and became the first tokamak to have more neutral beam heating than ohmic heating. It also was the first device to obtain an electron temperature of 2 keV. A tokamak was proposed

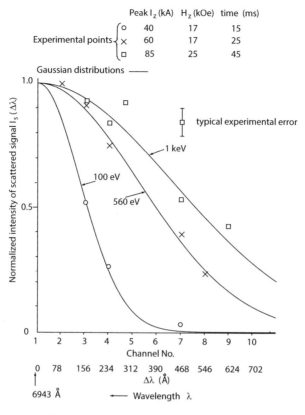

Fig. 1.7 Thomson scattering data from the T-3 tokamak, reported in 1969. From [27]. Used with permission, Springer Nature

by a group at MIT in Cambridge, Massachusetts which drew upon the strengths of the Francis Bitter Magnet Laboratory there to design a high-field device with 10 T fields at the center of the plasma. Italian physicist Bruno Coppi was a proponent of this high-field device, which he named "Alcator" from the Italian "Alto Campo Toro." The device was funded by the US Atomic Energy Commission in 1970 and was operational by 1972 but with some technical problems involving electrical breakdown in the magnet coils and poor vacuum. These problems were resolved by 1974 and good results for plasma performance were obtained. The early 1970s saw a proliferation of tokamaks worldwide. The French built a tokamak at Fontenay-aux-Roses called TFR (for "Tokamak Fontenay-aux-Roses") in 1973, and held performance records until 1976. The British built a dual mode stellarator/tokamak which was operating by 1972 and converted to an iron core T-3 type tokamak later. A machine at General Atomics in San Diego called Doublet was originally intended to be a multipole device, and had been built with in-house funding in 1968. It was converted into a tokamak during its lifetime and was replaced by a larger, noncircular cross section machine, Doublet II, which started running in 1972. A tokamak was online in Japan in 1972 called JFT-2. Germany built a machine called Pulsator which was online in 1973. It was an early demonstration of pellet fueling. The Soviets had a larger tokamak T-4 to provide better performance than their historic T-3 machine, replacing T-3 in 1971. It featured a carbon limiter instead of the more usual metal limiter. Another Soviet tokamak, T-8, came online in 1973, with a noncircular cross section as a new feature. Another Soviet tokamak, TO-1, came online in 1972. Its new feature was feedback control of the plasma position using variable currents in the vertical field coils. A machine built at Frascati in Italy was designed to study plasma turbulence. It was named TTF for "Frascati Turbulent Tokamak" and was operated from 1973 onwards.

At Princeton, work had begun on a machine to be called the Princeton Large Torus, or PLT. Funding was approved for this machine in the spring of 1972, and the machine was operational by 1975 [20]. This was the first tokamak to obtain a 5 keV electron temperature, and demonstrated neutral beam injection, ion cyclotron heating, and lower hybrid current drive. It was originally designed to be the first tokamak to carry 1.0 MA of toroidal current, although it was normally run below 700 kA. Many new diagnostics were developed and tested on PLT, and it paved the way for the next generation of DT-burning devices.

Meanwhile, 1973 and 1974 were notable as the years of the Arab oil embargo, which cut off supply of petroleum from various Middle Eastern oil-producing countries to the US, Britain, the Netherlands, Japan, and Canada. This gave an awakening to the US Congress and other parliamentary bodies as to the critical nature of the world energy supply. This in turn sparked a demand for the rapid production of a working D-T burning fusion reactor. Then fusion research went truly into the phase of Big Science. US president Richard Nixon announced that he would increase the budget for energy research to $10 billion for the next 5 years from 1974 forward. Both Princeton and Oak Ridge National Laboratory put in proposals for a D-T burning reactor experiment. Ultimately, the Princeton design was chosen over the Oak Ridge design. It came to be known as the Tokamak Fusion Test Reactor, or TFTR. An inside view of the TFTR machine is shown as Fig. 1.8. The cost was estimated to be $100 million. It became operational in 1982 and produced 10.7 MW of fusion power in 1994. The design work for the Joint European Torus, or JET, also started in 1973. The machine went into operation in 1983. An inside view of this machine is shown as Fig. 1.9. The major radii of these devices (2.65 m for TFTR and 3.0 m for JET) set them in quite another class from earlier tokamaks. Additionally, JET was designed with a D-shaped (as opposed to circular) cross section, which gave it some performance advantage over TFTR. JET is the current record-holder for the highest fusion power output at 16 MW, achieved in 1997. JET also achieved a steady power output of about 4 MW for almost 5 s.

While JET and TFTR were still in their first years of operation, another event was quietly happening. US President Ronald Reagan and Soviet Secretary Mikhail Gorbachev met in Geneva in November 1985. Among many points in a joint statement disclosed on November 21, both world leaders announced that they "emphasized the potential importance of the work aimed at utilizing controlled thermonuclear fusion for peaceful purposes and, in this connection, advocated the widest practicable

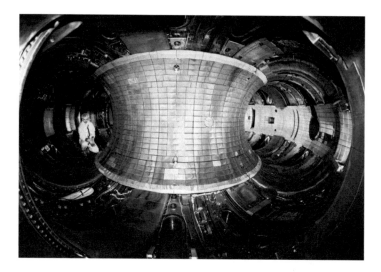

Fig. 1.8 The Tokamak Fusion Test Reactor (TFTR) at Princeton. Princeton Plasma Physics Laboratory

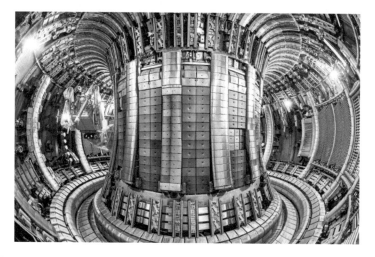

Fig. 1.9 Joint European Torus (JET) inside view ©EUROfusion

development of international collaboration in obtaining this source of energy, which is essentially inexhaustible, for the benefit of mankind." This can be marked as the conception of the International Thermonuclear Experimental Reactor, or ITER. An actual agreement was worked out at a summit meeting in Reykjavik on 12 October 1986, and a four-way agreement (with the European nuclear partnership EURATOM and Japan added) was reached before the end of 1986. The design activity started in 1988. China and Korea both joined the project in 2003, and India joined in 2005. However, there have been a few changes along the way. The original Conceptual

Fig. 1.10 Conceptual drawing of the ITER design, circa 2009. The ITER Organization

Design Activity (CDA) EURATOM resulted in a design which had an 8 m major radius, and was completed in 1990. It was to cost upwards of ten billion USD. The United States pulled out of the project in 1998, since the Congress had the view that this design was not an economical path to energy production at that point, and the low oil prices of that era contributed to a relatively low public opinion concerning nuclear fusion research. Another design activity started and that resulted in a smaller machine (6.2 m major radius), which would cost half as much to build. The US rejoined the ITER design activity in 2003. The expected performance of the machine is now $Q = 10$, with a 500 MW output. (The older design was to be an ignition machine, $Q = \infty$.) A conceptual drawing of the latest design is shown as Fig. 1.10.

1.2.2   Inertial Confinement Fusion (ICF)

The laser was first conceived by Charles Townes and Arthur Schawlow in 1958 [28]. In 1960, Theodore Maiman was the first to successfully demonstrate an optical laser using ruby [23, 24]. Prior to this result, John Nuckolls at Livermore had

Fig. 1.11 12-beam laser fusion target chamber used at Lawrence Livermore National Laboratory in the mid-1960s. From [26]. Courtesy of Lawrence Livermore National Laboratory

proposed heating and compressing a DT target capsule using some radiation source symmetrically applied to the capsule, and the process was simulated on Livermore's computers. An internal memo regarding this was circulated in 1958, but this was kept classified; the work was not published in the open literature until 1972 [25]. Others proposed such an approach to nuclear fusion, including Basov in 1964 [1]. A "4 π laser" system was built and tested in the mid-1960s using twelve lasers for about 20 J of total output (see Fig. 1.11. A laser fusion directorate was set up at Livermore in 1971. The first experimental laser built was called Long Path, and it was designed to put out 40–50 J in 10 ns[26]. This laser was the first in a long line of laboratory-built lasers using neodymium-doped glass as the lasing medium. Long Path was discovered to have a poor beam quality, and the neutron yields on its deuterated plastic targets were low. A succession of lasers were soon built, each more powerful than the last: Cyclops (1974), a one beam, 10 J neodymium-doped glass laser making 0.2 TW (later, a second beam was added to make 0.4 TW), followed by Janus (also in 1974), which could produce 1 TW output pulses. A significant technical hurdle had to be overcome at this point: the high electric fields in the lasing medium caused a nonlinear propagation characteristic which in turn would lead to high-intensity spots in the glass, leading to optical damage. This was overcome by using spatial filters, devices where the light was focused to a very small area and the non-smooth components were removed by passing the light through a pinhole. Following Janus was Argus, a two-beam system with 1 kJ and 2 TW per beam. Argus thus became a prototype for Shiva (1977), a 20-beam laser delivering 10.2 kJ to a target in 0.5–1.0 ns [26]. Figure 1.12 shows photos of these four early laser systems at Livermore.

   Shiva was a partial success: targets were shown to be compressing to 50–100 times liquid density, but that was still lower than what would be needed to ignite the target (around 1000–10,000 times liquid density). The cause of this failure to compress further was that fast electrons were being generated by a nonlinear laser–plasma interaction called stimulated Raman scattering (SRS), and these hot

Fig. 1.12 Four laser systems at Livermore from 1973 to 1977: (**a**) Cyclops, (**b**) Janus, (**c**) Argus, and (**d**) Shiva. From [26]. Lawrence Livermore National Laboratory

electrons would pre-heat the fuel in the capsule, causing the fuel to be on a higher pressure vs. density curve (its "adiabat") than originally designed for. Since the nonlinear force leading to SRS scales as $I\lambda^2$, where I is the intensity of the laser light and λ is the wavelength, a method of shifting the laser light to a shorter wavelength was sought. A technique for frequency shifting had been developed at the Laboratory for Laser Energetics (LLE) in Rochester, New York and this was to pass the leaser light through a crystal with nonlinear optical properties. The material used was potassium dihydrogen phosphate, or KDP. The laser light could be converted to its second or third harmonic (2ω or 3ω) compared to its original frequency ω, this is equivalent to saying that the new wavelength was $\lambda/2$ or $\lambda/3$, and thus as much as a $9\times$ increase in the threshold for SRS.

The Nova laser system was in the design phase starting in the late 1970s. A preliminary version was built called Novette with only two of Nova's planned ten beams, and it ran between 1983 and 1984. Nova came online in 1984 and was run until 1999. The Nova laser system could deliver 100 kJ at the Nd:glass initial wavelength of 1054 nm or about 45 kJ at the frequency-tripled wavelength of 351 nm. Typical yields on Nova experiments were about 10^{13} neutrons per shot, corresponding to tens of joules of fusion energy generated.

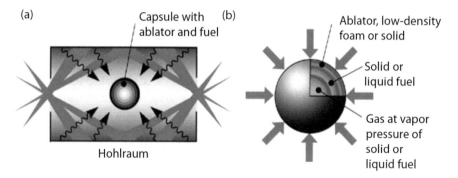

Fig. 1.13 Direct vs. indirect drive laser fusion, illustrating hohlraum concept. From [18]. (**a**) Indirect drive, (**b**) Direct drive

An important event happened in the US related to laser fusion in 1992. That was the declassification of the hohlraum concept through the US Department of Energy (DOE) [8]. The idea behind the hohlraum concept is that using lasers in the visible or ultraviolet wavelength range cannot work if the plasma generated around the target capsule is dense enough to reflect the laser light due to the frequency of the laser light being, at some point, lower than the plasma frequency there (see Sect. 3.1.1). Livermore's method of getting around this had been, for years, to focus the laser energy not on the capsule but on the inside surface of a thin metallic cylinder surrounding the capsule, called a "hohlraum" (German for "hollow room"). This metal surface heats up to temperatures in the 300 eV range, whereupon it emits a blackbody spectrum of soft X-rays, which are at sufficiently short wavelength to penetrate the target even at very high compressed densities. The hohlraum/capsule combination is said to use "indirect drive," whereas the application of laser light directly to the target without a hohlraum is called "direct drive." Figure 1.13 shows the two methods.

While some scientists felt that public disclosure of these formerly secret design ideas was a bad idea from the standpoint of potentially helping others design thermonuclear weapons, the US Department of Energy, with advice from the US National Academy of Sciences, argued that the hohlraum concept had been disclosed in the open literature by authors from France, Japan, Germany, and others. It also helped to explain why the Livermore laser fusion program did not have to stick to the "magic numbers" of the regular polyhedra (4, 6, 8, 12, and 20) for the number of beams on the system by symmetry: the current experiment at this time, Nova, had ten, and plans for the next machine called for 288 beams.

Another important de-classification happened before the hohlraum disclosure mentioned earlier, this one in September 1987 [7]. That was the disclosure that the US had designed some special nuclear weapons tests using devices that contained small capsules of DT fuel. These tests were called Centurion/Halite (one set of tests was done by Livermore and the other by Los Alamos, hence the two names). The X-ray flux from the bomb interacted with the small DT targets, and the yield of these targets was measured. This led to a discussion within the labs about what the laser energy might be for a successful "microfusion" experiment. (Note that 1.0 kiloton TNT = 4.18×10^{12} J, so laboratory fusion yields of a few megajoules are truly "microfusion" to weapon designers.) Estimates were that around 10 MJ of X-rays delivered to the target would give target gains of 10–100. The conversion of laser light into X-rays is at best about a 10% efficient process, so that would mean a laser with 100 MJ of output. This was considered impractical. However, computer simulations showed that a more carefully shaped pulse could achieve a higher compression factor with less laser energy and ensure a more symmetrical drive on the target, and Livermore proposed a 5 MJ laser for the LMF (for "Laboratory Micro-Fusion") call by the Department of Energy (DOE). The DOE asked for the National Academy of Sciences to review these plans, and they responded in 1990 with a recommendation that this plan was too ambitious. Livermore responded with a bid to build a 1 MJ facility, re-using as much as possible from the Shiva and Nova experiments. The experiment would be called "Nova Upgrade." The expected fusion yield was quoted as between 2 and 10 MJ, with an estimated construction cost of around $400 million, with construction to take place between 1995 and 1999.

However, another change happened about this time, and that was what would become the Comprehensive Nuclear Test Ban Treaty, which would stop all under-ground nuclear testing. This was the genesis of the Stockpile Stewardship and management program at the Department of Energy. This agency had the charge of confirming the validity of computer models used to design nuclear weapons by experimental tests. The Nova Upgrade project was considered too small to meet this need, and then the Nova Upgrade was re-designed and called the National Ignition Facility (NIF), with a $1 billion construction cost. Many delays and cost over-runs occurred, and the laser was brought online slowly between 2003 and 2010, at which time the laser system could deliver 1.8 MJ. A schematic of the overall facility is shown as Fig. 1.14. (To appreciate the scale of this facility, note that the target chamber is a 10 m diameter sphere shown near the bottom of the figure.) As of this writing (2018), the maximum target yield has been around 20 kJ, or about 1% of the laser energy entering the hohlraum.

Laser fusion programs have been around at other institutions for almost as long as Livermore's program. The Laboratory for Laser Energetics, mentioned earlier in regard to the KDP development, instituted a program in 1970. It currently houses the Omega laser, a 60-beam, frequency-tripled Nd:glass system with output up to about 40 kJ. This laser facility holds the record for the highest yield for a direct-drive shot of around 10^{14} neutrons, or about a 1% ratio of thermonuclear energy produced to laser energy on target. The GEKKO laser in Japan has been operational since 1983 and produces around 10 kJ at doubled Nd:glass frequency through twelve beams. It now has several beams which have been switched to short-pulse (femtosecond) operation, to explore the concept known as "fast ignition." The Laser Mégajoule (LMJ) facility in Bordeaux, France is similar to NIF, a 1.8 MJ frequency triples Nd:glass system. As of this writing, beam alignment experiments have been reported, but no target experiments have been disclosed.

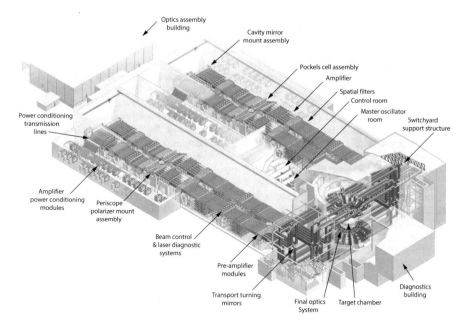

Fig. 1.14 Schematic of the national ignition facility. Lawrence Livermore National Laboratory

1.3 Some Final Comments

Fusion research has a rich history and has engaged a large number of scientists, engineers, government officials, and others over a period that can be traced back to the discovery of the basic nuclear reactions in the early 1930s. Contrary to what some critics have said, real progress has been made. As our knowledge of plasma physics matures, however, new challenges have appeared that were unknown to early researchers. The cost of a credible ignited fusion device is now considered to be in the tens of billions of dollars. And who wants to make this gamble? The answer is practically every scientifically advanced country in the world. The potential reward is simply too great to ignore. For those using this book as an introduction to the subject with an eye to getting involved in fusion research, it is the author's hope that you are inspired to the grand challenge of fusion, and that you have been given fair warning as to the refractory nature of the problem.

References

1. Basov, N.G., Krokhin, O.N.: Conditions for heating up of a plasma by the radiation from an optical generator. Zh. Eksp. Toer. Fiz. **46**, 171 (1964). English translation: Sov. Phys. JETP **19**, 123
2. Bennett, W.H.: Magnetically self-focussing streams. Phys. Rev. **45**, 890–897 (1934). http://link.aps.org/doi/10.1103/PhysRev.45.890

3. Bethe, H.A., Bacher, R.F.: Nuclear physics A. Stationary states of nuclei. Rev. Mod. Phys. **8**, 82–229 (1936). http://link.aps.org/doi/10.1103/RevModPhys.8.82
4. Bodin, H.A.B.: Evolution of the RFP. Plasma Phys. Controll. Fusion **30**(14), 2021 (1988). http://stacks.iop.org/0741-3335/30/i=14/a=006
5. Bodin, H.A.B., Keen, B.E.: Experimental studies of plasma confinement in toroidal systems. Rep. Prog. Phys. **40**(12), 1415 (1977). http://stacks.iop.org/0034-4885/40/i=12/a=001
6. Botts, T.E., Powell, J.R.: Waste management considerations for fusion power reactors. Nucl. Technol. **37**(2), 129–137 (1978). https://doi.org/10.13182/NT78-A31980
7. Broad, W.J.: Secret advance in nuclear fusion spurs a dispute among scientists. New York Times. http://www.nytimes.com/1988/03/21/us/secret-advance-in-nuclear-fusion-spurs-a-dispute-among-scientists.html?pagewanted=all&src=pm (21 March 1988)
8. Broad, W.J.: U.S. is starting to declassify H-bomb fusion technology. New York Times. http://www.nytimes.com/1992/09/28/us/us-is-starting-to-declassify-h-bomb-fusion-technology.html?pagewanted=all (28 September 1992)
9. Bromberg, J.: Fusion: Science, Politics, and the Invention of a New Energy Source. MIT Press, Cambridge (1982). https://books.google.com/books?id=ECOvgg7b3MQC
10. Buneman, O.: The Bennett pinch. In: Drummond, J.E. (ed.) Plasma Physics, p. 202. McGraw-Hill, New york (1961)
11. Conner, J.P., Hagerman, D.C., Honsaker, J.L., Karr, I.J., Mize, J.P., Osher, J.E., Phillips, J.A., Stovall, Jr., E.J. : Operational characteristics of the stabilized toroidal pinch machine, Perhapsatron S04. In: Proceedings of the Second United Nations International Conference on the Peaceful Uses of Atomic Energy, vol. 32, p. 297 (1958). http://www-naweb.iaea.org/napc/physics/2ndgenconf/data/Proceedings%201958/papers%20Vol32/Paper36_Vol32.pdf
12. Coombe, R.A., Ward, B.A.: The energy spectra of the neutrons from ZETA. J. Nucl. Energy. C Plasma Phys. Accel. Thermonucl. Res. **5**(5), 273 (1963). http://stacks.iop.org/0368-3281/5/i=5/a=301
13. Eddington, A.S.: The internal constitution of the stars. The Observatory **43**, 341–358 (1920). http://articles.adsabs.harvard.edu/cgi-bin/nph-iarticle_query?1920Obs\protect.\kern\fontdimen3\font.\kern\fontdimen3\font.\kern\fontdimen3\font.43..341E&data_type=PDF_HIGH&whole_paper=YES&type=PRINTER&filetype=.pdf
14. Editorial Board of Fizika Plazmy: Vitaly Dmitrievich Shafranov (in honor of his 80th birthday). Plasma Phys. Rep. **35**(12), 1068–1070 (2009). http://dx.doi.org/10.1134/S1063780X09120101
15. Gamow, G.: Zur quantentheorie des atomkernes. Z. Phys. **51**(3), 204–212 (1928). http://dx.doi.org/10.1007/BF01343196
16. Goldston, R.J., Glaser, A., Ross, A.F.: Proliferation risks of fusion energy: clandestine production, covert production, and breakout. In: 9th IAEA Technical Meeting on Fusion Power Plant Safety (2009). https://www.osti.gov/servlets/purl/962921-w4qgvc
17. Haines, M.G.: Historical perspective: fifty years of controlled fusion research. Plasma Phys. Controll. Fusion **38**, 643–656 (1996). https://doi.org/10.1088/0741-3335/38/5/001
18. Heller, A.: On target: designing for ignition. Science and Technology Review (1999). https://str.llnl.gov/str/September02/Haan.html
19. Heppenheimer, T.A.: The Man-Made Sun: The Quest for Fusion Power. Little, Brown, Boston (1983). See also http://history.nasa.gov/SP-4305/ch2.htm
20. Hosea, J., Goldston, R., Colestock, P.: The Princeton Large Torus (PLT). Nucl. Fusion **25**(9), 1155 (1985). http://stacks.iop.org/0029-5515/25/i=9/a=027
21. Kruskal, M., Tuck, J.L.: The instability of a pinched fluid with a longitudinal magnetic field. Proc. R. Soc. Lond. A: Math. Phys. Eng. Sci. **245**(1241), 222–237 (1958). https://doi.org/10.1098/rspa.1958.0079; http://rspa.royalsocietypublishing.org/content/245/1241/222
22. Lidsky, L.M.: The trouble with fusion. Technol. Rev. **86**(7), 32–44 (1983)
23. Maiman, T.H.: Optical and microwave-optical experiments in ruby. Phys. Rev. Lett. **4**, 564–566 (1960). https://doi.org/10.1103/PhysRevLett.4.564
24. Maiman, T.H.: Stimulated optical radiation in ruby. Nature **187**, 493–494 (1960). https://doi.org/10.1038/187493a0

25. Nuckolls, J., Wood, L., Thiessen, A., Zimmerman, G.: Laser compression of matter to super-high densities: thermonuclear (CTR) applications. Nature **239**, 139–142 (1972). http://dx.doi.org/10.1038/239139a0
26. Parker, A.: Empowering light–historic accomplishments in laser research. Science and Technology Review. https://str.llnl.gov/str/September02/September50th.html (2002)
27. Peacock, N.J., Robinson, D.C., Forrest, M.J., Wilcock P.D., Sannikov, V.V.: Measurement of the electron temperature by Thomson scattering in Tokamak T3. Nature **224**, 488–490 (1969). http://dx.doi.org/10.1038/224488a0
28. Schawlow, A.L., Townes, C.H.: Infrared and optical masers. Phys. Rev. **112**, 1940–1949 (1958). https://doi.org/10.1103/PhysRev.112.1940
29. Schröder, K.P., Connon Smith, R.: Distant future of the sun and earth revisited. Mon. Not. R. Astron. Soc. **386**(1), 155–163 (2008). https://doi.org/10.1111/j.1365-2966.2008.13022.x; http://mnras.oxfordjournals.org/content/386/1/155.abstract
30. Taylor, J.B.: Relaxation of toroidal plasma and generation of reverse magnetic fields. Phys. Rev. Lett. **33**, 1139–1141 (1974). http://link.aps.org/doi/10.1103/PhysRevLett.33.1139
31. Thonemann, P.C., Butt, E.P., Carruthers, R., Dellis, A.N., Fry, D.W., Gibson, A., Harding, G.N., Lees, D.J., McWhirter, R.W.P., Pease, R.S., Ramsden, S.A., Ward, S.: Controlled release of thermonuclear energy: production of high temperatures and nuclear reactions in a gas discharge. Nature **181**, 217 (1958). https://doi.org/10.1038/181217a0; http://www.nature.com/nature/journal/v181/n4604/pdf/181217a0.pdf

Chapter 2
Fusion Nuclear Reactions

2.1 Cross Sections and Reactivity

Suppose that one were to shoot a bullet into a bale of hay, and the bale of hay contains some rocks. The locations of the rocks within the bale of hay are unknown, but the number of rocks per unit volume n is known, and the average cross-sectional area of a rock facing the incoming bullet σ is known. We want to calculate the probability of the bullet striking a rock. We assume for simplicity that the bullet has a cross-sectional area of zero. Figure 2.1 shows the basic geometry involved. The incremental probability dP/dx of a collision is given by the density of the scatterers times the cross-sectional area σ, or

$$\frac{dP}{dx} = n\sigma \tag{2.1}$$

Then the reaction rate is given by

$$\frac{dP}{dt} = n\sigma\frac{dx}{dt} = n\sigma v \tag{2.2}$$

The application of this concept to nuclear cross sections is straightforward, except that the cross section can be a function of the velocity between the projectile and the scatterers, both of which can be moving. Thus we use the term "reactivity" to mean the average rate of collision per unit particle density $< \sigma v >$. Furthermore we can have a density of projectiles, which we now label as species i and the scatterers as species j. (At this point the designation of "projectile" and "scatterer" is no longer needed.) The reaction rate (the number of scattering events per unit volume per unit time) is then given by

$$R.R. = \frac{n_i n_j < \sigma_{ij} v_{rel} >}{1 + \delta_{ij}} \tag{2.3}$$

© Springer Nature Switzerland AG 2018
E. Morse, *Nuclear Fusion*, Graduate Texts in Physics,
https://doi.org/10.1007/978-3-319-98171-0_2

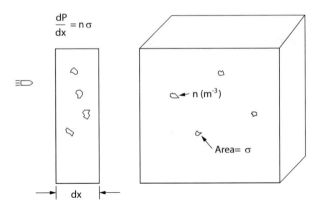

Fig. 2.1 Illustration of the basic cross section concept

Here the Kronecker delta appears: $\delta_{ij} = 1$ if $i = j$ and is zero otherwise. This needs to be added because if a species is interacting with itself, the collisions would be counted twice otherwise.

Sometimes what we want is not the total interaction rate, but rather the rate at which the scattering events lead to the projectile being scattered into some region in solid angle space. In the example above, the pertinent question is "will the shooter get hit with his own bullet?" The differential scattering cross section answers this question. In this case the reaction rate per projectile for a scattering into a region $\Omega = \int d\Omega$ in solid angle is

$$\frac{dP}{dt} = \int d\Omega \left(\frac{d\sigma}{d\Omega} \right) (\Omega, |\mathbf{v}_1 - \mathbf{v}_2|) |\mathbf{v}_1 - \mathbf{v}_2|, \qquad (2.4)$$

where the explicit dependence of the differential scattering cross section on the relative velocity $|\mathbf{v}_1 - \mathbf{v}_2|$ and scattering angle Ω are shown.

2.2   Solar Fusion Reactions

It is instructive to look at the primary fusion reactions on the sun in order to compare and contrast with terrestrial fusion reactions. In fact the term "fusion" should really only apply to the most basic nuclear reaction on the sun described here because it is the one where two heavy particles "fuse" together, leaving one heavy particle as a result. The other reactions studied for terrestrial applications all result in two heavy particles reacting to form two or more heavy particles coming out, so the term "fusion" does not strictly apply—these are "exchange" reactions. This also has led to countless media articles stating that "scientists are trying to re-create the fusion process in the sun on earth...", and this is simply inaccurate.

The most primitive fusion reaction provides the primary source of energy generation on the sun. It is given by

$$_1^1 H +_1^1 H \rightarrow_1 H^2 + e^+ + \nu_e \quad (Q = 1.44 \, \text{MeV}).$$

Here $_1^1 H$ is a proton and $_1^2 H$ is a deuteron, e^+ is a positron, and ν_e is an electron-neutrino. Because of the repulsive force on the protons caused by their positive charge, the reaction rate for this process is very small at low temperatures. Furthermore, the release of a positron and a neutrino shows that this reaction is a weak interaction, causing a reduction in reaction probability over interactions with a reaction using the strong force.

For this reaction to take place, one might think that the particle energies must be high enough for them to physically touch, treating the particles as billiard ball-like objects. A simple calculation shows that this requires very high energies. The potential energy generated by moving the two protons together, each having a radius of about 1.2 fm (femtometer or "fermi") is given by

$$U = \frac{e^2}{4\pi \epsilon_0 (r_p + r_p)} = \frac{1.44 \, \text{MeV} - \text{fm}}{2.4 \, \text{fm}} = 600 \, \text{keV} \tag{2.5}$$

Since the sun's core temperature (in plasma physics, always written in energy units, i.e. multiplied by Boltzmann's constant kT and then suppressing the k) is 1.335 keV (about 15.5 million Kelvin) we can see that the average energy per particle $(3/2)T$ is much less than this potential energy quoted above. However, quantum mechanics tells us that the protons are not discrete particles but rather must be regarded as having a wave function and have a probability distribution in space that includes places where they would not physically be allowed to exist under the laws of classical mechanics. A calculation of this process was done by Gamow [7] and this adds a "tunnelling factor" to the cross section, given by

$$P_G = \exp(-2\pi \eta(E)), \tag{2.6}$$

where

$$\eta(E) = \frac{Z_1 Z_2 e^2}{4\pi \epsilon_0 \hbar v}. \tag{2.7}$$

Here Z_1 and Z_2 are the atomic numbers of the colliding particles (both one in this case), $\hbar$ is the reduced Planck constant ($= 1.05 \cdot 10^{-34}$ kg s), ϵ_0 is the permittivity of free space ($8.85 \cdot 10^{-12}$ F m^{-1}), and v is the relative velocity of the particles in the center-of-mass frame and is given by

$$v = \left(\frac{2E}{\mu}\right)^{1/2}, \tag{2.8}$$

where μ is the reduced mass

$$\mu = \frac{M_1 M_2}{M_1 + M_2}. \tag{2.9}$$

This expression for P_G can be written more conveniently by introducing a "Gamow energy" E_G given by

$$E_G = 2\mu c^2 (\pi \alpha Z_1 Z_2)^2. \tag{2.10}$$

It is more convenient to work with the square root of this number $B_G \equiv E_G^{1/2}$. Then

$$P_G = \exp\left(-\left(\frac{E_G}{E}\right)^{1/2}\right) = \exp(-B_G/\sqrt{E}). \tag{2.11}$$

The overall cross section for the proton–proton fusion reaction is given by

$$\sigma(E) = \frac{S(E)}{E} P_G = \frac{S(E)}{E} \exp(-B_G/\sqrt{E}) \tag{2.12}$$

where the function $S(E)$ is a slowly varying function of energy given by

$$S(E) = S_0 + S'(0)E + \ldots \tag{2.13}$$

For proton–proton fusion, $S(0) \approx 4.0 \cdot 10^{-22}$ keV-b (1 barn=b=10^{-28} m^2) and $S'(0) = 4.48 \cdot 10^{-24}$ b [1]. (Here we can ignore the higher-order terms in $S(E)$, although they make about an 8% correction to the overall reactivity at the core temperature in the sun.) The primary reaction rate given in Eq. (2.3) applies, with the Kronecker delta taken as one for p-p fusion:

$$R.R(pp) = \frac{n_p^2}{2} < \sigma v > . \tag{2.14}$$

The average $< \sigma v >$, known as the reactivity, is obtained by averaging the cross section σ and the relative velocity v over the relative velocity distribution in the plasma. This is typically a maxwellian distribution, normalized to one:

$$f(v) = \left(\frac{\mu}{2\pi T}\right)^{3/2} \exp\left(-\frac{\mu v^2}{2T}\right) \tag{2.15}$$

This can be transformed into a distribution in center-of-mass energy E as follows:

$$f(E) = \frac{2}{\sqrt{\pi T^3}} E^{1/2} \exp\left(-\frac{E}{T}\right). \tag{2.16}$$

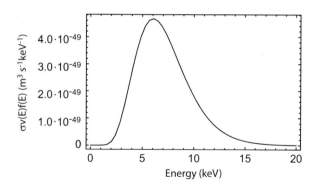

Fig. 2.2 Integrand for p-p-fusion cross section

Then the quantity $< \sigma v >$ is obtained from

$$< \sigma v > = \sqrt{\frac{8}{\pi \mu T^3}} \int_0^\infty E \sigma(E) \exp\left(-\frac{E}{T}\right) dE. \qquad (2.17)$$

Note that by the form that $\sigma(E)$ takes, $\propto \frac{1}{E}$ times the Gamow factor, the reactivity takes the form

$$< \sigma v > = \sqrt{\frac{8}{\pi \mu T^3}} \int_0^\infty S(E) \exp\left(-\left(\frac{E_G}{E}\right)^{1/2}\right) \exp\left(-\frac{E}{T}\right) dE. \qquad (2.18)$$

It is instructive to look at the integrand of Eq. (2.18) as a function of energy. The behavior of this integrand is shown in Fig. 2.2. Note that the Gamow factor is very small at the average energy of the particles ($(3/2)T \approx 2\,\text{keV}$), and the exponentially falling maxwellian gives rise to a peak in the integrand around 6 keV (called the "Gamow window"). Thus most fusion reactions come about in the "tail" of the distribution function. The location of the peak energy E_0 is given by

$$\frac{E_0}{T} = (\pi \alpha Z_1 Z_2 / \sqrt{2})^{2/3} \cdot \left(\frac{\mu}{T}\right)^{1/3} \qquad (2.19)$$

and the integrated cross section $< \sigma v >$ becomes, approximately [2]:

$$< \sigma v > = \frac{1.28 \cdot 10^{-41} \exp(-18.1/T)}{T^{2/3}} \text{m}^3 \text{s}^{-1} \qquad (2.20)$$

with T in keV. For a temperature of 1.335 keV, this gives a reactivity $< \sigma v > = 1.3639 \cdot 10^{-47}\,\text{m}^3\,\text{s}^{-1}$.

To find the specific power generation, we first must look at follow-on reactions caused by production of deuterons. These include:

$$
\begin{aligned}
{}^{2}\mathrm{H} + \mathrm{p} &\rightarrow {}^{3}\mathrm{He} + \gamma \\
{}^{3}\mathrm{He} + {}^{3}\mathrm{He} &\rightarrow {}^{4}\mathrm{He} + 2\mathrm{p} \\
{}^{3}\mathrm{He} + {}^{4}\mathrm{He} &\rightarrow {}^{7}\mathrm{Be} + \gamma \\
{}^{7}\mathrm{Be} + \mathrm{e}^{-} &\rightarrow {}^{7}\mathrm{Li} + \nu_{e} \\
{}^{7}\mathrm{Li} + \mathrm{p} &\rightarrow 2\,{}^{4}\mathrm{Be}
\end{aligned}
\tag{2.21}
$$

These reactions have much larger reactivities than the pp-fusion reaction, which forms the rate-limiting step. Ignoring the intermediate steps, which occur on two main branches, the overall process is

$$
4\mathrm{p} \rightarrow {}^{4}\mathrm{He} + Q
\tag{2.22}
$$

where the average $Q = 26.1\,\mathrm{MeV}$, or about 6.53 MeV per proton consumed. Thus the power density in the sun's core is given by

$$
P_{f}''' = \frac{n_{H}^{2}}{2} <\sigma v> Q/2.
\tag{2.23}
$$

The proton number density in the core of the sun, which has a mass density of $1.53 \cdot 10^{5}\,\mathrm{kg\,m^{-3}}$ and a hydrogen mass fraction X of 0.34, is given by

$$
n_{H} = \frac{X\rho}{m_{H}} = \frac{0.34 \cdot 1.53 \cdot 10^{5}}{1.67 \cdot 10^{-27}} = 3.115 \cdot 10^{31}\mathrm{m^{-3}},
$$

and this gives a fusion power density of $P_{f}''' = 279.4\,\mathrm{W\,m^{-3}}$. However, the density and temperature fall off as the distance from the center increases, so this value is about 1000 times larger than the total output of the sun divided by its volume (0.268 W per cubic meter). It might be said that what the sun lacks in efficiency it makes up for with size.

As the sun ages, it will run on higher-Z nuclear reactions such as the "CNO" cycle [2]. These other nuclear reactions will not be discussed here. Suffice it to say that less than 3% of the sun's energy comes from reactions other than those described here. The timescale for the sun to run out of protons is on the order of 10^{10} y.

2.3 Fusion Reactions for Terrestrial Energy Production

In stark contrast to the primary nuclear reaction on the sun are reactions using the strong nuclear force on the light elements. Here we define D to stand for ${}^{2}_{1}\mathrm{H}$, T to stand for ${}^{3}_{1}\mathrm{H}$, and α to stand for ${}^{4}_{2}\mathrm{He}$. The principal reactions are [13]:

$$
\begin{array}{llll}
\text{D} + \text{T} & \rightarrow & \text{n} + \alpha & (Q = 17.6\,\text{MeV}) \\
\text{D} + {}^3\text{He} & \rightarrow & \text{p} + \alpha & (Q = 18.3\,\text{MeV}) \\
\text{D} + \text{D} & \rightarrow & \text{p} + \text{T} & (Q = 4.03\,\text{MeV},\ B = 50\%) \\
 & \rightarrow & \text{n} + {}^3\text{He} & (Q = 3.27\,\text{MeV},\ B = 50\%) \\
\text{T} + \text{T} & \rightarrow & \alpha + 2\text{n} & (Q = 11.3\,\text{MeV}) \\
{}^3\text{He} + \text{T} & \rightarrow & \alpha + \text{p} + \text{n} & (Q = 12.1\,\text{MeV},\ B = 51\%) \\
 & \rightarrow & \alpha + \text{D} & (Q = 9.5\,\text{MeV},\ B = 43\%) \\
 & \rightarrow & {}^5\text{He} + \text{p} & (Q = 9.5\,\text{MeV},\ B = 6\%) \\
\text{p} + {}^6\text{Li} & \rightarrow & \alpha + {}^3\text{He} & (Q = 4.0\,\text{MeV}) \\
\text{p} + {}^7\text{Li} & \rightarrow & 2\alpha & (Q = 17.3\,\text{MeV},\ B = 20\%) \\
 & \rightarrow & {}^7\text{He} + \text{n} & (Q = -1.6\,\text{MeV},\ B = 80\%) \\
\text{D} + {}^6\text{Li} & \rightarrow & 2\alpha & (Q = 22.4\,\text{MeV}) \\
\text{p} + {}^{11}\text{B} & \rightarrow & 3\alpha & (Q = 8.7\,\text{MeV})
\end{array}
\tag{2.24}
$$

Note that in all cases where there are two particles in the exit channel, conservation of momentum requires that the energy split between the reaction products be inversely proportional to the masses of the particles. For the particles to each have momentum P, the energies will be $P^2/(2M_1)$ for particle 1 and $P^2/(2M_2)$ for particle 2, and thus

$$
E_1 = \frac{M_2}{M_1 + M_2} Q, \qquad E_2 = \frac{M_1}{M_1 + M_2} Q. \tag{2.25}
$$

For example, for the D-T reaction, the $Q = 17.60\,\text{MeV}$ energy release is divided such that the neutron gets $(4/5)Q = 14.08\,\text{MeV}$ and the alpha gets $(1/5)Q = 3.52\,\text{MeV}$. (For the case of three reaction products, a distribution of energy for each particle is found based on the more complex energy–angle relationship that develops.)

We will discuss the D(d, n)^{3}He D(d, p)^{3}H reactions first as they are the most straightforward and are similar in structure to the p-p fusion reaction just discussed. Reference [3] gives some details of the two reactions from the standpoint of the "astrophysical" factor $S(E)$ discussed earlier. In this work, the $S(E)$ factor is fit from measured data using a polynomial form

$$
S(E) = A1 + E(A2 + E(A3 + E(A4 + E\,A5))),
$$

with the center-of-mass energy E in keV, and $S(E)$ in units of keV-barns. The coefficients used are shown in Table 2.1. These shape functions have very small

Table 2.1 Astrophysical shape function $S(E)$ polynomial fit parameters

Reaction	A1	A2	A3	A4	A5
D(d, n)^{3}He	$5.3701 \cdot 10^4$	$3.3027 \cdot 10^2$	$-1.2706 \cdot 10^{-1}$	$2.9327 \cdot 10^{-5}$	-2.5151×10^{-9}
D(d, p)^{3}H	$5.5576 \cdot 10^4$	$2.1054 \cdot 10^2$	$-3.2638 \cdot 10^{-2}$	$1.4987 \cdot 10^{-6}$	1.8181×10^{-10}

From [3]

Table 2.2 D-D fusion reactivity fit parameters

Reaction	C1	C2	C3	C5
D(d, n)^{3}He	$5.4336 \cdot 10^{-18}$	$5.85778 \cdot 10^{-3}$	$7.68222 \cdot 10^{-3}$	$-2.964 \cdot 10^{-6}$
D(d, p)^{3}H	$5.6718 \cdot 10^{-18}$	$3.41267 \cdot 10^{-3}$	$1.99167 \cdot 10^{-3}$	$1.0506 \cdot 10^{-5}$

Adapted from [3]

coefficients for $A3$, $A4$, and $A5$, and neglecting these terms gives at most a 2% error in the reactivity calculation up to temperatures of 200 keV. Note that these shape functions are some twenty-five orders of magnitude larger than the p-p fusion shape function given earlier! Ref. [3] also gives a convenient fit to the integrated reactivities $< \sigma v >$ for a maxwellian distribution; for the D-D reaction, these are modeled as

$$< \sigma v >= C1 \cdot \theta \sqrt{\xi/(\mu c^2 T^3)} \exp(-3\xi) \qquad (2.26)$$

with

$$\theta = T \cdot \left[1 - \frac{C2\,T}{1 + T(C3 + C5\,T)} \right]^{-1} \qquad (2.27)$$

and

$$\xi = (E_G/(4\theta))^{1/3}. \qquad (2.28)$$

Table 2.2 shows the fit parameters. Note that temperatures are in keV and the reactivity is expressed in MKS units (m^3 s^{-1}).

Now we will discuss the D-T and D-^{3}He reactions. These reactions are more complicated because there are excited states in the compound nuclei involved in these reactions. In fact these reactions can be written as

$$
\begin{aligned}
\text{D} + \text{T} &\rightarrow {}^5\text{He}^* \rightarrow & \text{n} + \alpha & \quad (Q = 17.6\,\text{MeV}) \\
\text{D} + {}^3\text{He} &\rightarrow {}^5\text{Li}^* \rightarrow & \text{p} + \alpha & \quad (Q = 18.3\,\text{MeV})
\end{aligned} \qquad (2.29)
$$

A level diagram for the compound nucleus ^{5}He is shown as Fig. 2.3. This shows that an excited state with spin and parity $J^\pi = \frac{1}{2}^-$ is present at an energy 48 keV higher than the d+t rest mass. The Gamow penetration factor causes the experimentally measured cross section to peak at higher energies, however. The existence of these resonances at energies some tens of keV above the rest energies of D+T and D+^{3}He causes these reactions to have extraordinarily large cross sections at relatively modest energies. Also, the relatively rare emission of a 16.84 MeV gamma from the excited state to the ground state in ^{5}He has been useful in the past for development of a diagnostic of DT fusion reactions that is independent from neutron measurements [10, 11]. The astrophysical shape function for the DT reaction is shown as Fig. 2.4 (Note that in this figure, energies in the lab frame,

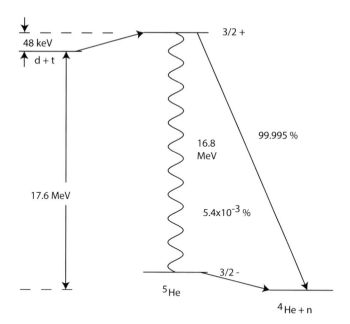

Fig. 2.3 Level structure of the ^{5}He compound nucleus

Fig. 2.4 Measured shape function $S(E)$ for the D-T fusion reaction. From [13]. Used with permission, Springer

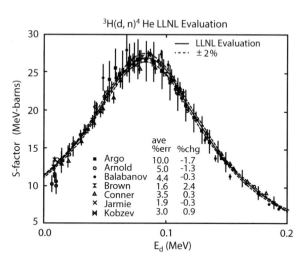

as opposed to the center-of-mass frame, are used. This causes the resonant peak to appear at $E_D = (2/1.2)\, E_{CM} \approx 80$ keV.) In the range of deuteron lab energies from $0 \rightarrow 200$ keV, this shape function can be modeled within a few percent by a simple Breit-Wigner type of resonance:

$$S(E) = \frac{S_0}{(E - E_0)^2 + \Gamma^2/4}, \tag{2.30}$$

Table 2.3 Parameters for
shape function $S(E)$ fit for
the D-T and D-^{3}He reactions

Coefficient	T(d,n)^{4}He	^{3}He(d,p)^{4}He
BG($\sqrt{\text{keV}}$)	34.3827	68.7508
A1	$6.927 \cdot 10^4$	$5.7501 \cdot 10^6$
A2	$7.454 \cdot 10^8$	$2.5226 \cdot 10^3$
A3	$2.050 \cdot 10^6$	$4.566 \cdot 10^1$
A4	$5.2002 \cdot 10^4$	0
B1	$6.38 \cdot 10^1$	$-3.1995 \cdot 10^{-3}$
B2	$-9.95 \cdot 10^{-1}$	$-8.5530 \cdot 10^{-6}$
B3	$6.981 \cdot 10^{-5}$	$5.9014 \cdot 10^{-8}$
B4	$1.728 \cdot 10^{-4}$	0

From [3]

with recommended values $E_0 = 48$ kev and $\Gamma = 74.5$ keV [12]. From [13] we can estimate $S_0 = 27$ Mev-b. At higher energies, a more accurate model must include other excited states in the compound nucleus, which requires an extended R-matrix theory [8, 14].

For the D+^{3}He reaction, the same Breit-Wigner form for the astrophysical shape function $S(E)$ given in Eq. (2.30) can be used with parameters $E_0 = 210$ keV, $\Gamma = 267$ keV, and $S_0 = 17$ MeV-b, again using data from [12] and [13]. The similarity of the nuclear structure of the compound nucleus ^{5}Li with ^{5}He is not surprising, since these are "mirror nuclei," where the proton number of one is the neutron number of the other, and vice versa. Reference [3] gives a multiparameter fit to the cross sections for these two reactions with an approximating polynomial-fraction:

$$S(E) = \frac{A1 + E(A2 + E(A3 + EA4))}{1 + E(B1 + E(B2 + E(B3 + EB4)))}, \tag{2.31}$$

with E in keV and $S(E)$ in eV-b. Parameters for this fit are given in Table 2.3.

Figure 2.5, left side, shows the energy dependence of the overall cross sections for the D-T, D-^{3}He, and the two D-D reaction branches. Note that the two branches of the D-D reaction have almost the same cross sections at low energies, but they diverge (about 25% different) at higher energies in the fractional-MeV range. Also note that the D-T and D-D reactions have more than a 100$\times$ different cross section near the D-T resonance (the resonant peak is shifted up to around 60 keV because of the Gamow factor), but as the energies approach MeV energies, the cross sections appear to be converging. Figure 2.5, right side, shows some other fusion reaction cross sections which require higher energies and generally yield lower reaction probabilities. Most of these come about from nonresonant S factors. The exception is p-^{11}B, which shows two resonances at around 150 and 550 keV. More recent studies of these reactions have attempted to take into account the "screening" process, where the effects of electrons near the nuclei in the cross section measurements can lower the Gamow barrier, causing discrepancies in the cross section data from what would occur in fully ionized fusion plasma [4–6].

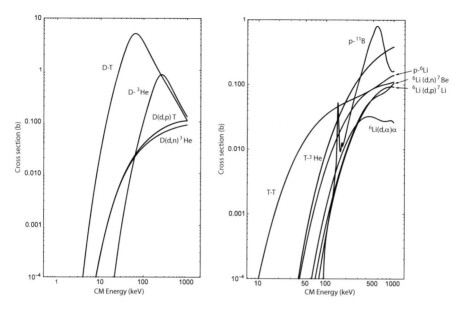

Fig. 2.5 Cross sections for fusion reactions among the light elements

The reactivities for these cross sections can be derived for maxwellian energy distribution using the integral expression given earlier, Eq. (2.17). The maxwellian-averaged reactivities are shown in Fig. 2.6.

Some manipulation of the basic cross section data can yield useful insight into the usability of these fusion reactions. First, in most magnetic fusion confinement concepts, there is a limit on the total pressure that can be confined. Generally this limit is caused by the propensity of the plasma to become unstable if this limit is exceeded. The plasma pressure is given by the total pressure of the electrons and ions. For conditions where the electrons and ions have roughly equal temperatures T, we can write

$$p = (n_e + n_i)T, \tag{2.32}$$

and in general plasmas are close to charge neutrality:

$$n_e = \sum_s n_i^{(s)} Z_s = n_i^{tot} \bar{Z} \tag{2.33}$$

where $\bar{Z}$ represents the average atomic number of the ions. Then the fusion power density P_f for a fifty–fifty mixture of two ion species can be written as

$$P_f = \frac{1}{4}\left(\frac{p}{1+\bar{Z}}\right)^2 \frac{<\sigma v>}{T^2} E_f \tag{2.34}$$

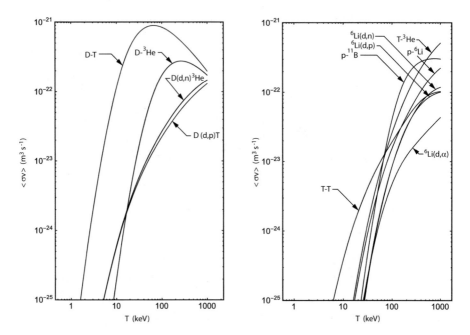

Fig. 2.6 Maxwellian-averaged reactivities $< \sigma v >$ for fusion reactions among the light nuclides

Note that the only temperature-dependent part is the fraction $< \sigma v > /T^2$, and so fusion power production is maximum when this parameter is optimized. Figure 2.7, left side, shows plots of the quantity $< \sigma v > /T^2$ for the most reactive fuels at low plasma temperatures, namely D-T and D-^{3}He. This plot shows that the optimum temperature for D-T is around 15 keV, and around 60 keV for D-^{3}He.

Another interesting situation is where a beam of particles of one species impinges on a target plasma made out of another species. In this case there are fusion reactions between the beam particles and the background plasma as well as fusion reactions between the target particles. For the case of a maxwellian energy distribution for the background (velocities $\mathbf{v'}$ mass m_b and temperature T) and a single energy component (velocity $\mathbf{v_0}$ for the beam, we have a relative velocity $\mathbf{v} = \mathbf{v_0} - \mathbf{v'}$ and then we have a beam-background reactivity given by [9]:

$$< \sigma v >_{beam} = \frac{2\beta}{\sqrt{\pi}} \frac{1}{v_0} \int_{v=0}^{\infty} \sigma(v) v^2 \exp(-\beta^2(v^2 + v_0^2)) \sinh(2\beta^2 v_0 v) dv.$$

(2.35)

Here $\beta = (m_b/(2T))^{1/2}$. A plot of this reactivity vs. beam energy E for the case of a deuteron beam entering a tritium target plasma with temperatures of 0, 10, 30, and 50 keV are shown in Fig. 2.7, right side. Note that at a temperature of zero the reactivity curve is the same as the raw DT fusion cross section shown in Fig. 2.5 with the conversion of the deuteron energy E_D into the equivalent center-of-mass energy: $E_{CM} = (3/5)E_D$. Also note that at high temperatures, the beam-plasma reactivity

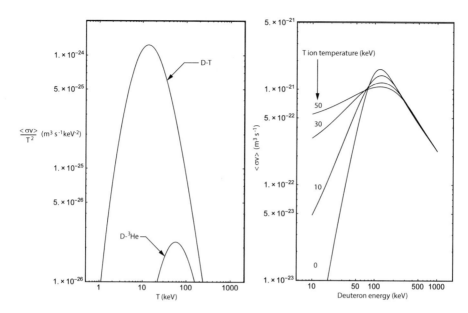

Fig. 2.7 Left: Plots of $< \sigma v > / T^2$ for D-T and D-^{3}He maxwellian plasma. Right: plots of $< \sigma v >$ for the interaction of an energetic particle beam of deuterons with a tritium plasma with a maxwellian energy distribution

actually drops because the resonant peak is "smeared out" by the distribution of relative velocities.

One should realize that in the beam-background reactivity calculation the beam particles will experience an energy loss as they suffer Coulomb collisions with the background particles. Therefore it is usually necessary to track the energy vs. time characteristic for the beam, and integrate the reactivity from Fig. 2.7 over the initial beam energy down to thermal energies.

There is no simple analog to this pressure-limited reactivity optimization in inertial fusion because temperature, density, volume, and the equation of state all vary over the history of the compressed plasma. However it is safe to say that lower ion temperatures, usually below 10 keV, have shown by computer simulation to give optimized yield for a given driver energy input. This will be discussed in detail in a later chapter.

2.3.1   Catalyzed D-D

An examination of the reactivities for the two branches of the D-D fusion reactions shows that in intermediate temperatures ($20 \rightarrow 100$ keV), the D-D reactions have a lower reactivity than the D-T and D-^{3}He reactions by a factor ≈ 100. Since the reaction products from the two branches of the D-D reaction contain either a T or a ^{3}He, we can "take credit" for the energy released in the subsequent D-T and D-^{3}He reactions. This is similar to the treatment given earlier for the solar p-p fusion energy release, since the rate-limiting step is the p-p reaction, with the subsequent deuterons (as well as the positron) being consumed quickly with more total energy release than the p-p reaction itself. Furthermore, the proton produced in the D-D reaction roughly 50% of the time has an energy of 14.7 MeV, considerably more than the other charged particles produced in the D-D reaction (an average of about 2.4 MeV per D-D reaction). Since the neutron interaction length is longer than the dimensions of the plasma, it is primarily the fast charged particles from fusion that heat the plasma, and so this 14.7 MeV proton greatly assists in the overall energy balance for the plasma. Hence the name "catalyzed" D-D. If we take the D(d,n)^{3}He and D(d,p)T to have equal reactivities, we have, for every three deuterons consumed,

$$
\begin{aligned}
3D \rightarrow \\
(1/2)n + (1/2)n(+8.265 \text{ MeV (uncharged)}) \\
+(1/2)p + (1/2)p + (1/2)^3\text{He} + (1/2)T + (1/2)\alpha \\
+(1/2)\alpha(+13.335 \text{ MeV (charged)}),
\end{aligned}
\tag{2.36}
$$

or a total yield of 21.6 MeV per three deuterons and a charged-particle fraction of 61.73 %. This is in contrast to the case for D-T fusion, where only 20% of the energy is released as charged particles.

References

1. Adelberger, E.G., Austin, S.M., Bahcall, J.N., Balantekin, A.B., Bogaert, G., Brown, L.S., Buchmann, L., Cecil, F.E., Champagne, A.E., de Braeckeleer, L., Duba, C.A., Elliott, S.R., Freedman, S.J., Gai, M., Goldring, G., Gould, C.R., Gruzinov, A., Haxton, W.C., Heeger, K.M., Henley, E., Johnson, C.W., Kamionkowski, M., Kavanagh, R.W., Koonin, S.E., Kubodera, K., Langanke, K., Motobayashi, T., Pandharipande, V., Parker, P., Robertson, R.G.H., Rolfs, C., Sawyer, R.F., Shaviv, N., Shoppa, T.D., Snover, K.A., Swanson, E., Tribble, R.E., Turck-Chièze, S., Wilkerson, J.F.: Solar fusion cross sections. Rev. Mod. Phys. **70**, 1265–1291 (1998). http://link.aps.org/doi/10.1103/RevModPhys.70.1265
2. Adelberger, E.G., García, A., Robertson, R.G.H., Snover, K.A., Balantekin, A.B., Heeger, K., Ramsey-Musolf, M.J., Bemmerer, D., Junghans, A., Bertulani, C.A., Chen, J.W., Costantini, H., Prati, P., Couder, M., Uberseder, E., Wiescher, M., Cyburt, R., Davids, B., Freedman, S.J., Gai, M., Gazit, D., Gialanella, L., Imbriani, G., Greife, U., Hass, M., Haxton, W.C., Itahashi, T., Kubodera, K., Langanke, K., Leitner, D., Leitner, M., Vetter, P., Winslow, L., Marcucci, L.E., Motobayashi, T., Mukhamedzhanov, A., Tribble, R.E., Nollett, K.M., Nunes, F.M., Park, T.S., Parker, P.D., Schiavilla, R., Simpson, E.C., Spitaleri, C., Strieder, F., Trautvetter, H.P., Suemmerer, K., Typel, S.: Solar fusion cross sections. II. The pp chain and cno cycles. Rev. Mod. Phys. **83**, 195–245 (2011). http://link.aps.org/doi/10.1103/RevModPhys.83.195
3. Bosch, H.S., Hale, G.: Improved formulas for fusion cross-sections and thermal reactivities. Nucl. Fusion **32**(4), 611 (1992). http://stacks.iop.org/0029-5515/32/i=4/a=I07
4. Engstler, S., Krauss, A., Neldner, K., Rolfs, C., Schröder, U., Langanke, K.: Effects of electron screening on the 3he(d, p)4he low-energy cross sections. Phys. Lett. B **202**(2), 179–

184 (1988). http://dx.doi.org/10.1016/0370-2693(88)90003-2; http://www.sciencedirect.com/science/article/pii/0370269388900032

5. Engstler, S., Raimann, G., Angulo, C., Greife, U., Rolfs, C., Schröder, U., Somorjai, E., Kirch, B., Langanke, K.: Test for isotopic dependence of electron screening in fusion reactions. Phys. Lett. B **279**(1), 20–24 (1992). http://dx.doi.org/10.1016/0370-2693(92)91833-U; http://www.sciencedirect.com/science/article/pii/037026939291833U

6. Engstler, S., Raimann, G., Angulo, C., Greife, U., Rolfs, C., Schröder, U., Somorjai, E., Kirch, B., Langanke, K.: Isotopic dependence of electron screening in fusion reactions. Z. Phys. A: Hadrons Nucl. **342**(4), 471–482 (1992). http://dx.doi.org/10.1007/BF01294958

7. Gamow, G.: Zur quantentheorie des atomkernes. Z. Phys. **51**(3), 204–212 (1928). http://dx.doi.org/10.1007/BF01343196

8. Lane, A.M., Thomas, R.G.: R-matrix theory of nuclear reactions. Rev. Mod. Phys. **30**, 257–353 (1958). http://link.aps.org/doi/10.1103/RevModPhys.30.257

9. Miley, G.H., Towner, H.: Reactivities for two-component fusion calculations. In: Schrack, R., Bowman, C., Protection, A.N.S.R., Division, S. (eds.) Nuclear Cross Sections and Technology: Proceedings of a Conference, Washington, D.C., 3–7 March 1975, Volume 2; Volume 13. NBS Special Publication, pp. 716–721. U.S. Department of Commerce/National Bureau of Standards, Washington, D.C./Gaithersburg (1975). https://books.google.com/books?id=uAv286X1KdcC

10. Moran, M.J.: Detector development for γ-ray diagnostics of d-t fusion reactions. Rev. Sci. Instrum. **56**(5), 1066–1068 (1985). http://dx.doi.org/10.1063/1.1138219; http://scitation.aip.org/content/aip/journal/rsi/56/5/10.1063/1.1138219

11. Moran, M.J.: The fusion diagnostic gamma experiment: a high-bandwidth fusion diagnostic of the National Ignition Facility. Rev. Sci. Instrum. **70**(1), 1226–1228 (1999). http://dx.doi.org/10.1063/1.1149338; http://scitation.aip.org/content/aip/journal/rsi/70/1/10.1063/1.1149338

12. Tilley, D., Cheves, C., Godwin, J., Hale, G., Hofmann, H., Kelley, J., Sheu, C., Weller, H.: Energy levels of light nuclei a=5, 6, 7. Nucl. Phys. A **708**(1–2), 3–163 (2002). http://dx.doi.org/10.1016/S0375-9474(02)00597-3; http://www.sciencedirect.com/science/article/pii/S0375947402005973

13. White, R.M., Resler, D.A., Warshaw, S.I.: Evaluation of charged-particle reactions for fusion applications. In: Qaim, S.M. (ed.) Nuclear Data for Science and Technology: Proceedings of an International Conference, held at the Forschungszentrum Jülich, Fed. Rep. of Germany, 13–17 May 1991, pp. 834–839. Springer, Berlin (1992). http://dx.doi.org/10.1007/978-3-642-58113-7_230

14. Wigner, E.P., Eisenbud, L.: Higher angular momenta and long range interaction in resonance reactions. Phys. Rev. **72**, 29–41 (1947). http://link.aps.org/doi/10.1103/PhysRev.72.29

Chapter 3
Collisions and Basic Plasma Physics

3.1 Plasma Properties

A plasma is more than just a collection of charged particles. The name plasma was coined by Irving Langmuir, a physicist working at General Electric in the 1920s, who found that ionized gases behaved with fluid-like properties which reminded him of blood plasma. The main properties which define plasma are: (1) The appearance of collective effects, such as waves and instabilities, (2) the screening of long-range electric fields by individual particles, and (3) the tendency towards charge neutrality.

3.1.1 Plasma Oscillations

Consider a box of plasma which is initially in a charge-neutral state $n_e = Zn_i$, with Z being the atomic number of the ions and n_e the electron density and n_i the ion density. Now suppose that the entire population of electrons is shifted down a distance δx as shown in Fig. 3.1. This creates a volume of purely positive charges on the top and a layer of purely negative charges on the bottom. The charge is given by $\pm n_e q_e dV = \pm n_e q_e A \delta x$, where A is the area of the top and bottom faces of the box. This charge separation creates an electric field $E = -V/d$ inside the box. The voltage on this "plasma capacitor" can be calculated from the capacitance law $V = Q/C$, where $C = \epsilon_0 A/d$. Then the factors A and d drop out and we have $E = -n q_e \delta x$. If we set up an equation of motion for the electrons, we have

$$m_e \frac{d^2 \delta x}{dt^2} = q_e E = -\frac{n_e q_e^2}{\epsilon_0} \delta x. \qquad (3.1)$$

© Springer Nature Switzerland AG 2018
E. Morse, *Nuclear Fusion*, Graduate Texts in Physics,
https://doi.org/10.1007/978-3-319-98171-0_3

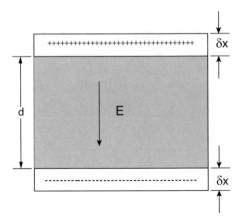

Fig. 3.1 Model for a simple plasma oscillation

With the substitution of $\omega_{pe}^2 \equiv n_e q_e^2/(m_e \epsilon_0)$, we have an equation describing simple harmonic motion:

$$\frac{d^2 \delta x}{dt^2} + \omega_{pe}^2 \delta x = 0. \tag{3.2}$$

This shows that unmagnetized cold plasmas can exhibit undamped oscillation at a frequency ω_{pe} called the plasma frequency. The linear frequency f_{pe} associated with this angular frequency ω_{pe} is just $\omega_{pe}/(2\pi)$. Substituting in physical constants m_e, q_e, and ϵ_0 gives, in MKS units

$$f_{pe} \approx 9\sqrt{n}. \tag{3.3}$$

For magnetic confinement fusion, plasma densities are around $10^{20}\,\mathrm{m}^{-3}$ and thus $f_{pe} \sim 90\,\mathrm{GHz}$. Inertial confinement plasma at $1000\times$ liquid DT density has a plasma frequency of around $7 \cdot 10^{16}\,\mathrm{Hz}$.

The plasma oscillation frequency given here has been thoroughly investigated as far back as 1929 and reported in a classic paper by Tonks and Langmuir [11]. Their derivation used a microscopic treatment rather than the slab model given above. They started with the continuity equation for the electrons (the ions are considered immobile because of their higher masses):

$$\frac{\partial n_e}{\partial t} + \nabla \cdot (n_e \mathbf{u}_e) = 0 \tag{3.4}$$

Integrating this over time for a small perturbation to the equilibrium density n_{e0} with no equilibrium flow ($\mathbf{u}_{e0} = 0$) gives

$$\delta n = -n_{e0} \nabla \cdot \boldsymbol{\xi}, \tag{3.5}$$

where $\boldsymbol{\xi}$ is the displacement of the fluid particles from equilibrium. Without loss of generality we can assume that $\boldsymbol{\xi}$ is in the $\hat{x}$ direction. Then we can write a Poisson equation as

$$\nabla \cdot \mathbf{E} = \frac{\partial E_x}{\partial x} = \rho / \epsilon_0 = +\frac{n_{e0} e}{\epsilon_0} \frac{\partial \xi_x}{\partial x}. \tag{3.6}$$

we can integrate this equation and find $E_x = n_{e0} e \xi_x$. We then find that there is a restoring force on the electrons $-e E_x$ which then gives an equation of motion for the electrons as

$$m_e \ddot{\xi}_x + \frac{n_{e0} e^2}{\epsilon_0} \xi_x = 0 \tag{3.7}$$

and thus

$$\ddot{\xi} + \omega_{pe}^2 \xi = 0, \tag{3.8}$$

the equation for simple harmonic motion at the plasma frequency, as in the previous case.

A third (and perhaps even more compelling) case for describing plasma oscillations arises from the description of the propagation of electromagnetic waves through a cold homogeneous unmagnetized plasma. Suppose that a plane wave is propagating through a plasma with electron density n_{e0}. This plane wave has a form

$$\begin{aligned} \mathbf{E}(\mathbf{x}, t) &= \hat{x} E_0 \exp\left(i(k_z z - \omega t)\right) \\ \mathbf{B}(\mathbf{x}, t) &= \hat{y} B_0 \exp\left(i(k_z z - \omega t)\right) \end{aligned} \tag{3.9}$$

Again treating the ions as being immobile, we find that the response of the plasma electrons is given by (dropping the $\mathbf{v} \times \mathbf{B}$ term, which is of order v/c times smaller):

$$m_e \frac{\partial \mathbf{v}_e}{\partial t} = -i \omega m_e \mathbf{v}_e = -e \mathbf{E} \tag{3.10}$$

The equation describing the wave propagation is found by taking the curl of the equation $\nabla \times \mathbf{E} = -\dot{\mathbf{B}}$ and applying another Maxwell equation $\nabla \times \mathbf{B} = \mu_0 \left(\mathbf{J} + \epsilon_0 \dot{\mathbf{E}}\right)$. This gives

$$\nabla \times \nabla \times \mathbf{E} = -\mu_0 \frac{\partial \mathbf{J}}{\partial t} - \mu_0 \epsilon_0 \frac{\partial^2 \mathbf{E}}{\partial t^2} \tag{3.11}$$

The current time derivative term $\dot{\mathbf{J}}$ is just $-i \omega n_{e0}(-e)(\mathbf{v}_e) = -(n_{e0} e^2 / m_e)\mathbf{E}$ and replacing $\nabla \times$ with $i \mathbf{k} \times$ (and noting that $\mathbf{k} \cdot \mathbf{E} = 0$ here), and using the identity $\mu_0 \epsilon_0 = 1/c^2$, turns Eq. (3.11) into

$$k^2 \mathbf{E} = -\frac{\omega_p e^2}{c^2} \mathbf{E} + \frac{\omega^2}{c^2} \mathbf{E}. \tag{3.12}$$

Then for a nonzero propagating electric field **E** we must have

$$\omega^2 = \omega_{pe}^2 + k^2 c^2 \qquad (3.13)$$

This shows that if $\omega < \omega_{pe}$ then **k** is imaginary, i.e. fields diminish exponentially ("evanescent"), and propagation is possible at $\omega > \omega_{pe}$ but with a wavenumber different from the free-space value. But more interestingly, the solution to this "dispersion relation" at **k** = 0 is just $\omega = \omega_{pe}$, and this implies that bulk plasma oscillation at this frequency is possible.

3.1.2 Debye Screening

In a 1923 paper [3], Peter Debye and Irwin Hückel put forward a theory for the electrostatic screening of ions in aqueous solutions. It was apparent to Langmuir and other early plasma researchers that this theory could be applied to plasma as well.

Suppose that an ion with charge $+Ze$ is put at position **r** = 0 in a plasma. If there were no other particles in the system, then the electrostatic potential of the ion would be given by

$$\Phi = \frac{+Ze}{4\pi \epsilon_0 r}. \qquad (3.14)$$

The long-range potential, falling off at this $\propto 1/r$ rate, tends to attract nearby electrons and repel nearby ions. This attraction and repulsion will cause some adjustment of the local electron and ion density. First we consider only the re-distribution of the electrons. If the electrons are in thermodynamic equilibrium, the local value of the electron density will follow a Boltzmann law:

$$n_e(r) = n_{e0} \exp\left[-U/T_e\right] = n_{e0} \exp\left[\frac{+|e|\Phi}{T_e}\right]. \qquad (3.15)$$

The re-distributed electron density represents a charge density given by $\rho_e = -|q_e|\,(n_e(r) - n_{e0})$. Then using Eq. (3.15) gives a Poisson equation

$$\nabla^2 \Phi = -\frac{\rho_e}{\epsilon_0} = \frac{+|q_e| n_{e0}\,(\exp(|q_e|\Phi/T_e) - 1)}{\epsilon_0}. \qquad (3.16)$$

This nonlinear partial differential equation is difficult to work with, but we can solve for Φ in places where $e\Phi \ll T_e$ by linearizing this equation by using $e^x - 1 \approx x$ for small x which yields

$$\nabla^2 \Phi - \frac{n_e q_e^2}{\epsilon_0 T} \Phi = 0. \tag{3.17}$$

Then defining $\lambda_{De} \equiv \left(\epsilon_0 T_e / (n_e q_e^2) \right)^{1/2}$ turns Eq. (3.17) into:

$$\nabla^2 \Phi - \frac{1}{\lambda_{De}^2} \Phi = 0. \tag{3.18}$$

which has the solution $\Phi(r) = (A/r) \exp(-r/\lambda_{De})$. Fitting this solution to the "undressed" potential as $r \to 0$ gives

$$\Phi = \frac{+Ze \exp(-r/\lambda_{De})}{4\pi \epsilon_0 r}. \tag{3.19}$$

Thus the electrostatic potential for the "dressed" charge falls off much faster than $1/r$ at distances greater than a Debye length λ_{De}. This length has a numerical value

$$\lambda_{De} = 7432 \, (T/n_e)^{1/2} \quad (T \text{ in eV}, n_e \text{ in m}^{-3}). \tag{3.20}$$

Thus for a typical magnetic fusion plasma ($T = 15 \, \text{keV}$, $n_e = 10^{20} \, \text{m}^{-3}$) the electron Debye length is about $90 \, \mu\text{m}$.

One can extend the single-species Debye length to multiple species by examining the screening term arising in Eq. (3.18). The constant term $1/\lambda_{De}^2$ can be replaced by $1/\lambda_{De}^2 + 1/\lambda_{Di1}^2 + 1/\lambda_{Di2}^2 + \ldots$ for all ion species present. In general it is easy to show that the overall Debye length λ_D is given by

$$\lambda_D = \lambda_{De} \left[\sum_s \frac{n_s}{n_e} \frac{T_e}{T_s} Z_s^2 \right]^{-1/2}. \tag{3.21}$$

Note that for a $Z = 1$ plasma with equal ion and electron temperatures, $\lambda_D = \lambda_{De}/\sqrt{2}$. Thus for a typical magnetic fusion plasma ($T = 15 \, \text{keV}$, $n_e = 10^{20} \, \text{m}^{-3}$) with $Z = 1$ and equal ion and electron temperatures, the overall $\lambda_D = 60 \, \mu\text{m}$.

An important parameter related to the Debye length is the quantity $n\lambda_D^3$, related to the number of particles in a sphere of radius λ_D. It is this quantity that tells us the propensity for the plasma to behave in a collective and fluid-like manner. In fact some authors define a plasma to be a collection of ions and electrons in a charge-neutral state with $n\lambda_D^3 \gg 1$. However that does little service to those who study "strongly coupled plasmas" ($n\lambda_D^3 = \mathcal{O}(1)$) and "nonneutral plasmas" (usually pure electron plasmas). Most fusion applications, including both magnetic and inertial confinement approaches, have $n\lambda_D^3 \gg 1$, although highly compressed, hot inertial fusion plasmas come close to being "strongly coupled."

3.2   Coulomb Collisions

An understanding of the basic collision process between two charged particles is essential to understanding many aspects of energy and momentum transport in plasma. Since the force between two charged particles (at distances small compared to the Debye length) is an inverse-square law, the mathematics of this interaction is identical to the Keplerian orbit problem in celestial mechanics, with the added feature that the forces can be attractive or repulsive, depending on the sign of the charges involved.

3.2.1   Coulomb Kinematics

First consider two charged pariclces of mass m_1 and m_2 and charges q_1 and q_2 and moving with velocities $\mathbf{v}_1(t)$ and $\mathbf{v}_2(t)$. The interactive force $\mathbf{F}_{12} = -\mathbf{F}_{21}$ and is given by

$$\mathbf{F}_{12} = -\mathbf{F}_{21} = \frac{q_1 q_2 (\mathbf{r}_1 - \mathbf{r}_2)}{4\pi\epsilon_0 |\mathbf{r}_1 - \mathbf{r}_2|^3}. \tag{3.22}$$

We can then combine the two Newtonian equations $\ddot{\mathbf{r}}_1 = \mathbf{F}_{12}/m_1$ and $-\ddot{\mathbf{r}}_2 = \mathbf{F}_{12}/m_2$ to obtain

$$\ddot{\mathbf{r}}_1 - \ddot{\mathbf{r}}_2 = \mathbf{F}_{12} \left(\frac{1}{m_1} + \frac{1}{m_2} \right). \tag{3.23}$$

We can then solve the equation of motion in the center-of-mass frame by defining $\mathbf{r} = \mathbf{r}_1 - \mathbf{r}_2$ and a reduced mass $\mu = (1/m_1 + 1/m_2)^{-1} = m_1 m_2/(m_1 + m_2)$ and treating a particle with mass μ and charge q_1 as being in motion around a fixed charge q_2. Figure 3.2 shows the geometry of the problem.

Fig. 3.2 Geometry of
Coulomb scattering problem

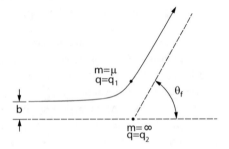

The equation of motion for particle 1 is then given by a radial equation and an equation representing the conservation of angular momentum:

$$\mu\ddot{r} - \mu r\dot{\theta}^2 - \frac{q_1 q_2}{4\pi\epsilon_0 r^2} = 0 \qquad (a)$$

$$\mu r^2 \dot{\theta} = P_\theta = \text{constant.} \quad (b)$$

(3.24)

(Note that the radial equation contains a term $\mu r\dot{\theta}^2$, which is a centrifugal term brought about by the non-rectilinear coordinate system.) We can then write $r = r(\theta)$ and then transform this system of equations to eliminate the time dependence by observing that, by the chain rule

$$\dot{r}(\theta) = \frac{dr(\theta)}{d\theta}\dot{\theta}, \ \ddot{r}(\theta) = \frac{d^2 r(\theta)}{d\theta^2} \cdot \dot{\theta}^2 + \frac{dr(\theta)}{d\theta} \cdot \ddot{\theta}. \tag{3.25}$$

Then using Eq. (3.24)(b), we have

$$\ddot{\theta} = -\frac{2 P_\theta \dot{r}}{\mu r^3}. \tag{3.26}$$

Substituting Eqs. (3.25) and (3.26) into Eq. (3.24)(a) gives

$$\frac{P_\theta^2}{2\mu r^4} \left(\frac{d^2 r}{d\theta^2} - \frac{2}{r} \left(\frac{dr}{d\theta} \right)^2 - r \right) - \frac{q_1 q_2}{4\pi\epsilon_0 r^2} = 0. \tag{3.27}$$

Now we make a variable substitution $u = 1/r$ and note that

$$\frac{dr}{d\theta} = -\frac{1}{u^2}\frac{du}{d\theta}$$

$$\frac{d^2 u}{d\theta^2} = \frac{2}{u^3} \cdot \left(\frac{du}{d\theta} \right)^2 - \frac{1}{u^2} \cdot \frac{d^2 u}{d\theta^2}$$

(3.28)

This yields a simple linear differential equation for u:

$$\frac{P_\theta^2}{\mu} \cdot \left(\frac{d^2 u}{d\theta^2} + u \right) + \frac{q_1 q_2}{4\pi\epsilon_0} = 0 \tag{3.29}$$

This has a solution

$$u = \frac{\mu q_1 q_2}{4\pi\epsilon_0 P_\theta^2} (e\,\cos(\theta - \theta_0) - 1) \tag{3.30}$$

The boundary conditions on $u(\theta)$ at the entrance angle of $\theta = \pi$ is $u(\pi) = 0$, meaning that the particle enters at $r = \infty$ from the left. Then $e\cos(\pi - \theta_0) = 1$. We find another boundary condition from noting that at the place where r is a minimum

(u is a maximum), the radial motion $\dot{r}$ is zero and the kinetic+potential energy there equals the kinetic energy of the inbound particle E_{in}. At this point

$$\frac{P_\theta^2}{2\mu} u_{max}^2 + \frac{q_1 q_2}{4\pi\epsilon_0} u_{max} - E_{in} = 0. \tag{3.31}$$

We can put this in standard quadratic form by assigning $A = P_\theta^2/(2\mu)$, $B = q_1 q_2/(4\pi\epsilon_0)$, and $C = -E_{in}$. Then the solution is

$$u_{max} = -\frac{B}{2A} + \sqrt{\left(\frac{B}{2A}\right)^2 + \frac{E_{in}}{A}}. \tag{3.32}$$

We can equate this to the solution for u given in Eq. (3.30) taken at the angle that maximizes u, namely $\theta = \theta_0$. Plugging in A and B in Eq. (3.30) gives

$$u_{max} = -\frac{B}{2A} + \frac{B}{2A}e. \tag{3.33}$$

Then we have

$$e = \sqrt{1 + \frac{4A E_{in}}{B^2}}. \tag{3.34}$$

We are now in a position to find a relation between the impact parameter b and the exit angle θ_f. Note that the next zero of u will occur at an angle equal to $\theta_f = \theta_0 + (\theta_0 - \pi) = 2\theta_0 - \pi$ by symmetry. Dividing by two gives $\theta_f/2 = \theta_0 - \pi/2$. Then $\tan^2(\theta_0) = \tan^2(\theta_f/2 + \pi/2) = \cot^2(\theta_f/2)$. The boundary condition at $\theta = \pi$ can be written as $e^2 = \sec^2\theta_0 = 1 + \tan^2(\theta_0)$. Equating this to the solution found for e in Eq. (3.34) then gives

$$\cot^2(\theta_f/2) = \frac{4A E_{in}}{B^2}. \tag{3.35}$$

And finally we set the angular momentum P_θ to the initial angular momentum $\mu v b$. After replacing the original parameters for A, B, and P_θ we obtain

$$b = \frac{q_1 q_2 \cot(\theta_f/2)}{4\pi\epsilon_0 \mu v^2}. \tag{3.36}$$

Note that the impact parameter for a 90° scatter in the CM frame (for $q_1 q_2 > 0$) is given by setting $\theta_f = \pi/2$:

$$b_{90} = \frac{q_1 q_2}{4\pi\epsilon_0 \mu v^2}. \tag{3.37}$$

A plot of the solution to the Coulomb scattering problem is shown as Fig. 3.3.

Fig. 3.3 Coulomb scattering for constant energy in the transformed frame and varying impact parameter b and $q_1 q_2 > 0$

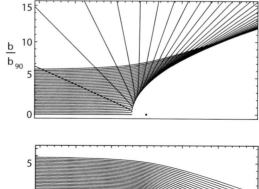

Fig. 3.4 Coulomb scattering for constant energy in the transformed frame with $q_1 q_2 < 0$

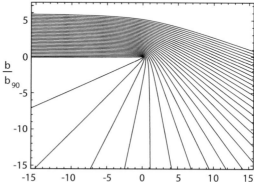

It is also worth noting that only one small change is necessary to accommodate the case where $q_1 q_2 < 0$ and that is to take e to be a negative number in Eq. (3.30). Then $\cos(\pi - \theta_0) = 1/e$ is a negative number, and thus $\theta_0 < \pi/2$. The exit angle θ_f is now is between $-\pi$ and 0, i.e. the particle more nearly encircles the fixed charge at $r = 0$. Equation (3.34) then becomes

$$e = -\sqrt{1 + \frac{4 A E_{in}}{B^2}}, \tag{3.38}$$

and the rest of the mathematics, which does not depend on the sign of e, proceeds the same way. Note especially that Eq. (3.36) stays exactly the same, as the $\cot(\theta_f/2)$ function gives a negative result for $q_1 q_2 < 0$. A plot of the trajectories for a negative $q_1 q_2$ is shown as Fig. 3.4. It is also worth noting that the same impact parameter for a $270°$ scatter for $q_1 q_2 < 0$ is obtained from Eq. (3.37) with the sign of $q_1 q_2$ reversed.

We can set $\theta_f = \theta$ in Eq. (3.36) now that the kinematics calculation is done. We can calculate the differential cross section $d\sigma/d\Omega$ by noting that $2\pi b\, db$ represents the differential cross-sectional area associated with a differential solid angle $2\pi \sin\theta\, d\theta$. This gives the result

$$\frac{d\sigma}{d\Omega} = \left| \frac{b}{\sin\theta} \frac{db}{d\theta} \right| = \left(\frac{q_1 q_2}{8\pi \epsilon_0 \mu v^2} \right)^2 \cdot \frac{1}{\sin^4(\theta/2)}. \tag{3.39}$$

This is sometimes referred to as the Rutherford cross section. It is important to note the $\sim 1/\theta^4$ dependence at small scattering angles, which means that interaction rates that are calculated using this cross section must be cut off at some minimum scattering angle. However, the Debye screening process discussed earlier provides just such a mechanism for calculation of a minimum scattering angle.

Another problem remains and that is that we have transformed the system into a different space from the "laboratory" frame. We have worked with a variable $\mathbf{r} = \mathbf{r}_1 - \mathbf{r}_2$. To get back to the real world, or lab frame, we must add to the position for $\mathbf{r}_1$ an amount equal to the part of $\mathbf{r} = \mathbf{r}_1 - \mathbf{r}_2$ belonging to $\mathbf{r}_1$, which is $m_2/(m_1 + m_2)\mathbf{r}$. Then the incremental velocity that gets added to the velocity of particle 1 due to the collision of particle 1 with particle 2 in the lab frame is

$$\Delta \mathbf{v}_1^L = |\mathbf{v}_1 - \mathbf{v}_2|^L \, (\hat{\mathbf{x}}(\cos\theta^T - 1) + \hat{\mathbf{y}}\sin\theta^T)\frac{m_2}{m_1 + m_2}. \tag{3.40}$$

The momentum change $\Delta \mathbf{p}_1$ associated with the collision is then $m_1 \Delta \mathbf{v} = \mu \Delta \mathbf{v}_T$ in the transformed system. This should not be a surprise, because exchange momenta are conserved under a Gallilean transformation.

We can now calculate the momentum loss of a test particle using the cross sections given here. For a distribution of background velocities $f(v_2)$, we have

$$\frac{d\mathbf{p}}{dt} = \int d\mathbf{v}_2 \int_{\theta_{min}}^{\pi} 2\pi \sin\theta d\theta f(\mathbf{v}_2)(\mathbf{v}_1 - \mathbf{v}_2)\left(\frac{q_1 q_2}{8\pi\epsilon_0\mu\,|\mathbf{v}_1 - \mathbf{v}_2|^2 \sin^2(\theta/2)}\right)^2$$
$$\times \mu\,|\mathbf{v}_1 - \mathbf{v}_2|\,(\cos\theta - 1). \tag{3.41}$$

Using trigonometric identities $1 - \cos\theta = 2\sin^2(\theta/2)$ and $\sin\theta = 2\sin(\theta/2)\cos(\theta/2)$ turns Eq. (3.41) into

$$\frac{d\mathbf{p}}{dt} = -\frac{q_1^2 q_2^2}{4\pi\epsilon_0^2\mu}\int d\mathbf{v}_2 \frac{(\mathbf{v}_1 - \mathbf{v}_2)f(\mathbf{v}_2)}{|\mathbf{v}_1 - \mathbf{v}_2|^3}\int_{\theta_{min}}^{\pi} d(\theta/2)\cot(\theta/2)$$
$$= \frac{q_1^2 q_2^2}{4\pi\epsilon_0^2\mu}\int d\mathbf{v}_2 \frac{(\mathbf{v}_1 - \mathbf{v}_2)f(\mathbf{v}_2)}{|\mathbf{v}_1 - \mathbf{v}_2|^3}\ln[\sin(\theta_{min}/2)]. \tag{3.42}$$

The minimum scattering angle is chosen to be the angle associated with an impact parameter $b = \lambda_D$, the Debye length. Using (3.37), we then have for the logarithmic term in Eq. (3.42), which we call the Coulomb logarithm $\ln\Lambda$:

$$\ln\Lambda = -\ln[\sin(\theta_{min}/2)] = \ln\left[\frac{4\pi\epsilon_0\mu\,|\mathbf{v}_1 - \mathbf{v}_2|^2\lambda_D}{q_1 q_2}\right]. \tag{3.43}$$

Here we have made the approximation $\sin(\theta_{min}/2) \approx \tan(\theta_{min}/2)$.

Because of the slow variation of $\ln\Lambda$ with relative velocity $|\mathbf{v}_1 - \mathbf{v}_2|$, $\ln\Lambda$ is usually taken outside the integral over background velocities and taken as a constant

Table 3.1 Coulomb logarithm for different species and temperature ranges

Collision type	Temperature range	$\ln \Lambda$
Electron–electron	All	
		$30.5 - 0.5 \ln n_e + 1.25 \ln T_e - \left[10^{-5} + (\ln T_e - 2)^2/16 \right]^{1/2}$
Electron–ion	$T_i m_e/m_i < T_e < 10 Z^2$ eV	$30 - 0.5 \ln n_e - \ln Z + 1.5 \ln T_e$
(=ion–electron)	$T_i m_e/m_i < 10 Z^2$ eV $< T_e$	$31 - 0.5 \ln n_e + \ln T_e$
	$T_e < T_i m_e/m_i$	$23 - 0.5 \ln n_i + 1.5 \ln T_i - 2.0 \ln Z$
		$+ \ln (m_i/m_p)$
Mixed ion–ion	All	
		$30 - \ln \left[\dfrac{Z_1 Z_2 (m_1 + m_2)}{m_1 T_2 + m_2 T_1} \right] - 0.5 \ln \left(\dfrac{n_1 Z_1^2}{T_1} + \dfrac{n_2 Z_2^2}{T_2} \right)$
Counter-streaming ions	$v_{thi1}, v_{thi2} < v_D < v_{the}$	
		$50 - \ln \left[\dfrac{Z_1 Z_2 (m_1 + m_2) m_p}{m_1 m_2 (v_D/c)^2} \right] - 0.5 \ln n_e + 0.5 \ln T_e$

Units are density in m^{-3} and temperature in eV. Adapted from [5]

with $|\mathbf{v}_1 - \mathbf{v}_2|^2 \approx 2T/\mu$. This gives (for charges $|q_1| = |q_2| = |q_e|$ and equal temperatures $T_1 = T_2$):

$$\ln \Lambda = \ln \left[\frac{12\pi/\sqrt{2} \epsilon_0^{3/2} T^{3/2}}{n^{1/2} |q_e|^3} \right] \approx 30 + 1.5 \ \ln \ T \ (\text{eV}) - 0.5 \ \ln n \ (\text{m}^{-3}) \quad (3.44)$$

However we can consider the functional form of $\ln \Lambda$ as being equivalent to $\ln(b_{max}/b_{min})$ for whatever b_{max} and b_{min} are most appropriate. In fact setting $b_{min} = b_{90}$ may be too small a value if this choice is less than half the deBroglie wavelength $\approx \hbar/(2\mu\bar{v})$ of the particles and a classical treatment of the scattering dynamics is inappropriate at this length scale. For electrons scattering on electrons, these two lengths become equal at temperatures around 10 eV and the deBroglie half-wavelength exceeds b_{90} at temperatures higher than that; for ions the temperature must be a factor m_i/m_e times higher, or about 35 keV (deuterium). Table 3.1 gives formulas for $\ln \Lambda$ for various reaction types and temperature ranges.

3.2.2 Momentum and Energy Loss for a Test Particle

We find the total momentum loss for a test particle (labeled 1) by summing Eq. (3.41) over all species i and obtain

$$\frac{dp_{1\parallel}}{dt} = - \sum_i \frac{q_1^2 q_i^2 \ln \Lambda_{1,i}}{4\pi \epsilon_0^2 \mu_{1,i}} \int d\mathbf{v}_i \frac{(\mathbf{v}_1 - \mathbf{v}_i) f(\mathbf{v}_i)}{|\mathbf{v}_1 - \mathbf{v}_i|^3}. \quad (3.45)$$

Here $\mu_{1,i}$ means the reduced mass $m_1 m_i/(m_1 + m_i)$ for the i-th component, and the summation is taken over all species (including the projectile species, if applicable). Similarly the Coulomb logarithm $\ln \Lambda_{1,1}$ might need to be calculated for each species for careful work.

The energy loss for a test particle is derived in a way analogous to the momentum loss. If we follow the energy of particle 1 before and after a collision, we can write [4]:

$$
\begin{aligned}
\delta W &= (m_1/2)|\mathbf{v}_1 + \Delta\mathbf{v}_1|^2 - (m_1/2)|\mathbf{v}_1|^2 = m_1\mathbf{v}_1 \cdot \Delta v_1 + m_1|\Delta\mathbf{v}_1|^2 \\
&= \frac{m_1(m_1\mathbf{v}_1 + m_2\mathbf{v}_2)\cdot\Delta\mathbf{v}_1 + m_1 m_2(\mathbf{v}_1 - \mathbf{v}_2)\cdot\Delta\mathbf{v}_1}{m_1 + m_2} + (m_1/2)|\Delta\mathbf{v}_1|^2 \\
&= m_1\mathbf{V}_{CM}\cdot\Delta\mathbf{v}_1 + \mu(\mathbf{v}_1 - \mathbf{v}_2)\cdot\Delta\mathbf{v}_1 + (m_1/2)|\Delta\mathbf{v}_1|^2 \\
&= m_1\mathbf{V}_{CM}\cdot(\Delta\mathbf{v}_1 - \Delta\mathbf{v}_2) + \mu^2/m_1((\mathbf{v}_1 - \mathbf{v}_2)\cdot(\Delta\mathbf{v}_1 - \Delta\mathbf{v}_2) \\
&\quad + (1/2)(\mathbf{v}_1 - \mathbf{v}_2)\cdot(\mathbf{v}_1 - \mathbf{v}_2)) \\
&= m_1\mathbf{V}_{CM}\cdot\Delta\mathbf{v}_1 \\
&= \mathbf{V}_{CM}\cdot\Delta\mathbf{p}_1 .
\end{aligned}
\tag{3.46}
$$

Then we can integrate over the impact parameter and relative velocity to obtain

$$
\frac{dW_1}{dt} = -\sum_i \frac{q_1^2 q_i^2 \ln \Lambda_{1,i}}{4\pi \epsilon_0^2 \mu_{1,i}} \int d\mathbf{v}_i \frac{f(\mathbf{v}_i)\mathbf{V}_{CM}\cdot(\mathbf{v}_1 - \mathbf{v}_i)}{|\mathbf{v}_1 - \mathbf{v}_i|^3}.
\tag{3.47}
$$

3.2.3   Energy and Momentum Loss in a Maxwellian Plasma

Equations (3.41) and (3.47) can be integrated for the case of a maxwellian background plasma. The background distribution function is taken to be

$$
f_i(\mathbf{v}_i) = n_i\left(\frac{m}{2\pi T_i}\right)^{3/2}\exp\left(-\frac{m_i v_i^2}{2T_i}\right) = n_i b_i^3 \pi^{-3/2}\exp\left(-b_i^2 v_i^2\right),
\tag{3.48}
$$

where $b_i \equiv (m_i/(2T_i))^{1/2}$. Then the momentum and energy loss are given by [12]:

$$
\begin{aligned}
\frac{dp_\parallel}{dt} &= \sum_i \frac{n_i q_1^2 q_i^2 b_i^2 \ln \Lambda_{1,i}}{4\pi \epsilon_0^2 \mu_{1,i}} \frac{H(x,0)}{x} \\
\frac{dW_1}{dt} &= \sum_i \frac{n_i q_1^2 q_i^2 b_i \ln \Lambda_{1,i}}{4\pi \epsilon_0^2 m_i} H(x, m_i/m_1)
\end{aligned}
\tag{3.49}
$$

Here

$$
x \equiv b_i v_1
\tag{3.50}
$$

Fig. 3.5 Plots of the function $H(x, y)$ for various values of y

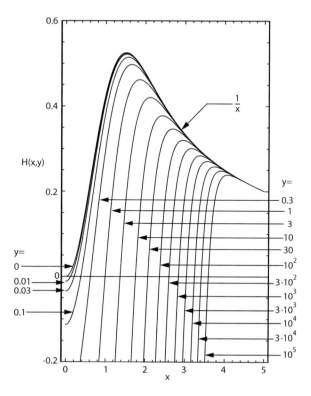

and

$$H(x, y) \equiv \frac{\text{erf}(x)}{x} - \frac{2}{\sqrt{\pi}}(1 + y)\exp\left(-x^2\right) \qquad (3.51)$$

A plot of the function $H(x, y)$ is shown as Fig. 3.5.

It is usually sufficient to work with asymptotic values for the function $H(x, y)$. For example, for a 3.5 MeV alpha particle released from a DT fusion event, the quantity $b_{\alpha/e} v_\alpha$ for slowing down on electrons is about 0.18. A Taylor series expansion of $H(x, y)$ around $x = 0$ gives

$$H(x, y) = -\frac{2y}{\sqrt{\pi}} + \frac{2x^2(3y + 2)}{3\sqrt{\pi}} + \frac{x^4(-5y - 4)}{5\sqrt{\pi}} + O\left(x^5\right). \qquad (3.52)$$

For the case of alphas on electrons, $y = m_e/m_\alpha \approx 1/7300$ and we can drop the y terms. Then to first order

$$H(x, y) \approx \frac{4x^2}{3\sqrt{\pi}}, \qquad (3.53)$$

which is within 3% of the actual value. This means that the alpha energy loss is
given by

$$\left. \frac{dW_\alpha}{dt} \right|_{\alpha/e} \approx -\frac{n_e q_\alpha^2 q_e^2 m_e^{1/2} \ln \Lambda v_\alpha^2}{6\sqrt{2}\pi^{3/2} \epsilon_0^2 T_e^{3/2}} \tag{3.54}$$

Conversely, for the alpha particles slowing down on ions, $x \gg 1$. For a 15 keV
ion temperature and a 3.5 MeV alpha, the quantity $x = b_{\alpha/DT} v_\alpha \approx 11.4$ (taking
an average mass of 2.5 proton masses) and the asymptotic value of $H(x) \approx 1/x$ at
large x, as can be seen from Fig. 3.5. Substituting this value for the function $H(x, y)$
gives the result

$$\left. \frac{dW_\alpha}{dt} \right|_{\alpha/DT} \approx -\frac{n_{DT} q_\alpha^2 q_e^2 \ln \Lambda}{4\pi \epsilon_0^2 m_{DT} v_\alpha} \tag{3.55}$$

Energy loss rates for an alpha particle in a 15 keV ($T_e = T_i$) DT plasma are shown
for electrons, deuterons, tritons, and a fictitious particle with $Z = 1$ and $m = 2.5m_p$
are shown in Fig. 3.6. The asymptotic values of these in accordance with Eqs. (3.54)
and (3.55) are shown as well. One can see that at the higher energy portion of the
alpha slowing down energy spectrum (above $\approx$600 keV), the electrons receive most
of the alpha energy, whereas the ions receive more energy in the lower portion.

Note in Fig. 3.6 that both the electron and ion energy loss rates approach zero
as the particle undergoing the slowing down process approaches the thermal energy
in the plasma. In the case of alphas slowing down on electrons, this happens where
$W_\alpha \approx (3/2)T_e$; for alphas on deuterium or tritium, the cutoff is closer to $W_\alpha \approx T_i$.

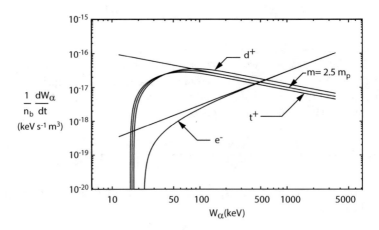

Fig. 3.6 Energy loss rates (in units of $(1/n_b)dE_\alpha/dt$) for an alpha particle in a D-T plasma with
temperature $T_e = T_i = 15$ keV, with asymptotic values shown

It is also interesting to note that using the asymptotic forms for dE/dt from Eqs. (3.54) and (3.55), we can see that the two rates are equal at a "critical" energy E_{crit} by equating these two terms and taking the $2/3$ power of each. This gives

$$E_{crit} = \left(\frac{3}{4}\sqrt{\pi}\right)^{2/3} \left(\frac{m_i}{m_e}\right)^{1/3} \frac{m_1}{m_i} T_e. \tag{3.56}$$

Then the overall energy loss rate can be written in terms of the electron loss rate as

$$\frac{dE_1}{dt} = \frac{dE_1}{dt}\bigg|_e \left[1 + \left(\frac{E_{crit}}{E}\right)^{3/2}\right]. \tag{3.57}$$

This means that the fraction of the energy going from test particle 1 to the ions can be found by integrating over the ratio of dE/dt going to the ions to the total, which is given by

$$f_i = \frac{1}{E_{max} - E_{min}} \int_{E_{min}}^{E_{max}} dE \frac{dE/dt|_i}{dE/dt|_i + dE/dt|_e}. \tag{3.58}$$

By substituting $x = E/E_{crit}$, and using the approximate form shown in Eq. (3.57), we have

$$f_i = \frac{1}{x_{max} - x_{min}} \int_{x_{min}}^{x_{max}} dx \frac{1}{1 + x^{3/2}}$$

$$= \frac{\left(-6\log\left(\sqrt{x}+1\right) + 3\log\left(x - \sqrt{x}+1\right) - 6\sqrt{3}\tan^{-1}\left(\frac{1-2\sqrt{x}}{\sqrt{3}}\right) + \sqrt{3}\pi\right)\bigg|_{x_{min}}^{x_{max}}}{9(x_{max} - x_{min})} \tag{3.59}$$

This analytical form, with $x_{min} = 0$, gives a result which is within 1% of the exact form. The results of a numerical integration to obtain the fractional energy to the ions and electrons are shown as Fig. 3.7.

3.2.4 Beam-Plasma Fusion with Slowing Down

We are now in a position to discuss the competition between Coulomb collisions and beam-background fusion reactions. This topic has relevance when the beam energy is somewhat above the D-T fusion cross section peak (in the laboratory frame, about 110 keV for fast d^+ ions on a t^+ target plasma). We want to find the ratio of fusion energy released per beam particle to the beam particle's initial energy. We can call this quantity $Q_{beam-plasma}$. Ignoring removal of the beam particles by fusion reactions, we have

Fig. 3.7 Fraction of 3.5 MeV
alpha energy to ions and
electrons for a 50–50 D-T
plasma with $T_e = T_i$

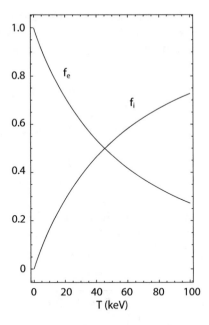

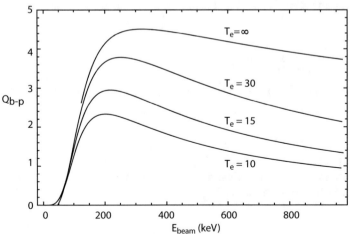

Fig. 3.8 Beam-plasma energy multiplication Q_{bp} as a function of temperature ($T_e = T_i$) for a d^+
beam on a tritium target plasma

$$Q_{beam-plasma} = \frac{1}{E_{beam}} \int_{E_{th}}^{E_{beam}} dE \frac{< \sigma_f v > (E_{beam}, T) E_f}{(1/n_i) dE/dt}, \qquad (3.60)$$

where $< \sigma_f v > (E_{beam}, T)$ was given in Eq. (2.35). Figure 3.8 shows the results
of this numerical integration for different beam energies and background plasma
temperatures.

3.2.5 Runaway Electrons

Another example of the application of the slowing down formulas is the study of
runaway electrons in a plasma subjected to an electric field [6]. Since in general at
high electron energies the electron drag force from both the ions and electrons scales
as $1/v_e^2$, a sufficiently fast electron can gain momentum from the electric field faster
than it loses it by drag. Recalling the momentum equation (Eq. (3.49)), we can write
the total drag force on the electron as

$$\frac{dp_\parallel}{dt} = -\frac{e^4 n_e \ln \Lambda}{4\pi \epsilon_0^2 T_e}\left[\frac{H(x,0)}{x} + \frac{1}{n_e m_e}\sum_i m_i n_i Z_i^2 \frac{H((m_i/m_e)^{1/2}x,0)}{2(m_i/m_e)^{1/2}x}\right],$$

(3.61)

where the index i is taken over all ion species present and $x = v_e/v_{th} = v_e/(2T_e/m_e)^{1/2}$. A plot of the normalized parallel friction (the part of the expression
inside the brackets), showing the contributions of the electrons and ions, is shown
as Fig. 3.9. We can then determine the fraction of the electron population that is

Fig. 3.9 Normalized drag for
runaway electron calculation

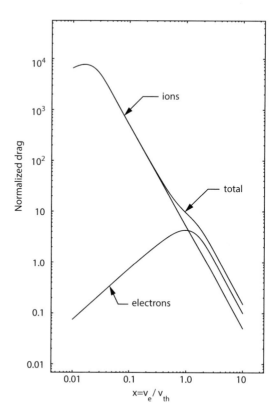

susceptible to runaway by finding the velocity of an electron where the electric force just matches the drag, i.e. where

$$eE = \left| \frac{dp_{\parallel}}{dt} \right|,$$
(3.62)

or when

$$E = \frac{e^3 n_e \ln \Lambda}{4\pi \epsilon_0^2 T_e} \left[\frac{H(x,0)}{x} + \frac{1}{n_e m_e} \sum_i m_i n_i Z_i^2 \frac{H((m_i/m_e)^{1/2} x, 0)}{2(m_i/m_e)^{1/2} x} \right]$$
(3.63)

A plot of the electric field required for runaway as a function of the dimensionless velocity $x = V_e/(2T_e/m_e)^{1/2}$ is shown as Fig. 3.10 for the case of a deuterium plasma with a density of 10^{20} m^{-3} and a temperature of $T_e = T_i = 15$ keV.

It is then a simple matter to find the fraction of the electrons in a thermal plasma whose velocity exceeds the runaway threshold. The one-dimensional electron distribution function normalized to one is given by

$$f_e(x) = \frac{1}{\sqrt{\pi}} e^{-x^2}$$
(3.64)

Fig. 3.10 Critical electric field for runaway vs. normalized velocity $v_e/(2T_e/m_e)^{1/2}$ for a deuterium plasma with $n_e = 10^{20}$ m^{-3} and a temperature of $T_e = T_i = 15$ keV

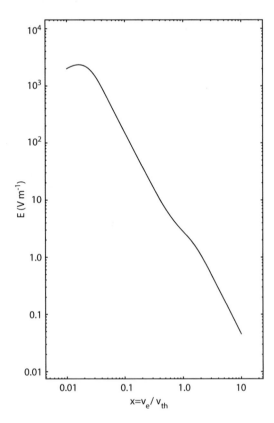

and then the fraction of electrons with normalized velocities above some critical value x_{crit} is given by

$$\frac{n_R}{n_e} = \frac{1}{\sqrt{\pi}} \int_{x_{crit}}^{\infty} dx \ e^{-x^2} = 2\text{erfc}(x_{crit}) \qquad (3.65)$$

Example: Runaway Fraction in a Magnetically Confined Plasma

As an example, for the density and temperature given earlier, an electric field of $0.5\,\text{V}\,\text{m}^{-1}$ gives a value of $x_{crit} = 3.017$ and this gives a runaway fraction $n_R/n_e = 9.9 \cdot 10^{-6}$. If we consider that the runaways pick up enough energy to reach a velocity near the speed of light, then they will be carrying a current density of $J_R \approx n_R ec \approx 50\,\text{kA}\,\text{m}^{-3}$. At this electric field, that amounts to an energy transfer from the electric field $\mathbf{J}_R \cdot \mathbf{E} \approx 25\,\text{kW}\,\text{m}^{-3}$. The runaway electrons can produce radiation by bremsstrahlung (radiation caused by collisions with ions, as will be explained in a subsequent chapter). Their orbital trajectories may allow them to collide with the walls of the plasma chamber, also producing radiation and causing large heat loads in unpredictable places.

The impact of runaway electrons is profoundly influenced by the plasma density, temperature, and the level of the electric field. A plasma with a density below $10^{19}\,\text{m}^{-3}$ and temperatures above $1\,\text{keV}$ will have a strong runaway population and will channel a great deal of energy loss into the runaway population. This is especially a problem at plasma startup, where densities tend to be low and the electric field required for breakdown of the initial gas fill can be quite large. This is mitigated somewhat by the drag on the electrons caused by collisions with neutral gas atoms and molecules.

The effects of runaway electrons include not only radiation loss from the plasma, but the potential threat of radiation exposure to personnel close to the reactor, if adequate shielding and interlocks are not in place. Some tokamak experiments in the 1970s produced several hundred rem per shot near the machine—an almost lethal dose. Furthermore, energetic photons (MeV range) can cause inelastic reactions in limiter and divertor materials such as tungsten and molybdenum. Specifically, $^{74}\text{W}(\gamma, n)$ and $^{74}\text{W}(\gamma, p)$, $^{42}\text{Mo}(\gamma, n)$, $^{42}\text{Mo}(\gamma, p)$, and $^{42}\text{Mo}(\gamma, \alpha)$ reactions have been studied for consideration of induced activity in the machine. Also, the photodissociation of deuterium $d(\gamma, n)p$, with a threshold of $2.2\,\text{MeV}$, can be a source of neutrons as well. In the early days of fusion research, experimenters often found neutrons present and would occasionally be deluded into thinking that these had been produced by fusion reactions, when in reality the neutrons probably came from the photon-induced reactions just described, with runaway electrons ultimately being the culprit.

3.3   Collisions in Velocity Space

Consider a distribution of particles in space with a variety of velocities $\mathbf{v}$. The distribution function $f(\mathbf{x}, \mathbf{v}, t)$ is defined so that $f(\mathbf{x}, \mathbf{v}, t)d^3\mathbf{x}d^3\mathbf{v}$ is the number of particles contained in the interval $(\mathbf{x}, \mathbf{x} + d\mathbf{x}) \otimes (\mathbf{v}, \mathbf{v} + d\mathbf{v})$. Without collisions, conservation of particles as this interval evolves implies that the total derivative on $f \equiv (d/dt)f = 0$. Invoking the chain rule gives:

$$\frac{df}{dt} = \frac{\partial f}{\partial t} + \mathbf{v} \cdot \nabla f + \frac{d\mathbf{v}}{dt} \cdot \frac{\partial f}{\partial \mathbf{v}} = 0. \tag{3.66}$$

Using Newton's law and considering only electromagnetic forces allows us to write $d\mathbf{v}/dt = (q/m)(\mathbf{E} + \mathbf{v} \times \mathbf{B})$, or

$$\frac{\partial f}{\partial t} + \mathbf{v} \cdot \nabla f + \frac{q}{m}(\mathbf{E} + \mathbf{v} \times \mathbf{B}) \cdot \frac{\partial f}{\partial \mathbf{v}} = 0. \tag{3.67}$$

This is called the Vlasov equation [13, 14]. Many problems involving waves and high-frequency instabilities can be addressed with this equation.

We now consider a collision process to be present which can modify the distribution function of species 1 by collisions with species i as before. This gives a modification to the Vlasov equation (Eq. (3.67) above)

$$\frac{\partial f_1}{\partial t} + \mathbf{v} \cdot \nabla f_1 + \frac{q}{m}(\mathbf{E} + \mathbf{v} \times \mathbf{B}) \cdot \frac{\partial f_1}{\partial \mathbf{v}} = \left(\frac{\partial f_1}{\partial t}\right)_{\text{coll}(1/i)}. \tag{3.68}$$

(Note that i can include species 1 as well as others.) Collisions are considered to be instantaneous transformations of the velocity vector $\mathbf{v}_1$ at a point, i.e. there is no spatial translation during the collision. For collisions resulting in an additional particle at velocity $\mathbf{v}_1$ after the collision, the primes denote the quantities before the collision and the unprimed symbols indicate the values after the collision. For particles being removed from velocity $\mathbf{v}_1$ by the collision, the primes and unprimed symbols are reversed in time order (allowed because of time reversal symmetry). Then the total collision rate is the sum of interactions placing particles into f_1 at velocity $\mathbf{v}_1$ minus collisions taking particles out of f_1 at velocity $\mathbf{v}_1$. Then the overall collisional time derivative on f_1 is given by the well-known Boltzmann collision integral:

$$\left(\frac{\partial f_1}{\partial t}\right)_{\text{coll}} = \sum_i \int |\Delta\mathbf{v}| \frac{d\sigma}{d\Omega}(|\Delta\mathbf{v}|, \Omega)[f_1(\mathbf{v}_1', t)f_i(\mathbf{v}_i', t)$$

$$- f_1(\mathbf{v}_1, t)f_i(\mathbf{v}_i, t)] d\Omega \, d^3\mathbf{v}_i'. \tag{3.69}$$

Now Eq. (3.69) can be formally rearranged to yield an equation of form (following Chandrasekhar [2]):

$$f_1(\mathbf{v}, t + \Delta t) = \int f_1(\mathbf{v} - \Delta\mathbf{v}, t) F(\mathbf{u} - \Delta\mathbf{u}, \Delta\mathbf{u}) d\Delta\mathbf{v} \qquad (3.70)$$

Now we can analyze this by expanding f_1 in a Taylor series about $\mathbf{v}$ and the transition probability $F(\mathbf{v} - \Delta\mathbf{v}, \Delta\mathbf{v})$ to obtain

$$
\begin{aligned}
f_1(\mathbf{v}, t) + \frac{\partial f_1}{\partial t} \Delta t &= \int d\Delta\mathbf{v} \left\{ f_1(\mathbf{v}, t) - \Delta\mathbf{v} \cdot \frac{\partial f_1}{\partial \mathbf{v}} + \frac{1}{2} \frac{\partial^2 f_1}{\partial \mathbf{v} \partial \mathbf{v}} : \Delta\mathbf{v}\Delta\mathbf{v} - \cdots \right\} \\
&\times \left\{ F(\mathbf{v}, \Delta\mathbf{v}) - \Delta\mathbf{v} \cdot \frac{\partial F}{\partial \mathbf{v}} + \frac{1}{2} \frac{\partial^2 F}{\partial \mathbf{v} \partial \mathbf{v}} : \Delta\mathbf{v}\Delta\mathbf{v} - \cdots \right\}.
\end{aligned}
$$

$$(3.71)$$

The expansion in powers of $\Delta\mathbf{v}$ is justified because Coulomb scattering is dominated by small-angle scattering events and thus $\Delta\mathbf{v}$ is small. Stopping the expansion at second order results in the well-known Fokker-Planck equation. We then look at averaged values for $\Delta\mathbf{v}$ and $\Delta\mathbf{v}\Delta\mathbf{v}$:

$$
\begin{aligned}
&< \Delta\mathbf{v} >= \int F(\mathbf{v}, \Delta\mathbf{v}) \Delta\mathbf{v} d\Delta\mathbf{v} \\
&< \Delta\mathbf{v}\Delta\mathbf{v} >= \int F(\mathbf{v}, \Delta\mathbf{v}) \Delta\mathbf{v}\Delta\mathbf{v} d\Delta\mathbf{v}.
\end{aligned}
$$

$$(3.72)$$

We also note that the transition probability must integrate to one, i.e. $\int F(\mathbf{v}, \Delta\mathbf{v}) d\Delta\mathbf{v} = 1$. We then take the limit of (3.71) as $\Delta t \to 0$ and obtain

$$\left(\frac{\partial f_1}{\partial t} \right)_{\text{coll}} = -\frac{\partial}{\partial \mathbf{v}} \cdot \left(f_1(\mathbf{v}) \frac{d < \Delta\mathbf{v} >}{dt} \right) + \frac{1}{2} \frac{\partial}{\partial \mathbf{v} \partial \mathbf{v}} : \left(f_1(\mathbf{v}) \frac{d < \Delta\mathbf{v}\Delta\mathbf{v} >}{dt} \right) - \cdots$$

$$(3.73)$$

Generally terms of order higher than two are dropped (these are smaller by a factor of $1/\ln \Lambda$). We then examine average collisional rates $d < \Delta\mathbf{v} > /dt >$ and $d < \Delta\mathbf{v}\Delta\mathbf{v} >$ as the sum of collisional integrals over the background distributions i as

$$\frac{d < \Delta\mathbf{v} >}{dt} = \sum_i \int |\Delta\mathbf{v}| \frac{d\sigma}{d\Omega} (|\Delta\mathbf{v}|, \Omega) \Delta\mathbf{v} f_i(\mathbf{v}_i, t)] d\Omega \, d^3\mathbf{v}_i' \qquad (3.74)$$

and

$$\frac{d < \Delta\mathbf{v}\Delta\mathbf{v} >}{dt} = \sum_i \int |\Delta\mathbf{v}| \frac{d\sigma}{d\Omega} (|\Delta\mathbf{v}|, \Omega) \Delta\mathbf{v}\Delta\mathbf{v} f_i(\mathbf{v}_i, t)] d\Omega \, d^3\mathbf{v}_i \qquad (3.75)$$

The integration of these expressions over solid angle $d\Omega$ is straightforward and leads to the Coulomb logarithm as discussed earlier. The integrals of $\Delta\mathbf{v}$ and $\Delta\mathbf{v}\Delta\mathbf{v}$ over solid angle then become

$$\int d\Omega\, \Delta\mathbf{v} \equiv \{\Delta\mathbf{v}\} = -\frac{Z_1^2 Z_i^2 e^4 \ln\Lambda}{4\pi\epsilon_0^2 \mu_{1i} m_1 |\mathbf{v}_1 - \mathbf{v}_i|^3}(\mathbf{v}_1 - \mathbf{v}_i)$$

$$\int d\Omega\, \Delta\mathbf{v}\Delta\mathbf{v} \equiv \{\Delta\mathbf{v}\Delta\mathbf{v}\} = \frac{Z_1^2 Z_i^2 e^4 \ln\Lambda}{4\pi\epsilon_0^2 m_1^2}\left(\mathbf{I} - \frac{(\mathbf{v}_1 - \mathbf{v}_i)(\mathbf{v}_1 - \mathbf{v}_i)}{|\mathbf{v}_1 - \mathbf{v}_i|^2}\right)$$

(3.76)

In Ref. [10], the authors note that the velocity-dependent terms given in Eq. (3.76) can be written in a form suggestive of the expressions appearing in electromagnetic potential theory. They noted that, using $u \equiv |\mathbf{v}_1 - \mathbf{v}_i|$ and $\Gamma_1 \equiv Z_1^2 Z_i^2 e^4 \ln\Lambda/(4\pi\epsilon_0^2 m_1^2)$:

$$\{\Delta\mathbf{v}\} = \Gamma_1 \frac{m_1}{\mu_{1i}} \frac{\partial}{\partial\mathbf{v}_1}\frac{1}{u}$$

$$\{\Delta\mathbf{v}\Delta\mathbf{v}\} = \Gamma_1 \frac{\partial^2 u}{\partial\mathbf{v}\partial\mathbf{v}}$$

(3.77)

We then obtain simplified expressions for the collision rates given in Eqs. (3.74) and (3.75):

$$\frac{d}{dt} < \Delta\mathbf{v} > = \Gamma_1 \frac{\partial h_1}{\partial\mathbf{v}}$$

$$\frac{d}{dt} < \Delta\mathbf{v}\Delta\mathbf{v} > = \Gamma_1 \frac{\partial^2 g_1}{\partial\mathbf{v}\partial\mathbf{v}}$$

(3.78)

with the "Rosenbluth potentials" defined as

$$h_1 = \sum_i \frac{m_1 + m_i}{m_i}\int dv_i' f_i(\mathbf{v}_i')|\mathbf{v} - \mathbf{v}_i|^{-1}$$

$$g_1 = \sum_i \int dv_i' f_i(\mathbf{v}_i')|\mathbf{v} - \mathbf{v}_i|.$$

(3.79)

We can then substitute the expressions from Eq. (3.78) into Eq. (3.73) to obtain

$$\frac{1}{\Gamma_1}\left(\frac{\partial f_1}{\partial t}\right)_{coll} = -\frac{\partial}{\partial\mathbf{v}}\cdot\left(f_1\frac{\partial h_1}{\partial\mathbf{v}}\right) + \frac{1}{2}\frac{\partial}{\partial\mathbf{v}\partial\mathbf{v}}:\left(f_1\frac{\partial g_1}{\partial\mathbf{v}\partial\mathbf{v}}\right)$$

(3.80)

The resultant fourth-order, nonlinear, three-dimensional partial differential equation for the evolution of the distribution function of a single species through collisions with itself and with other species may seem intractable at first. But often one is looking for solutions in a simplified geometry and with a distribution function which is close to a maxwellian to start with, and in fact a number of linearized, one-dimensional calculations with either a single species or a binary system where self-collisions are unimportant can be obtained from the Fokker-Planck equation with meaningful results.

3.4 Dynamics of Particles in Magnetic Fields

One of the two main approaches to obtaining fusion conditions involves confining the plasma particles in a magnetic field. This is called "magnetic confinement." In the alternative approach, the plasma is heated and compressed by photons. Magnetic fields are not strictly required in this approach, but they may be generated anyway due to certain instabilities, or magnetic fields may be induced on the target intentionally. In any case, the subject of the orbits of particles in magnetic fields is essential.

3.4.1 Cyclotron Motion

The simplest case to understand is when a particle is in a constant magnetic field $\mathbf{B} = \hat{z}\, B_0$. The equation of motion for the particle is written as

$$\begin{aligned} m\dot{v}_x &= qv_y B_0 \\ m\dot{v}_y &= -qv_x B_0 \\ m\dot{v}_z &= 0. \end{aligned} \tag{3.81}$$

The last equation shows that mv_z is a constant, i.e. the momentum P_z is an invariant. Differentiating the first two equations in time and substituting gives

$$\begin{aligned} \ddot{v}_x + (qB_0/m)^2 v_x &= 0 \\ \ddot{v}_y + (qB_0/m)^2 v_y &= 0. \end{aligned} \tag{3.82}$$

We define $\Omega \equiv qB/m$ as the cyclotron frequency. Equations (3.82) describe simple harmonic motion, but with coupling between the x and y motion. For an initial condition $v_x(0) = 0, x(0) = v_0/\Omega, v_y(0) = -v_0, y(0) = 0$, we have a solution

$$\begin{aligned} x(t) &= v_0/\Omega \cos(\Omega t) \\ y(t) &= -v_0/\Omega \sin(\Omega t). \end{aligned} \tag{3.83}$$

Thus particles with a positive charge rotate in a negative direction (clockwise) about a magnetic field line and negative charges (electrons) rotate in a positive (counterclockwise) sense. Note that constants of integration are not allowed to the v_x and v_y solutions, i.e. the particle is "pinned" to a magnetic field line. The radius of the rotational orbit is $\rho = v_0/|\Omega|$, called the gyroradius. For a 15 keV electron and ion (deuterium) temperature in a 4.0 T magnetic field, taking $v_{th} = (2T/m)^{1/2}$, gives $\rho_i \approx 100\,\mu\text{m}$ and $\rho_i \approx 6.2\,\text{mm}$. The smallness of these gyroradii compared to typical plasma sizes (in magnetic confinement systems) allows some general fluid approximations to describe low-frequency bulk motion of the plasma.

3.4.2   E × B *Drift*

We can modify the situation just described by adding a static electric field in a direction perpendicular to **B**. Without loss of generality, we add a constant electric field in the $\hat{y}$ direction, thus $\mathbf{E} = E_0 \hat{y}$. We take $E_0 \ll cB_0$ to avoid an unphysical solution for the classical treatment used here. The equations of motion become

$$
\begin{aligned}
m\dot{v}_x &= qv_y B_0 \\
m\dot{v}_y &= -qv_x B_0 + qE_0 \\
m\dot{v}_z &= 0.
\end{aligned}
\tag{3.84}
$$

Note that differentiating these equations in time gives the same set of equations as Eq. (3.82), except for a forcing term in the x component:

$$
\begin{aligned}
\ddot{v}_x + \Omega^2 v_x &= \Omega^2 E_0/B_0 \\
\ddot{v}_y + \Omega^2 v_y &= 0.
\end{aligned}
\tag{3.85}
$$

We can adapt the boundary conditions to retain the solution from before with superposition of the forcing term by setting $v_x(0) = E_0/B_0$, $x(0) = v_0/\Omega$, $v_y(0) = -v_0$, $y(0) = 0$, and then the solution is

$$
\begin{aligned}
x(t) &= v_0/\Omega \, \cos(\Omega t) + (E_0/B_0)t \\
y(t) &= -v_0/\Omega \, \sin(\Omega t).
\end{aligned}
\tag{3.86}
$$

A solution for this motion (with $\Omega = 1$, $v_0 = 1$, $E_0/B_0 = 0.1$) is shown as Fig. 3.11. We can generalize this result by noticing that if we had put the forcing term qE_0 into the x component of Eq. (3.84) instead of the y component, the drift would be in the $-\hat{y}$ direction, so we conclude that the drift can be written as

$$
\mathbf{v}_D = \frac{\mathbf{E} \times \mathbf{B}}{B^2}
\tag{3.87}
$$

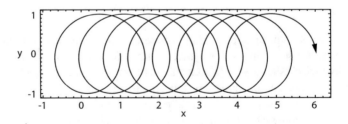

Fig. 3.11 Example of **E** × **B** drift with $\omega = 1$, $v_0 = 1$, and $E_0/B_0 = 0.1$

We notice a few things about this drift result. First, the drift does not depend on the magnitude or sign of the charge q, nor on the mass of the particle, so that all charged particles drift at the same rate, and no net current is generated in a charge-neutral plasma by the $\mathbf{E} \times \mathbf{B}$ drift. Second, the electic field $\mathbf{E}$ in this expression can be replaced by a generalized electric field due to some force $\mathbf{F}$ such that $\mathbf{E}$ is replaced by $\mathbf{F}/q$, giving a generalized drift

$$\mathbf{v}_D = \frac{\mathbf{F} \times \mathbf{B}}{q B^2}, \tag{3.88}$$

and third, neither $\mathbf{E}$ or the equivalent $\mathbf{F}/q$ can be so large that the drift velocity result is comparable to or larger than the speed of light.

3.4.3 Changing Magnetic Field: μ Conservation

Next we can study the case of a particle orbiting in a spatially homogenous magnetic field as before, but now we allow the magnetic field to change in time. An additional electric field must be added to satisfy Maxwell's equation, however. Since $\nabla \times \mathbf{E} = -\dot{\mathbf{B}}$, we must add an electric field to satisfy this. One choice of a self-consistent set of fields is (in cylindrical coordinates $\{r, \theta, z\}$):

$$\begin{aligned} \mathbf{B} &= \hat{z} \, B_0(t) \\ \mathbf{E} &= -\hat{\theta} \, (r/2) d B_0(t)/dt \end{aligned} \tag{3.89}$$

The equations of motion can be written in Cartesian geometry by noting that $\mathbf{E} \cdot \hat{x} = \sin\theta E_\theta = -r \sin\theta \dot{B}_0/2 = -(y/2)\dot{B}_0$ and $\mathbf{E} \cdot \hat{y} = \cos\theta E_\theta = -r\cos\theta \dot{B}_0/2 = -(x/2)\dot{B}_0$. Then the equations of motion become

$$\begin{aligned} m\dot{v}_x &= q v_y B_0(t) - qy/2\dot{B}_0 \\ m\dot{v}_y &= -q v_x B_0(t) - qx/2\dot{B}_0 \\ m\dot{v}_z &= 0. \end{aligned} \tag{3.90}$$

These equations no longer have a closed-form solution, but are rather simple to solve numerically. One such solution is shown here as Fig. 3.12. Note that since

$$\Delta W_\perp = \oint q\mathbf{E} \cdot d\boldsymbol{\ell}, \tag{3.91}$$

the energy of the particle increases when B increases. We note that

$$\oint \mathbf{E} \cdot d\boldsymbol{\ell} = -\frac{d}{dt} \int \mathbf{B} \cdot d\mathbf{S} \approx +\pi\rho^2 \frac{dB}{dt} \tag{3.92}$$

(The sign change is because positively charged particles circulate in a negative (left-handed) direction of gyration, and vice versa.) Note that $\rho = v_\perp / \Omega$ and that

$$\frac{dW_\perp}{dt} = \frac{\Omega}{2\pi} \oint q\mathbf{E} \cdot d\boldsymbol{\ell} = \left(\frac{\Omega}{2\pi}\right) \cdot q \cdot \left(\frac{\pi v_\perp^2}{\Omega^2} \frac{dB}{dt}\right), \qquad (3.93)$$

or

$$\frac{dW_\perp}{dt} = \frac{mv_\perp^2}{2}\left(\frac{1}{B}\frac{dB}{dt}\right). \qquad (3.94)$$

So that we have a relation

$$\frac{1}{W_\perp}\frac{dW_\perp}{dt} = \frac{1}{B}\frac{dB}{dt}, \qquad (3.95)$$

which implies that

$$\frac{d}{dt}\left(\frac{W_\perp}{B} \equiv \mu\right) = 0. \qquad (3.96)$$

The result is only approximate, because the particle does not have a closed orbit to evaluate the line integral $\oint \mathbf{E} \cdot d\boldsymbol{\ell}$, and changes in B must be small over a gyroperiod. We call this an "adiabatic" invariant. For the numerical example shown in Fig. 3.12, a plot of μ vs. time is shown as Fig. 3.13.

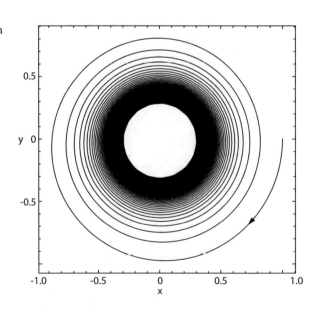

Fig. 3.12 Numerical solution for particle orbiting in an increasing magnetic field, $B_0(t) = 1 + \alpha t$, with $m = q = 1$ and $\alpha = 0.1$. Initial conditions are $x(0) = 1, y(0) = 0, v_x(0) = 0, v_y(0) = -1$

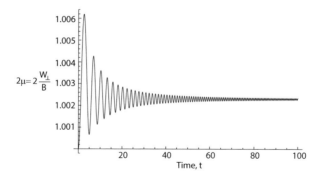

Fig. 3.13 Adiabatic invariant μ vs. time for the particle trajectory shown in Fig. 3.12

As a particle gyrates about a magnetic field line, it also conserves μ without there having to be a time-dependent magnetic field. As the particle visits different parts of space with a changing magnetic field, the changing magnetic field appears in the particle's frame of reference as a time-varying magnetic field, so that the invariance of μ happens in a spatially varying magnetic field as well as a time varying field. A condition for adiabatic invariance for a particle moving in a magnetic field, changing in time or not, is that changes in the field are gradual compared to the gyroradius of the particle in that field, which is equivalent to saying that changes in **B** in time in the particle's frame of reference are gradual over a gyroperiod. The robustness of the invariance of μ has been the subject of much theoretical analysis, and Kruskal proved in 1962 that μ was the lowest-order term in a series expression for some invariant quantity in powers of some smallness parameter ϵ related to the gyroradius-to-scale length ratio [8]. However, this result is not always applicable, and some systems can contain stochastic particle orbits for which no invariant at all can exist. The conditions for stochasticity in particle obits are given in the celebrated Kolomogorov-Arnold-Moser theorem [1, 7, 9].

Another Meaning of μ

Magnetism in solid materials is the result of circulating current loops in the material. This concept dates back to Ampere in 1820. The individual magnetic loop strength **m** is given by the current-area product of the loop IA and is normal to the loop in a right-handed sense. If we look at the orbit of a charged particle in a magnetic field, we see that $I = q\Omega/(2\pi)$ and $A = \pi\rho^2 = \pi v_\perp^2/\Omega^2$ so that

$$|\mathbf{m}| = \left(\frac{q\Omega}{2\pi}\right) \cdot \left(\frac{\pi v_\perp^2}{\Omega^2}\right) = \frac{m v_\perp^2}{2B} = \mu \tag{3.97}$$

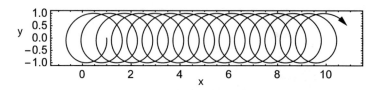

Fig. 3.14 Grad-B drift for a particle with $q = m = 1$, $B_z = 1 + \alpha y$, and $\alpha = -0.2$

Ions travel clockwise on magnetic field lines and electrons travel counterclockwise, so that **m** points opposite **B** and we write

$$\mathbf{m} = -\hat{b}|\mu| \equiv \overrightarrow{\mu} \tag{3.98}$$

3.4.4   Grad-B Drift

Now let B have a steady gradient in the y direction:

$$\mathbf{B} = \hat{z}(B_0 + \alpha y) \tag{3.99}$$

Here $\alpha\hat{y} = \nabla B$. The conservation of μ implies that we can think of the perpendicular energy μB as being "locked up," or as a potential energy Φ. This creates a force $\mathbf{F} = -\nabla\Phi = -\nabla(\mu B) = -\mu\nabla B$. Then applying the force-drift relation from Eq. (3.88), we have

$$v_{drift} = \frac{\mathbf{F} \times \mathbf{B}}{qB^2} = \frac{-\mu\nabla B \times \mathbf{B}}{qB^2} = \frac{mv_\perp^2 \mathbf{B} \times \nabla B}{2qB^3} \tag{3.100}$$

Figure 3.14 shows a solution for a trajectory with $q = m = 1$, $B_z = 1 + \alpha y$, and $\alpha = -0.2$.

Another interpretation of the grad-B drift is as follows. The local curvature of the particles orbit is proportional to the magnetic field at that point. From Fig. 3.14, one can see that the curvature of the particle trajectory is tighter on the bottom of the trajectory (negative y) than on the top. The net effect is that the orbit must make a cylcloidal-type trajectory—thus a geometrical argument not requiring knowledge of μ-conservation.

3.4.5   Curvature Drift

In a manner similar to the grad-B drift, a drift motion associated with a curved magnetic field can be seen by thinking of a particle's trajectory, spiraling on a curved magnetic field, as causing a centrifugal force to be acting on the particle. The general form for a centrifugal force is $mv_\parallel^2/R_c$, where R_c is the radius of curvature, and in the direction away from the center of the "osculating circle," i.e. the best-fitting circle to the curve at this point. If we use the formula from differential geometry for the curvature, we have $\boldsymbol{\kappa} = \hat{\mathbf{b}} \cdot \nabla \hat{\mathbf{b}}$ as the curvature vector, with $\hat{\mathbf{b}}$ representing the unit vector in the direction of $\mathbf{B}$, i.e. $\hat{\mathbf{b}} = \mathbf{B}/B$. We note that $|\boldsymbol{\kappa}| = 1/R_c$ pointing towards the center of the osculating circle, so that the centrifugal force is $-mv_\parallel^2\boldsymbol{\kappa} = -mv_\parallel^2\hat{\mathbf{b}} \cdot \nabla \hat{\mathbf{b}}$. Then using the force-drift relation (Eq. (3.88)), we have

$$\mathbf{v}_{drift} = \frac{-mv_\parallel^2(\hat{\mathbf{b}} \cdot \nabla \hat{\mathbf{b}}) \times \mathbf{B}}{qB^2} \tag{3.101}$$

After some vector manipulation (invoking Maxwell's equation $\nabla \cdot \mathbf{B} = 0$) we find that

$$(\hat{\mathbf{b}} \cdot \nabla \hat{\mathbf{b}}) \times \mathbf{B} = \frac{(\mathbf{B} \cdot \nabla \mathbf{B}) \times \mathbf{B}}{B^2}, \tag{3.102}$$

so that the curvature drift can be written as

$$\mathbf{v}_{drift} = \frac{-mv_\parallel^2(\mathbf{B} \cdot \nabla \mathbf{B}) \times \mathbf{B}}{qB^4}. \tag{3.103}$$

An example of this drift is given here. Suppose that we have a solenoidal, azimuthally symmetric magnetic field given (in cylindrical coordinates $\{r, \theta, z\}$) by

$$\mathbf{B} = B_0\hat{\theta} \tag{3.104}$$

(Note that this field has $\nabla \cdot \mathbf{B} = 0$, which is not only a requirement from Maxwell but also necessary to get the correct answer from particle dynamics.) We start with a particle with initial conditions $x(0) = 10$, $y(0) = 0$, $v_x(0) - 1$, $v_y(0) = -1$, $z(0) = 0$, $v_z(0) = 0$, and take $q = m = B_0 = 1$. The results of this numerical integration are shown in Fig. 3.15. Note the upward drift of the particle in the $+\hat{z}$ direction. This magnetic field has $\nabla B = 0$ and therefore this orbit demonstrates the curvature drift in a field with no grad-B drift. If we had chosen a more typical field where $\mathbf{B} = \hat{\theta}B_0(r_0/r)$, such as the vacuum field inside a toroidal field coil, there would be both grad-B and curvature drifts, in the same direction for the same particle, but opposite in sign for particles of opposite charge.

3.4.6 The Magnetic Mirror Effect

As mentioned earlier, we can treat the perpendicular energy $W_\perp = \mu B$ as being "locked up" and being taken away from the total energy $E = W_\parallel + W_\perp$. Thus a

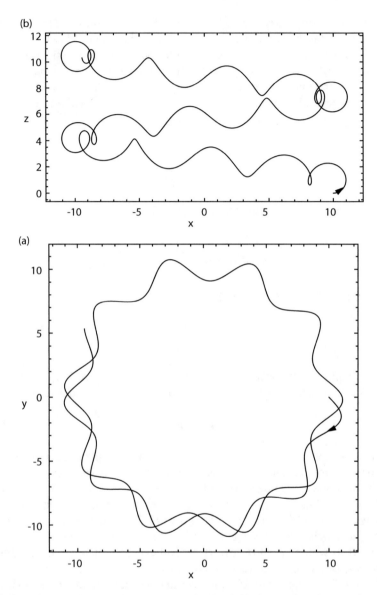

Fig. 3.15 Trajectory of particle in (**a**) the xy plane and (**b**) the xz plane for a magnetic field $\mathbf{B} = B_0\hat{\theta}$

particle launched at some position with magnetic moment μ at magnetic field B, then the particle cannot venture past a place where the magnetic field B is such that $W_\parallel = E - \mu B = 0$. A particle moving in the direction of increasing B may thus bounce off, or be "mirrored," at this point.

Suppose that we look at the initial angle θ_0 that the velocity vector $\mathbf{v}$ makes with the magnetic field, i.e. $\cos\theta_0 = v_\parallel/v$. Then the value of μ is given by $\mu = W_\perp/B = E\sin^2\theta_0/B_0$. Since μ and E are both conserved, the local value of the angle θ is given by $\sin^2\theta = \sin^2\theta_0(B/B_0)$. Since $\sin^2\theta \le 1$, then we predict reflection if

$$\sin^2\theta_0 \ge \frac{B_0}{B_{max}}, \tag{3.105}$$

where B_{max} is the maximum field along the path that the (unmirrored) particle would take. The angle θ_0 defines a "loss cone" in velocity space so that particles inside the loss cone pass through the maximum field position, and particles with a greater pitch angle are reflected.

A simple demonstration of the mirror effect is given here. Consider a magnetic field of the form (in cylindrical coordinates $\{r, \theta, z\}$):

$$\mathbf{B} = \hat{z}(B_0 + \alpha z) - \hat{r}(\alpha/2)r. \tag{3.106}$$

Note that this field has $\nabla \cdot \mathbf{B} = 0$. Now we launch a particle with initial conditions $x(0) = 1$, $y(0) = 0$, $z(0) = 0$, $v_y(0) = -1$, $v_x(0) = 0$, $v_z(0) = 1$ and we take $q = m = B_0 = 1$ and take $\alpha = 0.1$. The orbit of this particle is shown as Fig. 3.16. Note that for this initial condition, $\theta_0 = \pi/4$, and we expect that mirroring will take place where the magnetic field (on axis) is doubled, which is where $z = 10$, and the numerical results show precisely that behavior.

"Mirror machines" were an early approach to magnetically confined fusion devices. They consisted of a pair of circular coils in a Helmholtz-type arrangement, but spaced so that the "mirror ratio" (the ratio of the field at the midsection to the field in the bore of the coils) was at least about 1:2. This allowed a good fraction of the ion population to be trapped by the mirror effect. The electrons would also be lost through the loss cone mechanism, but the buildup of an electrostatic confining potential would equalize their loss to the ion loss. Confinement was determined by the rate at which ions would be scattered into the loss cone. It was found experimentally that losses were fairly high, and certain instabilities further increased the loss rate. Many embellishments were added, such as magnetic coils with more sophisticated shapes to reduce bulk instabilities and directing the output side of one mirror to another mirror with a long, weaker field in between (called the "tandem mirror"), but ultimately the mirror program was phased out in most countries.

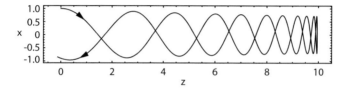

Fig. 3.16 Trajectory of a particle in an increasing magnetic field, showing "mirroring" at the point $z = 10$

But one cannot ignore the mirror effect entirely. In toroidal devices such as the tokamak, the appearance of rising and falling levels of magnetic field along a particle trajectory gives rise to trapped particles, causing some deleterious effects (as well as some beneficial effects). So one might say that the defunct mirror program has come to haunt the rival tokamak program!

3.4.7   Polarization Drift

Consider the case where a *time-dependent* electric field appears in the $\hat{y}$ direction and a static field $\mathbf{B}_0$ is in the $\hat{z}$ direction. The equations of motion, as before, are:

$$
\begin{aligned}
m\frac{dv_x}{dt} &= qv_y B_z \\
m\frac{dv_y}{dt} &= -q(v_x B_z - E_y(t)) \\
m\frac{dv_z}{dt} &= 0
\end{aligned}
\tag{3.107}
$$

However, the procedure used to derive the $\mathbf{E} \times \mathbf{B}$ drift by differentiating this set of equations and substituting brings in a new term because $\mathbf{E}$ now has a time derivative. The new solution for initial condition $v_x(0) = v_0$, $vy(0) = -v_0$, $v_z(0) = v_{z0}$, $x(0) = v_0/\Omega$ is

$$
\begin{aligned}
v_x(t) &= -v_0 \sin \Omega t + \frac{E_y}{B} \\
v_y(t) &= -v_0 \cos \Omega t + \frac{m}{q B^2}\dot{E}_y \\
v_z(t) &= v_{z0}
\end{aligned}
\tag{3.108}
$$

The new term is the polarization drift:

$$
v_p = \frac{m\dot{\mathbf{E}}_\perp}{q B^2}
\tag{3.109}
$$

Notice that unlike the $\mathbf{E} \times \mathbf{B}$ drift, the polarization drift has a charge dependence. There is thus an overall current from the polarization drift:

$$
\mathbf{J}_p = n_i q_i \mathbf{v}_{pi} + n_e q_e \mathbf{v}_{pe} = \frac{n_e m_e + n_i m_i}{B}\frac{d}{dt}\left(\frac{\mathbf{E}}{B}\right)
\tag{3.110}
$$

Note that $\mathbf{J}_p$ is in the same direction as $\mathbf{E}$ but has a time dependence related to the time derivative of $\mathbf{E}$, similar to the displacement current $\epsilon_0\, d\mathbf{E}/dt$ appearing in Maxwell's equation, and so it appears as an adjustment to the dielectric properties of the plasma.

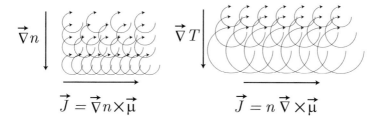

$$\vec{J} = \vec{\nabla}n \times \vec{\mu} \qquad \vec{J} = n\,\vec{\nabla} \times \vec{\mu}$$

Fig. 3.17 Magnetization current caused by density and temperature gradients

3.4.8 Magnetization Current

While not, strictly speaking, a drift term, the magnetization current describes the way that currents can flow in a plasma due to gradients in temperature and density which cause an imbalance in the local cancellation of current by the gyrating particles. Figure 3.17 illustrates the basic concept. We draw the particle trajectories as nearby circles in the plasma. If there are no temperature or density gradients, no net current exists because there is a direct cancellation of the current carried by each particle from its nearest neighbor. However, a density gradient affects the population of neighboring particles, and the cancellation is not exact. Furthermore, a temperature gradient affects the magnetic moments of the particles involved since $<\mu>=<W_\perp/B>=T/B$.

The magnetic moment μ is also the current-area product for the particle, as discussed in Sect. 3.4.3. The magnetization $\mathbf{m} = -\hat{\mathbf{b}}\mu$ reflects the current loop with the correct "handedness," and the overall magnetization density $\mathbf{M} = n\mathbf{m}$. Then we can write

$$\mathbf{J}_m = \nabla \times \mathbf{M}, \qquad (3.111)$$

which includes the two parts shown in Fig. 3.17.

3.4.9 The Plasma Current

We will now demonstrate that combining the drift-induced currents and the magnetization current leads to the same result for the perpendicular current density $\mathbf{J}_\perp$ as one would get from the fluid equilibrium equation $\nabla p = \mathbf{J} \times \mathbf{B}$. Assume a maxwellian distribution for all of the particles in the plasma and note that $< mv_\perp^2/2 >= T$ (since there are two degrees of freedom perpendicular to $\mathbf{B}$) and that $< mv_\parallel^2 >= T$ (one degree of freedom parallel to $\mathbf{B}$). The grad-B and curvature drifts then combine to yield

$$J_{drift} = nq v_{drift} = -\frac{nq <mv_{\parallel}^2> (\mathbf{B} \cdot \nabla)\mathbf{B} \times \mathbf{B}}{qB^4} + \frac{nq <mv_{\perp}^2> \mathbf{B} \times \nabla B}{2qB^3}$$

$$= (nT) \left[-\frac{(\mathbf{B} \cdot \nabla)\mathbf{B} \times \mathbf{B}}{B^4} + \frac{\mathbf{B} \times \nabla(B^2/2)}{B^4} \right]$$

$$= (p) \left[-\frac{(\mathbf{B} \cdot \nabla)\mathbf{B} \times \mathbf{B}}{B^4} - \frac{\nabla(B^2/2) \times \mathbf{B}}{B^4} \right].$$

$$(3.112)$$

Now look at the magnetization current:

$$\mathbf{J}_m = \nabla \times \mathbf{M} = \nabla \times (n\boldsymbol{\mu}) = -\nabla \times \left(\frac{n\hat{b} <W_{\perp}>}{B} \right) = -\nabla \times \left(\frac{nT\mathbf{B}}{B^2} \right).$$

$$= -\frac{\nabla p \times \mathbf{B}}{B^2} - p\frac{\nabla \times \mathbf{B}}{B^2} - 2p\frac{\mathbf{B} \times \nabla B}{B^3}.$$

$$(3.113)$$

Now add the two currents together:

$$\left(\mathbf{J}_m + \mathbf{J}_{drift} \right)_{\perp} =$$

$$\left[-\frac{\nabla p \times \mathbf{B}}{B^2} - p\frac{\nabla \times \mathbf{B}}{B^2} - 2p\frac{\mathbf{B} \times \nabla B}{B^3} + p\left(-\frac{(\mathbf{B} \cdot \nabla)\mathbf{B} \times \mathbf{B}}{B^4} - \frac{\nabla(B^2/2) \times \mathbf{B}}{B^4} \right) \right]_{\perp}$$

$$= -\frac{\nabla p \times \mathbf{B}}{B^2}.$$

$$(3.114)$$

(For the last step, we used a vector identity:

$$(\nabla \times \mathbf{B}) \times \mathbf{B} = -\nabla \left(\frac{B^2}{2} \right) + (\mathbf{B} \cdot \nabla)\mathbf{B}.) \qquad (3.115)$$

This shows that the drift and magnetization terms all add up so that $\mathbf{J} \times \mathbf{B} = \nabla p$, as one might expect from fluid theory. Thus the plasma is "smart enough" to set up drifts and magnetization currents to attain the proper force equilibrium. Notice, however, that the current $\mathbf{J}_{\parallel}$ is not addressed here and this is not derivable from the drift model. This goes back to being able to freely adjust v_z for each particle's initial velocity along $\mathbf{B}$. Other physics is required to obtain solutions for the parallel current in the plasma, which is typically nonzero.

Problems

3.1 Find the slowing down rate dE/dt of a 3.52 MeV alpha particle (He^{2+}) in a DT plasma of 10^{20} m^{-3} density and 10 keV temperature ($T_e = T_i$) using Eqs. (3.54) and (3.55). Calculate the individual losses to the electrons and ions. Assume the background ions have a mass of 2.5 proton masses.

3.2 A plasma has a 6.0 T magnetic field in the $+\hat{z}$ direction with a gradient $dB_z/dx = 0.4\,\mathrm{T\,m^{-1}}$. Find the drift velocity of a deuteron with $W_\perp = 10\,\mathrm{keV}$ in meters per second. Give the vector direction of this drift.

3.3 For the above problem, what value of an electric field would cancel out this drift exactly? Give magnitude, direction, and sign. Is it possible to cancel the drift of electrons at the same time as the ions?

3.4 During re-entry, the Shuttle spacecraft becomes incommunicado on its 432 MHz telemetry radio channel with the Space Center. Find the density of plasma surrounding the spacecraft at the time when this first starts happening.

3.5 Find the overall (electron+ion) Debye length in a laser-fusion plasma at a density of $250\,\mathrm{g\,cm^{-3}}$ and at $10\,\mathrm{keV}$ temperature. Calculate $\ln \Lambda$ for this case. Can one use binary collision theory with a small-angle cutoff in this case?

References

1. Arnol'd, V.I.: Proof of a theorem of A.N. Kolmogorov on the invariance of quasi-periodic motions under small perturbations of the Hamiltonian. Russ. Math. Surv. **18**(5), 9–36 (1963). http://stacks.iop.org/0036-0279/18/i=5/a=R02
2. Chandrasekhar, S.: Stochastic problems in physics and astronomy. Rev. Mod. Phys. **15**, 1–89 (1943). http://link.aps.org/doi/10.1103/RevModPhys.15.1
3. Debye, P., Hückel, E.: The theory of electrolytes. I. Lowering of freezing point and related phenomena. Phys. Z. **24**, 185–206 (1923). http://electrochem.cwru.edu/estir/hist/hist-12-Debye-1.pdf
4. Dolan, T.: Fusion Research. Fusion Research, vol. 1. Elsevier Science, Amsterdam (1982). https://books.google.com/books?id=AahkPAAACAAJ
5. Huba, J.D.: NRL Plasma Formulary. Naval Research Lab, Washington, DC (2016). http://www.nrl.navy.mil/ppd/sites/www.nrl.navy.mil.ppd/files/pdfs/NRL_FORMULARY_16.pdf
6. Knoepfel, H., Spong, D.: Runaway electrons in toroidal discharges. Nucl. Fusion **19**(6), 785 (1979). http://stacks.iop.org/0029-5515/19/i=6/a=008
7. Kolmogorov, A.: On the conservation of conditionally periodic motions under small perturbation of the Hamiltonian. Dokl. Akad. Nauk SSSR **98**, 527–530 (1954). Engl. transl.: Stochastic Behavior in Classical and Quantum Hamiltonian Systems, Volta Memorial Conference, Como, 1977. Lecture Notes in Physics, vol. 93. Springer, Berlin (1979), pp. 51–56. http://dx.doi.org/10.1007/BFb0021737, ISBN 978-3-540-35510-6
8. Kruskal, M.: Asymptotic theory of Hamiltonian and other systems with all solutions nearly periodic. J. Math. Phys. **3**(4), 806–828 (1962). http://dx.doi.org/10.1063/1.1724285; http://scitation.aip.org/content/aip/journal/jmp/3/4/10.1063/1.1724285
9. Moser, J.: On invariant curves of area-preserving mappings of an annulus. Nachr. Akad. Wiss. Göttingen Math.-Phys. Kl. II, **1**, 1–20 (1962)
10. Rosenbluth, M.N., MacDonald, W.M., Judd, D.L.: Fokker-Planck equation for an inverse-square force. Phys. Rev. **107**, 1–6 (1957). http://link.aps.org/doi/10.1103/PhysRev.107.1
11. Tonks, L., Langmuir, I.: Oscillations in ionized gases. Phys. Rev. **33**(2), 195–211 (1929). http://www.columbia.edu/~mem4/ap6101/Tonks_Langmuir_PR29.pdf
12. Trubnikov, B.A.: Particle interaction in a fully ionized plasma. In: Leontovich, M.A. (ed.) Reviews of Plasma Physics, vol. 1, pp. 105–204. Consultants Bureau, New York (1965)

13. Vlasov, A.A.: On vibration properties of electron gas. J. Exp. Theor. Phys. (in Russian) **8**, 291 (1938)
14. Vlasov, A.A.: The vibrational properties of an electron gas. Sov. Phys. Usp. **10**(6), 721–733 (1968). https://doi.org/10.1070/PU1968v010n06ABEH003709; http://ufn.ru/en/articles/1968/6/a/. (English translation of [13])

Chapter 4
Energy Gain and Loss Mechanisms in Plasmas and Reactors

4.1 Bremsstrahlung

Bremsstrahlung (German for "braking radiation") is the most basic radiative energy loss in a plasma, and the most unavoidable. It is caused by the collisions of electrons with ions in the plasma. (Note that electron–electron collisions do not produce bremsstrahlung because of the symmetrical motion of the particles during a collision, and ions do not produce bremsstrahlung in collisions with other ions to any appreciable extent because of their larger mass.) The guiding principle to calculating bremsstrahlung is to look at Larmor's formula, which is a classical expression for the energy lost to radiation by an accelerating electron. In the non-relativistic limit, it is given by:

$$P_{rad} = \frac{e^2}{6\pi\epsilon_0}\frac{\dot{v}^2}{c^3}. \tag{4.1}$$

Consider a simplified model where an energetic electron (mass m_e) encounters an ion (with charge $+Ze$) with a fairly large impact parameter b, so that its trajectory past the ion can be taken as a straight line. We keep track of the total acceleration of the electron, however, even though we approximate the trajectory as rectilinear. Then the acceleration of the electron is given by

$$\dot{v} = \frac{Ze^2}{4\pi\epsilon_0 m_e r^2} = \frac{Ze^2}{4\pi m_e \epsilon_0 (b^2 + (vt)^2)} \tag{4.2}$$

Here we have taken the time t such that $t = 0$ is the time of closest approach. The integrated radiation loss power is then given by

$$E_{rad} = \int P_{rad}\,dt = \frac{2Z^2 e^6}{3(4\pi\epsilon_0)^3 m_e^2 c^3}\int_{-\infty}^{\infty}\frac{1}{\left(b^2 + (vt)^2\right)^2}\,dt = \frac{\pi Z^2 e^6}{3(4\pi\epsilon_0)^3 m_e^2 c^3 b^3 v} \tag{4.3}$$

© Springer Nature Switzerland AG 2018
E. Morse, *Nuclear Fusion*, Graduate Texts in Physics,
https://doi.org/10.1007/978-3-319-98171-0_4

To get the overall specific radiation loss, we multiply by the number of scatterers n_Z and now multiply by the incremental area $2\pi b \, db$ of impact parameter and the velocity v of the incoming particles, and integrate over the impact parameters b and a maxwellian distribution function

$$f_e(v) = n_e \left(\frac{m_e}{2\pi T_e} \right)^{3/2} \exp\left(-m_e v^2 / (2T_e) \right).$$

We still end up with a divergent integral $\int db/b^2$ to integrate over. However, we recognize that we are using classical physics, and we are not taking account of quantum effects which happen on a length scale of a deBroglie (reduced) wavelength $\Delta x = \hbar/(m_e v)$, so we cut off the impact parameter integration at $b_{min} = \hbar/(m_e v)$. We then get a total bremsstrahlung rate per unit volume of

$$P_{rad} = \frac{n_e n_Z Z^2 e^6 Z^2 \sqrt{\dfrac{T_e}{m_e^3}}}{12\sqrt{2\pi}c^3 \epsilon_0^3 h} = 6.57 \times 10^{-37} n_e n_Z Z^2 T_e^{1/2} (\text{keV}) \, \text{Wm}^{-3}, \qquad (4.4)$$

where the last equality takes n_e and n_Z in MKS units.

This result can be refined by taking a more realistic look at the actual particle trajectories. In a 1923 paper, Kramers [17] derived a result for the frequency-dependent bremsstrahlung arising from an electron–ion interaction using a detailed trajectory of the electron and ion from collision theory, keeping the hyperbolic orbits as in the preceding chapter. He then integrated this over impact parameters down to the deBroglie wavelength as before, and obtained the following result for the integrated bremsstrahlung:

$$P_{brem} = \frac{32\pi}{3} \left(\frac{2\pi}{3} \right)^{1/2} \frac{e^6}{(4\pi\epsilon_0)^3 h m_e c^3} \left(\frac{T_e}{m_e} \right)^{1/2} Z^2 n_e n_i \bar{g}$$

$$= 4.83 \times 10^{-37} n_e n_Z Z^2 T_e^{1/2} (\text{keV}) \bar{g}, \qquad (4.5)$$

with $\bar{g} = 1$ for the semi-classical theory of Kramers. A full quantum-mechanical treatment was done by Sommerfeld [29] and put in a more convenient form by Biedenharn [2].

The quantum correction is as follows (see [14]). We define a Gaunt factor for free–free transitions g_{ff} as the dimensionless quantity to multiply the semi-classical expression for the radiation emitted, as a function of the electron energy before and after the photon emission. (In this treatment, we use the non-relativistic quantum theory so that $E_{e^-} = \hbar^2 k^2/(2m)$.) We write down normalized values for the initial and final inverse wavenumbers k_i and k_f by using

$$\eta = \frac{Z e^2}{4\pi\epsilon_0 \hbar v} = \frac{Z}{k a_0}, \qquad (4.6)$$

where a_0 is the Bohr atomic radius (0.51 Å). Conservation of energy requires that $E_f = E_i - h\nu$, where $h\nu$ is the energy of the emitted photon. This can be written as

$$\frac{1}{\eta_f^2} = \frac{1}{\eta_i^2} - \frac{h\nu}{Z^2}. \tag{4.7}$$

Here the photon energy $h\nu$ is in Rydbergs (1 Ry=13.6 eV). The quantum-mechanical correction to the classical expression for the emitted radiation, as a function of the normalized inverse wavenumbers η_i and η_f, is then given by overlap integrals over the Coulomb wave function for the electron before and after the emission of a photon, which can be evaluated using a hypergeometric function:

$$g_{ff} = \frac{2\sqrt{3} I(0, \eta_i, \eta_f) \left((2\eta_f^2 \eta_i^2 + \eta_f^2 + \eta_i^2) I(0, \eta_i, \eta_f) - 2\eta_f \sqrt{\eta_f^2 + 1} \eta_i \sqrt{\eta_i^2 + 1} I(1, \eta_i, \eta_f) \right)}{\pi \eta_f \eta_i}, \tag{4.8}$$

where

$$I(\ell, \eta_i, \eta_f) =$$

$$\frac{4^\ell \left(\dfrac{\eta_f - \eta_i}{\eta_f + \eta_i} \right)^{i\eta_f + i\eta_i} e^{\frac{\pi |\eta_i - \eta_f|}{2}} \left(\dfrac{\eta_f \eta_i}{(\eta_f - \eta_i)^2} \right)^{\ell+1} \left| \Gamma(i\eta_f + \ell + 1)\Gamma(i\eta_i + \ell + 1) \right|}{\Gamma(2\ell + 2)}$$

$$\times\ _2F_1 \left(-i\eta_f + \ell + 1, -i\eta_i + \ell + 1; 2\ell + 2; -\frac{4\eta_f \eta_i}{(\eta_i - \eta_f)^2} \right). \tag{4.9}$$

(See [1] for the properties of the hypergeometric functions.) We then can integrate the Gaunt factor g_{ff} over the distribution of electron energies and photon energies. We make normalized variables by letting

$$u = \frac{h\nu}{T} \quad \text{and} \quad \gamma^2 = \frac{Z^2}{T(\text{Ry})}.$$

Then the overall averaged Gaunt factor $\bar{g}$ is given by:

$$\bar{g} = \int_0^\infty du \left[\int_u^\infty e^{-x} g_{ff} \left(\frac{\gamma}{\sqrt{x}}, \frac{\gamma}{\sqrt{x-u}} \right) dx \right]. \tag{4.10}$$

Note that by replacing the dummy variable x with $x + u$, we arrive at a more convenient form for integration:

$$\bar{g} = \int_0^\infty \int_0^\infty dx\, du\, e^{-(x+u)} g_{ff} \left(\frac{\gamma}{\sqrt{x+u}}, \frac{\gamma}{\sqrt{x}} \right). \tag{4.11}$$

We can also find an asymptotic form for g_{ff} as $\gamma \rightarrow 0$, i.e. at temperatures $T_e \gg 1$ Ry. We note that taking a limit of the hypergeometric function $_2F_1$:

$$_2F_1(1, 1; 2; -x) = \frac{\ln(1 + x)}{x}. \tag{4.12}$$

The Gaunt factor at high temperature becomes

$$\lim_{\epsilon \rightarrow 0} g_{ff}(\epsilon, \epsilon x) = \frac{2\sqrt{3}}{\pi}\left[\frac{1}{4}\ln\left(\frac{(1 + \sqrt{x})^2}{(1 - \sqrt{x})^2}\right)\right], \tag{4.13}$$

and using

$$\int_0^{\infty}\int_0^{\infty} e^{-(s+x)}\ln\left(\frac{(\sqrt{s + x} + \sqrt{x})^2}{(\sqrt{x} - \sqrt{s + x})^2}\right)\,ds\,dx = 4 \tag{4.14}$$

gives

$$\lim_{T_e \rightarrow \infty} \bar{g} = \frac{2\sqrt{3}}{\pi}. \tag{4.15}$$

A plot of $\bar{g}$ as a function of electron temperature is shown in Fig. 4.1. Note that at high temperatures, the asymptotic value for $\bar{g}$ is $2\sqrt{3}/\pi = 1.10266$, as shown above. This results in the bremsstrahlung as frequently quoted in the literature:

$$P_{brem} = 5.35 \times 10^{-37} n_e n_Z Z^2 T_e^{1/2}(\text{keV}). \tag{4.16}$$

As a final refinement, relativistic corrections also become important if the electron temperature is a substantial fraction of the electron rest energy $m_e c^2 = 511$ keV. Then Eq. (4.16) is modified to read [9]:

$$P_{brem} = 5.35 \times 10^{-37} n_e n_Z Z^2 T_e^{1/2}(\text{keV})\left(1 + 5.1 \times 10^{-3}T_e(\text{keV})\right). \tag{4.17}$$

For a multi-species plasma at non-relativistic temperatures, the bremsstrahlung power is found by summing over all ionic charge states Z, giving:

$$P_{brem} = 5.35 \times 10^{-37} n_e T_e^{1/2}(\text{keV})\left(\sum_Z n_Z Z^2\right). \tag{4.18}$$

The Z^2 dependence shows that a small amount of a high-Z impurity can greatly increase the bremsstrahlung power emitted from the plasma. For example, suppose that molybdenum ($Z = 42$) is in a plasma at a concentration f of 0.1% of the

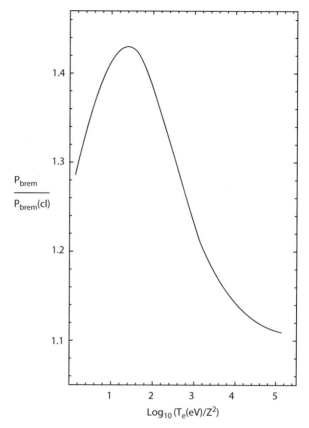

Fig. 4.1 Multiplication factor $\bar{g}$ for quantum Gaunt factor corrections to the semi-classical bremsstrahlung calculation. See [14]

DT fuel density. Then the ratio of the bremsstrahlung power in the "dirty" plasma to that in the "clean" case is found by noticing that in addition to the $\sum n_Z Z^2$ term, the electron density also increases over the "clean" DT case because of charge neutrality. So we have

$$\frac{n_e(\text{dirty})}{n_e(\text{clean})} = 1 + fZ \quad \text{and} \quad \frac{\sum_Z n_Z Z^2(\text{dirty})}{\sum_Z n_Z Z^2(\text{clean})} = 1 + fZ^2 \quad (4.19)$$

Thus the overall ratio of the "dirty" to "clean" bremsstrahlung power is

$$\frac{P_{brem}(\text{dirty})}{P_{brem}(\text{clean})} = (1 + fZ)\left(1 + fZ^2\right), \quad (4.20)$$

which for the example given here gives a ratio $P_{brem}(\text{dirty})/P_{brem}(\text{clean}) = 2.88$, or a 188% increase in the bremsstrahlung radiation.

4.2   Line and Recombination Radiation

The treatment of radiation loss given in the previous section covers only the case of a free electron interacting with an ion in such a way that it remains a free particle after the encounter, i.e. free–free interactions. This is totally adequate for hydrogen and other light ions, which tend to be completely stripped of bound electrons. As the atomic number increases, the energy required to strip all of the electrons from the atom also increases. Figure 4.2 shows the average charge of an oxygen atom as a function of electron temperature. When electrons remain on the oxygen ion ($Z < 8$), then the bound electrons can radiate on the quantized transition energies for the atomic system that the ion is in. This is called line radiation. Figure 4.3 shows the radiation parameter (see below) from oxygen ions at various temperatures. By comparing these two figures as the electron temperature increases, we can see that the radiation is the highest at those temperatures where the average number of electrons is in transition. The two peaks in Fig. 4.3 correspond to those temperatures where the oxygen ion is transitioning from having four bound electrons down to having two bound electrons (i.e., being helium-like) at around 20 eV, and then around 200 eV, where the ions are starting to become fully stripped. The ionization energies for oxygen for O^+ through O^{8+} are 13.62, 35.12, 54.94, 77.41, 113.90, 138.12, 739.29, and 871.4101 eV, respectively. Notice that for the higher ionization states, each two ionization energies are roughly paired together. This is because the last two electrons are in a $1s$ state and the two before that are in a $2s$ state.

4.2.1   Basic Concepts

The most important concept in discussing atomic physics for plasmas is the degree of ionization of the species in the plasma. For hydrogen, the energy required to strip the (only) electron from the ground state is given by

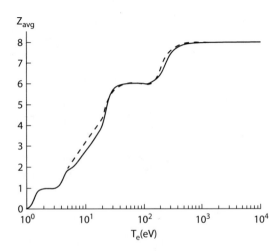

Fig. 4.2 Average charge state z for oxygen in an optically thin plasma as a function of electron temperature. Solid line: data from [23], dashed line: data from [25]. From [23]. Used by permission, Springer

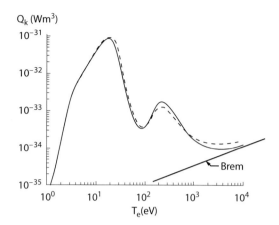

Fig. 4.3 Radiation parameter Q_k for oxygen in an optically thin plasma as a function of electron temperature. Solid line: data from [23], dashed line: data from [25]. From [23]. The contribution from bremsstrahlung is shown. Used by permission, Springer

$$E^0_\infty(1,0) = 13.6 \, \text{eV} \tag{4.21}$$

Here the notation $E^{Z^*}_{n_f}(n,l)$ describes the energy to effect a transition of an atom in charge state Z^* from level n with principal angular quantum number l into final state n_f. Thus $n_f = \infty$ represents an electron completely stripped from the atom. Equation (4.21) can be extended to a "hydrogenlike" atom with nuclear charge Z and charge state $Z^* = Z - 1$, i.e. an atom with just one bound electron as

$$E^{Z-1}_\infty(1,0) = 13.6 Z^2 \, \text{eV}. \tag{4.22}$$

This result is arrived at by noting that writing down Schrodinger's equation with the potential term $-e^2/(4\pi\epsilon_0 r)$ replaced with $-Ze^2/(4\pi\epsilon_0 r)$ and then replacing the radial coordinate r with $r' = r/Z$, and the equations become self-similar with the energy units scaled by $\{E', V'\} = \{E/Z^2, V/Z^2\}$. However, this result is not exact for heavy atoms such as uranium, because the electron spatial scale involves the electron spending time inside the nucleus, and also relativistic effects become important.

The degree to which atoms with intermediate Z become fully stripped depends on the relationship of the energy to strip the last electron 13.6 Z^2 and the electron temperature of the plasma. Clearly if $T_e \gg 13.6 \, Z^2$, then that species will be fully stripped. In this case, the only contribution to the plasma radiation from this species is bremsstrahlung as discussed earlier. But even for the case of oxygen ($Z = 8$, 13.6 $Z^2 \approx 870 \, \text{eV}$), the situation is not so clear for places near the edge of a confined plasma, where low temperatures may result in partially stripped oxygen ions. In the case of uranium ($Z = 92$, 13.6 $Z^2 = 115 \, \text{keV}$), full stripping would rarely be the case.

Some radiative processes can be quite complex, and in recent times codes have been developed to carefully account for the details of the quantum levels of partially

Fig. 4.4 Radiation parameter
Q_k for selected elements.
Data from [3]

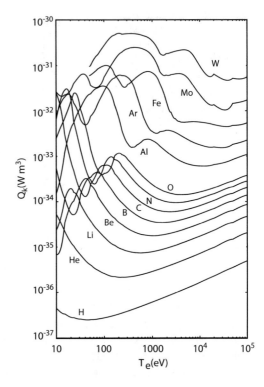

stripped ions and transition rates for the various reactions. Some general physical details are given here. Usually, the overall radiation loss per unit volume is a function of the plasma electron temperature and of the impurity species k involved, so one can construct a radiation coefficient Q_k for each species, so that

$$P_{rad} = n_e \sum_k Q_k n_k \tag{4.23}$$

A plot of the overall radiation parameter Q_k for various elements is shown in Fig. 4.4.

4.2.2 Radiation in Optically Thin Plasmas

The case where the plasma is optically thin and excitation is dominated by electron collisions with partially stripped atoms and de-excitation is primarily radiative is called coronal equilibrium. Atoms which are not fully stripped will undergo electron impact excitations, along with other processes, which can result in excited states in these species. In the case of optically thin plasma, the radiation loss from partially stripped atoms can be severe. The equilibrium population of excited states is determined by the balance of the collisional excitation rate and the decay rate to the less excited states, with three additional features: (1) the partially stripped (or fully stripped) atom can trap an electron in a bound state (this is called recombination radiation), (2) the excited state can become de-excited by an electron collision, and (3) an electron can ionize the partially stripped atom to a higher charge state. The recombination process just mentioned might also involve the excitation of another bound electron simultaneously with the attachment of the formerly free electron, and this is called dielectronic recombination. In general, recombination radiation loss is less than line radiation except at plasma temperatures close to that required for full stripping, but bremsstrahlung dominates at higher temperatures. At thermonuclear temperatures, ($T_e > 10\,\text{keV}$), line radiation becomes a serious issue for $Z \geq 26$ (iron and higher), whereas edge temperatures in the hundreds of eV can radiate severely from carbon and oxygen impurities. We will first look at line radiation, followed by recombination radiation in the subsequent section.

Line Radiation

The radiative de-excitation rate R_{nm} for a transition $m \to n$ is described using the Einstein A coefficient:

$$R_{nm}^{rad} = A_{nm} n_m, \tag{4.24}$$

and the collisional excitation rate R_{mn}^{coll} (for the transition $n \to m$ between these two states is given by

$$R_{mn}^{coll} = n_e n_n < \sigma^{ex}(n \to m) v > \tag{4.25}$$

(Note that here we use the notation of [10], where the order of the subscripts is (final, initial); however, this notation is not universal and may lead to confusion.) We consider the excited states to be weakly populated compared to the ground state and thus take $n_m \ll n_0$. Setting these rates equal gives an expression for the density of the excited state n_m and thus the radiative power density $n_m E_{0m} A_{0m}$:

$$n_0 n_e < \sigma^{ex} v >= n_m A_{0m} \tag{4.26}$$

The Einstein A_{nm} coefficient is the rate of spontaneous emission between the excited state m and de-excited (not necessarily ground) state n of the partially stripped ion and is given by (with the transition energy ΔE_{nm} in eV):

$$A_{nm} = 2\frac{r_0 \omega^2}{c}|f_{nm}| = 2\frac{e^2 \omega^2}{4\pi \epsilon_0 m_e c^3}|f_{nm}| = 4.3 \times 10^7 f_{mn} \frac{g_n}{g_m} (\Delta E_{nm})^2\, \text{s}^{-1}. \tag{4.27}$$

Here f_{nm} is the oscillator strength for the transition, which compares the probability of the transition to a simple harmonic oscillator with the same quantized levels. It is derived from the coupling integral for the electromagnetic plane wave with the two states n and m and is given by

$$f_{nm} = \frac{2}{3}\frac{m_e}{\hbar^2}(E_n - E_m)\sum_{i=1,3} |<n|x_i|m>|^2. \tag{4.28}$$

The statistical factors g_n and g_m are the number of independent states at each level, e.g. $2J + 1$ for states labeled with angular momentum quantum number J.

As an example, here is a calculation of an oscillator strength for a transition in a hydrogenlike atom. Suppose that the initial state is a $2p$ state ($n = 2$, $\ell = 1$, $m_n = 0$) and the final state is the $1s$ ground state ($n = 1$, $\ell = 0$, $m_n = 0$). The wavefunctions associated with these states are (using a reduced radius r as the real radial coordinate divided by the Bohr radius a_0):

$$\Psi_1 = 2\exp(-r)Y_{00}(\theta, \phi) = 2\exp(-r)/\sqrt{4\pi}$$
$$\Psi_2 = \frac{r\exp(-r/2)}{\sqrt{24}}Y_{10}(\theta, \phi) = \frac{r\exp(-r/2)}{\sqrt{24}}\sqrt{\frac{3}{4\pi}}\cos\theta \tag{4.29}$$

For this combination of initial and final states, the only nonzero term in $<n|x_i|m>$ is the z term, i.e. $<n|z|m> = <n|r\cos\theta|m>$. The radial integration gives

$$R_{10}^{21} = \int_0^\infty dr\,R_{10}\cdot r^3\cdot R_{20} = \frac{128}{81}\sqrt{\frac{2}{3}}, \tag{4.30}$$

and the angular integration gives

$$I_\Omega = \left(\sqrt{\frac{3}{4\pi}}\right)\left(\frac{1}{\sqrt{4\pi}}\right)\int_0^{2\pi} d\phi\int_0^\pi d\theta\cos^2\theta\sin\theta = \frac{1}{\sqrt{3}} \tag{4.31}$$

Using the Bohr radius $a_0 = 4\pi\epsilon_0\hbar^2/(me^2)$ and $E_1 - E_0 = -3/4$ Ry$=(-3/4)e^4 m_e/(2\cdot(4\pi\epsilon_0)^2\hbar^2)$ makes all the dimensions go away, and we are left with

$$f_{21} = (2/3)(-3/4)(1/2)\left(\frac{128}{81}\sqrt{\frac{2}{3}}\right)^2\cdot\frac{1}{3} = -\frac{8192}{59049} = -0.138732. \tag{4.32}$$

Note that oscillator strengths are negative for emission events such as this one, and positive for absorption events, due to the sign change in the term $E_n - E_m$. A table of oscillator strengths for hydrogenlike atoms is given in [10]. Note that the oscillator strengths for different magnetic quantum numbers M_{NL} for a given level

N, L are the same, but oscillator strengths for the reverse process differ in sign and by the statistical weights:

$$g_n f_{nm} = -g_m f_{mn} \qquad (4.33)$$

Equation (4.27) shows that, for transitions in the keV energy range, and taking emitted photons in the keV range such that $\mathbf{k} \cdot \mathbf{r} = 0.25$, we approximate $f_{nm} \approx 0.1$ and then the Einstein A coefficient is approximately 10^{13} for the "average" keV-range transition, thus an average optical half-life on the order of less than a picosecond.

Now we look at collisional excitation rates. The excitation cross section for an electron inelastic reaction producing a promotion of a bound electron in state n to state m is given by

$$\sigma_{mn}(E_{e^-}) = \frac{8\pi^2}{\sqrt{3}} a_0^2 f_{mn} \frac{E_H^2}{\Delta E_{mn} E} \bar{g} = 2.36 \times 10^{-17} \frac{f_{mn} g(n,m)}{E_{e^-} \Delta E_{nm}} \, m^2. \qquad (4.34)$$

Here $g(n,m)$ is the Gaunt factor for this transition which, as in the bremsstrahlung case, gives a correction for the semi-classical result to the quantum-mechanical treatment. Typical Gaunt factors are around 1.0 for atoms and 0.2 for ions. It should be noticed that this formula only holds for the energetically possible case where $E_{e^-} \geq E_{mn}$. Integrating Eq. (4.34) over the maxwellian distribution above this energy cutoff gives the collisional excitation rate per ion $X_{mn} = n_e <\sigma_{ex} v>$ as:

$$\begin{aligned} X_{mn} &= 16\pi \left(\frac{2\pi E_H}{3m_e}\right)^{1/2} a_0^2 \frac{f_{nm} <g(n,m)> n_e}{\Delta E_{nm} T_e^{1/2}} \exp\left(-\frac{E_{nm}}{T_e}\right) \\ &= 1.6 \times 10^{-11} \frac{f_{nm} <g(n,m)> n_e}{\Delta E_{nm} T_e^{1/2}} \exp\left(-\frac{E_{nm}}{T_e}\right). \end{aligned} \qquad (4.35)$$

Here T_e is in eV, and a_0 is the Bohr radius ($= 5.3 \times 10^{-11}$ m). We can then use the result of Eqs. (4.35) and (4.26) to obtain an overall expression for the power loss by line radiation from a single line at energy ΔE_{nm} for a maxwellian plasma:

$$\begin{aligned} P_{nm}^{line} &= 32\pi \left(\frac{\pi}{3}\right)^{1/2} \left(\frac{E_H}{T_e}\right)^{1/2} \left(\frac{E_H^2}{\hbar}\right) a_0^3 f_{nm} \bar{g}_m n_e n_0 \left(\frac{\Delta E_{nm}}{\Delta E_{n0}}\right)^3 \exp\left(-\frac{\Delta E_{n0}}{T_e}\right) \\ &= 1.6 \times 10^{-32} \frac{f_{nm} \bar{g}_m n_e n_0}{T_e^{1/2}} \left(\frac{\Delta E_{nm}}{\Delta E_{n0}}\right)^3 \exp\left(-\frac{\Delta E_{n0}}{T_e}\right) \end{aligned} \qquad (4.36)$$

Here (in the second line with the numerical coefficient) T_e is in keV and n_0 means the ion density in the ground state at this charge state. Without studying the dimensionless quantities at this point, notice that the coefficient out front (1.6×10^{-32}) is many orders of magnitude larger than the coefficient in the bremsstrahlung equation (5.35×10^{-37}), and this shows that at intermediate temperatures where there are many partially stripped ions, line radiation can dominate.

Recombination Radiation

The atomic processes involved at low values of $T_e/(13.6Z^2)$ resulting in radiation loss include ionization, radiative recombination ($e^- + A^{Z+} \rightarrow A^{(Z-1)+} + h\nu$), and dielectronic recombination ($e^- + A^{Z+} \rightarrow A^{**(Z-1)+} \rightarrow A^{(Z-1)+} + h\nu + h\nu$, autoionization($A^{**(Z-1)+} \rightarrow A^{Z+} + e^-$, the inverse process to dielectronic recombination). There can also be electron collisional de-excitation and three-body recombination, but these processes carry an additional factor of the electron density in the rate equations, and thus are negligible in low-density plasmas. If radiative recombination α_R and electron impact ionization S dominate, then we have a simple rate equation for charge state Z:

$$\frac{dn_Z}{dt} = n_e \left[-n_Z S(Z) - \alpha_R(Z)n_Z + S(Z-1)n_{Z-1} + \alpha_R(Z+1)n_{Z+1} \right].$$

$$(4.37)$$

The ionization rate, following Ref. [21], is given (approximately) by

$$S = 8\pi \left(\frac{2E_H}{\pi m_e} \right)^{1/2} a_0^2 \left(\frac{E_H}{E_\infty} \right)^{3/2} \beta^{1/2} \left[0.69 E_1(\beta) \right]. \qquad (4.38)$$

Here $\beta = E_\infty/T_e$, where E_∞ is the ionization energy for each level from the ground state, and E_1 is the exponential integral:

$$E_1(\beta) = \int_\beta^\infty \frac{e^{-x}}{x} dx. \qquad (4.39)$$

Figure 4.5 shows the collisional ionization rate for the various charge states of oxygen.

An expression for the recombination rate for hydrogenlike ions is given in Ref. [28]:

$$\alpha = \frac{2^6}{3} \left(\frac{\pi}{3} \right)^{1/2} \alpha^4 c a_0^2 Z \left(\frac{E_\infty^Z}{T_e} \right)^{1/2} \left[0.4288 + \frac{1}{2} \ln \left(\frac{E_\infty^Z}{T_e} \right) + 0.469 \left(\frac{E_\infty^Z}{T_e} \right)^{-1/3} \right]$$

$$= 5.2 \times 10^{-20} Z \left(\frac{E_\infty^Z}{T_e} \right)^{1/2} \left[0.4288 + \frac{1}{2} \ln \left(\frac{E_\infty^Z}{T_e} \right) + 0.469 \left(\frac{E_\infty^Z}{T_e} \right)^{-1/3} \right]$$

$$(4.40)$$

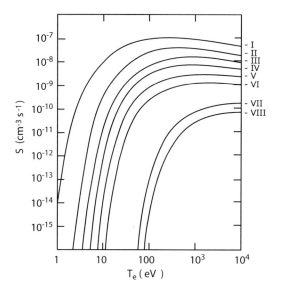

Fig. 4.5 Collisional ionization rate coefficient for oxygen. Note that spectroscopic notation is used here, e.g. OIII=O^{2+}. From [21]. Used by permission, Springer

Dielectronic recombination expressions are more complex because they involve more detailed information about the atomic levels involved and the transition rates for the competing branches in the system. However, Ref. [10] gives a rough estimate of the relative rates energy loss from dielectronic recombination (dr) and direct recombination (fb) as

$$\frac{P_{dr}}{P_{fb}} \approx \frac{\sqrt{3}\pi^2 E_H}{16\alpha T_e} \approx 1.5 \times 10^2 \frac{E_H}{T_e}, \tag{4.41}$$

and thus the rates would become equal at around a 2 keV electron temperature. Dielectronic recombination greatly exceeds direct recombination at lower temperatures and medium-Z ionic species. Figure 4.6 shows a calculation of the components of the radiation loss for iron ($Z = 26$) as a function of electron temperature for coronal equilibria at temperatures of interest for fusion experiments.

Transient Behavior

The non-equilibrium radiation from optically thin plasmas can exceed the levels after equilibration occurs. This happens because the populations of one-electron and two-electron atoms at temperatures that will ultimately lead to almost fully stripped ions will be larger than equilibrium values before the collisional processes have time to push the ion populations into higher charge states. Figure 4.7 shows an example calculation of oxygen atoms being introduced to thermal electrons

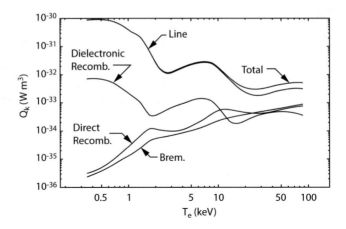

Fig. 4.6 Radiation parameter Q_k for Fe ($Z = 26$) as a function of electron temperature. Line radiation, dielectronic recombination radiation, direct recombination radiation, bremsstrahlung, and the total radiative loss are shown. Data from [5]

Fig. 4.7 Radiation from a plasma with an oxygen impurity as a function of time, showing the transient effects. From [11]

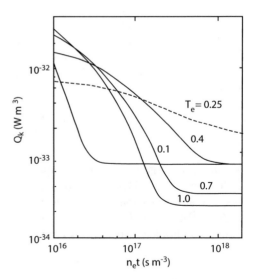

at various temperatures. The radiation parameter is plotted as a function of time. One can see that radiation levels exceed coronal equilibrium values by orders of magnitude for characteristic times $n_e t$ of less than $10^{17}\,\mathrm{m}^{-3}$ s, i.e. 1 ms in a magnetic fusion plasma with a density of $10^{20}\,\mathrm{m}^{-3}$. This timescale is set by the characteristic electron ionization cross sections given in Eq. (4.38). These rates for the last two ionization stages are around $7.1 \times 10^{-18}\,\mathrm{m}^3\mathrm{s}^{-1}$ for $O^{6+} \rightarrow O^{7+}$, $3.5 \times 10^{-18}\,\mathrm{m}^3\,\mathrm{s}^{-1}$ for $O^{7+} \rightarrow O^{8+}$, and $9.16 \times 10^{-19}\,\mathrm{m}^3\,\mathrm{s}^{-1}$ for $O^{8+} \rightarrow O^{7+}$ recombination at $T_e = 300\,\mathrm{eV}$. To illustrate the process, suppose that ionization proceeds rapidly to O^{6+} and then proceeds to O^{7+} and O^{8+} through the coupled equations (ignoring dielectronic recombination, and using the recombination term from Eq. (4.40) only for the transition from O^{8+} to hydrogenlike O^{7+}):

$$\frac{dN_8}{dt} = \lambda_2 N_7 - \lambda_3 N_8$$

$$\frac{dN_7}{dt} = \lambda_1 N_6 - \lambda_2 N_7 + \lambda_3 N_8 \qquad (4.42)$$

$$\frac{dN_6}{dt} = -\lambda_1 N_6.$$

Here $\lambda_1 = \lambda_{6+\rightarrow7+} = n_e < \sigma_{6+\rightarrow7+}v >$, $\lambda_2 = \lambda_{7+\rightarrow8+} = n_e < \sigma_{7+\rightarrow8+}v >$, and $\lambda_3 = \lambda_{8+\rightarrow7+} = n_e\alpha_{8+\rightarrow7+}$. The equations have a solution for $N_6(0) = N_0$, $N_7(0) = 0$, $N_8(0) = 0$:

$$N_6(t) = N_0 \exp(-\lambda_1 t)$$

$$N_7(t) = N_0 \frac{e^{-\lambda_1 t}\left(\lambda_1\lambda_2 + \lambda_1\lambda_3 - \lambda_2\lambda_3 - \lambda_3{}^2\right) - e^{-(\lambda_2+\lambda_3)t}\lambda_1\lambda_2 + \lambda_2\lambda_3 - \lambda_1\lambda_3 + \lambda_3{}^2}{(\lambda_2+\lambda_3)(-\lambda_1+\lambda_2+\lambda_3)}$$

$$N_8(t) = N_0 - N_7(t) - N_6(t)$$

$$(4.43)$$

If we take a density $n_e = 10^{19}\,\text{m}^{-3}$, then $\lambda_1 = 71\,\text{s}^{-1}$, $\lambda_2 = 35\,\text{s}^{-1}$, and $\lambda_3 = 9.16\,\text{s}^{-1}$, we obtain a plot of the density fractions vs. time as shown in Fig. 4.8. This simple model is qualitatively similar to Figure 2 in Ref. [11], where a full model with nine atomic states was used and recombination (with and without dielectronic

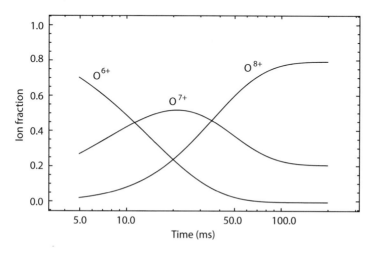

Fig. 4.8 Ion charge state fractions in oxygen vs. time for a simple rate model. $T_e = 300\,\text{eV}$, $n_e = 1.0 \times 10^{19}\,\text{m}^{-3}$

recombination) was calculated. Note that with this model, the asymptotic values of N_6, N_7, and N_8 are

$$N_6(\infty) = 0, \quad N_7(\infty) = N_0 \frac{\lambda_3}{\lambda_2 + \lambda_3} = 0.209, \quad N_8(\infty) = N_0 \frac{\lambda_2}{\lambda_2 + \lambda_3} = 0.791,$$

(4.44)

in agreement with the asymptotic values found in Ref. [11].

4.2.3   Radiation in Optically Thick Plasmas

Optically thick plasmas re-absorb some of the radiation generated locally before it escapes the plasma. Optical thickness is determined by the overall absorption coefficient for photons in the plasma:

$$\kappa_{tot} = \kappa_{line} + \kappa_{cont} + \kappa_R + \kappa_{Thom}$$

(4.45)

where κ_{line}, κ_{cont}, κ_R, and κ_{Thom} are the attenuation and absorption coefficients for line absorption, continuum absorption, Rayleigh scattering, and Thomson scattering, respectively. The radiation attenuation coefficient results in an intensity variation $\propto \exp(-\kappa_{tot}x)$ as radiation is transmitted from its source in the plasma through a travel distance x. For a characteristic plasma size L, "optically thick" means that $\kappa_{tot}L > 1$. (Note, however, that in most laser fusion applications, opacity (κ') is defined as the mass density times the cross sections, and thus has units of $cm^2\,g^{-1}$, and then the optical thickness parameter $\tau = \kappa L = \kappa'(\rho L)$.)

Since a cross section σ can be associated with each process, we can write $\kappa = n_\alpha \sigma$ for each process, so that plasmas become optically thick if the density increases for a fixed size. However, this also means that collisional de-excitation rates, which scale with the square of density, can also compete with spontaneous emission, thus changing the level densities in the plasma. At high densities, level populations are controlled almost entirely by collisions, and thus are accurately modeled by statistical thermodynamics. In true thermodynamic equilibrium, however, the electron and ion populations should have the same temperature, and the photon population follows a blackbody spectrum with the same temperature. This is almost never obtained in laboratory plasmas, however. Typically the electrons and ions have different temperatures unless the temperature gradients are weak and the characteristic confinement time is greater than the electron–ion equipartition time, and that is one of the longest timescales in a plasma. Usually at higher densities and adequate timescales a condition known as "local thermodynamic equilibrium," or LTE, prevails. Usually the radiation output is below the blackbody level consistent with the electron temperature, but the electrons are in a near-maxwellian distribution, and the atomic levels are filled according to thermodynamic considerations.

A necessary condition for LTE to hold is that collisional depopulation rates be much larger, say, a factor of ten, than the radiative recombination rate. Reference [10] gives the density required for this to hold as:

$$n_e \gtrsim \frac{5}{8\sqrt{\pi}} \left(\frac{\alpha}{a_0}\right)^3 z^7 (E_2/z^2 E_H)^3 (T_e/(z^2 E_H))^{1/2}$$

$$\approx 3.9 \times 10^{23} z^7 (T_e/(z^2 E_H))^{1/2} \text{ m}^{-3}. \tag{4.46}$$

Here z is the charge state of the ion. This formula loses its accuracy when other than one-electron atoms are involved.

Conditions for LTE are easily met by many current laser fusion experiments. As an example, a recent paper [18] describes an experiment at the National Ignition Facility (NIF) where a polystyrene (CH) target was illuminated with an X-ray photon spectrum from a hohlraum, yielding an average temperature of 86 eV and density of 6.74 g cm^{-3}; Eq. (4.46) gives an electron density of 1.5×10^{28} m^{-3} as the threshold for LTE, whereas the inferred electron density (using an experimentally inferred $< z >$ for the carbon ions of 4.92) of 1.85×10^{30} m^{-3} in this experiment, is two orders of magnitude greater. However, at these densities, departures from standard ionization models are observed, and describing the correct ionization model for warm dense matter such as this example will require further effort.

The LTE condition also has a timescale limitation, just as in the coronal equilibrium case. Given a typical collisional ionization (and de-population) reactivity $< \sigma v > \approx 10^{-17}$ m^3 s^{-1} for each level, we expect collisions to arrange the level populations into their thermal equilibrium values on a timescale $n_e \tau_{LTE} \approx 10^{18}$ m^{-3} s or faster. In Ref. [18] cited above, the plasma was measured at 480 ps after shock coalescence, giving $n_e t \approx 9 \times 10^{20}$ m^{-3} s, or almost two orders of magnitude longer than the LTE condition.

Radiation Transport in Optically Thick Plasmas

If we define a spectral intensity $I_\nu(\mathbf{r}, \hat{\Omega}, t)$ indicating the energy per unit time per unit area per unit frequency per unit solid angle, then we can write down an equation for the transport of radiation, including spontaneous emission, stimulated emission, and absorption as

$$\frac{1}{c}\frac{\partial I_\nu}{\partial t} + \hat{\Omega} \cdot \nabla I_\nu = \rho \eta_\nu' \left(1 + \frac{c^2}{2h\nu^3} I_\nu\right) - \rho \kappa_\nu' I_\nu. \tag{4.47}$$

Here κ' and η' are the opacity and emissivity, respectively, in units of per unit mass. The famous Einstein B_{21} stimulated emission coefficient is given by the second term inside the parentheses. For true thermodynamic equilibrium (ions, electrons, and photons all at the same temperature), the intensity function is given by the Planckian function:

$$I_{vp} = \frac{2hv^3}{c^2} \frac{1}{\exp(hv/T) - 1} \tag{4.48}$$

An equilibrium solution with no gradients thus has a left-hand side of zero for Eq. (4.47), and then this requires that

$$\eta_v = k_v \exp(-hv/T), \tag{4.49}$$

known as Kirchoff's radiation law. This becomes a statement that "absorption equals emissivity" at each photon energy (frequency) with a (different but appropriate) set of units chosen for the two parameters. Since the blackbody spectrum is not specifically required, this expression holds for any LTE system, whether or not the intensity spectrum is a true blackbody spectrum.

Saha Equilibrium

If the conditions for LTE are met, one expects that the ionization states in the atomic systems to be filled according to the laws of statistical thermodynamics. If we consider a reaction of the form

$$A^{(n-1)+} \rightleftharpoons A^{n+} + e^-, \tag{4.50}$$

then using Maxwell-Boltzmann statistics we can write an equilibrium density of states as

$$\frac{n_{Z+}}{n_{(Z-1)+}} = \frac{g_e g_Z}{g_{Z-1}} \exp\left(-\frac{E_{Z+} - E_{(Z-1)+}}{T_e}\right) \tag{4.51}$$

Here the g-factors are the usual degeneracy factors of states for the ionic species involved. But what do we choose for the degeneracy factor of the free electron? Planck answered this question in 1924 [24]. He found that if the electron existed in a box of volume $V = 1/n_e$, the reciprocal of the density of the electrons, then the degeneracy factor for the electron is found by counting the number of quantized states for the electron in this box, weighted by the Maxwell-Boltzmann probability that each level is filled. The number of quantized levels per unit momentum wavenumber $k = p/\hbar$ is given by the Rayleigh-Jeans number (with an additional factor of two for the two spin states of the electron):

$$N(k)dk = \frac{1}{\pi^2} V \cdot k^2 dk. \tag{4.52}$$

The electron energy $E_{e^-} = \hbar^2 k^2/(2m_e)$, and the electrons are also in a Maxwell-Boltzmann distribution such that the occupancy factor is

$$W(k) = \exp\left(-\frac{\hbar^2 k^2}{2m_e T_e}\right). \tag{4.53}$$

Then integrating over k gives

$$g_{e^-} = \frac{V}{\pi^2} \cdot \int_0^\infty k^2 \exp\left(-\frac{\hbar^2 k^2}{2m_e T_e}\right) dk = \frac{2}{n_e}\left(\frac{m_e T_e}{2\pi \hbar^2}\right)^{3/2}. \tag{4.54}$$

Then the equilibrium densities are given by the Saha equation [26]:

$$
\begin{aligned}
\frac{n_e n_{Z+}}{n_{(Z-1)+}} &= \frac{2g_Z}{g_{Z-1}}\left(\frac{m_e T_e}{2\pi \hbar^2}\right)^{3/2} \exp\left(-\frac{E_{Z+} - E_{(Z-1)+}}{T_e}\right) \\
&= \frac{2g_Z}{g_{Z-1}a_0^3}\left(\frac{T_e}{4\pi E_H}\right)^{3/2} \exp\left(-\frac{E_I((Z-1) \to Z)}{T_e}\right)
\end{aligned}
\tag{4.55}
$$

Note especially that in the second form of (4.55) that the ion fraction n_Z/n_{Z-1} has the usual Boltzmann factor times the dimensionless parameter $1/(n_e a_0^3)$, and this number is large at low densities, so that the relative population at higher ionization levels is favored even at fairly low temperatures compared to the ionization energy. As the parameter $n_e a_0^3$ becomes larger, the ionization levels change from their low-density values as well, due to interaction of the electrostatic fields between the atoms. Generally, the ionization energy drops as the density increases, allowing an increase in the population of states of higher ionization over that calculated with the values of ionization energies found at lower densities.

NLTE Models

The most general model for plasma equation of state, ionization, and radiation transport is the non-LTE (NLTE) model. Here both collisional and radiative contributions to the excitation and de-excitation of each state are used. In most currently used codes, "detailed configuration accounting" (DCA) methods are applied, meaning that many atomic levels are tracked individually, with energy levels, oscillator strengths, collisional cross sections, and autoionization rates calculated using systematic models [27]. Typically hundreds of atomic levels are kept, and timesteps as short as 10^{-16} s are sometimes required. Figure 4.9 shows a comparison of the radiation models outlined above with a state-of-the art NLTE model used in the LASNEX code.

Fig. 4.9 Comparison of average charge state $< Z >$ for gold using coronal model, LTE model, and non-LTE model of a laser-induced ablation plasma with the LASNEX code. From [22]. Used with permission, Springer

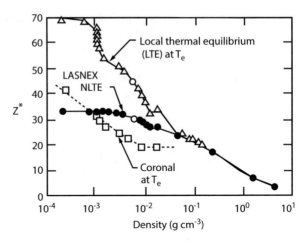

4.3 Charge Exchange

Charge exchange processes such as

$$\underline{D}^+ + D \rightarrow \underline{D} + D^+ \tag{4.56}$$

(here the underline represents an energetic particle) can cause an ion in the interior of a magnetically confined plasma with a substantial energy to be replaced by an ion which was originally a neutral particle from the outside of the hot plasma, typically at $0 \rightarrow 10\,\text{eV}$ energies. The source of these slow atoms can either be by recycling of neutral gas from the walls, especially from the limiter or divertor (typically at the wall temperature of about $0.025 \rightarrow 0.05\,\text{eV}$), or by Franck-Condon neutrals [6, 15, 16] formed by the dissociation of molecular deuterium:

$$D_2^+ + e^- \rightarrow e^- + D^+ + D^0 \tag{4.57}$$

with energies in the range from $0 \rightarrow 15\,\text{eV}$. Figure 4.10 shows a plot of the cross section and reactivity for a typical atomic reaction producing Franck-Condon neutrals. Because the D_2^+ are created by impact ionization into relatively long-lived vibrational states v, the reaction rates shown here are averaged over the population of vibrational states.

A single charge exchange event in the warm plasma represents an almost total energy loss of $(3/2)T_i$ on average. Figure 4.11 shows the Maxwellian-averaged $< \sigma_{cx} v >$ for a hydrogen plasma. (Note that the cross sections depend on the (fast) ion thermal velocity, so the hydrogen reactivity at half the deuterium temperature is used for the deuterium reactivity.) At higher temperatures, Fig. 4.11 shows that proton impact ionization occurs at a faster rate than charge exchange, and this is a less severe energy loss mechanism, as all resultant particles are charged and thus trapped in the plasma.

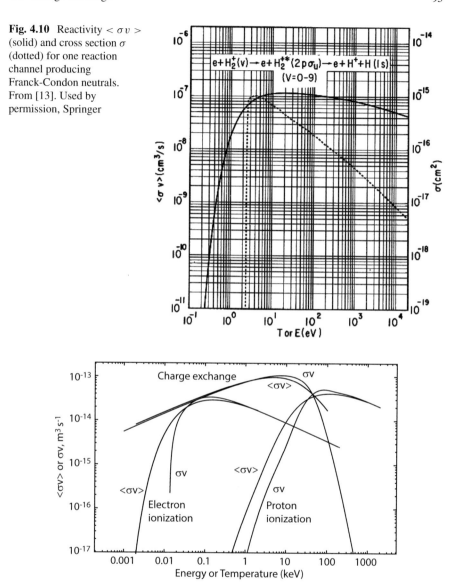

Fig. 4.10 Reactivity $< \sigma v >$ (solid) and cross section σ (dotted) for one reaction channel producing Franck-Condon neutrals. From [13]. Used by permission, Springer

Fig. 4.11 Charge exchange, electron impact ionization, and ion impact ionization for atomic hydrogen targets. Maxwellian average $< \sigma v >$ and reaction rates σv shown. Data from ALLADIN database, compiled by IAEA [12]

Historically, charge exchange by Franck-Condon neutrals formed a barrier from reaching high densities and temperatures in early open-ended magnetic confinement experiments at Livermore and Oak Ridge [4]. The balance of loss by charge exchange and injection of particles from an external source could produce an S-shaped curve in plasma density vs. injection current, with a rather high threshold to

Fig. 4.12 Density vs. injection current due to charge exchange. The densities jump from low values (solid line) to stable, high values as the injection current exceeds some limiting value. From [7]. Oak Ridge National Laboratory

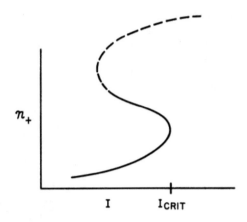

n_+

I I_{CRIT}

a transition from modest densities to high density [7] (see Fig. 4.12). It was through the development of large neutral-beam sources and radiofrequency heating methods that plasmas were capable of "burning through" the charge exchange barrier. Once the plasma is brought to higher densities and temperatures, the mean free path for the neutrals drops significantly, and losses to charge exchange in the center of the plasma become very small. For example, a 10 eV Franck-Condon deuterium atom in a plasma with an electron density of 10^{20} m^{-3} and an electron temperature of 100 eV, where the electron impact ionization reactivity is around 3×10^{-14} m^3 s^{-1}, will have a mean free path of $v\tau = v/(n < \sigma v >) \approx 7$ mm.

4.4 Synchrotron Radiation

An electron spiraling about a magnetic field experiences a perpendicular acceleration $\dot{v}_\perp = \Omega_{ce} v_\perp$, with $\Omega_{ce} = eB/m_e$. Applying Larmor's formula for radiation power, we have

$$P_{rad} = \left(\frac{2}{3}\right) \frac{e^2}{4\pi\epsilon_0 c^3} \dot{v}^2 = \left(\frac{2}{3}\right) \frac{e^4 B^2}{4\pi\epsilon_0 m_e^2 c^3} v_\perp^2 \qquad (4.58)$$

Using an average $< v_\perp^2 >= 2T_e/m_e$ and an electron density n_e then gives a volumetric synchrotron radiation power

$$P_{synch} = \frac{n_e e^4 B^2 T_e}{3\pi\epsilon_0^2 m_e^3 c^3} = 6.21 \cdot 10^{-17} B^2 n_e T_e \,(\text{keV}) \qquad (4.59)$$

This is a surprisingly high value for the radiation loss. If we consider a magnetically confined fusion plasma with an electron density of 10^{20} m^{-3} in a 4.0 T magnetic field at an electron temperature of 15 keV, this gives a synchrotron radiation loss of 1.49 MW m^{-3}, about four times the alpha heating power in a

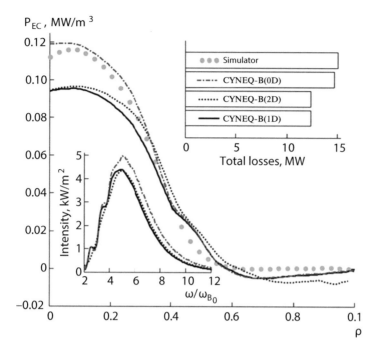

Fig. 4.13 Calculation of cyclotron radiation losses for ITER. From [19]. Used by permission, Springer

DT plasma at this temperature and density (380 kW m^{-3}), so ignition would be impossible. However this calculation assumes that there is no re-absorption of the radiated electromagnetic energy, and also ignores reflection of this radiation back into the plasma. The frequencies involved in this synchrotron emission are harmonics of the electron cyclotron frequency, and therefore in the range of 100–500 GHz at typical magnetic field strengths. Metallic surfaces are highly reflective at these frequencies, and the plasma is very absorptive at these frequencies. Rather sophisticated synchrotron loss calculations have emerged, including the broadening of the cyclotron frequencies due to spatially varying magnetic fields, relativistic effects on the emitted spectrum, wall resistivity, reflection optics, and other factors. Figure 4.13 shows a state-of-the-art calculation for synchrotron losses expected in ITER.

4.5   Energy Balance and the Lawson Criterion

A young British engineer named J. D. Lawson entertained some thoughts about the overall power balance in a fusion reactor and wrote a report on his analysis in 1955 [20]. His overall analysis was quite simple and included a fusion reactor with some

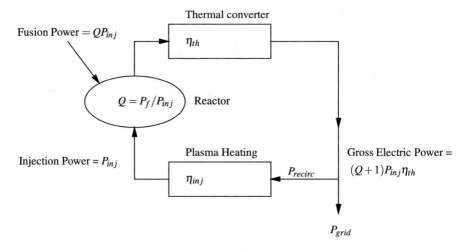

Fig. 4.14 Overall energy balance for fusion power plant

injected power, with the reactor creating fusion power at a ratio to the injection power R (which we now call Q), and a thermal energy conversion step where heat taken out of the reactor through its walls is converted into electricity (or some other type of "high-quality" energy), some of which is recirculated in the reactor system to provide for the injected power. Figure 4.14 shows a schematic of the power flow through the reactor system.

One of Lawson's early observations was that the thermal conversion step has an efficiency determined by thermodynamics and is always less than the Carnot efficiency for such a device. In fact, the efficiency of large turbines used for electrical power generation with water (steam) as a working fluid has not changed in the 60 years since Lawson's report and is roughly $\eta_{th} = 1/3$. So the most optimistic case possible for a fusion power plant to just break even, and not send any electricity out on the grid, can be arrived at by assuming that $\eta_{inj} = 1$ and that the recirculating power equals the gross electric power generated. That is to say

$$(Q+1)\eta_{th} = 1 \qquad \text{("breakeven")}, \qquad (4.60)$$

which for $\eta_{th} = 1/3$ gives $Q = 2$ as the energy breakeven criterion. Thus $Q = 2$ is often referred to as the Lawson criterion. (It is almost universal in the popular press to refer to "energy breakeven" as being the point where the fusion power equals the injection power, or $Q = 1$ as being the important milestone, and sometimes this is referred to as "scientific feasibility," although there is nothing "scientific" or "feasible" about that condition.)

Lawson went one step further and used the expression for the energy balance inside the plasma in the reactor, rather than the external loop representing the overall power plant as shown above. He observed that there are two energy gain terms in the plasma: the injected power $P_{inj} = P_f/Q$ and the self-heating of the plasma by charged particles made in the fusion process. We can label this term as $f_c P_f$, where f_c is the fraction of the fusion energy release coming from charged particles (and also multiplied by the fraction of that quantity actually absorbed in the plasma rather than escaping to the walls or the hot gas outside the confines of the burning plasma). The loss terms present include radiation from the plasma, usually given as a fraction χ_R times the fusion power. This will be explored in detail later, but it can be as low as a few percent in the more optimistic scenarios. Then there is loss from the plasma by heat conduction, which on a per unit volume basis can be written as the plasma thermal energy $\frac{3}{2}(n_e T_e + n_i T_i)$ representing the thermal energies in the electrons and the ions, divided by an energy confinement time τ_E. If we assume that the plasma is a 50–50 mixture of D and T and that $n_e = n_i \equiv n$ and that $T_e = T_i \equiv T$, we end up with an expression

$$\frac{P_f}{Q} + f_c P_f = P_f \chi_R + \frac{3nT}{\tau_E} \tag{4.61}$$

We note that the fusion power per unit volume is given by, again for a 50–50 mixture of D-T:

$$P_f = \frac{n^2}{4} < \sigma v > E_f \tag{4.62}$$

Substituting into Eq. (4.61) and rearranging, we have

$$n\tau_E = \frac{3T}{(1/4) < \sigma v > E_f(1/Q + f_c - \chi_R)} \tag{4.63}$$

In some cases it is more logical to replace the radiation loss term $(1/4) < \sigma v > E_f \chi_R$ in the expanded denominator of Eq. (4.63) with the equivalent quantity P_{rad}/n^2, where P_{rad} is the radiative power per unit volume. Radiation loss from the plasma can come from radiative electron–ion collisions, called bremsstrahlung, radiation caused by the centripetal acceleration of electrons in a magnetic field, called synchrotron radiation and described earlier, and by atomic processes in plasmas with impurities, called line and recombination radiation. These processes were discussed in detail earlier in this chapter. For all of these except synchrotron radiation, $P_{rad} \propto n^2$ because the radiation is produced by a two-body collisional process. The simplest (and most unavoidable) of these radiation losses is bremsstrahlung. This is caused by binary collisions between electrons and ions which produce photons with energy comparable to the plasma temperature, as shown earlier. For each ionic species with atomic number Z, the bremsstrahlung energy produced per unit volume P_{brem} is

$$P_{brem} = n_e n_Z Z^2 b T_e^{1/2} \tag{4.64}$$

with the coefficient $b = 5.35 \cdot 10^{-37} \, \text{W m}^3 \, \text{keV}^{-1/2}$. For a D-T plasma with no impurities, $Z = 1$ and then Eq. (4.63) reads

$$n\tau_E = \frac{3T}{(1/4) < \sigma v > E_f(1/Q + f_c) - bT^{1/2}}. \tag{4.65}$$

We can also model the effect of impurities with a simple trick. Let us suppose that all of the plasma ions have an "effective" charge Z_{eff}, not necessarily an integer. Then the total number of ions in a charge-neutral plasma will be n_e/Z_{eff}. Typically the impurities present have an atomic number $Z \gg Z_{eff}$, so that $n_e \approx n_{DT}$ with the addition of the impurity. (For example, a 0.1% iron impurity adds an additional $(0.001)26^2 = 0.676$, or 67.6% increase in the radiation loss, while only decreasing n_{DT}/n_e from unity to $1 - (.001)(26) = 0.974$.) Then using the imaginary quantity Z_{eff}, we have

$$P_{rad} = n_e n_{Z_{eff}} Z_{eff}^2 bT_e^{1/2} = n_e^2 bT_e^{1/2} Z_{eff}$$

and then assuming that $n_e \approx n_{DT}$ gives for Eq. (4.63):

$$n\tau_E = \frac{3T}{(1/4) < \sigma v > E_f(1/Q + f_c) - bT^{1/2}Z_{eff}}. \tag{4.66}$$

A plot of $n\tau_E$ vs. temperature is shown as Fig. 4.15. Note that since $f_c = 0.2$ for D-T (with all alpha energy retained in the plasma), the ratio of $n\tau_E$ for a reactor with $Q = 5$ and that for an "ignited" reactor ($Q = \infty$, no external plasma heating required) is only a factor of two, ignoring radiation losses. Also note that the minimum $n\tau_E$ for each Q-value is at a temperature of about 28 keV, which is not the temperature for optimum power density of around 15 keV, as described earlier. At temperatures around 10 keV or less, the slope of the $n\tau_E$ vs. temperature curves for fixed Q are around -1 on a log-log plot such as Fig. 4.15, which has motivated the notion of using the "triple product" $n\tau_E T$ as a metric for fusion plasma performance. Figure 4.16 shows a plot of $n\tau_E T$ vs. temperature for the same Q values as before. Now the flat part of each curve is around 10 keV, which is a temperature more likely to be seen in current and future fusion experiments. What is generally quoted as "The Lawson criterion" is the value of $n\tau_E$ for $Q = 2$ for D-T at $T = 10$ keV, or

$$n\tau_E \geq 10^{20} \text{m}^{-3}\text{s}. \tag{4.67}$$

4.5.1 Advanced Energy Recovery Cycles

Embellishments on the basic energy conversion cycle analyzed by Lawson are possible. One possibility is that plasma exhaust in the form of charged particles may be put through an energy recovery system that extracts energy from the

Fig. 4.15 Lawson parameter
$n\tau_E$ vs. temperature T for a
D=T plasma with $Z_{eff} = 2$

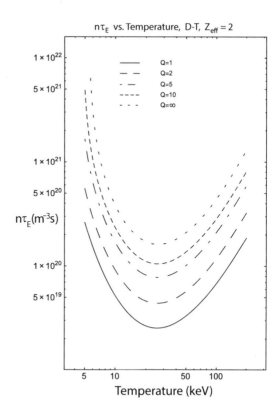

charged particles directly without the need of a thermal conversion step with its relatively low efficiency. With favorable conditions, efficiencies up to around 90% are possible. Furthermore, the energy not collected in the direct converter may be directed to the thermal converter if the temperatures of the coolant streams from the direct conversion equipment are compatible with the output temperature from the blanket. Another possibility is that the blanket, whose main purpose in DT fusion reactors is to provide tritium in a closed fuel cycle, may also contain other materials which can interact with the neutrons coming out of the burning plasma and provide an additional energy multiplication. Examples of these materials are the ^{6}Li which needs to be there anyway for tritium breeding and has a nuclear reaction with neutrons which is exothermic ($Q_{reaction} = 4.8$ MeV), manganese in the structural steels used in the blanket, and actinide breeding materials such as ^{232}Th and ^{238}U. (One may also consider fissionable isotopes such as ^{235}U and ^{239}Pu, but one might wonder whether it would be faster and cheaper to build a fission reactor, since the fission–fusion hybrid system obtained this way would probably have the same safety risk as the fission reactor, with the additional cost and availability issues with the hybrid.) Thus one might define a multiplication parameter M such that the thermal energy created by each neutron E_n might be treated as creating ME_n units of energy in the blanket and being presented in the thermal conversion cycle. The radiation

Fig. 4.16 Triple product $n\tau_E T$ vs. temperature for a D-T plasma with $Z_{eff} = 2$

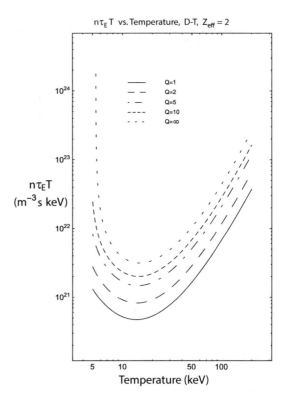

from the plasma would be converted to thermal energy in the blanket without any multiplication factor. Figure 4.17 shows a schematic of the more sophisticated energy recovery system. If we define a second radiation parameter Ψ_R as the fraction of the fusion energy leaving the plasma as charged particles, we have

$$\Psi_R = \frac{f_c P_f - P_{rad}}{P_f} = f_c - \chi_R, \tag{4.68}$$

and then the energy leaving the plasma as charged particles is given by $P_{inj} + f_c P_f \Psi_R$. The electrical output of the direct converter is given by this quantity times the direct converter's efficiency, or $\eta_{DC}(P_{inj} + f_c P_f \Psi_R)$. The waste heat from the direct converter is then $(1 - \eta_{DC})(P_{inj} + f_c P_f \Psi_R)$. The total heat presented to the thermal converter is this amount plus the non-charged particle power (including neutrons) $(1 - \Psi_R) P_f$ and the extra power due to the multiplication M, or $f_n(M - 1)P_f$. So the total output gross electric power P_G is given by:

$$P_G = \eta_{DC}\left(P_{in} + \Psi_R P_f\right)$$
$$+ \eta_{th}\left[(1 - \Psi_R) P_f + (M - 1) f_n P_f + (1 - \eta_{DC})\left(P_{in} + \Psi_R P_f\right)\right] \tag{4.69}$$

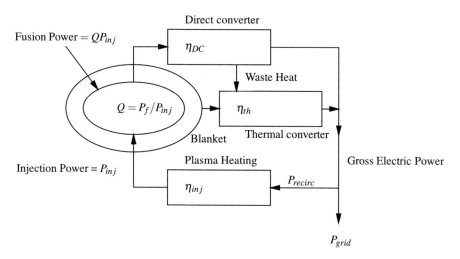

Fig. 4.17 Energy balance for fusion power plant with direct conversion and blanket multiplication

The total power output from the reactor, including blanket multiplication, is just the input power plus the fusion power with the neutron portion multiplied by M, or $P_{in} + P_f(f_c + Mf_n)$. Dividing this into Eq. (4.69), and noting that $P_f/P_{in} \equiv Q$ gives an effective efficiency for the power conversion equipment:

$$\eta_{eff} = \eta_{th} + \frac{\eta_{DC}\,(1 - \eta_{th})\,(1 + Q\Psi_R)}{1 + (f_c + f_n M)\,Q} \tag{4.70}$$

The quantity η_{eff} can be used in place of η_{th} for the calculation of the overall plant efficiency η_0, which is defined as the ratio of the power to the grid P_{grid} to the nuclear power produced P_{nuc}. One can see that for the simple power balance shown in Fig. 4.14, where the nuclear power is all from fusion reactions, that

$$P_{grid} \equiv \eta_0 P_f = (Q + 1)\eta_{th} P_{in} - P_{in}/\eta_{in} = \left(1 + \frac{1}{Q}\right) P_f \eta_{th} - \frac{1}{Q\eta_{in}} P_f, \tag{4.71}$$

and thus

$$\eta_0 = \eta_{th} - \frac{1}{Q}\left(\frac{1}{\eta_{in}} - \eta_{th}\right). \tag{4.72}$$

Since the quantity multiplying $1/Q$ in Eq. (4.72) is always positive, $\eta_0 \leq \eta_{th}$. For the advanced recovery cycle, we substitute η_{eff} for η_{th} to obtain

$$\eta_0 = \eta_{eff} - \frac{1}{Q}\left(\frac{1}{\eta_{in}} - \eta_{eff}\right) \tag{4.73}$$

Note that as the blanket energy multiplication M becomes large, this expression reverts to $\eta_0 = \eta_{th}$, i.e. the plant behaves as if it is a fission reactor.

4.5.2 The Lawson Concept Applied to Inertial Fusion

Inertial fusion is based on the idea that a capsule containing a D-T (or some other fuel) mixture can be compressed and heated by a short pulse energy source such as a laser or ion beam. It is interesting to note how the development of the Lawson criterion above can be modified to apply to this situation. The key to understanding this is to look at what to use for τ_E in Eq. (4.67).

Suppose that we have a compressed sphere, radius R_0 of cold D-T fuel which has an external source of energy impinging on its surface. The surface immediately heats up to fusion burn temperature and an expansion wave is formed which moves in towards the center of the sphere. If the density is not so high that Fermi degeneracy is an issue, this expansion wave has a velocity given by the sound speed [8]:

$$c_s = \left(\frac{\gamma p}{\rho} \right)^{1/2} = \left(\frac{\gamma_e T_e + \gamma_i T_i}{m_i} \right)^{1/2}. \tag{4.74}$$

Typically the values of γ are those of a monatomic gas, i.e. 5/3, assuming adiabatic conditions apply. (In some cases, however, the electrons behave isothermally, and then $\gamma_e = 1$.) Thus the speed of sound in the plasma in typical conditions is $c_s = \sqrt{10/3}(T_i/m_i)^{1/2}$. We assume that a constant fraction of the fusion fuel burns as the burn wave progresses inward, and that the fuel behind the burn wave is ejected radially outward as is required by momentum balance. The average mass $< m >$ involved in the fusion burn is given by the mass density times instantaneous volume $4\pi/3 R(t)^3$ integrated over time and divided by the time for the burn wave to reach the center of the sphere R_0/c_s:

$$< m > = \frac{1}{R_0/c_s} \int_0^{R_0/c_s} (4/3)\rho\pi R(t)^3 dt = \frac{1}{R_0/c_s} \int_0^{R_0/c_s} (4/3)\pi\rho \left(R_0 - c_s t \right)^3 dt$$

$$= (4/3)\pi\rho R_0^3 \cdot \frac{1}{4} = \frac{m_0}{4} \tag{4.75}$$

Then the effective disassembly time is given by

$$\tau_e = \frac{R}{4c_s}. \tag{4.76}$$

and thus the Lawson criterion becomes

$$n\tau_E = \frac{n_{DT} R}{4c_s} = \frac{\rho R \cdot N_A}{4c_s M_{DT}}.$$

It is usual to work in cgs units with ρR, and plugging in $N_A = 6.02 \cdot 10^{23}$ (g-mole)$^{-1}$, and using $T_e = T_i = 10\,\text{keV}$ gives

$$n\tau_e = 5.2 \cdot 10^{14}\,\text{cm}^{-3}\,\text{s} \cdot \rho R(\text{cgs}) \tag{4.77}$$

The MKS Lawson criterion given earlier in Eq. (4.67) becomes $n\tau_E \geq 10^{14}\,\text{cm}^{-3}\,\text{s}$. Because of the time-dependent nature of the fusion burn, the coefficient in Eq. (4.77) given above is usually dropped by a factor of five, and then the Lawson equivalent for inertial fusion is given simply as

$$\rho R \geq 1\,\text{g}\,\text{cm}^{-2}. \tag{4.78}$$

The total fusion yield from the target is also predicted by the parameter ρR. Since the total number of fusions is given by $N_f = (n^2/4) <\sigma v> \tau_e$ and two particles are burned per reaction, the fuel burnup fraction becomes

$$f_B = \frac{\Delta M_f}{M_f} = \frac{n\tau_e <\sigma v>}{2} = \frac{\rho R <\sigma v> N_A}{8 c_s M_{DT}} = \rho R \frac{<\sigma v>}{k\sqrt{T}} \equiv \frac{\rho R}{H_B}, \tag{4.79}$$

where k is a constant($= 1.19376 \cdot 10^{-15}$ (cgs units, with T in keV). A plot of H_B vs. temperature for D-T is shown as Fig. 4.18. The fusion yield is given by

$$Y = M_f f_B Y''' \tag{4.80}$$

where Y''' is the specific yield in the fuel ($=3.39 \cdot 10^{11}\,\text{J}\,\text{g}^{-1}$ for D-T).

The simple model for burnup can be adapted to add some additional physics features. Firstly, the model does not account for fuel depletion at high burnup rates. Secondly, the model does not include the energy amplification possible at lower initial temperatures below 20 keV. More sophisticated computer models have shown

Fig. 4.18 Burnup parameter H_B vs. temperature for D-T

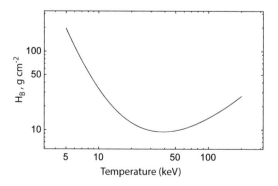

lower values for H_B (around $6 \rightarrow 10\,\mathrm{g\,cm^{-2}}$ at $8\,\mathrm{keV}$). Thus a more realistic model for burnup might be (first suggested in [8]):

$$f_B = \frac{\rho R}{\rho R + H_B} \qquad (4.81)$$

with H_B between 6 and $11\,\mathrm{g\,cm^{-2}}$ in most cases with $\rho R > 1\,\mathrm{g\,cm^{-2}}$.

Problems

4.1 Neon ($Z = 10$) gas puffing has been suggested as a means of controlling the radiative output of a plasma to prevent disruptions. Find the increase in bremsstrahlung radiation when a plasma with $10^{20}\,\mathrm{m^{-3}}$ density of DT fuel ions has injected into it neon ($Z = 10$) gas to a ratio of one neon atom for every twenty DT fuel ions. If this plasma were at $15\,\mathrm{keV}$ temperature, could it still be in ignition ($Q = \infty, n\tau_E < \infty$)?

4.2
Draw curves of the Lawson product $n\tau_E$ vs. temperature for $Q = 1, 2, 5, 10$ and ∞ for a 50–50 D-T plasma with a $Z_{eff} = 2.5$. Use a temperature range of 5–200 keV. Use the reactivity $< \sigma v >$ from the fit algorithm provided on the web page. Use log-log coordinates and provide a legend, labeled axes, and a title for your plot. Avoid the use of color, use one-point lineweights, and use a sans serif font for the labels. Also provide a table of $n\tau_E$ for these Q-values at temperatures $5 \rightarrow 20\,\mathrm{keV}$ by 1 keV increments and then every 5 keV going up to 200 keV.

4.3 (a) Find the maximum concentration of carbon ($Z = 6$) impurity that can be tolerated and still allow the DT plasma to achieve ignition ($Q = \infty$) at any temperature, assuming $n\tau_E = \infty$. Assume that the carbon is fully stripped. (b) Now find the same answer for $Q = 10$.

4.4 Find the maximum concentration of tungsten impurity (Z=74) which can be tolerated to give impurity radiation equal to one-third of the alpha heating power at $T_i = T_e = 15\,\mathrm{keV}$ in a DT plasma. Use the data in Fig. 4.4.

4.5 Using the burnup formula for inertial fusion, Eq. (4.80), with $H = 10$, find the yield from a $160\,\mu\mathrm{g}$ load of DT.

4.6 This example shows the need for compression in order to "miniaturize" the inertial confinement fusion process for industrial-scale energy production. Suppose that we have two targets, each achieving a ρR of one. One is at liquid DT density of $0.25\,\mathrm{g\,cm^{-3}}$ and the other is compressed to $10{,}000\times$ liquid density. Find the yield for both cases again using Eq. (4.80) with $H = 10$. For the first case, express your answer in kilotons as well as MJ ($1\,\mathrm{kT} = 4.18 \times 10^{12}\,\mathrm{J}$).

References

1. Abramowitz, M.: Handbook of Mathematical Functions, With Formulas, Graphs, and Mathematical Tables. Dover Publications, New York (1974)
2. Biedenharn, L.C.: A note on Sommerfeld's bremsstrahlung formula. Phys. Rev. **102**, 262–263 (1956). http://link.aps.org/doi/10.1103/PhysRev.102.262
3. Chung, H.K., Chen, M., Morgan, W., Ralchenko, Y., Lee, R.: FLYCHK: generalized population kinetics and spectral model for rapid spectroscopic analysis for all elements. High Energy Dens. Phys. **1**(1), 3–12 (2005). http://dx.doi.org/10.1016/j.hedp.2005.07.001; http://www.sciencedirect.com/science/article/pii/S1574181805000029
4. Colchin, R.: Target plasma trapping. Nucl. Fusion **11**(4), 329 (1971). http://stacks.iop.org/0029-5515/11/i=4/a=002
5. Davis, J., Jacobs, V.L., Kepple, P.C., Blaha, M.: Radiative cooling of tokamak plasmas due to multiply-charged Fe impurity ions. J. Quant. Spectrosc. Radiat. Transf. **17**, 139–147 (1977)
6. Dunn, G.H., Van Zyl, B.: Electron impact dissociation of h_2^+. Phys. Rev. **154**, 40–51 (1967). http://link.aps.org/doi/10.1103/PhysRev.154.40
7. Fowler, T.K.: Effect of energy degradation on the critical current in an OGRA-type device. Oak Ridge National Laboratory Technical Report, ORNL-3037 (1960)
8. Fraley, G.S., Linnebur, E.J., Mason, R.J., Morse, R.L.: Thermonuclear burn characteristics of compressed deuterium-tritium microspheres. Phys. Fluids **17**(2), 474–489 (1974). http://dx.doi.org/10.1063/1.1694739; http://scitation.aip.org/content/aip/journal/pof1/17/2/10.1063/1.1694739
9. Ginzburg, V.L.: Important elementary processes in cosmic-ray astrophysics and X-ray astronomy. In: DeWitt, C., Schatzman, E., Veron, P. (eds.) High Energy Astrophysics, vol. 1. Gordon and Breach, New York (1967)
10. Griem, H.R.: Principles of Plasma Spectroscopy. Cambridge University Press, Cambridge (1997). Cambridge Books Online, http://dx.doi.org/10.1017/CBO9780511524578
11. Hopkins, G.R., Rawls, J.M.: Impurity radiation from medium density plasmas. Nucl. Technol. **36**(2), 171–186 (1977)
12. International Atomic Energy Agency: ALLADIN database. https://www-amdis.iaea.org/ALADDIN/ (2016)
13. Janev, R.K., Langer, W.D., Post, D.E., Evans, K.: Electron impact collision processes. In: Elementary Processes in Hydrogen-Helium Plasmas: Cross Sections and Reaction Rate Coefficients, pp. 17–114. Springer, Berlin (1987). http://dx.doi.org/10.1007/978-3-642-71935-6_2
14. Karzas, W.J., Latter, R.: Electron radiative transitions in a coulomb field. Astrophys. J. Suppl **6**, 167 (1961). https://doi.org/10.1086/190063
15. Kieffer, L.J., Dunn, G.H.: Dissociative ionization of H_2 and D_2. Phys. Rev. **158**, 61–65 (1967). http://link.aps.org/doi/10.1103/PhysRev.158.61
16. Kieffer, L.J., Dunn, G.H.: Dissociative ionization of H_2 and D_2. Phys. Rev. **164**, 270–270 (1967). http://link.aps.org/doi/10.1103/PhysRev.164.270.5
17. Kramers, H.A.: Xciii. On the theory of x-ray absorption and of the continuous x-ray spectrum. Philos. Mag. Ser. 6 **46**(275), 836–871 (1923). https://doi.org/10.1080/14786442308565244
18. Kraus, D., Chapman, D.A., Kritcher, A.L., Baggott, R.A., Bachmann, B., Collins, G.W., Glenzer, S.H., Hawreliak, J.A., Kalantar, D.H., Landen, O.L., Ma, T., Le Pape, S., Nilsen, J., Swift, D.C., Neumayer, P., Falcone, R.W., Gericke, D.O., Döppner, T.: X-ray scattering measurements on imploding CH spheres at the National Ignition Facility. Phys. Rev. E **94**, 011202 (2016). http://link.aps.org/doi/10.1103/PhysRevE.94.011202
19. Kukushkin, A.B., Minashin, P.V., Polevoi, A.R.: Impact of magnetic field inhomogeneity on electron cyclotron radiative loss in tokamak reactors. Plasma Phys. Rep. **38**(3), 187–196 (2012). DOI 10.1134/S1063780X12030038. http://dx.doi.org/10.1134/S1063780X12030038

20. Lawson, J.D.: Some criteria for a power producing thermonuclear reactor. Technical report, Atomic Energy Research Establishment, Harwell, Berkshire (1955). https://www.euro-fusion. org/wpcms/wp-content/uploads/2012/10/dec05-aere-gpr1807.pdf
21. Lotz, W.: Electron-impact ionization cross-sections and ionization rate coefficients for atoms and ions from hydrogen to calcium. Z. Phys. **216**(3), 241–247 (1968). http://dx.doi.org/10. 1007/BF01392963
22. More, R.M.: Atoms in dense plasmas. In: Briand, J.P. (ed.) Atoms in Unusual Situations, pp. 155–215. Springer US, Boston (1986). http://dx.doi.org/10.1007/978-1-4757-9337-6_7
23. Morozov, D.K., Baronova, E.O., Senichenkov, I.Y.: Impurity radiation from a tokamak plasma. Plasma Phys. Rep. **33**(11), 906–922 (2007). http://dx.doi.org/10.1134/S1063780X07110037
24. Planck, M.: Zur quantenstatistik des bohrschen atommodells (On the quantum statistics of Bohr's atom model). Ann. Phys. **380**, 673–684 (1924)
25. Post, D., Jensen, R., Tarter, C., Grasberger, W., Lokke, W.: Steady-state radiative cooling rates for low-density, high-temperature plasmas. At. Data Nucl. Data Tables **20**(5), 397– 439 (1977). http://dx.doi.org/10.1016/0092-640X(77)90026-2; http://www.sciencedirect.com/ science/article/pii/0092640X77900262
26. Saha, M.N.: LIII. Ionization in the solar chromosphere. Philos. Mag. Ser. 6 **40**(238), 472–488 (1920). http://dx.doi.org/10.1080/14786441008636148
27. Scott, H., Hansen, S.: Advances in NLTE modeling for integrated simulations. High Energy Dens. Phys. **6**(1), 39–47 (2010). http://dx.doi.org/10.1016/j.hedp.2009.07.003; http://www. sciencedirect.com/science/article/pii/S1574181809000834
28. Seaton, M.J.: Radiative recombination of hydrogenic ions. Mon. Not. R. Astron. Soc. **119**(2), 81–89 (1959). https://doi.org/10.1093/mnras/119.2.81; http://mnras.oxfordjournals. org/content/119/2/81.abstract
29. Sommerfeld, A.J.F.: Atombau und Spektrallinien, vol. 2. Ungar, New York (1953)

Chapter 5
Magnetic Confinement

5.1 Fluid Equations for Plasma: MHD Model

Equation (3.68) in Chap. 3 gives an expression for the evolution of a distribution function $f_i(\mathbf{x}, \mathbf{v}, t)$ for each component in a plasma. The generality of this equation prevents its general solution. However, there are many situations where this equation for the evolution of the plasma can be put in a much simpler form by making a few assumptions about the plasma. This allows the description of the plasma by a set of fluid equations. A fluid model is strictly only valid when the following conditions apply [11]: (1) the rates of change of plasma conditions with time are slow compared to collision processes, (2) (in magnetic confinement systems) the scale length for plasma parameters to vary is long compared to the electron and ion gyroradii, or the scale length is long compared to the mean free path in unmagnetized systems, (3) the number of particles in a Debye sphere (see Chap. 3 is large, or equivalently, $\ln \Lambda \gg 1$, (4) average drift velocities are less than the characteristic ion thermal speed, (5) the electron gyroradius ρ_e is large compared to the Debye length, and (6) velocity-space "microinstabilities" are negligible.

While all of these criteria might not be met in a particular application, the resulting fluid model may faithfully reproduce certain aspects of a plasma's behavior but miss others. This is especially true where the fluid equations generated are closely linked to absolute conservation laws for conservation of particles, momentum, energy, and others. In general, the fluid equations are derived by taking moments of the Vlasov-Boltzmann equations with varying powers in velocity. Writing Eq. (3.68) in the simplified form $\mathscr{D} f = \mathscr{C} f$, where $\mathscr{D} = \partial/\partial t + \mathbf{v} \cdot \nabla + (q/m)\,(\mathbf{E} + \mathbf{v} \times \mathbf{B}) \cdot \partial/\partial \mathbf{v}$ and $\mathscr{C}$ is the Boltzmann collision operator, the moment equations have the characteristic form

$$\int d\mathbf{v}\, \mathbf{v}^n (\mathscr{D} f - \mathscr{C} f) = 0. \tag{5.1}$$

© Springer Nature Switzerland AG 2018
E. Morse, *Nuclear Fusion*, Graduate Texts in Physics,
https://doi.org/10.1007/978-3-319-98171-0_5

In this equation, $\mathbf{v}^n$ may represent a scalar such as $v^2 = \mathbf{v} \cdot \mathbf{v}$ or it may represent a tensor form such as $\mathbf{vv}$.

The zeroth moment of the Vlasov-Boltzmann equation takes a particularly simple form, since $\int d\mathbf{v} Cf = 0$ (there is no "teleportation" of particles instantaneously in space through collisions, although this point might be revisited if there is a source of particles (such as charge exchange) or a sink of particles (nuclear fusion!). Ignoring these effects for now, we write the zeroth moment equation as [11]:

$$\frac{dn_\alpha}{dt} + \nabla \cdot (n_\alpha \mathbf{u}_\alpha) = 0, \tag{5.2}$$

where d/dt is the Lagrangian derivative

$$\frac{d}{dt} = \frac{\partial}{\partial t} + \mathbf{u}_\alpha \cdot \nabla \tag{5.3}$$

and the average velocity $\mathbf{u}_\alpha$ is given by

$$\mathbf{u}_\alpha = \frac{1}{n_\alpha} \int d\mathbf{v}_\alpha f_\alpha \mathbf{v}_\alpha. \tag{5.4}$$

Equation (5.2) shows a recurring feature of the system of moment equations: each moment equation requires the definition of a new moment of the distribution function f_α at one higher power in velocity. While the first few such moments will have an obvious physical meaning, eventually they will not, and some sort of closure method must be employed to end this vicious cycle.

The next moment equation, the first moment with the vector $\mathbf{v}_\alpha$, is given by [11]:

$$m_\alpha n_\alpha \frac{d\mathbf{u}_\alpha}{dt} = -\nabla p_\alpha - \nabla \cdot \mathbf{P}_\alpha + Z_\alpha e n_\alpha \left[\mathbf{E} + \mathbf{u}_\alpha \times \mathbf{B} \right] + \mathbf{R}_\alpha. \tag{5.5}$$

Here $p_\alpha = n_\alpha T_\alpha$, and the temperature T_α is derived from the second scalar moment, using only the random part of $\mathbf{v}_\alpha$:

$$T_\alpha = \frac{1}{3} \frac{1}{n_\alpha} \int d\mathbf{v}_\alpha \, m_\alpha \, |\mathbf{v}_\alpha - \mathbf{u}_\alpha|^2 \, f_\alpha \mathbf{v}_\alpha. \tag{5.6}$$

The terms $\mathbf{P}_\alpha$ and $\mathbf{R}_\alpha$ are moments of the collision operator on f_α with the reduced velocity $\mathbf{v}_\alpha - \mathbf{U}_\alpha$. The term $\mathbf{P}_\alpha$ represents the viscous force on the species α caused by collisions with itself, and the term $\mathbf{R}_\alpha$ represents drag force with the other species. The former is unwieldy to work with in general, but the drag term R_α can be used to derive an Ohm's law for the plasma, especially in the direction parallel to the magnetic field. This moment equation represents conservation of momentum.

The second moment of the Vlasov-Boltzmann hierarchy is the equation describing conservation of thermal energy. It is given by [11]:

$$\frac{3}{2}n_\alpha \frac{T_\alpha}{dt} + p_\alpha \nabla \cdot \mathbf{u}_\alpha = -\nabla \cdot \mathbf{q}_\alpha - \mathbf{P}_\alpha : \nabla \mathbf{u}_\alpha + Q_\alpha. \tag{5.7}$$

Here the heat flux $\mathbf{q}_\alpha$ is defined by

$$\mathbf{q}_\alpha = \int d\mathbf{v}_\alpha \frac{m_\alpha}{2} |\mathbf{v}_\alpha - \mathbf{u}_\alpha|^2 (\mathbf{v}_\alpha - \mathbf{u}_\alpha), \tag{5.8}$$

and Q_α represents the heating caused by inter-species collisions, including the frictional heating such that total thermal energy is conserved. For one ionic species present, it takes the form

$$Q_e = -Q_i - \mathbf{R} \cdot (\mathbf{u}_e - \mathbf{u}_i). \tag{5.9}$$

Note that the second moment of the Vlasov-Boltzmann equation has a simple thermodynamic explanation. The change in internal energy of the plasma is $dU_\alpha = d\left[(3/2)n_\alpha T_\alpha\right]$, which is represented by the first term. (That n_α appears outside the derivative is because parts of the zeroth and first moment equations have been factored out, a persistent trait of the moment equations). The next term represents the pdV work being done on the plasma by compression or expansion, since the term $\nabla \cdot \mathbf{u}_\alpha$ is the rate of change of the volume of each differential volume in the plasma. The terms in $\mathbf{q}$, Q, and $\mathbf{P} : \nabla \mathbf{u}$ all represent some type of heat transfer in or out of each species α, with the $\mathbf{P} : \nabla \mathbf{u}$ term representing the viscous heating of the species. Thus in a general sense, the first law of thermodynamics is shown by Eq. (5.7); that is

$$dU = dQ - pdV. \tag{5.10}$$

5.1.1 The MHD Limit

As a subset of the general set of fluid equations, we can make the magnetohydrodynamic, or "MHD" approximation. The general conditions for validity of the multi-species transport equations given above are usually satisfied in magnetically confined fusion plasma with one exception: the approximation of high collisionality typically does not hold. In fact, collisions are very infrequent, and thus mean free paths for collisions along $\mathbf{B}$ are tens of kilometers. The idea behind the MHD model is to isolate those physical processes that actually might require the assumption of high collisionality, and eliminate those degrees of freedom for modeling the system.

The overall continuity equation is formed by summing the contributions to the mass and mass flow:

$$\rho = m_e n_e + m_i n_i \approx m_i n_i$$

$$\rho \mathbf{u} = n_e m_e \mathbf{u}_e + m_i n_i \mathbf{u}_i \approx n_i m_i \mathbf{u}_i$$

$$\frac{d\rho}{dt} + \rho \nabla \cdot \mathbf{u} = m_i \frac{d\mathbf{n}_i}{dt} + m_e \frac{d\mathbf{n}_e}{dt} + n_e m_e \nabla \cdot \mathbf{u}_e + n_i m_i \cdot \nabla \mathbf{u}_i \tag{5.11}$$

$$\approx m_i \frac{d\mathbf{n}_i}{dt} + n_i m_i \nabla \cdot \mathbf{u}_i = 0.$$

Next we note that Maxwell's equation for $\nabla \times \mathbf{B}$ is

$$\nabla \times \mathbf{B} = \mu_0 e(n_i \mathbf{u}_i - n_e \mathbf{u}_e) + \mu_0 \epsilon_0 \frac{\partial \mathbf{E}}{\partial t}. \tag{5.12}$$

This allows fast electromagnetic waves to propagate, which complicates a description limited to low frequencies. We thus take the limit of $\epsilon_0 \to 0$ to eliminate these fast electromagnetic waves. This, however, forces a limit to occur when evaluating Poisson's equation:

$$\nabla \cdot \mathbf{E} = \frac{e(Z_i n_i - n_e)}{\epsilon_0} \tag{5.13}$$

so that we must have charge neutrality

$$n_e = Z_i n_i. \tag{5.14}$$

This in turn provides an expression for the current $\mathbf{J} = e(n_i Z_i \mathbf{u}_i - n_e \mathbf{u}_e)$, which is just the electromagnetic force term showing up when we sum the electron and ion momentum equations from Eq. (5.5). Note that the electric field terms $Z_\alpha e n_\alpha \mathbf{E}$ for the ion and electron cancel when we add the $\alpha = e$ and $\alpha = i$ versions of Eq. (5.5) together, and also the interspecies drag terms $\mathbf{R}_e$ and $\mathbf{R}_i$ cancel exactly (Newton's First Law). This then gives the "single-fluid" version of the momentum equation:

$$\rho \frac{d\mathbf{u}}{dt} = -\nabla p + \mathbf{J} \times \mathbf{B} + \nabla \cdot (\mathbf{P}_i + \mathbf{P}_e). \tag{5.15}$$

It is not clear in general how the last term in Eq. (5.15) can be removed, as it scales with the collisional times τ_i and τ_e. However, it is found that many problems involving bulk plasma motion in magnetic fields are modeled quite well by the MHD momentum equation with this term removed. It has been shown that the perpendicular components of these terms are small in a collisionless approximation [4], implying that the high-collisionality result also applies at low collisionality for perpendicular components in the momentum equation. Thus the MHD model equation of motion is simplified to:

$$\rho \frac{d\mathbf{u}}{dt} = -\nabla p + \mathbf{J} \times \mathbf{B}. \tag{5.16}$$

5.2 MHD Equilibria

5.2.1 General Stationary Equilibria

We now want to examine Eq. (5.16) for equilibrium solutions where $\partial/\partial t = 0$. We first must notice that d/dt and $\partial/\partial t$ are not the same thing, so that the time-independent version of Eq. (5.16) is

$$\rho \mathbf{u} \cdot \nabla \mathbf{u} = -\nabla p + \mathbf{J} \times \mathbf{B}, \tag{5.17}$$

i.e., one may include plasmas with a steady flow $\mathbf{u} \neq 0$ as a class potentially having equilibrium solutions. However, since the generalized Ohm's law $\mathbf{J} = \sigma(\mathbf{E} + \mathbf{u} \times \mathbf{B})$, and the plasma conductivity is very large, these solutions typically have $\mathbf{u}$ parallel to $\mathbf{B}$, with the possible inclusion of a component along the toroidal angle in axisymmetric systems (see Ref. [10]). Rotation in tokamak plasmas has been observed experimentally [6] and may have some impact on equilibrium considerations. For the case of a rotation in toroidal systems, the centrifugal term on the left in Eq. (5.17) has a magnitude $\sim \rho v_\phi^2/R$, and the first term on the right $\sim p/a$ where a is the minor radius so that the ratio of these terms is of order $(a/R)v_\phi^2/(p/\rho)$, which is $\gamma(a/R)(v_\phi^2/c_s^2) = \gamma(a/R)M^2$, where γ is the adiabatic index ($(5/3)$ in most situations) and M is the Mach number v/c_s. Thus for subsonic flows in typical tokamaks ($R/a \approx 3$), the left-hand side of Eq. (5.17) is small compared to the other terms. However, the situation is not so clear for the poloidal component of flows at Mach numbers exceeding one.

5.2.2 Static Equilibria

A static equilibrium has $\mathbf{u} = 0$ everywhere in addition to $\partial/\partial t = 0$. We then have

$$\nabla p = \mathbf{J} \times \mathbf{B}, \tag{5.18}$$

the most basic equation of MHD theory. We note that $\nabla \times \mathbf{B} = \mu_0 \mathbf{J}$ (no displacement current, this is called the "pre-Maxwell" equation), so we can also write

$$\nabla p = \frac{\nabla \times \mathbf{B} \times \mathbf{B}}{\mu_0}, \tag{5.19}$$

and invoking a vector identity

$$\nabla \times \mathbf{B} \times \mathbf{B} = -\nabla\left(\frac{B^2}{2}\right) + \mathbf{B} \cdot \nabla \mathbf{B} \tag{5.20}$$

The right-most term in (5.20) is worth exploring further. Note that

$$\mathbf{B} \cdot \nabla \mathbf{B} = B\hat{b} \cdot (B\hat{b}) = B^2\hat{b} \cdot \nabla b + B\hat{b}\hat{b} \cdot \nabla B = B^2\kappa + \hat{b}\nabla_{\parallel}(B^2/2). \qquad (5.21)$$

Here $\kappa = \hat{b} \cdot \nabla\hat{b}$ is the curvature vector of differential geometry [15]. Also note that $\nabla_{\parallel}$ means the component of the gradient along $\mathbf{B}$. Using the vector identity of Eq. (5.20) again, this time with $\hat{b}$ substituted for $\mathbf{B}$, shows that $\kappa = \hat{b} \cdot \nabla\hat{b}$ is perpendicular to $\hat{b}$, since $\nabla |\hat{b}|^2 = 0$. Then (5.21) can be used to construct only perpendicular components:

$$\nabla_{\perp}\left[p + \frac{B^2}{2\mu_0}\right] = \frac{B^2}{2\mu_0}\kappa, \qquad (5.22)$$

and the parallel force equation is just $\nabla_{\parallel}p = 0$, which is obvious from Eq. (5.18). Equation (5.22) implies that in a simple equilibrium where the magnetic field lines are straight (but not necessarily aligned with one another) one has a simple relationship

$$\nabla\left[p + \frac{B^2}{2\mu_0}\right] = 0. \qquad (5.23)$$

This brings up the concept of "magnetic pressure," where the quantity $p + B^2/(2\mu_0)$ is a constant, and the plasma can be thought of as "digging a well" in the magnetic field. It is instructive to show a numerical example: a 4.0 T magnetic field has a magnetic pressure equal to $6.3662 \cdot 10^6$ Pa, or more than 60 atmospheres. By contrast, a plasma with a density of $10^{20}\,\mathrm{m}^{-3}$ and a temperature of 15 keV has a kinetic pressure $p = 2nT = 0.48 \times 10^6$ Pa. The dimensionless parameter β is defined by

$$\beta = \frac{p}{\dfrac{B^2}{2\mu_0}}. \qquad (5.24)$$

Here we take B to be the magnetic field when $p = 0$. (Many different versions of β exist, however. In tokamaks B is taken as the toroidal field at the location of the magnetic center of the plasma, but in vacuum.) The quantity β can also mean a local value or a mean value for the plasma, and it can be quoted as a pressure ratio for only one component of $\mathbf{B}$, such as a "poloidal beta" β_{pol} and a "toroidal beta" β_{tor}.

For a slab-type plasma equilibrium with no field line curvature, it is clear that $0 \le \beta \le 1$, but other geometries with magnetic curvature can have higher values of β. Typically, however, the maximum value of β for any type of confinement device is determined not by arbitrary geometrical factors but by the appearance of instabilities at some threshold value of β. Since the nuclear power in a fusion reactor scales as β^2, the highest stable value for β is typically the design goal. Additionally, since the fusion reactivity is a strong function of temperature, there can be ways of optimizing the density and temperature profiles to optimize the nuclear power produced within a set pressure profile.

5.2.3 A Simple Slab MHD Equilibrium Model

Suppose that we have a slab of plasma of infinite extent in y and z and ranging from $-1/2$ to $1/2$ in the x-directions. Suppose that we have $B_z = 1.0$ outside the domain of the plasma and that the current inside the plasma is given by

$$J_y(x) = \begin{cases} -2x, & |x| \le 1/2 \\ 0, & |x| > 1/2 \end{cases}.$$

(5.25)

For simplicity we take $\mu_0 = 1$. Then integrating the curl equation for $\mathbf{B}$ gives $\nabla \times \mathbf{B} = -\hat{y}(d/dx)B_z(x)$, and thus $B_z(x) = x^2 + C$. We match the boundary conditions at $x = \pm 1/2$ by choosing $C = 3/4$. Then

$$B_z(x) = \begin{cases} x^2 + 3/4, & |x| \le 1/2 \\ 1, & |x| > 1/2 \end{cases}.$$

(5.26)

Since $\nabla p = \mathbf{J} \times \mathbf{B}$ we have a cubic form for dp/dx:

$$\frac{dp(x)}{dx} = \begin{cases} -2x\left(x^2 + 3/4\right), & |x| \le 1/2 \\ 0, & |x| > 1/2 \end{cases}.$$

(5.27)

Integrating this gives the pressure, and the constant of integration is chosen so that $p(\pm 1/2) = 0$. This gives

$$p(x) = \begin{cases} \dfrac{7 - 24x^2 - 16x^4}{32}, & |x| \le 1/2 \\ 0, & |x| > 1/2 \end{cases}.$$

(5.28)

Figure 5.1 illustrates this magnetic equilibrium. Note that this equilibrium has a peak value of beta of $7/16 = 0.4375$, which is not very descriptive of magnetic confinement fusion plasma beta, but then, neither are infinite slabs. But this model simply illustrates the concept of magnetic pressure in systems where the field lines are nearly straight.

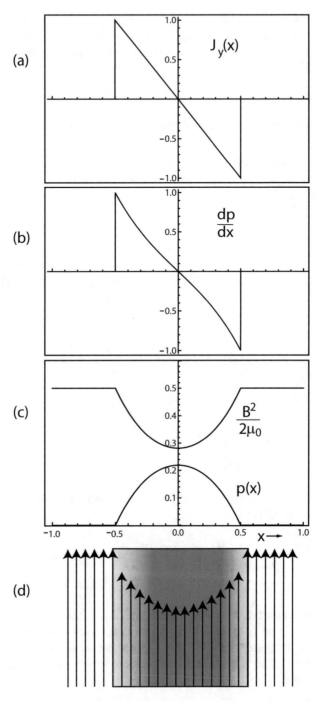

Fig. 5.1 Slab model MHD equilibrium. (**a**) current profile $J_y(x)$, (**b**) pressure gradient $dp(x)/dx$, (**c**) magnetic pressure $B^2/(2\mu_0)$ and kinetic pressure $p(x)$, (**d**) density plot of pressure with vector plot of the magnetic field $\mathbf{B}(x)$. Note that μ_0 has been taken as 1.0 for simplicity

5.2.4 The z-Pinch

The z-pinch was one of the earliest plasma configurations studied [3, 5] and is pictured in Fig. 1.2. We now examine its MHD equilibrium properties.

We consider an azimuthally symmetric current $J_z(r)$ flowing in the z-direction. If we make a surface of integration as a circle of radius r inside the column, then the current enclosed in this surface $I(r)$ is given by

$$I(r) = \int_0^r 2\pi r' J_z(r')dr',$$ (5.29)

and then by Ampere's law, the azimuthal magnetic field B_θ is given by

$$B_\theta(r) = \frac{\mu_0 I(r)}{2\pi r}.$$ (5.30)

Differentiating Eq. (5.29) gives an expression for J_z in terms of I_z:

$$J_z(r) = \frac{I'(r)}{2\pi r},$$ (5.31)

which in turn gives

$$\frac{dp}{dr} = -J_z(r)B_\theta(r) = -\frac{\mu_0 I(r)I'(r)}{4\pi^2 r^2}.$$ (5.32)

We now ask what the average pressure $< p >$ is across the column, which has a radial extent from $r = 0$ to $r = a$. This is given by

$$< p >= \frac{\int_0^a 2\pi r p(r)dr}{\int_0^a 2\pi r dr} = \frac{2}{a^2}\int_0^a r p(r)dr,$$ (5.33)

and now we integrate by parts to obtain

$$< p > = -\frac{2}{a^2}\int_0^a \frac{r^2}{2}p'(r) = \frac{1}{a^2}\int_0^a \frac{\mu_0 I(r)I'(r)}{4\pi^2} = \left(\frac{\mu_0 I(a)}{2\pi a}\right)^2 \cdot \frac{1}{2\mu_0}$$

$$= \frac{B_\theta(a)^2}{2\mu_0}.$$ (5.34)

So the remarkable feature of the z-pinch is that its *average* beta $< \beta >$, computed using the magnetic field at the edge of the column $B_\theta(a)$, is one. There is no mathematical limit on the peak pressure at the center of the column compared to

Fig. 5.2 Toroidal field vs. R
in a toroidal, tokamak-like
plasma, illustrating
paramagnetic and
diamagnetic behavior

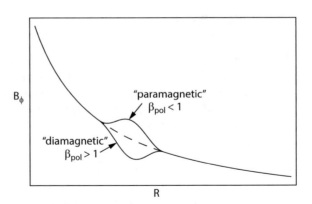

the edge magnetic field. (This is easy to understand when one considers that we
could define the boundary a great distance away, with little current flowing in the
outer regions, and then $B_\theta(a) \propto 1/a$.) Equation (5.34) is often referred to as the
Bennett Pinch theorem, although this explicit form does not appear in Bennett's
original 1934 paper [3].

While violent instabilities are frequently observed in z-pinches, it was found
that adding a stabilizing magnetic field in the z-direction (which does not affect
the pressure balance) could suppress instabilities. Then making the system toroidal
and inducing the current by magnetic induction makes the plasma into a tokamak.
We can define a "poloidal beta" as being the value of beta as calculated for the z-
pinch as before but using only the poloidal magnetic field in the calculation, i.e.
$\beta_{pol} = p/(B_\theta(a)^2/(2\mu_0))$. A tokamak with no currents flowing poloidally will
have $< \beta_{pol} >= 1$ and might be called a "Bennett Pinch tokamak." In general,
however, poloidal currents can flow in the plasma, and this can reduce or increase
the toroidal field inside the plasma. Since the vacuum toroidal field shows a $1/R$
dependence (just like the magnetic field surrounding a wire on the axis), departures
from this $1/R$ dependence are called "diamagnetic" if the toroidal field is reduced
and "paramagnetic" if it increases, causing $< \beta_{pol} > > 1$ in the former case and
$< \beta_{pol} > < 1$ in the latter case. This is shown in Fig. 5.2.

5.3   Confinement in Axisymmetric Systems

5.3.1   Toroidal and Poloidal Fields

Let ϕ represent the angular coordinate in cylindrical geometry, and R and Z are the
radial and vertical directions, and let $\{R, \phi, Z\}$ form a right-handed triple. Suppose
that we have symmetry in ϕ: $\partial/\partial\phi[\dots] = 0$ for all quantities. The magnetic field
B can be decomposed into a toroidal field pointing in the $\hat{\phi}$ direction and a poloidal

field lying in the $\{R, Z\}$ plane. Now look at the poloidal field. Since $\nabla \cdot \mathbf{B} = 0$, (and $\nabla \cdot \mathbf{B}_\phi$ is automatically zero), we can define a stream function Ψ such that:

$$\mathbf{B}_{pol} = \frac{\nabla \Psi \times \hat{e}_\phi}{R} = \nabla \times (A_\phi \hat{e}_\phi). \qquad (5.35)$$

Notice that $\Psi = R A_\phi$, where $\mathbf{A}$ is the vector potential. The physical meaning of Ψ is that the poloidal flux $\hat{\Psi} = \int \mathbf{B}_{pol} \cdot dS = 2\pi \Psi$ for any integration over a region S with a boundary a circle with $R =$ constant and $Z =$ constant. (Using this definition means that Ψ is really $1/(2\pi)$ of the flux; however, Ψ is often referred to as "the flux function.")

Similarly, in a symmetrical static equilibrium $\nabla \cdot \mathbf{J} = 0$ so that we can define a stream function for the poloidal current:

$$\mathbf{J}_{pol} = \frac{\nabla F \times \hat{e}_\phi}{\mu_0 R} \qquad (5.36)$$

Note, however, that $\mathbf{J}_{pol} \times \mathbf{B}_{pol} = 0$, because otherwise it would cause a pressure gradient to exist in the $\hat{e}_\phi$ direction. Therefore $F = F(\psi)$, i.e. it is a function of the poloidal flux only. Furthermore, since $\mathbf{J}_{pol} = 1/\mu_0 \nabla \times \mathbf{B}_\phi$, we can use Eq. (5.36) to write

$$F(\Psi) = R B_\phi. \qquad (5.37)$$

In order to find an equation for the overall force balance,

$$\nabla p = \mathbf{J} \times \mathbf{B} = \mathbf{J}_{tor} \times \mathbf{B}_{pol} + \mathbf{J}_{pol} \times \mathbf{B}_{tor},$$

we must find one more quantity and that is $\mathbf{J}_{tor}$. That is found using the Maxwell equation $\mathbf{J}_{tor} = (1/\mu_0) \nabla \times \mathbf{B}_{pol}$. Applying the curl to Eq. (5.35) gives

$$\mathbf{J}_\phi = -\frac{\hat{e}_\phi}{\mu_0} \frac{\Delta^* \Phi}{R}, \qquad (5.38)$$

where $\Delta^* \Phi$ is a "modified Laplacian"

$$\Delta^* \Psi = R \frac{\partial}{\partial R} \left(\frac{1}{R} \frac{\partial \Psi}{\partial R} \right) + \frac{\partial^2 \Psi}{\partial Z^2}. \qquad (5.39)$$

Then the force balance is obtained from

$$\nabla p = p'(\psi) \nabla \psi = \mathbf{J} \times \mathbf{B} = \left(\frac{1}{\mu_0} \right) \left[-\frac{F(\psi) F'_\psi \nabla \psi}{R^2} - \frac{\Delta^* \psi \nabla \psi}{R^2} \right]$$

This has the form of a scalar equation times $\nabla \psi$, so we can write

$$\Delta^* \psi = -\mu_0 R^2 p'(\psi) - F(\psi)F'(\psi) \tag{5.40}$$

This is the Grad-Shafranov equation [8, 12, 13].

5.3.2 Solov'ev Equilibria

Closed form solutions to (5.40) are relatively uncommon due to its nonlinear nature and because Ψ is both a dependent variable and a coordinate for the functions p and F. However, there is a remarkable set of solutions found by Solov'ev [14] by restricting the functions $p(\Psi)$ and $F(\Psi)$ to simple forms which result in a linear equation in Ψ. The forms assumed were

$$p'(\Psi) = -a \tag{5.41}$$

and

$$F(\Psi)F'(\Psi) = -bR_0^2. \tag{5.42}$$

Here R_0 is a constant representing the position of the magnetic axis. Then Eq. (5.40) has a general solution

$$\Psi(R, Z) = \frac{1}{8}(a - c)\left(R^2 - R_0^2\right)^2 + \frac{1}{2}Z^2\left(bR_0^2 + cR^2\right). \tag{5.43}$$

As a simple example, setting $b = 0$, $a = -10$, $R_0 = 1/\sqrt{2}$, $c = -2$ and then adding $1/4$ to this solution (adding a simple constant to the flux does not affect the solution) gives the well-known "Hill's Vortex" solution, mapping closed flux surfaces into a sphere of radius one:

$$\Psi(R, Z) = R^2\left(1 - R^2 - Z^2\right). \tag{5.44}$$

This solution is shown as Fig. 5.3.

A more intuitive version of the Solov'ev has been shown by Goedbloed et al. [7]. By working in normalized variables:

$$x = \frac{R - R_0}{\epsilon R_0}$$

$$y = \frac{Z}{\epsilon R_0}, \tag{5.45}$$

Fig. 5.3 Hill's vortex
solution to the
Grad-Shafranov equation

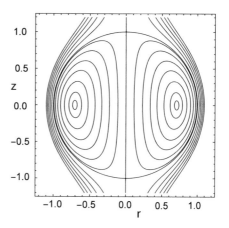

an equilibrium solution can be written down:

$$\Psi(x, y) = \left(x - \frac{\epsilon}{2}\left(1 - x^2\right)\right)^2 + \left(1 - \frac{\epsilon^2}{4}\right)[1 + \epsilon\tau x(2 + \epsilon x)]\left(\frac{y^2}{\sigma^2}\right). \quad (5.46)$$

With this form, there are only three free parameters: ϵ, σ, and τ. They each have
some physical meaning. ϵ is the inverse aspect ratio, with $x = -1/\epsilon$ being the
location of the symmetry axis. σ is related to the elongation of the plasma in the Z-
direction. τ is a triangularity parameter, with negative values making a "nose" on the
inside midplane last closed flux surface and positive values making some triangular
pointiness on the outside. Contour plots for constant ϵ and σ are shown in Fig. 5.4.
One notices that the structure of the X-points changes as the triangularity parameter
τ changes; the transition to two x-points from one happens at $\tau = \pm 1$. Note that the
Hill's vortex solution can be found with the transformed coordinates of Eq. (5.46)
with $\epsilon = 1$, $\tau = 1$, and $\sigma = \sqrt{3}$ and is roughly similar to the equilibrium shown in
Fig. 5.4c.

5.3.3 Solution with Whittaker Functions

Similar to the Solov'ev solution above, it is also possible to find analytic solutions to
the Grad-Shafranov equation with a pressure p and toroidal field function F derived
from quadratic functions of the flux (see Ref. [9], based on earlier work by Ref. [2]):

$$p(\Psi) = p_{axis}\left(\frac{\Psi}{\Psi_{axis}}\right)^2$$

$$F^2(\Psi) = R_0^2 B_0^2 \left[1 + b_{axis}\left(\frac{\Psi}{\Psi_{axis}}\right)^2\right]. \quad (5.47)$$

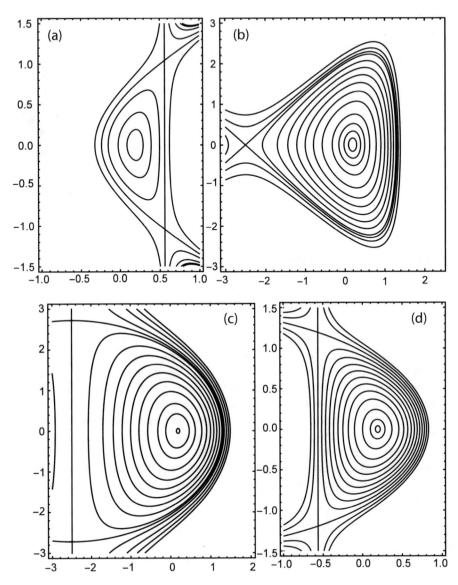

Fig. 5.4 Flux contour plots for Solov'ev equilibria, all with $\epsilon = 0.4$ and $\sigma = 1.4$. (**a**) $\tau = -2.0$, (**b**) $\tau = -0.5$, (**c**) $\tau = 1.0$, and (**d**) $\tau = 2.5$. See Ref. [7]

Here B_0 is the toroidal magnetic field in vacuum taken at the geometric center of the plasma at $R = R_0$ and $Z = 0$. The parameter b_{axis} determines the "paramagnetic" or "diamagnetic" nature of the plasma: if $b_{axis} > 0$, then the plasma is paramagnetic and $< \beta_{pol} > < 1$; if $b_{axis} < 0$, then the plasma is diamagnetic and $< \beta_{pol} > > 1$. (Note that in vacuum, $B_\phi \propto 1/R$ and so $F =$constant.) With these choices for the free functions $p(\Psi)$ and $F^2(\Psi)$, the Grad-Shafranov equation becomes

$$\Delta^* \Psi = -\frac{R_0^2 B_0^2}{\Psi_{axis}^2} \left(b_{axis} + \beta_{axis} \frac{R^2}{R_0^2} \right) \Psi, \tag{5.48}$$

where $\beta_{axis} = 2\mu_0 p_{axis}/B_0^2$. This equation is a linear partial differential equation. Then the flux Ψ and R and Z coordinates are transformed by taking $\Psi = \Psi_0 \psi$, $R^2 = R_0^2 x$, and $Z = ay$, where a is the minor radius $= \epsilon R_0$. Then the transformed Grad-Shafranov equation is

$$4\epsilon^2 x \frac{\partial^2 \psi}{\partial x^2} + \frac{\partial^2 \psi}{\partial y^2} + (\alpha x + \gamma)\psi = 0, \tag{5.49}$$

where

$$\gamma = \left(\frac{a R_0 B_0}{\Psi_{axis}} \right) b_{axis}, \tag{5.50}$$

and

$$\alpha = \left(\frac{a R_0 B_0}{\Psi_{axis}} \right) \beta_{axis}. \tag{5.51}$$

This equation can be solved by separation of variables:

$$\psi = \sum_m X(\rho) Y(y), \tag{5.52}$$

with a simple linear scaling $x = -i \left(\epsilon/\sqrt{\alpha} \rho \right)$. The equation for the Y_m term is given by

$$\frac{d^2 Y_m}{dy^2} + k_m^2 Y_m = 0, \tag{5.53}$$

which has a general solution

$$Y_m(y) = c_m \cos(k_m y) + d_m \sin(k_m y). \tag{5.54}$$

The coefficients d_m vanish for an equilibrium with up-down symmetry, but can be retained for equilibria that are asymmetric, such as plasmas with a single-null divertor.

The equation for $X_m(\rho)$ is given by

$$\frac{d^2 X_m}{d\rho^2} + \left[-\frac{1}{4} + \frac{\lambda_m}{\rho} \right] X_m = 0, \tag{5.55}$$

where the separation constant λ_m is given by

$$\lambda_m = -i \frac{\gamma - k_m^2}{4\epsilon \sqrt{\alpha}}. \tag{5.56}$$

The solutions to Eq. (5.55) are given in terms of Whittaker functions (see Ref. [1] for a description of these functions):

$$X_m(\rho) = a_m W_{\lambda_m,\mu}(\rho) + b_m M_{\lambda_m,\mu}(\rho).$$ (5.57)

Here μ takes on a value of 1/2. Since ρ and λ_m are defined as purely imaginary numbers, we note that $M_{\lambda_m,\mu}(\rho)$ is purely imaginary, but $W_{\lambda_m,\mu}(\rho)$ is arbitrarily complex. We therefore select the imaginary part of $X_m(\rho)$ to represent our solution, as both the real and imaginary parts of $X_m(\rho)$ solve Eq. (5.55) separately, because it is a differential equation with real coefficients, but only the imaginary part has two independent solutions (with one containing a logarithmic term with singular behavior at $\rho = 0$).

The authors of Ref. [9] found that, for typical symmetric tokamak equilibria, fairly realistic equilibrium fluxes could be fitted with a sum of three terms:

$$\psi(\rho, y) = \sum_{m=1}^{3} \left[a_m W_{\lambda_m,\mu}(\rho) + b_m M_{\lambda_m,\mu}(\rho) \right] \cos(k_m y).$$ (5.58)

This means that six unknown expansion coefficients are involved. Therefore, six boundary conditions are sought. Note that the y-wavenumbers k_m also have to be chosen. The authors of Ref. [9] found empirically that choosing $k_1 = 0$, k_2 as an imaginary number of order unity (generating cosh functions), and choosing $k_3 = \pi/\kappa$, where κ is the elongation b/a, were good choices for a broad spectrum of equilibria. The boundary conditions used include four zero-flux conditions at the inside midplane $(R, Z) = (R - a, 0)$, outside midplane $(R, Z) = (R + a, 0)$, and the top at a position $(R, Z) = (R_0 - \delta a, \kappa a)$ (with δ being the triangularity parameter), and the requirement that at this position the flux is a maximum, i.e. the gradient of the flux in R is zero there. (See Fig. 5.5 for the geometrical details.) This gives a set of equations:

$$\Psi(R_0 + a, 0) = 0,$$
$$\Psi(R_0 - a, 0) = 0,$$
$$\Psi(R_0 - \delta a, \kappa a) = 0,$$
$$\partial \Psi / \partial R(R_0 - \delta a, \kappa a) = 0,$$ (5.59)

plus an expression for the curvature on the inboard side:

$$\frac{1}{R_c} = \frac{\Psi_{ZZ}(R_0 - a, 0)}{\Psi_R(R_0 - a, 0)},$$ (5.60)

and finally, two coupled equations that define the maximum flux (at the magnetic axis R_{axis}) to be normalized to one:

$$\Psi_R(R_{axis}, 0) = 0,$$
$$\Psi(R_{axis}) = 1.$$ (5.61)

Fig. 5.5 Tokamak geometry.
From [9]

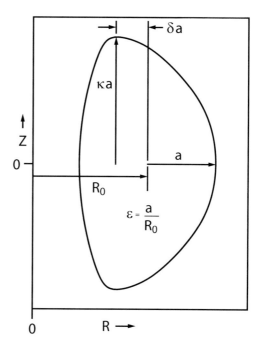

The equations in Eqs. (5.59), (5.60), and (5.61) are then written in terms of the Whittaker functions and their derivatives as:

$$a_1 W_1(\rho+) + b_1 M_1(\rho_+) + a_2 W_2(\rho_+) + b_2 M_2(\rho_+) + a_3 W_3(\rho_+) + b_3 M_3(\rho_+) = 0,$$

$$a_1 W_1(\rho_-) + b_1 M_1(\rho_-) + a_2 W_2(\rho_-) + b_2 M_2(\rho_-) + a_3 W_3(\rho_-) + b_3 M_3(\rho_-) = 0,$$

$$a_1 W_1(\rho_\delta) + b_1 M_1(\rho_\delta) + (a_2 W_2(\rho_\delta) + b_2 M_2(\rho_\delta)) \cos(k_2\kappa)$$
$$+ (a_3 W_3(\rho_\delta) + b_3 M_3(\rho_\delta)) \cos(k_3\kappa) = 0,$$

$$a_1 W_1'(\rho_\delta) + b_1 M_1'(\rho_\delta) + (a_2 W_2'(\rho_\delta) + b_2 M_2')(\rho_\delta) \cos(k_2\kappa)$$
$$+ (a_3 W_3'(\rho_\delta) + b_3 M_3'(\rho_\delta)) \cos(k_3\kappa) = 0,$$

$$\left(\frac{1-\epsilon}{2\epsilon\rho_-} \right)$$
$$\times \left[\frac{k_2^2(a_2 W_2(\rho_-) + b_2 M_2(\rho_-)) + k_2^3(a_3 W_3 + b_3 M_3)}{a_1 W_1'(\rho_-) + b_1 M_1'(\rho_-) + a_2 W_2'(\rho_-) + b_2 M_2'(\rho_-) + a_3 W_3'(\rho_-) + b_3 M_3'(\rho_-)} \right] = \frac{a}{R_c},$$

$$a_1 W_1'(\rho_a) + b_1 M_1'(\rho_-) + a_2 W_2'(\rho_a) + b_2 M_2'(\rho_-) + a_3 W_3'(\rho_a) + b_3 M_3'(\rho_a) = 0,$$

$$a_1 W_1(\rho_a) + b_1 M_1(\rho_a) + a_2 W_2(\rho_a) + b_2 M_2(\rho_a) + a_3 W_3(\rho_a) + b_3 M_3(\rho_a) = 1.$$

$$(5.62)$$

Here the argument ρ is given by

$$\rho = \frac{i\sqrt{\alpha}R^2}{R_0^2 \epsilon}, \tag{5.63}$$

with $\epsilon = a/R_0$, the inverse aspect ratio, and α is the normalized β on axis:

$$\alpha = \left(\frac{aR_0 B_0^2}{\Psi_{axis}}\right) \beta_{axis} \tag{5.64}$$

with B_0 being the field at $R = R_0$ and Ψ_{axis} being the poloidal flux on axis. Note that the position of the magnetic axis (labeled "a" above) is only known after solving the set of coupled equations, and it is always outward of the nominal major radius R_0, a phenomenon known as the "Shafranov shift."

Examples of solutions for the plasma equilibrium in ITER-like geometry ($R_0 = 6.2$m, $B_0 = 5.3$ T, $\epsilon = 0.32$, $\kappa = 1.8$, $\delta = 0.45$, $\alpha = 4.48$, $\beta_{axis} = 9.25\%$, $< \beta >= 2.1$ %, $I = 10.1$ MA) are shown in Fig. 5.6 for various choices for the inboard curvature R_c. Note that parts (a) and (d) are close to the actual design conditions. In Fig. 5.6d, terms in $\sin(k_2 Z/a)$ and $\sin(k_3 Z/a)$ are added to generate an asymmetrical equilibrium with only one singular X-point at the bottom of the plasma: this is called the "single-null divertor" type of equilibrium, which is the actual design basis for ITER. Figure 5.7 shows the pressure and toroidal current profiles for the solution shown in Fig. 5.6a.

5.3.4   The Safety Factor q

The magnetohydrodynamic stability of toroidal devices can be quite complex, and it is the subject of the next chapter. However, it is possible to give a brief overview at this point. Let us return to the cylindrical approximation for a tokamak. We "unwrap" the torus and make a straight cylinder with length $2\pi R$ and radius a. Figure 5.8 illustrates the idea. If we then unwrap the torus poloidally as well, we are left with a rectangle of length $2\pi R$ and width $2\pi a$. Now the magnetic field components B_ϕ (toroidal) and B_θ appear as straight lines on this rectangular mapping. We define the safety factor q as being the number of toroidal transits required to give one poloidal transit. Since the aspect ratio of the rectangular box (horizontal/vertical) is equal to R/a, it is easy to show that

$$q(a) = \frac{a}{R} \frac{B_\phi}{B_\theta(a)}. \tag{5.65}$$

We can extend this concept to arbitrary minor radii r such that

$$q(r) = \frac{r}{R} \frac{B_\phi}{B_\theta(r)}, \tag{5.66}$$

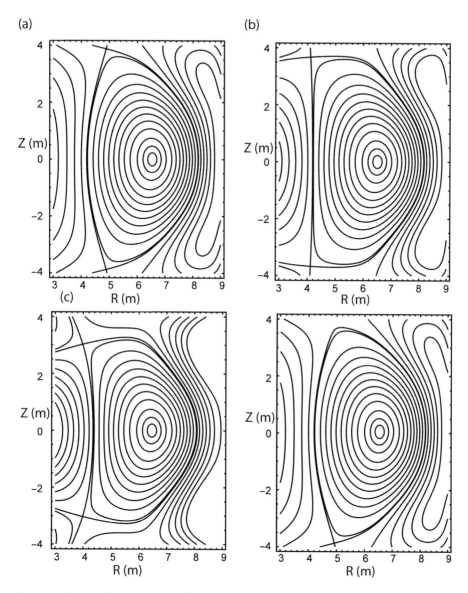

Fig. 5.6 MHD equilibrium for ITER-like parameters, using the Whittaker function method shown in [9]. (**a**) Inboard curvature > 0, (**b**) Inboard curvature = 0, (**c**) Inboard curvature < 0, (**d**) same as (**a**) but modified for single null divertor

with R taken to mean the major radius of the magnetic axis (as seen above, not necessarily at the geometric center of the plasma). We also take B_ϕ to mean the toroidal field at the magnetic axis, but without allowing modifications to the magnetic field by the presence of plasma. Equation (5.66) shows that the local value

Fig. 5.7 Pressure and
toroidal current profiles for
the equilibrium shown in
Fig. 5.6a

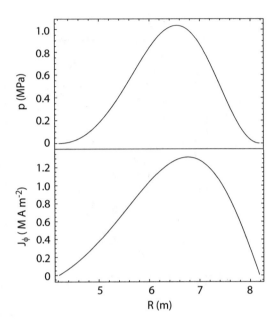

Fig. 5.8 Cylindrical tokamak
approximation, illustrating
safety factor concept

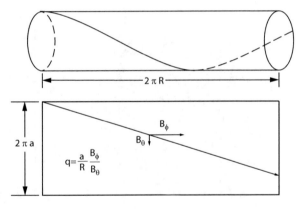

of $q(r)$ does depend on the poloidal field at r, which is a function of the current
density profile (which we will call $J_z(r)$ in the cylindrical approximation). Invoking
Ampere's law:

$$\oint \mathbf{B} \cdot \mathbf{d}\ell = \mu_0 \int \mathbf{J} \cdot \mathbf{d}S, \qquad (5.67)$$

gives

$$2\pi r B_\theta(r) = \mu_0 \int_0^r 2\pi r' dr' J_z(r'). \qquad (5.68)$$

Fig. 5.9 q profile for simple current model

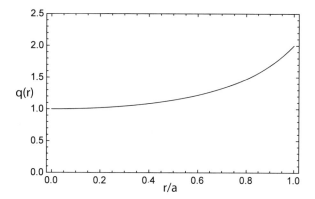

Suppose that we have a current profile given by

$$J_z(r) = J_0\left[1 - \left(\frac{r}{a}\right)^2\right].$$

(5.69)

Then integrating, and ignoring constants, we have

$$B_\theta(r) \propto \frac{r}{a} - \frac{1}{2}\left(\frac{r}{a}\right)^3,$$

(5.70)

and then the safety factor becomes

$$q(r) = \frac{q(0)}{1 - \frac{1}{2}\left(\frac{r}{a}\right)^2}$$

(5.71)

As we shall see later, a requirement for the stability of the plasma is that $q(r) > 0$ everywhere. We also see for this current model $q(a) = 2q(0)$, so the minimum $q(a) = 2.0$. A graph of this q profile is shown as Fig. 5.9.

For the general noncircular tokamak equilibrium, including poloidal currents, we take as a refinement in the simple expression for the safety factor given above by writing a more general form:

$$q(\psi) = \frac{d\chi(\psi)}{d\hat{\psi}},$$

(5.72)

where $\chi(\psi)$ is the toroidal flux $\int \mathbf{B}_\phi \cdot d\mathbf{S}$ and $\hat{\psi}$ is the "real" poloidal flux, i.e. 2π times the flux function Ψ being used thus far. If we look at an incremental volume element between two neighboring flux lines separated by an incremental flux $d\psi$, then by Eq. (5.35) we have the perpendicular distance dx between these neighboring flux lines is given by

$$dx = \frac{d\Psi}{|\nabla\Psi|} = \frac{d\Psi}{RB_{pol}}.$$

(5.73)

We also note that the toroidal field $B_\phi = F(\psi)/R$ as shown earlier. Then integrating along the poloidal flux line ℓ gives an expression for the safety factor, again only a function of ψ:

$$q(\psi) = \frac{d\chi}{d\hat{\Psi}_{pol}} = \frac{1}{2\pi} \oint \frac{F(\psi)}{R^2 B_{pol}} d\ell. \qquad (5.74)$$

Unfortunately, however, evaluating the safety factor in this way means that we have to return to the "real world" of coordinates R and Z to evaluate this integral, because we have to know the position and incremental length of the flux line for any particular value of Ψ. The most numerically stable way to do this is to adopt a coordinate system with a center point at the magnetic axis $\{R, Z\} = \{R_{axis}, 0\}$ and then assign a radial vector of length $r(\Psi, \theta)$ at an angle θ with respect to the outboard midplane such that

$$R(\Psi, \theta) = R_{axis} + r(\Psi, \theta) \cos \theta$$
$$Z(\Psi, \theta) = r(\Psi, \theta) \sin \theta. \qquad (5.75)$$

The radial vector's length $r(\Psi, \theta)$ can then be found with a root-finding algorithm at an array of angles θ (from 0 to π for the symmetrical equilibria and from 0 to 2π for the asymmetrical ones), and then an interpolation function can be constructed for each value of Ψ needed. Then the length element $d\ell$ can be constructed via Pythagoras:

$$d\ell = \sqrt{r(\Psi, \theta)^2 + (\partial [r(\Psi, \theta)] / \partial \theta)^2} d\theta \qquad (5.76)$$

For the analytic models given above, the poloidal field can be obtained by taking R and Z derivatives of the flux function, and then a q-profile can be constructed using an interpolation function. The q-profiles obtained in this way are typically well-behaved functions of flux. Figure 5.10 shows the q-profile obtained in this way

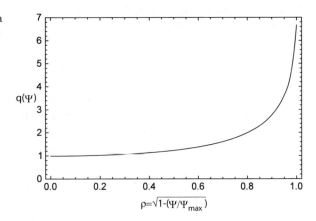

Fig. 5.10 Safety factor q as a function of normalized radius $\sqrt{1 - \Psi/\Psi_{axis}}$ for the equilibrium shown in Fig. 5.6a

for the equilibrium shown in Fig. 5.6a. Notice that here we have plotted the safety factor as a function of a "normalized radius"

$$\rho = \sqrt{1 - \psi/\psi_{max}} \qquad (5.77)$$

Note that the generalized expression for the safety factor and the normalized radius ρ revert to the safety factor formula for the straight circular approximation as a function of r/a and $\rho = r/a$, respectively.

Note also that once the framework has been set up to evaluate q, other important quantities are readily calculated. One is the total toroidal current:

$$I_{tor} = \frac{1}{\mu_0} \oint B_{pol} d\ell, \qquad (5.78)$$

where the line integral is taken at $\Psi = 0$, and another is the plasma volume as a function of flux:

$$V(\psi) = \int_{R_{min}}^{R_{max}} 2\pi R dR \left(Z_+(\Psi) - Z_-(\Psi)\right). \qquad (5.79)$$

Obtaining a functional fit for $V(\Psi)$ then allows a calculation of the average beta $<\beta>$:

$$<\beta> = \frac{\int p(\Psi) dV(\psi)}{V(0) B_0^2/(2\mu_0)}. \qquad (5.80)$$

Problems

5.1 Find the pressure represented by a 4.0 T magnetic field. If this field consisted of straight lines, what would be the maximum pressure in a slab of plasma confined by this field, if the field decreases to 3.5 T in the plasma? What density of 10 keV (temperature) electrons and DT ions could be confined in this case?

5.2 Find the total toroidal current in TFTR ($R = 3.1$m, $A = 3$, $B_{tor} = 5.2$ T, circular bore) if the safety factor at the edge $q(a) = 4.0$.

5.3 For the above problem, what would the average pressure p be if $< \beta_{pol} >= 1$?

5.4 The safety factor at the center in TFTR would sometimes fall below 1. For the edge safety factor as given above, find the minimum value of the Wesson current model ν which would fit this result. Draw a graph of the toroidal current profile.

5.5 From the textbook website, download the file "Guazatto-FreidbergEQ.nb." Run the program using *Mathematica* for the parameter "kGuaz" equal to your birthday

(day of month) divided by 100, and negative if your birthday is on an odd day. Plot your result as a contour plot of the flux. Solve for the exact locations of the x-points.

References

1. Abramowitz, M.: Handbook of Mathematical Functions, With Formulas, Graphs, and Mathematical Tables. Dover Publications, New York (1974)
2. Atanasiu, C.V., Günter, S., Lackner, K., Miron, I.G.: Analytical solutions to the Grad-Shafranov equation. Phys. Plasmas **11**(7), 3510–3518 (2004). http://dx.doi.org/10.1063/1.1756167; http://scitation.aip.org/content/aip/journal/pop/11/7/10.1063/1.1756167
3. Bennett, W.H.: Magnetically self-focussing streams. Phys. Rev. **45**, 890–897 (1934). http://link.aps.org/doi/10.1103/PhysRev.45.890
4. Bowers, E., Haines, M.G.: Application of finite Larmor radius equations for collisionless plasmas to a θ pinch. Phys. Fluids **14**(1), 165–173 (1971). http://dx.doi.org/10.1063/1.1693267; http://scitation.aip.org/content/aip/journal/pof1/14/1/10.1063/1.1693267
5. Buneman, O.: The Bennett pinch. In: Drummond, J.E. (ed.) Plasma Physics, p. 202. McGraw-Hill, New York (1961)
6. Garofalo, A.M., Solomon, W.M., Lanctot, M., Burrell, K.H., DeBoo, J.C., deGrassie, J.S., Jackson, G.L., Park, J.K., Reimerdes, H., Schaffer, M.J., Strait, E.J.: Plasma rotation driven by static nonresonant magnetic fields. Phys. Plasmas **16**(5), 056119 (2009). http://dx.doi.org/10.1063/1.3129164; http://scitation.aip.org/content/aip/journal/pop/16/5/10.1063/1.3129164
7. Goedbloed, J.P., Keppens, R., Poedts, S.: Advanced Magnetohydrodynamics. Cambridge University Press, Cambridge (2010)
8. Grad, H., Rubin, H.: Hydromagnetic equilibria and force-free fields. In: Proceedings of the Second International Conference on the Peaceful Uses of Atomics Energy, vol. 31, p. 190 (1958)
9. Guazzotto, L., Freidberg, J.P.: A family of analytic equilibrium solutions for the Grad-Shafranov equation. Phys. Plasmas **14**(11), 112508 (2007). http://dx.doi.org/10.1063/1.2803759; http://scitation.aip.org/content/aip/journal/pop/14/11/10.1063/1.2803759
10. Guazzotto, L., Betti, R., Manickam, J., Kaye, S.: Numerical study of tokamak equilibria with arbitrary flow. Phys. Plasmas **11**(2), 604–614 (2004). http://dx.doi.org/10.1063/1.1637918; http://scitation.aip.org/content/aip/journal/pop/11/2/10.1063/1.1637918
11. Huba, J.D.: NRL plasma formulary. Naval Research Lab., Washington, DC (2016). http://www.nrl.navy.mil/ppd/sites/www.nrl.navy.mil.ppd/files/pdfs/NRL_FORMULARY_16.pdf
12. Shafranov, V.D.: On magnetohydrodynamical equilibrium configurations. Sov. Phys. JETP **6**, 545–554 (1958)
13. Shafranov, V.D.: Plasma equilibrium in a magnetic field. In: Leontovich, M.A. (ed.) Reviews of Plasma Physics, vol. 2, p. 103. Consultants Bureau, New York (1966)
14. Solov'ev, L.S.: Hydrodynamic stability of closed plasma configurations. In: Leontovich, M.A. (ed.) Reviews of Plasma Physics, Volume 2, vol. 6, pp. 239–331. Consultants Bureau, New York (1975)
15. Struik, D.: Lectures on Classical Differential Geometry. Addison-Wesley Series in Mathematics. Addison-Wesley, Boston (1961). https://books.google.com/books?id=TIXg4hSXmFAC

Chapter 6
Magnetohydrodynamic Stability

6.1 Fluid Perturbations in Plasma

The theory of magnetohydrodynamics (MHD) has several different formulations depending on the assumptions made about the plasma's behavior. The most general fluid description takes the plasma to be nonlinear, resistive, viscous, contain equilibrium flow, and have certain gyrokinetic effects. In its most general form, MHD modeling has little to recommend it over direct kinetic modeling of the plasma using the Vlasov equation with a collision operator. However, MHD theory in its simplest form remains a useful tool for studying the low-frequency waves and instabilities present in magnetically confined plasmas of interest for nuclear fusion. The early history of fusion research depended heavily on the simplified MHD models to explore the behavior of the early experiments. Departures from ideal (infinite conductivity) MHD behavior in the experiments also could be directly identified and the underlying physics causing these departures was easier to identify because of the knowledge base from ideal MHD theory. In this chapter, we will start with ideal linear MHD behavior of static equilibria and then identify the changes brought about by more sophisticated models.

6.1.1 Linearized Perturbations to Stationary ($u_0 = 0$) Equilibrium

First we start with the static equilibrium as explored in Chap. 5:

$$\rho \frac{d\boldsymbol{u}_0}{dt} = 0 = -\nabla p_0 + \boldsymbol{J}_0 \times \boldsymbol{B}_0. \tag{6.1}$$

© Springer Nature Switzerland AG 2018
E. Morse, *Nuclear Fusion*, Graduate Texts in Physics,
https://doi.org/10.1007/978-3-319-98171-0_6

We now make perturbations to this system, using the following forms:

$$\rho = \rho_0 + \delta\rho$$
$$\boldsymbol{B} = \boldsymbol{B}_0 + \delta\boldsymbol{B}$$
$$\boldsymbol{u} = \delta\boldsymbol{u}$$
$$p = p_0 + \delta p$$

(6.2)

We now look at the equation of motion

$$\rho\frac{d\boldsymbol{u}}{dt} = -\nabla p + \boldsymbol{J} \times \boldsymbol{B}.$$

(6.3)

and subtract Eq. (6.1) from this, leaving only the perturbed component:

$$\rho_0\frac{d\delta\boldsymbol{u}}{dt} + \delta\rho\frac{d\delta\boldsymbol{u}}{dt} = -\nabla\delta p + \delta\boldsymbol{J} \times \boldsymbol{B}_0 + \delta\boldsymbol{J}_0 \times \delta\boldsymbol{B} + \delta\boldsymbol{J} \times \delta\boldsymbol{B}.$$

(6.4)

We think of the perturbation as being small compared to the equilibrium values of the components (if they exist). Then we linearize Eq. (6.4) by dropping the terms containing products of two perturbation terms, in this case the last term on the left and the last term on the right, giving

$$\rho_0\frac{d\delta\boldsymbol{u}}{dt} = -\nabla\delta p + \delta\boldsymbol{J} \times \boldsymbol{B}_0 + \boldsymbol{J}_0 \times \delta\boldsymbol{B}.$$

(6.5)

In order to find closure to this equation we will find expressions for the perturbations listed in Eq. (6.2) in terms of only one basic variable. We take as our basic variable a fluid displacement vector $\boldsymbol{\xi}$, given by:

$$\boldsymbol{\xi} = \int_{-\infty}^{t} \delta\boldsymbol{u}(t')dt'.$$

(6.6)

We take $\boldsymbol{\xi}$ to have grown very slowly from $t = -\infty$ to its present (still infinitesimal) value to avoid issues of the mathematical consistency of the equation of motion as an initial-value problem. Let us first look at eliminating $\delta\rho$ from the equation of motion. We do this by invoking the continuity equation:

$$\frac{\partial\rho}{\partial t} + \nabla \cdot (\rho\boldsymbol{u}) = 0.$$

(6.7)

Note that $\partial\rho_0/\partial t = 0$ and $\boldsymbol{u}_0 = 0$, so the linearized equation is just

$$\frac{\partial\delta\rho}{\partial t} + \nabla \cdot (\rho_0\delta\boldsymbol{u}) = 0.$$

(6.8)

Integrated, this becomes

$$\delta\rho + \nabla \cdot (\rho_0 \boldsymbol{\xi}) = 0, \tag{6.9}$$

or

$$\delta\rho + \rho_0 \nabla \cdot \boldsymbol{\xi} + \boldsymbol{\xi} \cdot \nabla \rho_0 = 0 \tag{6.10}$$

Now we look at the perturbed pressure δp. We use the adiabatic law for pressure:

$$\frac{d}{dt} p\rho^{-\gamma} = 0. \tag{6.11}$$

Note that $\partial\rho_0/\partial t = 0$ and $\partial p_0/\partial t = 0$ and $\boldsymbol{u}_0 = 0$. Also note that d/dt appears in this equation rather than $\partial/\partial t$, which is to say that the quantity $p\rho^{-\gamma}$ (related to the specific entropy) is conserved for a volume element moving in the fluid, which is to say "in a Lagrangian sense" rather than "in an Eulerian sense." Then the linearized expression for the adiabatic law, using $d/dt = \partial/\partial t + \delta\boldsymbol{u} \cdot \nabla$:

$$\left(\frac{\partial}{\partial t} + \boldsymbol{u} \cdot \nabla \right) p\rho^{-\gamma} = 0. \tag{6.12}$$

Integrating over time, the linearized equation is just

$$\rho_0^{-\gamma}\delta p - \gamma p_0 \delta\rho\rho_0^{-\gamma-1} + \boldsymbol{\xi} \cdot \nabla(p_0\rho_0^{-\gamma}) = 0. \tag{6.13}$$

We can then eliminate $\delta\rho$ using the continuity equation and then obtain

$$\delta p = -\boldsymbol{\xi} \cdot \nabla p_0 - \gamma p_0 \nabla \cdot \boldsymbol{\xi}. \tag{6.14}$$

Now we look at the electromagnetic force terms $\delta\boldsymbol{J} \times \boldsymbol{B}_0$ and $\boldsymbol{J}_0 \times \delta\boldsymbol{B}$. We take the plasma to be a perfectly conducting fluid. Then

$$\boldsymbol{E} + \boldsymbol{u} \times \boldsymbol{B} = 0. \tag{6.15}$$

Since $\boldsymbol{E}_0 = 0$ we have (leaving out the second-order term $\delta\boldsymbol{u} \times \delta\boldsymbol{B}$):

$$\delta\boldsymbol{E} = -\delta\boldsymbol{u} \times \boldsymbol{B_0} \tag{6.16}$$

and since $\nabla \times \delta\boldsymbol{E} = -\partial\delta\boldsymbol{B}/\partial t$, we have upon integrating over time:

$$\delta\boldsymbol{B} = \nabla \times (\boldsymbol{\xi} \times \boldsymbol{B}_0) \tag{6.17}$$

We now are in a position to write Eq. (6.5) entirely in terms of the fluid displacement vector $\boldsymbol{\xi}$, yielding:

$$\rho_0 \frac{\partial^2 \boldsymbol{\xi}}{\partial t^2} = \delta \boldsymbol{J} \times \boldsymbol{B}_0 + \boldsymbol{J}_0 \times \delta \boldsymbol{B} - \nabla \delta p \qquad (6.18)$$

$$= \frac{1}{\mu_0} [\nabla \times \nabla \times (\boldsymbol{\xi} \times \boldsymbol{B}_0) \times \boldsymbol{B}_0 + (\nabla \times \boldsymbol{B}_0) \times (\nabla \times (\boldsymbol{\xi} \times \boldsymbol{B}_0))] + \nabla [\boldsymbol{\xi} \cdot \nabla p_0 + \gamma p_0 \nabla \cdot \boldsymbol{\xi}]$$
$$(6.19)$$

This ideal MHD momentum equation is the fundamental building block for examining a broad class of low-frequency waves and instabilities in magnetically confined plasma. At first, it might appear a bit daunting to use, because of the complicated dependence on the equilibrium magnetic field $\boldsymbol{B}_0$ and its curl. It therefore is important to recognize some of the simple physical phenomena described by this equation under simplified conditions and geometries.

6.1.2 Special Solutions to the Ideal MHD Equation

First we look at a plasma with no magnetic field and no currents: $\boldsymbol{J}_0 = \boldsymbol{B}_0 = 0$. Then $\nabla p_0 = 0$ and Eq. (6.18) becomes

$$\rho_0 \frac{\partial^2 \boldsymbol{\xi}}{\partial t^2} = \gamma p_0 \nabla (\nabla \cdot \boldsymbol{\xi}). \qquad (6.20)$$

This is an ordinary sound wave with a velocity $c_s = (\gamma p_0/\rho_0)^{1/2}$. We take $\gamma = 5/3$ by treating the plasma as having particles with three degrees of freedom in space but without rotational or vibrational states such as found in ordinary gases. (Sometimes it is more appropriate to treat the electrons as isothermal rather than adiabatic, however.) For a DT plasma with a temperature of 15 keV, this gives $c_s = 1.38426 \times 10^6 \, \text{m s}^{-1}$. (Note that the result does not actually depend on density.)

Next we look at a case where a constant magnetic field $\boldsymbol{B}_0$ is present but the pressure is zero and thus $\boldsymbol{J}_0 \propto \nabla \times \boldsymbol{B}_0 = 0$. Then $\nabla p_0 = 0$ and Eq. (6.18) becomes

$$\rho_0 \frac{\partial^2 \boldsymbol{\xi}}{\partial t^2} = \frac{1}{\mu_0} (\nabla \times \nabla \times (\boldsymbol{\xi} \times \boldsymbol{B}_0)) \times \boldsymbol{B}_0 \qquad (6.21)$$

Let $\boldsymbol{B}_0 = B_0 \hat{z}$ and $\boldsymbol{\xi} = \xi(z)\hat{y}$. Then we obtain a wave equation

$$\rho_0 \frac{\partial^2 \xi}{\partial t^2} = \frac{B_0^2}{\mu_0} \frac{\partial^2 \xi}{\partial z^2} \qquad (6.22)$$

This is called an Alfvén wave and it has a wave velocity given by

$$v_A = \left(\frac{B_0^2}{\mu_0 \rho_0} \right)^{1/2}.$$ (6.23)

Sample Calculation of Alfvén Wave Speed
For a DT plasma with density of $10^{20}\,\mathrm{m}^{-3}$ in a 5.3 T magnetic field (ITER conditions), this gives $v_A = 7.31716 \times 10^6\,\mathrm{m\,s}^{-1}$, almost six times faster than the sound wave. (Unlike the sound wave, this result does depend on density.)

Having $v_A \gg c_s$ is almost always the situation in fusion plasma, since taking $p = \beta B^2/(2\mu_0)$ gives

$$\frac{c_s}{v_A} = \left(\frac{\gamma \beta}{2} \right)^{1/2},$$ (6.24)

so that if β is less than one, $v_A > c_s$. As a result, most MHD wave and instability action has more of an " Alfvénic" and less of a "sonic" character in fusion plasma, as the faster wave usually controls the propagation channel.

It is also interesting to contrast the nature of the two waves found above. In the first case, the perturbation ξ points in the same direction as the wave propagation direction and the wave depends on the compression $\nabla \cdot \xi$ being nonzero. This wave is thus a longitudinal, compressional wave. For the Alfvén wave, $\nabla \cdot \xi = 0$ and the wave is traveling along B with a perturbation $\xi \perp B$. (In fact, one can see from the expression $\delta B = \nabla \times (\xi \times B)$ that the component of ξ parallel to B never matters for the propagation of Alfvén waves.) While there are other solutions to Eq. (6.18) for waves in a zero-pressure plasma that are not propagating along B (the so-called fast magnetosonic wave), Alfvén waves usually refer to the incompressible solution given above. The Alfvén wave can be thought of as similar to the propagation of energy along a tensioned string such as found in musical instruments. On a stretched string, the wave velocity is $c_T = (T/\rho_m)^{1/2}$, where T is the tension and ρ_m is the mass per unit length. In the case of Alfvén waves, the "tension" is given by the Maxwell stress $T = B^2/\mu_0$ per unit area and the mass density goes from a one-dimensional form to a three-dimensional form. It is in this sense that the rather pessimistic analogy of magnetic confinement fusion being similar to attempting to confine jelly with rubber bands has come about!

6.2 MHD Energy Principle

For plasma geometries with some complexity, Eq. (6.18) becomes difficult to work with as it stands. It is a hyperbolic partial differential equation in three space dimensions. However, the force operator representing the right-hand side, written as a dependent function of ξ, can be written as

$$F(\boldsymbol{\xi}) = \boldsymbol{J}_0 \times \nabla \times (\boldsymbol{\xi} \times \boldsymbol{B}_0) + \frac{1}{\mu_0}[\nabla \times \nabla \times (\boldsymbol{\xi} \times \boldsymbol{B}_0)] \times \boldsymbol{B}_0$$

$$+ \nabla (\boldsymbol{\xi} \cdot \nabla p_0 + \nabla \gamma p_0 \nabla \cdot \boldsymbol{\xi}). \tag{6.25}$$

and has the remarkable property that $F(\boldsymbol{\xi})$ is self-adjoint; that is to say

$$\int d^3x \ \boldsymbol{\chi}^* \cdot \ F(\boldsymbol{\xi}) = \int d^3x \ \boldsymbol{\xi}^* \cdot \ F(\boldsymbol{\chi}) \tag{6.26}$$

for any two vector functions $\boldsymbol{\xi}$ and $\boldsymbol{\chi}$. This means that any eigenfunctions Λ of the equation $F(\boldsymbol{\xi}) = \Lambda\boldsymbol{\xi}$ are real. If we introduce a time dependence $\exp(i\omega t)$ for the left-hand side of Eq. (6.18), we have

$$- \rho\omega^2\boldsymbol{\xi} = F(\boldsymbol{\xi}), \tag{6.27}$$

and thus

$$- \omega^2 = \int d^3x \ \boldsymbol{\xi}^* \cdot F(\boldsymbol{\xi}) \bigg/ \int d^3x \ \rho \ \boldsymbol{\xi}^* \cdot \boldsymbol{\xi}. \tag{6.28}$$

We then make the quantity

$$\delta W = -(1/2) \int d\boldsymbol{x} \ \boldsymbol{\xi}^* \cdot F(\boldsymbol{\xi}), \tag{6.29}$$

which represents the perturbed energy going into the plasma by the perturbation $\boldsymbol{\xi}$. Then we have

$$\omega^2 = \delta W \bigg/ \int d\boldsymbol{x} \ \rho\boldsymbol{\xi}^* \cdot \boldsymbol{\xi}/2. \tag{6.30}$$

The self-adjointness property now implies that we can invoke the Ritz variational principle. This means that if we treat $\boldsymbol{\xi}$ as a member of a normalized vector space of trial functions such that $\int d\boldsymbol{x} \ \rho\boldsymbol{\xi}^* \cdot \boldsymbol{\xi}/2 = 1$, then ω^2 is minimized by an actual solution to (6.27). If that minimum value of ω^2 is greater than zero, then the time dependence $\exp(i\omega t)$ is purely oscillatory and the system is stable. If $\omega^2 < 0$ for any trial function, then there will be an actual solution that will have a time dependence of a growing exponential and the system is unstable.

Some steps can be taken to simplify Eq. (6.28). The first step is an integration by parts. This yields

$$\delta W = \frac{1}{2} \int d^3x \left[\frac{|\delta\boldsymbol{B}|^2}{\mu_0} - \boldsymbol{\xi}^* \cdot [\boldsymbol{J}_0 \times \delta\boldsymbol{B} + \nabla (\boldsymbol{\xi} \cdot \nabla p)] + \gamma p|\nabla \cdot \boldsymbol{\xi}|^2 \right]$$

$$- \frac{1}{2} \int d\hat{S} \cdot \boldsymbol{\xi}^* \left(\gamma p\nabla \cdot \boldsymbol{\xi} - \frac{\boldsymbol{B} \cdot \delta\boldsymbol{B}}{\mu_0} \right). \tag{6.31}$$

The surface energy contribution is the result of the integration by parts. Equation (6.31) can be simplified by dividing $\boldsymbol{\xi}$ into components parallel to and perpendicular to the equilibrium magnetic field: $\boldsymbol{\xi} = \boldsymbol{\xi}_\perp + \boldsymbol{\xi}_\parallel \boldsymbol{b}$. Some manipulation then shows that $\boldsymbol{\xi}_\parallel^* \boldsymbol{b} \cdot [\boldsymbol{J}_0 \times \delta\boldsymbol{B} + \nabla (\boldsymbol{\xi} \cdot \nabla p)] = 0$. Also, $\boldsymbol{\xi} \cdot \nabla p = \boldsymbol{\xi}_\perp \cdot \nabla p$. Integrating this term by parts adds another part to the surface term, so that Eq. (6.31) becomes

$$\delta W = \frac{1}{2} \int d^3x \left[\frac{|\delta\boldsymbol{B}|^2}{\mu_0} - \boldsymbol{\xi}_\perp^* \cdot \boldsymbol{J}_0 \times \delta\boldsymbol{B} + \left(\boldsymbol{\xi}_\perp \cdot \nabla p\right) \nabla \cdot \boldsymbol{\xi}_\perp^* + \gamma p |\nabla \cdot \boldsymbol{\xi}|^2 \right]$$
$$- \frac{1}{2} \int d\hat{S} \cdot \boldsymbol{\xi}^* \left(\gamma p \nabla \cdot \boldsymbol{\xi} - \frac{\boldsymbol{B} \cdot \delta\boldsymbol{B}}{\mu_0} + \boldsymbol{\xi}_\perp \cdot \nabla p \right). \tag{6.32}$$

The surface term is required if there is a vacuum region surrounding the plasma. A rather complicated set of matching equations then appear in order to evaluate the surface energy correctly. However, these boundary conditions can be absorbed into the expression for δW by exploiting the equilibrium boundary condition $[\![p + B^2/(2\mu_0)]\!] = 0$, where the symbol $[\![Q]\!]$ indicates the change in quantity Q from vacuum to plasma at the surface of the plasma. Then we can rewrite the overall expression for δW as

$$\delta W = \delta W_F + \delta W_S + \delta W_V, \tag{6.33}$$

with δW_F as before, and the surface term becoming

$$\delta W_S = \frac{1}{2} \int d\hat{S} \cdot \left[\!\left[\nabla \left(\frac{B_0^2}{2\mu_0} + p \right) \right]\!\right] \xi_\perp^2. \tag{6.34}$$

The vacuum perturbed energy is simply

$$\delta W_{vac} = \int dV \frac{\delta B^2}{2\mu_0}. \tag{6.35}$$

6.2.1 The Intuitive Form of δW

An alternative form for δW was derived by Furth [13] which highlights the two driving forces for instability in plasma, namely parallel current and pressure gradients by a rearrangement that has four positive definite terms and two potentially negative terms. We look at our previous equation for δW_F:

$$\delta W_F = \frac{1}{2} \int dV \left[\frac{\delta B^2}{\mu_0} - \boldsymbol{\xi}^* \cdot \boldsymbol{J} \times \delta\boldsymbol{B} + (\boldsymbol{\xi} \cdot \nabla p)(\nabla \cdot \boldsymbol{\xi}) + \gamma p (\nabla \cdot \boldsymbol{\xi})^2 \right] \tag{6.36}$$

Now we divide $\delta \boldsymbol{B}$ into parallel and perpendicular components $\delta \boldsymbol{B}_{\parallel}$ and $\delta \boldsymbol{B}_{\perp}$. Then the first two terms in Eq. (6.36) can then be decomposed as

$$|\delta \boldsymbol{B}|^2 = |\delta \boldsymbol{B}_{\perp}|^2 + |\delta \boldsymbol{B}_{\parallel}|^2, \tag{6.37}$$

and

$$\boldsymbol{\xi}^* \cdot \boldsymbol{J} \times \delta \boldsymbol{B} = J_{\parallel}(\boldsymbol{\xi}_{\perp}^* \times \boldsymbol{b}) \cdot \delta \boldsymbol{B}_{\perp} + \delta B_{\parallel} \boldsymbol{\xi}_{\perp}^* \cdot \boldsymbol{J} \times \boldsymbol{b}, \tag{6.38}$$

where $J_{\perp}$ and $\delta B_{\parallel}$ can be written as

$$J_{\perp} = \frac{\boldsymbol{b} \times \nabla p}{B}, \tag{6.39}$$

and

$$\begin{aligned}
\delta B_{\parallel} &= \boldsymbol{b} \cdot \nabla \times (\boldsymbol{\xi}_{\perp} \times \boldsymbol{B}) \\
&= \boldsymbol{b} \cdot (\boldsymbol{B} \cdot \nabla \boldsymbol{\xi}_{\perp} - \boldsymbol{\xi}_{\perp} \cdot \nabla \boldsymbol{B} - \boldsymbol{B} \nabla \cdot \boldsymbol{\xi}_{\perp}) \\
&= -B(\nabla \cdot \boldsymbol{\xi}_{\perp} + 2\boldsymbol{\xi}_{\perp} \cdot \kappa) + (\mu_0/B)\boldsymbol{\xi}_{\perp} \cdot \nabla p.
\end{aligned} \tag{6.40}$$

Substituting these into Eq. (6.36) then gives

$$\begin{aligned}
\delta W_F = \frac{1}{2} \int_P dV \Big[&|\delta \boldsymbol{B}|^2/\mu_0 + (B^2/\mu_0)|\nabla \cdot \boldsymbol{\xi}_{\perp} + 2\boldsymbol{\xi}_{\perp} \cdot \kappa|^2 + \gamma p|\nabla \cdot \boldsymbol{\xi}|^2 \\
&-2(\boldsymbol{\xi}_{\perp} \cdot \nabla p)(\kappa \cdot \boldsymbol{\xi}_{\perp}^*) - J_{\parallel}(\boldsymbol{\xi}_{\perp}^* \times \boldsymbol{b}) \cdot \delta \boldsymbol{B}_{\perp} \Big].
\end{aligned} \tag{6.41}$$

In this form, one can see the emergence of four positive terms and two potentially negative ones. The first potentially negative term represents potential instability due to the curvature of the magnetic field (unfavorable if $\kappa \cdot \nabla p > 0$), and possible current-driven instabilities due to parallel currents in the plasma. Thus MHD instabilities are thought of as being "pressure-driven" or "current-driven." Examples of the former are peeling and ballooning modes, and internal and external kinks caused by a too-low safety factor (meaning a relatively large $J_{\parallel}$) for the latter.

What also becomes clear in this form for δW is that the quantity $\xi_{\parallel}$ only appears in one term, and that is the positive definite term $\gamma p |\nabla \cdot \boldsymbol{\xi}|^2$ representing energy stored in the compression of the plasma. This term can be minimized with respect to $\xi_{\parallel}$ separately from the other terms, and this yields a minimizing condition (see [11]):

$$\boldsymbol{B} \cdot \nabla (\nabla \cdot \boldsymbol{\xi}) = 0. \tag{6.42}$$

In cases where the operator $\boldsymbol{B} \cdot \nabla$ can be inverted (such as on a magnetic field line that does not close on itself), this expression simplifies to

$$\nabla \cdot \boldsymbol{\xi} = 0, \tag{6.43}$$

and then the compressibility term can be dropped. In other cases, such as the $m = 0$ (axisymmetric) mode in Z-pinches, or rational surfaces in tokamaks where $q = m/n$ for some integers m and n, then Eq. (6.42) applies. In general, however, most analyses of tokamak stability assume that the term $\gamma p \, |\nabla \cdot \boldsymbol{\xi}|^2$ is set to zero, which is equivalent to saying that the plasma is incompressible.

6.3   MHD Stability of Tokamaks

We will discuss MHD stability in tokamaks with models of escalating complexity. As a first step, we treat the tokamak equilibrium in the large aspect ratio limit, which is to say we treat the inverse aspect ratio $\epsilon = a/R$ as a small parameter and assume that the poloidal beta β_{pol} is of order unity. See [31]. We assume that the plasma has a circular cross section. We take the safety factor q to be of order unity, and thus the ratio of the poloidal field to the toroidal field $B_\theta/B_\phi \sim \epsilon$ and the toroidal beta $\beta \sim \epsilon^2$. Then the pressure balance equation

$$\frac{dp}{dr} + \frac{d}{dr}\frac{B_\phi^2}{2\mu_0} + \frac{1}{\mu_0}\frac{B_\theta}{r}\frac{d}{dr}(r B_\theta) = 0 \tag{6.44}$$

requires that $dB_\phi/dr \sim \epsilon^2 B_\phi/a$ and $J_\theta \sim \epsilon J_\phi$. Analyzing the components of $\delta\boldsymbol{B} = \nabla \times (\boldsymbol{\xi} \times \boldsymbol{B})$ in this ordering shows that $\delta B_r \sim \delta B_\theta \gg \delta B_\phi$. Adopting the incompressibility condition $\nabla \cdot \boldsymbol{\xi} = 0$ and taking the modes to have a form $\exp(i(m\theta - n\phi))$ then gives

$$\xi_\theta = -\frac{i}{m}\frac{d}{dr}(r\xi_r), \tag{6.45}$$

which allows the expression for δW to be written in terms of a single component ξ_r, which we then drop the subscript and define $\xi_r = \xi$. The perturbed magnetic fields are given by

$$\delta B_r = -\frac{im B_\phi}{R}\left(\frac{n}{m} - \frac{1}{q}\right)\xi \tag{6.46}$$

and

$$\delta B_\theta = -\frac{B_\phi}{R}\frac{d}{dr}\left[\left(\frac{n}{m} - \frac{1}{q}\right)r\xi\right] \tag{6.47}$$

After some algebraic manipulation and an integration by parts, contribution to δW by the plasma, δW_F, is given by

$$
\delta W_F = \frac{\pi^2 B_\phi^2}{\mu_0 R} \left\{ \int_0^a \left[\left(r \frac{d\xi}{dr} \right)^2 + (m^2 - 1)\xi^2 \right] \left(\frac{n}{m} - \frac{1}{q} \right)^2 r \, dr \right.
$$
$$
\left. + \left[\frac{2}{q_a} \left(\frac{n}{m} - \frac{1}{q_a} \right) + \left(\frac{n}{m} - \frac{1}{q_a} \right)^2 \right] a^2 \xi_a^2 \right\}.
$$

$$(6.48)$$

The contribution to δW from the vacuum magnetic energy is derived by matching the perturbed magnetic fields δB_r and δB_θ at the plasma edge $r = a$ with a solution for the magnetic field in a vacuum, which is given by $\delta \boldsymbol{B} = \nabla \psi$, where ψ is a solution to a Laplace equation $\nabla^2 \psi = 0$, with a boundary condition that $\delta B_r = 0$ at the position of a conducting wall, taken to be at $r = b$. If we make the quantity λ to be a function of the radius ratio a/b such that

$$
\lambda = \frac{1 + (a/b)^{2m}}{1 - (a/b)^{2m}},
$$

$$(6.49)$$

then the total perturbed energy δW is obtained by a slight modification to Eq. (6.48), giving:

$$
\delta W_F = \frac{\pi^2 B_\phi^2}{\mu_0 R} \left\{ \int_0^a \left[\left(r \frac{d\xi}{dr} \right)^2 + (m^2 - 1)\xi^2 \right] \left(\frac{n}{m} - \frac{1}{q} \right)^2 r \, dr \right.
$$
$$
\left. + \left[\frac{2}{q_a} \left(\frac{n}{m} - \frac{1}{q_a} \right) + (1 + m\lambda) \left(\frac{n}{m} - \frac{1}{q_a} \right)^2 \right] a^2 \xi_a^2 \right\}.
$$

$$(6.50)$$

Note that λ has a minimum value of one when the wall is at infinity and a value of infinity when the wall is touching the plasma, in which case $\xi(a) = 0$. Thus moving the wall away from the plasma is destabilizing, since the perturbed energy contains a positive definite term multiplied by λ. In fact, this expression for δW given in Eq. (6.50) predicts complete stability for all modes if there is a tightly fitting conducting wall around the plasma. For a wall at infinity, Eq. (6.50) predicts instability for plasmas with shallow q profiles where $q_a < m/n$. Since the minimum value of n is one, narrow bands of instability can exist for safety factors q_a somewhat less than any integer. The region of stability can be determined by solving the eigenmode equation associated with Eq. (6.50) taken as a variational principle. This yields a second-order equation in ξ, given by

$$
\frac{d}{dr} \left[(\rho \omega^2 - F^2) r \frac{d}{dr} (r\xi) \right] - \left[m^2 \left(\rho \omega^2 - F^2 \right) - r \frac{dF^2}{dr} \right] \xi = 0.
$$

$$(6.51)$$

Here ρ is the plasma density, ω is the frequency, and $F = (m - nq)B_\theta/(r\mu_0)^{1/2}$. The boundary conditions at $r = 0$ and $r = \infty$ are that $\xi \propto r^{m-1}$ at $r = 0$ and

$$\frac{d}{dr}(r\xi) = \frac{m(m - nq)^2}{(\mu_0\rho\omega^2 a^2 B_\theta^2) - (m - nq_a)^2}\left(\lambda - \frac{2}{m - nq_a}\right)\xi \qquad (6.52)$$

at $\xi = a$. The q profile and the position of the wall affect the stability boundary. Wesson [30] took a simple current model:

$$J_\phi = J_0\left(1 - \left(\frac{r}{a}\right)^2\right)^\nu, \qquad (6.53)$$

which produces a safety factor profile

$$q(r) = q_0\frac{(\nu + 1)\ (r/a)^2}{1 - \left(1 - \frac{r^2}{a^2}\right)^{\nu+1}} \qquad (6.54)$$

and has the property that

$$\frac{q_a}{q_0} = \nu + 1. \qquad (6.55)$$

Equations (6.51) and (6.52) can then be solved numerically, resulting in the stability diagram shown in Fig. 6.1 for the case $b/a \to \infty$. More complex stability problems involving noncircular geometry are the domain of large computer codes, such as GATO and ERATO. An example of a finite element stability calculation for a JET-like equilibrium is shown in Fig. 6.2.

6.3.1 Internal Kink Mode

Suppose we look at Eq. (6.50) for an $m = 1$, $n = 1$ mode with a conducting wall boundary condition $\xi(a) = 0$. Then Eq. (6.50) becomes

$$\delta W = \frac{\pi^2 B_\phi^2}{\mu_0 R}\int_0^a\left[\left(r\frac{d\xi}{dr}\right)^2 + (m^2 - 1)\xi^2\right]\left(\frac{n}{m} - \frac{1}{q}\right)^2 r\,dr \qquad (6.56)$$

Now we supply a mode such as shown in Fig. 6.3. This mode is constant over the part of the plasma $r < r_1$ with $q < 1$ but has $d\xi/dr = 0$ there, so that both terms in the integrand vanish. We then allow for a region between $r = r_1 - \delta$ and $r = r_1$ where the perturbation becomes zero, and the perturbation remains zero out

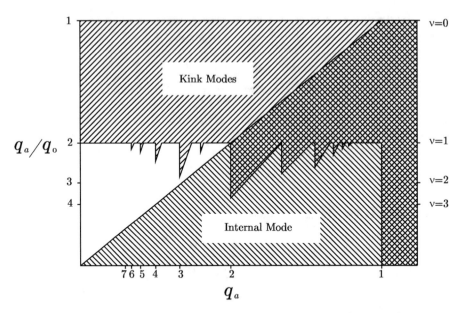

Fig. 6.1 Diagram of stable and unstable regions as a function of edge safety factor $q(a)$ and current model exponent ν. After [30]

to the wall. Then in the limit of $\delta \rightarrow 0$, $\delta W = 0$ and the plasma is marginally stable. However, a marginally stable result means that one has to go beyond the cylindrical approximation to the next order in the inverse aspect ratio parameter $\epsilon = a/R$ and add in another term representing the modification of δW due to toroidal effects. This term is negative for values of poloidal beta $\beta_{pol} > 0.3$, and is proportional to $1 - q(0)$. So except for the case of very low β_{pol}, it appears that q must be greater than one everywhere to avoid this mode. The criterion $q(0) > 1$ is thus shown as a stability boundary in Fig. 6.1; with the choice of variables for this diagram, $q(0) > 1$ appears as a straight line.

Experimentally, $q(0) < 1$ has been seen in some tokamak discharges, and sometimes reaching down as low as $q(0) = 0.7$. However, the emergence of $q(0) < 1$ is correlated with the appearance of sawtooth oscillations [27]. These oscillations are observed by the presence of X-ray emissions which have this sawtooth time-dependent shape. Figure 6.4 shows the X-ray signal from areas on both sides of the $q = 1$ surface, showing the reversal in the waveform. Apparently some plasma rearrangement happens at the time of the sawtooth crash, and thus there is some nonlinear mechanism for the saturation of this mode. The depth of the safety factor depression also seems to be influenced by velocity flow in the plasma, as well as its shear.

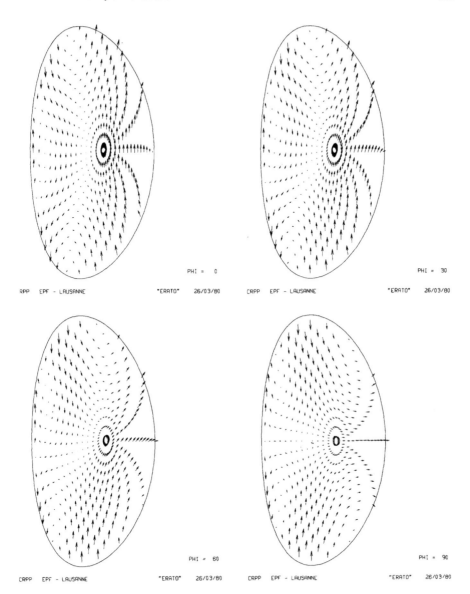

Fig. 6.2 Output from ERATO finite element MHD code. From [15]. Used with permission, Elsevier

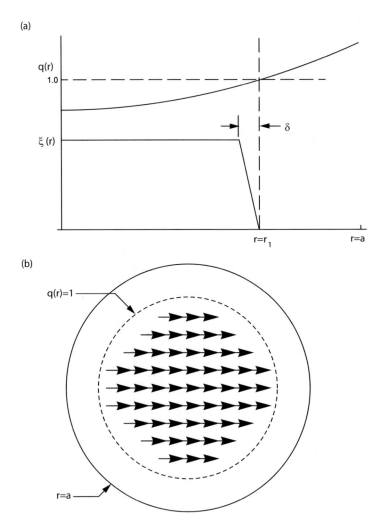

Fig. 6.3 (a) Internal $m = 1$, $n = 1$ kink mode in a plasma with $q(r) = 1$ at some radius r_1 in cylindrical approximation. Critical eigenmode, with $\xi(r_1) = 0$ where $q(r_1) = 1$. (b) Vector plot of the displacement vector $\boldsymbol{\xi}$

6.4   Resistive MHD

6.4.1   *Magnetic Reconnection*

The ideal MHD theory, with its assumption that conductivity is infinite, does not allow for the possibility of magnetic field line reconnection. One can see this by looking at the generalized Ohm's law for a conducting fluid:

$$\boldsymbol{J} = \sigma(\boldsymbol{E} + \boldsymbol{u} \times \boldsymbol{B}). \tag{6.57}$$

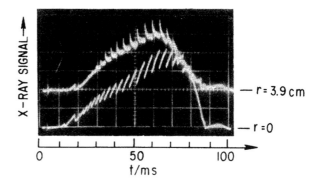

Fig. 6.4 Sawtooth X-ray signal observed in the ST tokamak. From [27]. Used with permission, American Physical Society

With $\sigma \rightarrow \infty$, Eq. (6.57) becomes $\boldsymbol{E} + \boldsymbol{u} \times \boldsymbol{B} = 0$, and if one takes the curl of this equation and integrates over a surface S one obtains, taking $\Phi = \int \boldsymbol{B} \cdot d\boldsymbol{S}$:

$$\frac{\partial \Phi}{\partial t} + \oint \boldsymbol{u} \cdot (d\boldsymbol{\ell} \times \boldsymbol{B}) = 0. \tag{6.58}$$

This shows that magnetic flux is convected along with the fluid, and that if any two fluid elements are connected by a magnetic field line at one time, they remain connected for all times. Thus ideal MHD has a feature that the topology of the magnetic fields cannot change.

However, the presence of a finite amount of resistivity can eliminate that feature of the MHD fluid description of the plasma. Suppose that we have a plasma with a magnetic field pointing in the $+\hat{y}$ direction at some distance away from an X-point type magnetic null at $x = 0$, $y = 0$, and in the $-\hat{y}$ direction at some (positive) distance away. Figure 6.5 shows the basic situation. The original analysis of this model has its origins in papers by Sweet [26] and Parker [24].

We can represent the two-dimensional magnetic field $\boldsymbol{B} = \hat{x}B_x + \hat{y}B_y$ with a stream function ψ such that

$$\boldsymbol{B} = \nabla \psi \times \hat{z}. \tag{6.59}$$

We assume that the plasma is incompressible and that at $|x| \gg \delta$ there is a flow towards the X-point with velocity $v = \pm \hat{x} u_0$. The upstream magnetic field is $\boldsymbol{B} = \pm \hat{y} B_0$ on the two sides of the X-point for $|x| \gg \delta$. Inside a column of radius δ, the magnetic field takes a sharp turn into the $\pm \hat{x}$ direction and is transported by a flow field $\pm v_0 \hat{y}$. Conservation of mass across the footprint of the reconnection region requires that $u_0 l = v_0 \delta$, typically with $l \gg \delta$. Inside the reconnection region, we allow for the finite resistivity. Near the X-point, $E_z \rightarrow 0$ by symmetry. This gives, by (6.57):

$$u_0 B_0 = \eta J \approx \eta \frac{B_0}{\mu_0 \delta}. \tag{6.60}$$

Fig. 6.5 Magnetic
reconnection schematic. lines
coming in from the left and
right represent the flow field,
and the other lines are the
magnetic field. From [32]

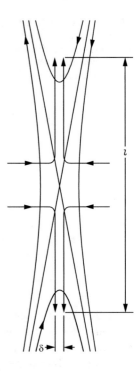

Typically the kinetic energy $(1/2)\rho u_0^2$ outside of the reconnection zone is much
smaller than the magnetic energy $B_0^2/(2\mu_0)$, so that force balance requires that
$(\partial/\partial x)\left(p + B^2/(2\mu_0)\right) = 0$ so that

$$\frac{B_0^2}{2\mu_0} = p_{max} - p_0, \qquad (6.61)$$

where p_{max} is the pressure at the X-point (where $B = 0$), and p_0 is the upstream
pressure. Integrating the steady-state fluid equation $\rho v_y(\partial/\partial y)v_y = -(\partial/\partial y)p$
between $y = 0$ and the end of the reconnection region at $y = \pm l/2$ gives

$$\rho v_0^2/2 = p_{max} - p_0 = B_0^2/(2\mu_0), \qquad (6.62)$$

with the last identity obtained from (6.61). Thus

$$v_0 = \left(\frac{B_0^2}{\mu_0 \rho}\right)^{1/2} = v_A, \qquad (6.63)$$

where v_A is the Alfvén wave speed. Using this solution for v_0 then gives a solution
for the Alfvénic Mach number

$$\frac{u_0}{v_0} \equiv M_0 = \left(\frac{\eta \sqrt{\rho}}{B_0 l \sqrt{\mu_0}} \right)^{1/2} \equiv S_0^{-1/2} \tag{6.64}$$

and for the aspect ratio A:

$$\frac{\delta}{l} \equiv A^{-1} = S_0^{-1/2} \tag{6.65}$$

In these equations a dimensionless number S_0 has been defined, called the Lundquist number

$$S_0 = \frac{B_0 l \sqrt{\mu_0}}{\eta \sqrt{\rho}} = \frac{v_A l}{\eta *}. \tag{6.66}$$

Here η^* is the magnetic diffusivity η/μ_0. Note that $v_A l$ and v^* have units of $(\text{length})^2(\text{time})^{-1}$. The Lundquist number relates the ratio of the transport of magnetic energy by Alfvénic effects to dissipation by resistivity. For magnetic fusion plasma , this number is in the range $10^5 \rightarrow 10^{10}$, showing that over the bulk of the plasma, resistivity is negligible.

Example: Lundquist Number in ITER

Consider the case of ITER, with $B_\phi = 5.3$ T, and DT plasma at a number density of 10^{20} m^{-3}. Using DT (average mass 2.5 proton masses) gives $\rho = 10^{20} \cdot 2.5 \cdot 1.67 \times 10^{-27} = 4.175 \times 10^{-7}$ kg m^{-3}. Take a magnetic scale length $l = 1.0$ m. Use the Braginskii resistivity

$$\eta = 2.8 \times 10^{-8} T_e (\text{keV})^{3/2} Z_{eff},$$

For $T_e = 10$ keV and $Z_{eff} = 2$ this gives $\eta = 1.77 \times 10^{-9}$ Ω-m. Then $S = B_0 l \sqrt{\mu_0}/(\eta \sqrt{\rho}) = 5.12 \times 10^9$.

But in those places where the magnetic field vanishes, such as in the current sheet example above, resistivity must be included. In this current sheet example given above, it is the fate of all field lines to be swept into the reconnection region and emerge connected to a field line from the other side. This is thus a fundamental example of how resistivity breaks topology. However, the Sweet-Parker reconnection theory presented here gives an estimate for the reconnection rate which is inconsistently slow compared to the observation of the reconnection rates actually seen in astrophysical observations and in laboratory fusion experiments [33]. A good analysis of subsequent modifications to the resistive MHD theory in attempts at greater self-consistency and closer agreement with observations is given in Ref. [3]. Since then, research into magnetic reconnection has included analyses of plasmoid formation at the current sheet, once it was generally accepted that the current sheet is intrinsically unstable. A recent review of the nonlinear MHD of reconnection is given in [20]. There have been several attempts to incorporate non-MHD phenomena, such as the effects of the finite ion Larmor radius and the growth of whistler waves and ion acoustic waves into the theory, but so far nothing has been very successful in predicting the fast timescale of sawtooth crashes in tokamaks.

6.4.2   The Resistive Tearing Mode ($m \neq 1$)

The previous section was devoted to the explanation of the reconnection of magnetic field lines through a resistive layer containing an X-point, i.e. a place where $\boldsymbol{B} = 0$. The situation in tokamaks and other toroidal devices is somewhat different in that the overall magnetic field is never zero. However, perturbations can form near rational surfaces where $m - nq = 0$ with a resonant structure such that the localized fluid displacement has a structure $\boldsymbol{\xi} \propto \exp(i(m\theta - n\phi))$, i.e. it has a structure in resonance with the equilibrium field at $q = m/n$. In ideal MHD, instabilities can form with perturbations localized to these rational surfaces, they are stabilized by magnetic shear when

$$\frac{r B_\phi^2}{8\mu_0}\left(\frac{q'}{q}\right)^2 > \left(-p'\right)\left(1 - q^2\right). \tag{6.67}$$

This is called the Mercier criterion, and is satisfied if $q > 1$ everywhere and the pressure profile is monotonic. The situation in resistive MHD is more complicated.

It is useful to "unwrap" the tokamak geometry into a straight cylinder and to select a coordinate system where the new coordinates are $x = r$, $z \propto m\theta - n\phi$, and a third coordinate $y \propto m\phi + n\theta$ orthogonal to both. Figure 6.6 shows the components of the magnetic field as the radial coordinate passes through the place where $m - nq = 0$. Thus in this geometry, the xy projection has a look similar to the one shown in Fig. 6.6, with a reversed magnetic field B_y through $x = 0$. Even though the plasma is strongly magnetized, a reconnection process can take place near the resonant surface.

Fig. 6.6 projection of magnetic field components through a sheared magnetic field, with a rational surface at x = 0

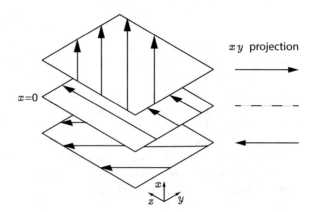

Now we return to our standard $\{r, \theta, \phi\}$ geometry for a moment. We attempt to construct a perturbed current with a toroidal and poloidal structure $\propto \exp(i(m\theta - n\phi))$. To do this, we note that since $\boldsymbol{J} \times \boldsymbol{B} = \nabla p$ and thus $\nabla \times \boldsymbol{J} \times \boldsymbol{B} = 0$, and since $\cdot \boldsymbol{J} = \nabla \cdot \boldsymbol{B} = 0$, we have

$$\boldsymbol{B} \cdot \nabla \boldsymbol{J} - \boldsymbol{J} \cdot \nabla \cdot \boldsymbol{B} = 0. \tag{6.68}$$

Now we invoke an ordering of the components of $\boldsymbol{J}$ and $\boldsymbol{B}$ in a large aspect ratio tokamak. Noting that $B_\theta \sim \epsilon B_\phi$ and $J_\theta \sim \epsilon J_\phi$, where $\epsilon = r/R$ is taken to be a small number, and forming perturbed quantities $\{\delta B_r, \delta B_\theta, \delta B_\phi\}$ and $\{\delta J_r, \delta J_\theta, \delta J_\phi\}$, we find that $\delta B_\phi \sim \epsilon \delta B_\theta \sim \delta B_r$ and $\delta J_r \sim \epsilon \delta J_\theta \sim \epsilon \delta J_\phi$. Setting $\epsilon \to 0$ then gives

$$\delta(\boldsymbol{B} \cdot \nabla J_\phi) = 0. \tag{6.69}$$

In order to assure that $\nabla \cdot \delta \boldsymbol{B} = 0$, we introduce a stream function ψ such that

$$\delta B_r = -\frac{1}{r}\frac{\partial \psi}{\partial \theta} \quad \text{and} \quad B_\theta = \frac{\partial \psi}{\partial r}. \tag{6.70}$$

We then find $\delta \boldsymbol{J}$ through Ampére's law:

$$\mu_0 \delta J_\phi = \nabla^2 \psi = \frac{1}{r^2}\frac{\partial}{\partial r} r \frac{\partial \psi}{\partial r} + \frac{1}{r^2}\frac{\partial^2 \psi}{\partial \theta^2} \tag{6.71}$$

Plugging this into Eq. (6.69), which can be rewritten as $\boldsymbol{B} \cdot \nabla \delta \boldsymbol{J} + \delta \boldsymbol{B} \cdot \nabla \boldsymbol{J} = 0$, then gives

$$\frac{1}{\mu_0}\left(\frac{mB_\theta}{r} - \frac{nB_\phi}{R}\right)\nabla^2 \psi - \frac{m}{r}\frac{dj_\phi}{dr}\psi = 0. \tag{6.72}$$

Note that upon dividing by the term $mB_\theta/r - nB_\phi$ this gives an equation which links second derivatives of ψ with a term linear in ψ with a singular coefficient at $m - nq = 0$. Thus Eq. (6.72) cannot be used near the resonant surface and a treatment including resistivity is required in a region surrounding the rational surface. Retaining the resistivity in the generalized Ohm's law (Eq. (6.57)) and taking the radial component gives:

$$-\frac{\partial B_r}{\partial t} + \boldsymbol{B} \cdot \nabla v_r = -\frac{\eta}{\mu_0}\nabla^2 B_r. \tag{6.73}$$

Taking all quantities to have spatial and time dependence $\propto \exp(\gamma t + i(m\theta - n\phi))$ and taking $B_r = -im\psi/r$ from Eq. (6.70) gives

$$\gamma \psi + B_\theta \left(1 - \frac{nq}{m}\right) v_r = \frac{\eta}{\mu_0}\nabla^2 \psi. \tag{6.74}$$

Since the resistive layer is assumed to be narrow, the radial second derivative in ∇^2 dominates, so that we can rewrite Eq. (6.74) as

$$\gamma \psi + B_\theta \left(1 - \frac{nq}{m} \right) v_r = \frac{\eta}{\mu_0} \frac{d^2 \psi}{dr^2}. \tag{6.75}$$

We now note that the term $1 - nq/m$ in Eq. (6.75) is zero at the resonant surface and we can make a first-order Taylor expansion

$$1 - \frac{nq}{m} = - \left. \frac{q'}{q} \right|_{r=r_s} s, \qquad \text{where} \quad s = r - r_s. \tag{6.76}$$

We then integrate Eq. (6.75) across the resistive layer, noting that ψ is slowly changing, but $d\psi/dr$ can be discontinuous across the layer because of the singularity in the "outer" ideal MHD equation, (6.72). This integration then yields

$$\Delta' = \frac{\mu_0 \gamma}{\eta} \int \left(1 - \frac{B_\theta q'}{q} - s \frac{v_r}{\psi} \right) ds. \tag{6.77}$$

Here Δ' represents the jump in the logarithmic derivative across the resistive layer:

$$\Delta' = \frac{1}{\psi} \left[\left. \frac{d\psi}{dr} \right|_{r=r_s+\delta/2} - \left. \frac{d\psi}{dr} \right|_{r=r_s-\delta/2} \right]. \tag{6.78}$$

The remaining task is to find v_r/ψ. From the equation of motion

$$\rho \frac{d\boldsymbol{v}}{dt} = \boldsymbol{J} \times \boldsymbol{B} - \nabla p,$$

we take the curl to obtain

$$\nabla \times \boldsymbol{J} \times \boldsymbol{B} = \nabla \times \rho \frac{\partial \boldsymbol{v}}{\partial t}. \tag{6.79}$$

We solve for the ϕ component of Eq. (6.79), yielding an equation for dv_θ/dr and then invoking the continuity equation for an incompressible fluid $\nabla \cdot \boldsymbol{v} = 0$:

$$\left(\nabla \times \rho \frac{\partial \boldsymbol{v}}{\partial t} \right)_\phi = \gamma \rho \frac{dv_\theta}{dr}$$

$$= \frac{i \gamma \rho}{m} r \frac{d^2 v_r}{dr^2}. \tag{6.80}$$

The equilibrium equation $J \times \nabla p = 0$, and its curl, as discussed above, is modified to include the inertial term:

$$\rho \frac{\partial u}{\partial t} = -\nabla p + J \times B,$$

and inserting the result for the curl of the equation of motion, with the inertial term added, from Eq. (6.80), gives:

$$\frac{dv_r}{ds^2} - \left(\frac{B_\theta^2 m^2 q'^2}{\rho \eta \gamma r^2 q^2} \right) s^2 v_r = - \left(\frac{B_\theta m^2 q'}{\rho \eta r^2 q} \right) s\psi - \frac{m^2}{\rho \gamma r^2} \frac{d J_\phi}{dr} \psi. \qquad (6.81)$$

Since ψ is roughly constant inside the resistive layer, We note that the forcing terms to this second-order ODE for ψ contains both odd and even terms in the length variable s, but only the odd terms in s contribute to the integral in Eq. (6.77). So we solve for v_r using only the first term on the right in Eq. (6.81). We can define some new variables for simplicity:

$$d = \left(\frac{\rho \eta r^2 q^2}{B_\theta^2 m^2 q'^2} \right)^{1/4},$$

$$x = \frac{s}{d}, \quad \text{and} \quad y = -\frac{\rho \gamma r^2 q}{m^2 B_\theta q' d^3} \frac{v_r}{\psi}. \qquad (6.82)$$

Then Eq. (6.81) becomes

$$\frac{d^2 y}{dx^2} = -y(1 - xy), \qquad (6.83)$$

and Eq. (6.77) becomes

$$\Delta' = \frac{\mu_0 \gamma d}{\eta} \int_{-\infty}^{+\infty} (1 - xy)dx. \qquad (6.84)$$

Equation (6.81) has a solution [1]:

$$\begin{aligned}
y &= \frac{x}{2} \int_0^1 dx' \exp(-x^2 x'/2) \left(1 - x'^2 \right)^{-1/4} \\
&= \frac{1}{2} \sqrt{\frac{\pi}{2}} \sqrt{x} \, \Gamma\left(\frac{3}{4} \right) \left(I_{1/4}\left(\frac{x^2}{2} \right) - L_{1/4}\left(\frac{x^2}{2} \right) \right) \\
&= \frac{\sqrt{x} \, \Gamma\left(\frac{3}{4} \right)}{\sqrt{2\pi}} \int_0^\infty dx' \frac{J_{1/4}(xx')}{(x')^{1/4} \left(x'^2 + 1 \right)}.
\end{aligned} \qquad (6.85)$$

Fig. 6.7 Solution for $y(x) \propto v_r/\psi$ in the resistive layer

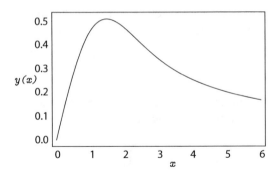

Here $I_{1/4}$ and $L_{1/4}$ are the modified Bessel function and modified Struve function, respectively. As both of these become exponentially large for a large argument, this expression is not numerically stable at large values of x. The last expression, known as the Sonin-Gubler identity (see Ref. [29, p. 426]), is quite stable numerically. A graph of this function is shown as Fig. 6.7. At large values of x, $y \approx 1/x + 2/x^5$, and thus the integrand $1 - xy$ rapidly goes to zero. Using this result, the expression for Δ' becomes

$$\Delta' = \frac{\mu_0 \gamma d}{\eta} \int (1 - xy)dx = \frac{\mu_0 \gamma d}{\eta} \sqrt{2} \left(\Gamma \left(\frac{3}{4} \right) \right)^2 = 2.12365 \frac{\mu_0 \gamma d}{\eta}. \tag{6.86}$$

Equation (6.86) clearly shows that a resonant surface where $\Delta' > 0$ is unstable and one with $\Delta' < 0$ is stable. Equation (6.86) can then be regrouped to give an expression for the growth rate γ:

$$\gamma = \frac{1}{2^{2/5} \left(\Gamma(3/4) \right)^{8/5}} \left(\left(\frac{\eta}{\mu_0} \right)^{3/5} \left(\frac{m B_\theta}{(\mu_0 \rho)^{1/2}} \frac{q'}{qr} \right)^{3/5} \right)_{r=r_s} \Delta'^{4/5}. \tag{6.87}$$

The numerical factor $1/(2^{2/5}(\Gamma(3/4))^2) = 0.54737$. By defining an Alfvén time τ_A and a resistive diffusion time τ_R as

$$\tau_A = \frac{a}{B_\phi/(\mu_0 \rho)^{1/2}} \quad \text{and} \quad \tau_R = \frac{\mu_0 a^2}{\eta}, \tag{6.88}$$

we can write the growth rate as

$$\gamma = \frac{0.55}{\tau_R^{3/5} \tau_A^{2/5}} \left(n \frac{a}{R} \frac{a q'}{q} \right)^{2/5} \left(a \Delta' \right)^{4/5}. \tag{6.89}$$

Example: Resistive Growth Times on ITER

To give a numerical example, we look at parameters relevant to ITER. Using a density of $10^{20}\,\mathrm{m}^{-3}$ of DT with a mass of 2.5 proton masses, with minor radius $a = 2.1\,\mathrm{m}$ and a toroidal field B_ϕ of 5.3 T gives an Alfvén time of $\tau_A = 2.87 \times 10^{-7}$ s. We take the plasma resistivity to be

$$\eta = \frac{2.8 \times 10^{-8}}{Z} T_e(\mathrm{keV})^{-3/2}, \tag{6.90}$$

which for an effective $Z = 2$ and $T_e = 15\,\mathrm{keV}$ gives $\eta = 9.64 \times 10^{-10}\,\Omega\mathrm{m}$, and then this gives a resistive decay time $\tau_R = 5748$ s. If we assume that the shear $q'a = 3.0$ at the $q = 2$ surface with mode numbers $(m, n) = (2, 1)$, and take $\Delta' = 15.0\,\mathrm{m}^{-1}$ for the outer MHD mode solution, we find $\gamma \approx 15\,\mathrm{s}^{-1}$, or a characteristic exponential growth period of 67 ms.

The "in-between" timescale that resistive MHD represents clearly differentiates this type of mode from an ideal MHD mode, which can have a growth period of less than a microsecond.

One should note, however, an inconsistency in the simple ohmic resistive-layer theory and that is the predicted value for the resistive layer thickness d. For the conditions given above and taking $r = 1.0\,\mathrm{m}$, we calculate $d = 0.25\,\mathrm{mm}$. This is smaller than the ion Larmor radius by a factor of about ten, and thus the one-fluid resistive MHD model breaks down.

6.4.3   Magnetic Islands

We now revisit the transformed geometry of Fig. 6.6 in order to look at the magnetic structure at the resonant surface and inside the resistive layer. In ideal MHD we have that the component of the perturbed magnetic field perpendicular to the magnetic surface $\delta B_\perp$ must be zero, since in ideal MHD, $\delta B_\perp = \nabla \times \boldsymbol{\xi} \times \boldsymbol{B}_0 = \boldsymbol{B} \cdot \nabla \boldsymbol{\xi} = 0$. In resistive MHD, however, $\delta \boldsymbol{B}$ is derived from the magnetic diffusion equation, including resistivity, Eq. (6.57). Using a flux function such that $\boldsymbol{B}_\perp = \hat{z} \times \nabla \psi$, we see that

$$B_y = B_\theta \left((1 - \frac{n}{m} q(r)) \right) = - \left(B_\theta \frac{q'}{q} \right)_{r=r_s} x, \tag{6.91}$$

and thus one component of the flux is given by the integral of this quantity:

$$\psi(x) = - \left(B_\theta \frac{q'}{q} \right)_{r=r_s} \frac{x^2}{2} + \tilde{\psi}(y) \tag{6.92}$$

The radial (x) component of the perturbed magnetic field is derived from $\tilde{\psi}(y)$ in Eq. (6.92) above; for a helical perturbation we can choose $\delta B_r = -\partial\tilde{\psi}/\partial y = \hat{B}_r \sin ky$. We then integrate this to obtain $\tilde{\psi}(y) = (\hat{B}_r/k)\cos(ky)$. Then we have

$$\psi(x) = -\left(B_\theta \frac{q'}{q}\right)_{r=r_s} \frac{x^2}{2} + \frac{\hat{B}_r}{k}\cos ky. \tag{6.93}$$

This can be re-written as

$$\psi(x) = -\left(B_\theta \frac{q'}{2q}\right)_{r=r_s}\left(x^2 - \frac{w^2}{8}\cos ky\right). \tag{6.94}$$

with

$$w = 4\left(\frac{q\hat{B}_r}{kq'B_\theta}\right)^{1/2} \tag{6.95}$$

A contour plot of the flux is shown as Fig. 6.8. The parameter w is chosen so that $x = \pm w/2$ at the separatrix at $y = 0$, i.e. the full width of the island is w.

We can map this island into circular geometry (as a poloidal projection) by replacing ky in Eq. (6.94) with $y = r_s\theta$ and $k = m/r_s$, i.e. the angular dependence is $\cos(m\theta)$ and the island width is

$$w = 4\left(\frac{rq\hat{B}_r}{mq'B_\theta}\right)^{1/2}_{r=r_s}. \tag{6.96}$$

A plot of the island in circular poloidal geometry is shown as Fig. 6.9. Note that more than one resonant surface can support an island: for example, there can be simultaneous island formation on the $m/n = 2/1$ and $3/2$ surfaces. When these islands grow to finite size, there tends to be a stochastic region where magnetic fields are no longer confined to a two-dimensional surface, but rather tend to fill a volume. The electron heat conduction can get very large when this happens, which in turn tends to modify the current profile and thus the safety factor profile. The nonlinear

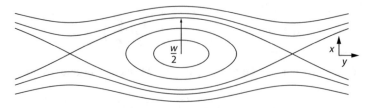

Fig. 6.8 Magnetic island in transformed geometry

Fig. 6.9 Magnetic island in circular geometry

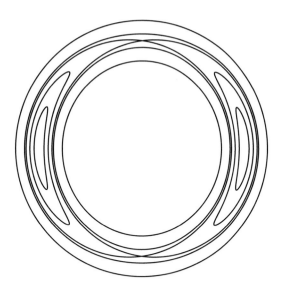

ramification of this is that the degraded confinement during island overlap tends to terminate the island growth and return the plasma to a more stable state, but then improved heat confinement can start the process over. Nearby modes are often characterized by (m, n) and $(m + 1, n + 1)$ quantum numbers such as 4/3 being paired with 5/4 or 3/2 and 2/1.

6.4.4 Resistive MHD for m = n = 1 Mode

Somewhat different physics is involved for the case where the $q = 1$ surface is inside the plasma. If the ideal MHD stability theory finds modes where $\delta W_{MHD} < 0$, then instability occurs with growth rates on an Alfvén timescale. If, however, the resulting δW_{MHD} is small but slightly positive, the resistive MHD treatment causes a finite growth rate.

We start by examining the shape of the trial function for the ideal MHD mode with $m = n = 1$. Since δW can be written as [31]:

$$\delta W = \frac{\pi^2 B_\phi^2}{\mu_0 R_0} \int_0^a \left[\left(r \frac{d\xi}{dr} \right)^2 + \left(m^2 - 1 \right) \xi^2 \right] \left(\frac{n}{m} - \frac{1}{q} \right)^2 r \, dr + O(\epsilon)^2. \quad (6.97)$$

We can construct a mode for $n = m = 1$ which will have $\delta W \to 0$ as shown in Fig. 6.3 earlier, i.e. $\xi(r < r_1) = \xi_0$, $\xi(r > r_1) = 0$. (Higher-order terms in (6.97), shown here as $\propto \epsilon^2$, will modify the expression for δW somewhat. The $m = n = 1$ mode with $q < 1$ over a region of the plasma will be ideal MHD-stable only if the poloidal beta is below some threshold value.) We then proceed with a set of resistive MHD equations similar to Eq. (6.81), working with normalized variables and dropping a small term in dJ_ϕ/dr:

$$\tilde{\eta}\frac{d^2\tilde{\psi}}{dx^2} = \tilde{\gamma}\tilde{\psi} + \tilde{\gamma}x\tilde{\xi}$$

$$\tilde{\gamma}^2\frac{d^2\tilde{\xi}}{dx^2} = x\frac{d^2\tilde{\psi}}{dx^2}.$$

(6.98)

Here $\tilde{\eta} = \tau_H/\tau_R$, $\tilde{\psi} = -\psi/(q'(r_1)B_\theta(r_1))$, $\tilde{\xi} = v_r/(\gamma r_1)$, $x = (r - r_1)/r_1$, and $\tilde{\gamma} = \gamma\tau_H$. Here the characteristic MHD growth period is given by

$$\tau_H = \left(\frac{\sqrt{\mu_0\rho}}{q'B_\theta}\right)_{q=1}$$

and the resistive time τ_R is taken to mean

$$\tau_R = \left(\frac{\mu_0 r^2}{\eta}\right)_{q=1}.$$

The solution to these equations is given by

$$\xi = \frac{1}{2}\xi_0\left(1 - \text{erf}\left(\frac{x}{\sqrt{2}\tilde{\eta}^{1/3}}\right)\right).$$

(6.99)

A comparison of the radial displacement eigenfunction ξ_r for the ideal and resistive MHD $m = 1$ mode is shown as Fig. 6.10. For the resistive $m = 1$ mode at neutral ideal MHD stability, the growth rate is given by:

$$\gamma = \tau_H^{-2/3}\tau_R^{-1/3}.$$

(6.100)

For a case where the MHD perturbed energy $\delta W > 0$ for the $m = 1$, $n = 1$ mode, the resistive effects become similar to the resistive modes for $m \geq 2$ as outlined earlier. In this case the radial eigenfunction has an odd part similar to Fig. 6.7 shown earlier. For $\delta W_{MHD} > 0$, the MHD growth rate $\gamma_{MHD}\tau_H < 0$. For large negative γ_H, the normalized growth rate $\hat{\lambda} \equiv \gamma\tau_H/\epsilon^{3/2}$, with $\epsilon \equiv \tau_H/\tau_R$, has an asymptotic limit

$$\hat{\lambda} = \left(-8\frac{\Gamma(5/4)}{\Gamma(-1/4)}\right)^{4/5}\left|\hat{\lambda}_H\right|^{-4/5} = 1.3679\left|\hat{\lambda}_H\right|^{-4/5}.$$

(6.101)

Here λ_H is given by a matching condition across the resistive layer. It can be written in terms of the ideal MHD δW for the mode [1]:

$$\lambda_H = -\pi\frac{\delta W_{min}|_{r=r_s}\ \mu_0}{\rho(\xi_\infty r B_\theta q')^2|_{r=r_s}}.$$

(6.102)

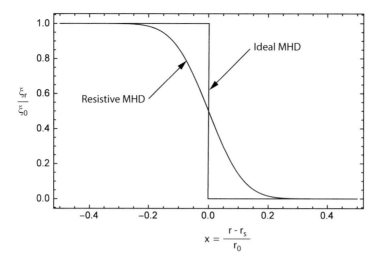

Fig. 6.10 Ideal and resistive radial displacement eigenfunctions for the $m = 1$ mode at neutral ideal MHD stability ($\delta W_{MHD} = 0$)

An expression for the growth rate of the resistive $m = n = 1$ mode for all regimes given above, $\delta W < 0$, $\delta W \approx 0$, and $\delta W > 0$, was given in [7]. With the definitions of λ_H and $\hat{\lambda}$ as given above, and with $\hat{\lambda}_H = \lambda_H/\epsilon^{3/2}$, Ref. [7] finds the following transcendental equation for $\hat{\lambda}$:

$$\frac{\hat{\lambda}}{\hat{\lambda}_H} = \frac{\hat{\lambda}^{9/4}}{8} \frac{\Gamma\left(\dfrac{\hat{\lambda}^{3/2} - 1}{4}\right)}{\Gamma\left(\dfrac{\hat{\lambda}^{3/2} + 5}{4}\right)}. \tag{6.103}$$

It is relatively straightforward to show that $\hat{\lambda} = 1$ for $\lambda_H = 0$, which translates back to $\gamma = \tau_H^{-2/3}\tau_R^{-1/3}$ at neutral MHD stability, and that for large MHD growth rate $\gamma_{MHD} \gg 0$, the growth rate asymptotically reaches $\gamma = \gamma_H$. For the case of ideal MHD stability with $\gamma_{MHD} \ll 0$, the limiting form given in Eq. (6.101) is also obtained, which translates back to

$$\gamma = K\tau_H^{-2/5}\tau_R^{-3/5}|\gamma_{MHD}\tau_H|^{-4/5},$$

with $K = 1.3679$ as before. An examination of the parameter Δ' used earlier shows that

$$\Delta' r_s = \frac{\pi}{\lambda_H},$$

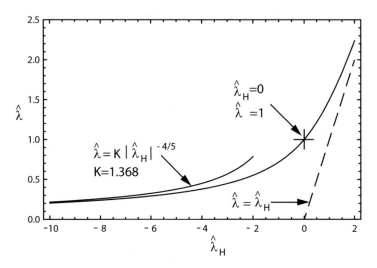

Fig. 6.11 Normalized growth rate $\hat{\lambda}$ vs. normalized MHD growth rate $\hat{\lambda}_H$, with asymptotic forms shown. From [7]

so that the growth rate can also be written as

$$\gamma = K/\pi^{4/5}\tau_H^{-2/5}\tau_R^{-3/5}(\Delta' r_s)^{4/5} = 0.55\tau_H^{-2/5}\tau_R^{-3/5}(\Delta' r_s)^{4/5},$$

exactly the same as Eq. (6.89) for the $m \neq 1$ modes.

A plot of the general solution for the normalized growth rate $\hat{\lambda}$ vs. $\hat{\lambda}_H$ is shown as Fig. 6.11, along with the two asymptotic forms and the special case at $\gamma_H = 0$. Figure 6.12 shows a plot of the same growth rates in the non-normalized form $\gamma\tau_H$ vs. $\gamma_{MHD}\tau_H$ by Lundquist number $S = \tau_R/\tau_H$.

6.4.5 Disruptions

Disruptions are MHD events resulting in complete termination of a discharge. These can result in physical damage to the machine because of the high mechanical stress that they can impart to the machine, along with thermal loads which can melt holes in limiters and other plasma-facing components.

6.4.6 Hugill Diagram

A convenient way to show the operational boundaries was formulated by Hugill et al. [10]. The Hugill diagram shows the operational space for tokamaks as a function

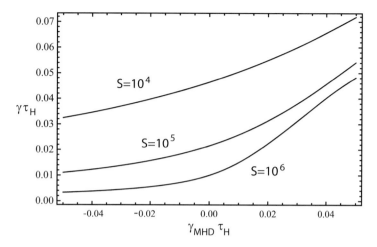

Fig. 6.12 Growth rates for the resistive $m = n = 1$ mode for different values of the Lundquist number $S = \tau_R/\tau_H$

Fig. 6.13 Hugill diagram for TFTR. From [21]. Used with permission, Springer

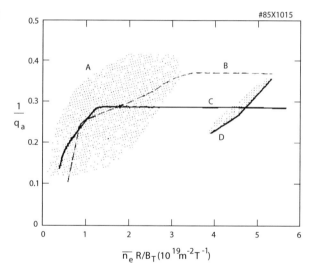

of the inverse safety factor at the edge q_a and a normalized density parameter $\bar{n}R/B_\phi$ first introduced by Murakami et al. [22]. The upper boundary corresponds to an edge safety factor dropping below 2.0 and is associated with current profiles susceptible to $m = 2$ kink and tearing modes. The lower boundary corresponds to the limit where the localized heat flux passing through the outer parts of the plasma cannot sustain the energy loss by radiation, resulting in a radiative collapse of the plasma. The overall boundary moves to the right (greater maximum $\bar{n}$) when external heating such as neutral beam injection is applied. Figure 6.13 shows the Hugill diagram for disruption-free, full-size ($R \approx 2.55$ m, $a \approx 0.8$ m) discharges on

TFTR. In this chart, the shaded area A is the operating space for deuterium gas-puff discharges with both ohmic and neutral beam heating. The dashed line B shows the time evolution of a single shot in helium, and C shows the time history of a shot with the injection of a large pellet (2.1×10^{21} atoms) and curve D shows the time history of a shot with several smaller pellets (7×10^{20} atoms).

Other events can trigger disruptions as well. If the plasma is unstable to vertical displacements, then the plasma can merge with an upper or lower wall, and return currents can flow through the wall. Also, metallic debris can fall into the plasma, causing considerable radiation and ultimately a radiative collapse. Additionally, at low electron densities, the thermal component of the plasma can disappear in a disruption, leaving behind a ring of runaway electrons, losing energy slowly in the form of hard X-rays.

6.4.7 Mode Locking

Resistive MHD modes in tokamaks typically appear with a real frequency caused by the presence of plasma rotation. When these modes grow to finite amplitude, the frequency of the mode is seen to drop. The coupling of momentum and energy out of the plasma is caused by the finite resistivity of the wall. The characteristic resistive time for the metallic shell surrounding the plasma is given by [23]:

$$\tau_V = \mu_0 \sigma b \delta / 2, \tag{6.104}$$

where δ is the wall thickness and σ is the conductivity. The characteristic frequency of the resistive MHD mode is in the range of $1 \rightarrow 10\,\text{kHz}$ in medium-sized tokamaks such as JET and TFTR, corresponding to a (toroidal) Mach number of a few percent.

A simple theoretical model for mode locking was given in [23]. Here the island width w is tied to the perturbed magnetic field at the island through

$$w = 4 \left. \left((q/q') |\Psi| / B_\theta \right)^{1/2} \right|_{r=r_s}. \tag{6.105}$$

We can then make a simple approximation for the time-dependent island width w by taking

$$\Delta'(w) = \Delta'(0)(1 - w/w_0) \tag{6.106}$$

which has a solution

$$w(t) = w_s \left(1 - \exp\left(t/\tau_s \right) \right). \tag{6.107}$$

Here w_s is the saturated island size. We can find an expression for the rate of change of the rotation rate $\omega(t)$ in the plasma by accounting for the angular momentum loss to the walls, which is determined by the wall resistivity and the distance between the wall (at $r = b$) and the plasma edge (at $r = a$), as well as the difference between the location of the resonant surface r_s in the plasma and the plasma edge. First, we can calculate the boundary condition on the perturbed flux at $r = a$. This is given by

$$\left[\frac{\psi'}{\psi}\right]_{r=a} = -\frac{m}{a}\frac{1 + f(a/b)^{2m}}{1 - f(a/b)^{2m}}. \tag{6.108}$$

Assuming that the wall thickness $\delta \ll 1/(\omega \tau_V)$, the frequency-dependent term f has the familiar form of the transfer function for an R-L filter circuit:

$$f = \frac{1}{1 + im/(\omega \tau_V)}. \tag{6.109}$$

The contributions to eddy currents in the wall leading to momentum exchange can then be written in a suggestive form

$$\frac{d\omega}{dt} = \frac{aJ}{2\mu_0}\mathrm{Im}\left(\psi^*\psi'\right)_{r=a}, \tag{6.110}$$

where J is defined as

$$J = \frac{m^2}{\int \rho r^3 dr} + \frac{n^2}{R^2 \int \rho r dr}, \tag{6.111}$$

representing a sort of aggregated moment of inertia.

To solve for the term $\mathrm{Im}\left(\psi^*\psi'\right)$ given in Eq. (6.110), one must make use of the frequency-dependent parameter f given above. This yields an equation for $d\omega/dt$:

$$\frac{d\omega}{dt} = \frac{mJ}{\mu_0}\phi_\infty^2(r_s)(r_s/b)^{2m}g\mathrm{Im}f. \tag{6.112}$$

Here g is another geometrical factor, which involves the outer solution for ψ from $r = r_s$ to $r = a$, and this requires knowledge of the current distribution. However, if the current density is small in this region, then a simplified version for g is obtained:

$$g = \frac{1 - (r_s/b)^{2m}}{\left|1 - f(r_s/b)^{2m}\right|^2}. \tag{6.113}$$

For the case where $r_s \ll b$, we can take $g \approx 1$ and then Eq. (6.112) becomes

$$\frac{d\omega}{dt} = -c\frac{1}{\tau_a^2}\frac{\omega\tau_V}{\omega^2\tau_V^2 + m^2}\left(\frac{w}{a}\right)^4, \qquad (6.114)$$

where

$$c = \frac{m^2}{256}\left(\frac{r_s}{b}\right)^{2m}\left(\frac{aq'}{q}\right)_{r_s}^2 \qquad (6.115)$$

and

$$\tau_A^2 = \mu_0/\left(Ja^2B_\theta^2(r_s)\right) \qquad (6.116)$$

with J defined in (6.111).

For this simplified set of assumptions, an analytic solution for $\omega(t)$ can be found:

$$\frac{\omega(t)}{\omega(0)} = (m/\tau_V)\sqrt{W\left(\tau_V^2/m^2\exp\left(a_4x^4 + a_3x^3 + a_2x^2 + a_1x + a_0\right)\right)}, \qquad (6.117)$$

where $x = \exp(-t/\tau_S)$ and

$$a_4 = \frac{c\tau_S\tau_V w_0^4}{2a^4m^2\tau_A^2}$$

$$a_3 = -\frac{8c\tau_S\tau_V w_0^4}{3a^4m^2\tau_A^2}$$

$$a_2 = \frac{6c\tau_S\tau_V w_0^4}{a^4m^2\tau_A^2} \qquad\qquad (6.118)$$

$$a_1 = -\frac{8c\tau_S\tau_V w_0^4}{a^4m^2\tau_A^2}$$

$$a_0 = -\frac{2c\tau_S\tau_V w_0^4}{a^4m^2\tau_A^2} + \frac{6a^4\tau_A^2\tau_V^2 + 25c\tau_S\tau_V w_0^4}{6a^4m^2\tau_A^2}.$$

Here $W(x)$ is the product logarithmic function of Lambert, defined so that $W(xe^x) = x$. (Note that although this expression appears to be dimensionally incorrect, the unit system of time is carried in a normalizing function inside the argument of the W-function, and in fact any time units can be used.) Figure 6.14 shows a solution using this form for $\omega(t)$, along with plots of the perturbed poloidal field $\tilde{B}_\theta$ and the island width $w(t)/a$ versus time. Some refinements of the theory have also been done to allow for more realistic modeling of mode locking. One example is the allowance for a non-smooth wall and the presence of error field errors which break symmetry caused by the finite number of toroidal field coils.

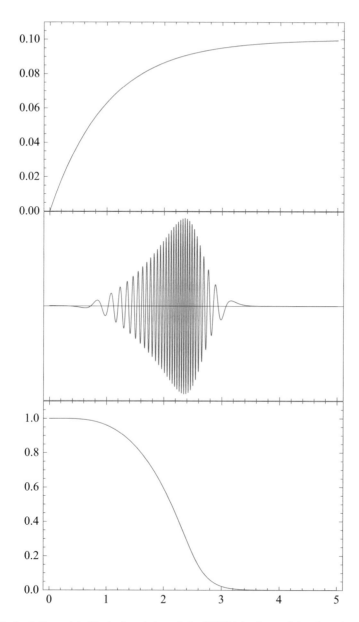

Fig. 6.14 Analytic model of locked mode in resistive MHD island growth in tokamak

6.4.8 Magnetic Stochasticity

The simple models for magnetic islands given above focus on the progression of a single resonant surface in the plasma. Typically, however, more than one resonance

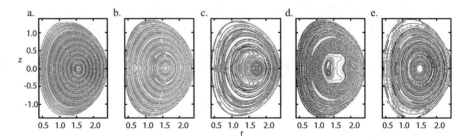

Fig. 6.15 Resistive MHD simulation of the CDX tokamak, showing developing island structure, growth, and healing of stochastic regions during the buildup and crash of a sawtooth mode. (**a**) Initial state, t=1266.17. (**b**) Island growing, t=1660.70. (**c**) Nonlinear phase, t=1795.61. (**d**) After first crash, t=1839.86. (**e**) Flux surfaces recovered, t=2094.08. (Times normalized to Alfvén time). From [4]. ©IOP Publishing. Reproduced with permission. All rights reserved

can come into play: for example, a 3/2 island may be forming along with a 2/1 island, and perhaps a 4/3 island at the same time. After some growth time, the island widths are such that the islands create a more complex environment than the simple outer-solution models allow for. The topology of the magnetic field lines outside of the islands is affected. Good flux surfaces start to disappear, and a single magnetic field line, traced around toroidally, appears to fill an area rather than form an isolated surface. The terms "ergodic" or "stochastic" are applied in this situation, perhaps with less mathematical rigor than applied to the formal definition of these concepts. The general idea, however, is that if a relationship in R and Z no longer holds for the position of a magnetic field line for each transit toroidally, then the magnetic field line appears to randomly filling a surface. Plots of the position of a magnetic field line as it passes through a poloidal surface are called Poincare plots, or puncture plots. Generally both islands with closed magnetic surfaces exist alongside with stochastic regions, and the regions of stochasticity grow with increasing island size.

Figure 6.15 shows a resistive MHD simulation of a small tokamak with an evolving island structure. Here one can see that as the islands grow, regions of magnetic stochasticity appear. The magnetic stochasticity present will effectively "short out" temperature gradients in the stochastic regions, since the heat conduction coefficients for the electrons are very different in the parallel (to B) and perpendicular directions. (In fact, they differ by a factor of $(\omega_{ce}\tau_e)^2$ in the classical Braginskii heat conduction model, and this can be on the order of $10^{10} \rightarrow 10^{16}$, with the parallel conduction being larger.) The flattening of temperature gradients directly affects the q profile in the plasma, and temperature fluctuations from stochastic field lines provide feedback to the island growth. The rapid removal of electron heat from the plasma may be linked to the sharp crash in sawtooth oscillations.

The modeling of resistive MHD dynamics in a real 3-D equilibrium of tokamaks with noncircular geometry is now largely the domain of codes run on supercomputers, and even these codes must make physics approximations in order to get results in a finite amount of time. However, the topological features of island overlap and

Fig. 6.16 Chirikov-like mapping of Goldston and Rutherford [14] with $\Delta = 0.5, 1.0, 1.4$, and 2.0 (in ascending order). Mappings contain forty random starting points

the buildup of stochastic regions can be shown to be roughly equivalent to what one can see with simple models. One such simple model, due to Goldston and Rutherford [14], uses a mapping of two coordinates x and y from one cycle to the next. The mapping is as follows:

$$y_{n+1} = x_n + y_n$$
$$x_{n+1} = x_n + \Delta \cos(y_{n+1}). \tag{6.119}$$

Here the variables x and y are taken to be modulo 2π, and Δ is an adjustable parameter, which roughly translates to the island width for the resistive MHD island formation in tokamaks. Figure 6.16 shows some maps generated with this algorithm for various values of Δ. (These were generated with forty individual starting points and 10,000 iterations.) For the case of small Δ, most of the trajectories lie on a single line, which can be thought of a good flux surface. The large island on the bottom left of each figure represents lines encircling the fixed point at $x = 0$, $y = \Pi/2$. As the parameter Δ is increased, more complex island structures emerge, as well as modest regions of stochasticity. Further increases in Δ, above 1.0, lead to larger regions of stochasticity, although the large island is never destroyed. This mapping is similar to the "Chirikov mapping," or "standard mapping" of [5], given by

$$x_{n+1} = x_n + k \sin y_n$$
$$\tag{6.120}$$
$$y_{n+1} = y_n + x_{n+1}.$$

Both of these mappings have the property of being area-preserving, which means that the determinant of the Jacobian transformation matrix

$$\det J = \frac{\partial x_{n+1}}{\partial x_n} \frac{\partial y_{n+1}}{\partial y_n} - \frac{\partial y_{n+1}}{\partial x_n} \frac{\partial x_{n+1}}{\partial y_n}$$

is one. This is required to represent divergence-free magnetic fields.

6.5 Neoclassical Tearing Modes

A refinement of the linear growth model for the tearing mode is to add the effects of trapped particles to the perturbed currents in the magnetic island. The effects of trapped particles on transport in general is called neoclassical transport theory and is the subject of a subsequent chapter. Here we focus on just one aspect of neoclassical transport theory and that is the effect of the trapped-particle perturbed current (called the "bootstrap current") on the stability and growth rates due to the presence of these trapped particles. The fraction of trapped particles in the plasma is given roughly by $\varepsilon^{1/2}$, where ε is the inverse aspect ratio. Reference [12] gives an expression for the rate of change of the island width w as a function of Δ' as:

$$\frac{\tau_R}{r^2} \frac{dw}{dt} = \Delta',\tag{6.121}$$

where $\tau_R = \mu_0 r^2/\eta$ as before. The addition of the neoclassical effect is to add a second term [16, 25]:

$$\frac{\tau_R}{r^2} \frac{dw}{dt} = \Delta' + \epsilon^{1/2} \frac{L_q}{L_p} \frac{\beta_p}{w},\tag{6.122}$$

where the characteristic lengths L_p and L_q are given by $L_q = q/(dq/dr)$ and $L_p = -p/(dp/dr)$. (Note the negative sign in the expression for L_p, i.e. it is positive for a typical pressure profile with $dp/dr < 0$.) The unstable modes outside

Fig. 6.17 Experimentally observed neoclassical tearing modes on TFTR, with comparison to the modified Rutherford theory. (**a**) $m/n = 3/2$ island. (**b**) $m/n = 4/3$ island. From [6]

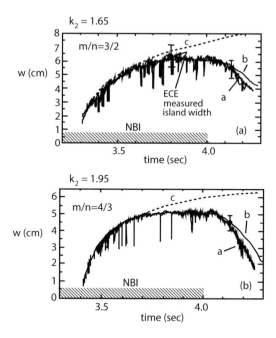

these new (usually more stringent) stability limits are called Neoclassical tearing modes, or NTMs.

Since stable equilibria have $\Delta' < 0$, there is no island growth unless the poloidal beta β_p exceeds some threshold value. Also, the form of this equation suggests that there is a mode saturation amplitude w_{sat} at which the mode no longer grows if $\Delta' < 0$. (Refinements on the theory also use an updated Δ' as the mode grows, because the outer solution changes as the island gets larger.) Figure 6.17 shows the experimentally observed island growth for both a 3/2 and a 2/1 mode on TFTR. Note that in both cases shown, when the neutral beam injection (NBI) is turned off, β_p drops, and thus the island shrinks, as predicted by Eq. (6.122).

Several factors can affect the currents flowing in the island. In addition to the neoclassical bootstrap effect, there is a "polarization effect," where the $\boldsymbol{E} \times \boldsymbol{B}$ drifts of the electrons and ions, equal to each other in a smooth magnetic field, become imbalanced by the larger ion gyroradius, which affects the average magnetic and electric fields seen by the ions. There is another trick that can be used to advantage, and that is electron cyclotron current drive (ECCD). Radiofrequency power in the electron cyclotron band (about 140 GHz in ITER) can be directed at the island in order to drive currents opposite to the bootstrap current perturbation, thus reversing the island growth. The relatively slow growth rate for the resistive islands allows for a set of movable mirrors to be used to aim the ECCD RF energy at the island, and these have been built into the ITER design.

6.6   Edge Localized Modes (ELMs)

A new mode of operation of a tokamak was first discovered on the ASDEX tokamak
[28]. The new mode, called the "H mode" (for "high confinement time," as opposed
to the "L" mode), was reached by increasing the injected power into the plasma
past some threshold value, at which point the energy confinement time more than
doubled in some cases. Figure 6.18 shows some of the details. Upon further analysis,
it was found that the new mode of operation involved the formation of a pedestal at
the outer plasma regions where the plasma density and temperature changed rapidly
towards the lower density and temperature seen on the region outside the closed flux
surfaces. While the confinement conditions are generally desirable for operation in
H mode, there was a price to pay: the emergence of bursts of D_α light (L_α, $\lambda =$
$1215.67 \, \text{Å}$) at the edge of the plasma, indicating some transient bursts of cold gas
interacting with warm plasma being ejected from inside the closed flux. Magnetic
signals also confirmed the correlation of these bursts with perturbations with high-m
mode numbers, in the range of $8 \rightarrow 12$. Furthermore, there were reductions in the
central density associated with these bursts. This magnetic activity has been labeled
"Edge localized modes," or ELMs. The current ITER design basis is based on a
scaling law called "ELMy H mode."

The sharp gradients at the outer plasma layer is thought to excite MHD instability
through ∇p-driven and ∇J-driven modes. Computer simulation of the plasma using
a resistive MHD code seems to support this assumption [2, 34]. Figure 6.19 shows
the result of one such simulation, showing the ballooning/peeling=type mode at the
plasma outer boundary.

Since the early observation of ELMs, further observations have resulted in the
modes being divided into different categories [34]. These were first noted on DIII-D

Fig. 6.18 Plot of
confinement time and density,
showing H mode and L mode
regimes. From [28]

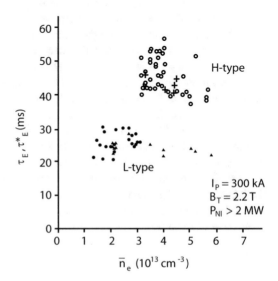

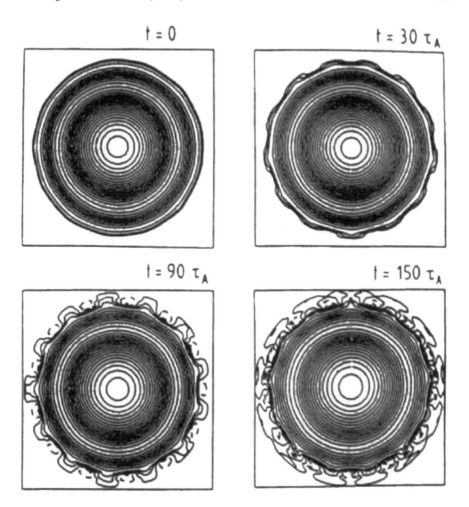

Fig. 6.19 Simulation of edge = localized modes in a resistive MHD code with circular equilibria. From [2]

[8]. The "Type 1" ELMs have no precursor and have a rotation frequency increasing with injected power. These modes are probably ideal MHD instabilities, combining the features of an ideal ballooning mode and a global MHD mode such as an ideal kink. The "dithering" or "grassy" ELM seems to represent a plasma which is transitioning back and forth between an L-mode and an H-mode; these happen near the L-H transition injected power threshold. "Type III" ELMs show a magnetic precursor with $n = 5 \to 10$ and $m = 10 \to 15$. These go away for sufficiently high edge temperatures, which suggests that they are connected to a resistive MHD mode. The candidate modes for this ELM type are resistive ballooning modes and a "peeling mode."

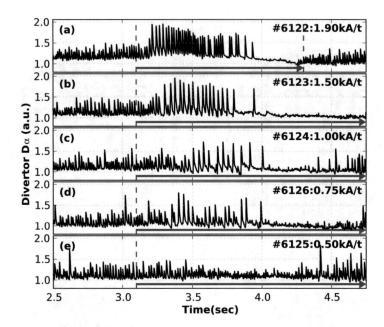

Fig. 6.20 ELM suppression on KSTAR. From [17]

Recent studies have concentrated on methods to suppress ELMs. Early attempts at ELM suppression were done at DIII-D [9], using "resonant magnetic perturbations" (RMPs) by placing some rectangular coils with a variety of toroidal and poloidal periodicity in an attempt to lock with the mode and suppress the ELM. A more recent attempt was done at the KSTAR tokamak in South Korea using an $n = 1$ coil configuration [17]. Figure 6.20 shows the results of the application of these ELM suppression coils on KSTAR: a clear mode suppression threshold is seen at about 1.2 KA-turn. Other ELM suppression techniques have been to dither the vertical field coil in order to move the divertor strike point around, and to "ergodicise" the x-point in the plasma. Moving into the reactor regime will require that if ELMs exist at any time during the discharge, the heat load caused by these ELMs must be spread over a finite area, lest damage in the walls and divertor occur from the enormous heat loads.

Another area of concern is the potential for the ELMs to lock with a neoclassical tearing mode (NTM) inside the plasma. This locking can lead to enhanced transport from the plasma. A recent study of this phenomenon was undertaken at DIII-D and was facilitated by a relatively new technique called motional Stark effect, or MSE [18]. Here the MSE allows a non-perturbative way of measuring the magnetic fields in the plasma, thus allowing a direct knowledge of the q profile and the degree of "flux pumping" associated with island formation. The ability to measure the degree of flux-pumping and mode locking may lead to the ability to control the ELM events through modification of the NTM island growth with ECCD or other means.

Fig. 6.21 Plasma
configuration for a
one-dimensional plasma
supported against gravity by a
magnetic field. For
Problem 6.1

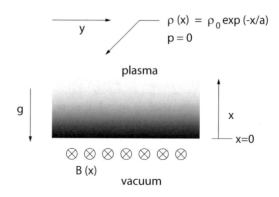

Problems

6.1 Consider a slab model of a plasma supported in equilibrium against gravity by a uniform magnetic field $\mathbf{B} = B\mathbf{e}_z$ as shown in the accompanying Fig. 6.21.

a. Write down an equation for the equilibrium force balance, with gravity $\mathbf{g} = -\hat{x}g$ and $p = 0$ everywhere.
b. Make a graph of the magnetic field $B_z(x)$ for all values of x, using $g = 1, a = 1$, $\rho_0 = 0.1$, $\mu_0 = 1$, and $B_0(\infty) = 1.0$.
c. Derive the one-dimensional ideal MHD eigenvalue equation describing linear stability. Assume that the perturbations vary as $\boldsymbol{\xi} = (\hat{x}\xi_x(x) + \hat{y}\xi_y(x))\exp(-i\omega t + ik_y y + ik_z z)$. For simplicity assume incompressional displacements, i.e. $ik_y\xi_y(x) + \xi'_x(x) = 0$.
Note that the basic momentum equation is given by

$$\rho\frac{\partial^2 \boldsymbol{\xi}}{\partial t^2} = \mathbf{J}_0 \times \delta\mathbf{B} + \delta\mathbf{J} \times \mathbf{B} + \delta\rho\mathbf{g} \equiv \mathbf{F}(\boldsymbol{\xi})$$

d. Form the perturbed energy δW from

$$\delta W = \frac{1}{2}\omega^2 \int d^3x \rho(x)\boldsymbol{\xi}^* \cdot \boldsymbol{\xi} = -\frac{1}{2}\int d^3x \boldsymbol{\xi}^* \cdot \mathbf{F}(\boldsymbol{\xi})$$

and show that this leads to an equation of form

$$\omega^2 = k_z^2 v_A^2 + \dots$$

where the other terms do not depend on k_z, and thus ω^2 will be minimized by taking $k_z = 0$. (Note that $v_A = B_0(x)/\sqrt{\mu_0\rho(x)}$.)
e. Show that the remaining terms in the integrand $-\boldsymbol{\xi}^* \cdot \mathbf{F}(\boldsymbol{\xi})$ evaluate to

$$-\boldsymbol{\xi}^* \cdot \mathbf{F}(\boldsymbol{\xi}) = \rho g(\boldsymbol{\xi}^* \cdot \boldsymbol{\xi}' + \boldsymbol{\xi}^{*'} \cdot \boldsymbol{\xi})$$

and thus show that for $\xi_x(x) = \xi_0 \exp(-|\mathrm{Im}\, k_x|x)$, we have

$$\omega^2 = -g|\mathrm{Im}\, k_x|,$$

recovering the classic Rayleigh-Taylor instability. (See [19].)

6.2 Given a tokamak ($B_\phi = 5.3$ T, $n_i = 10^{20}$ m^{-3}, R = 6.2 m, $A = 3$, $q(0.9a) = 3$, 50-50 D-T, circular cross section). Assume that the $q = 2$ surface in this tokamak is located at $r = 0.5a$ and that $\Delta' = -40.0$ m^{-1} there. Find the threshold value of β_{pol} for growth of a neoclassical tearing mode. Take $L_p = L_q$ and $w = 1$ cm.

6.3 A circular tokamak with a toroidal current density given by

$$J_\phi = J_0 \left(1 - \left(\frac{r}{a}\right)^2\right)^\nu$$

has a safety factor q profile given by

$$q(r) = -\frac{(\nu+1)q(0)(r/a)^2}{\left(1 - (r/a)^2\right)^{\nu+1} - 1}$$

Assume that $q(0) = 0.95$ and $\nu = 2.5$.

a. Find the location of the $q = 1$ and $q = 2$ surfaces in this tokamak as fractions of the minor radius a.

b. Assume that the electron temperature $T_e = 10$ keV throughout, $a = 2.0$ m, $A = 3$, and $B_\phi = 5.3$ T. Suppose that $\Delta'a = +10.0$ at the $q = 1$ surface and $\Delta'a = +5.0$ at the $q = 2$ surface. Find the resistive island exponential growth rates at both locations using Eq. (6.89). Ignore neoclassical effects.

c. Starting with each island having a half-width $w/2 = 1.0$ cm, find the time before the two islands will touch.

6.4 Using a tokamak with the same conditions as in Problem 3.2, but with $\delta W = 0$ for the $m = 1$ mode at the same radius as before, calculate the resistive kink instability growth rate using Eq. (6.100). (Note the different definition of τ_R for this treatment). Compare this value for the growth rate obtained in Problem 3.2. Also compare this value to the ideal MHD growth rate for $\delta W < 0$, taking $\gamma \tau_H = 0.4$.

6.5 Use the standard mapping of Chirikov (Eq. (6.120)) to develop plots similar to those shown in Fig. 6.16 for the Rutherford-Goldston mapping. Use the same values for k as used for Δ in Fig. 6.16, i.e. $k = \{0.5, 1.0, 1.4, 2.0\}$. Use 10,000 iterations for forty random initial points. Compare the results of this mapping with the Rutherford-Goldston mapping.

6.6 Given a tokamak ($B_\phi = 5.3$ T, $n_i = 10^{20}$ m^{-3}, R = 6.2 m, $A = 3$, $q(0.9a) = 3$, 50-50 D-T, circular cross section). Assume that the $q = 2$ surface in this tokamak is located at $r = 0.5a$ and that $\Delta' = -40.0$ m^{-1} there. Find the threshold value of β_{pol} for growth of a neoclassical tearing mode. Take $L_p = L_q$ and $w = 1$ cm.

References

1. Ara, G., Basu, B., Coppi, B., Laval, G., Rosenbluth, M., Waddell, B.: Magnetic reconnection and $m = 1$ oscillations in current carrying plasmas. Ann. Phys. **112**(2), 443–476 (1978). http://dx.doi.org/10.1016/S0003-4916(78)80007-4; http://www.sciencedirect.com/science/article/pii/S0003491678800074

2. ASDEX Team: The H-mode of ASDEX. Nucl. Fusion **29**(11), 1959 (1989). http://stacks.iop.org/0029-5515/29/i=11/a=010

3. Biskamp, D.: Nonlinear Magnetohydrodynamics. Cambridge Monographs on Plasma Physics. Cambridge University Press, Cambridge (1997). https://books.google.com/books?id=OzFNhaVKA48C

4. Breslau, J.A., Park, W., Jardin, S.C.: Massively parallel modeling of the sawtooth instability in tokamaks. J. Phys. Conf. Ser. **46**(1), 97 (2006). http://stacks.iop.org/1742-6596/46/i=1/a=014

5. Chirikov, B.V.: Research Concerning the Theory of Nonlinear Resonance and Stochasticity. Preprint N 267, Institute of Nuclear Physics, Novosibirsk (1969). Engl. Trans.: CERN Trans. 71–40 (1971)

6. Chang, Z., Callen, J.D., Fredrickson, E.D., Budny, R.V., Hegna, C.C., McGuire, K.M., Zarnstorff, M.C.a.: Observation of nonlinear neoclassical pressure-gradient–driven tearing modes in TFTR. Phys. Rev. Lett. **74**, 4663–4666 (1995). http://link.aps.org/doi/10.1103/PhysRevLett.74.4663

7. Coppi, B., Galvão, R., Pellat, R., Rosenbluth, M., Rutherford, P.: Resistive internal kink modes. Sov. J. Plasma Phys. **2**, 533 (1976)

8. Doyle, E.J., Groebner, R.J., Burrell, K.H., Gohil, P., Lehecka, T., Luhmann Jr., N.C., Matsumoto, H., Osborne, T.H., Peebles, W.A., Philipona, R.: Modifications in turbulence and edge electric fields at the L-H transition in the DIII-D tokamak. Phys. Fluids B: Plasma Phys. **3**(8), 2300–2307 (1991). http://dx.doi.org/10.1063/1.859597

9. Evans, T., Moyer, R., Watkins, J., Osborne, T., Thomas, P., Becoulet, M., Boedo, J., Doyle, E., Fenstermacher, M., Finken, K., Groebner, R., Groth, M., Harris, J., Jackson, G., Haye, R.L., Lasnier, C., Masuzaki, S., Ohyabu, N., Pretty, D., Reimerdes, H., Rhodes, T., Rudakov, D., Schaffer, M., Wade, M., Wang, G., West, W., Zeng, L.: Suppression of large edge localized modes with edge resonant magnetic fields in high confinement diii-d plasmas. Nucl. Fusion **45**(7), 595 (2005). http://stacks.iop.org/0029-5515/45/i=7/a=007

10. Fielding, S., Hugill, J., McCracken, G., Paul, J., Prentice, R., Stott, P.: High-density discharges with gettered torus walls in DITE. Nucl. Fusion **17**(6), 1382 (1977). http://stacks.iop.org/0029-5515/17/i=6/a=020

11. Freidberg, J.: Ideal Magnetohydrodynamics. Modern Perspectives in Energy Series. Plenum Publishing Company, New York (1987). https://books.google.com/books?id=UMDvAAAAMAAJ

12. Furth, H.P., Killeen, J., Rosenbluth, M.N.: Finite-resistivity instabilities of a sheet pinch. Phys. Fluids **6**(4), 459–484 (1963). http://aip.scitation.org/doi/abs/10.1063/1.1706761

13. Furth, H., Killeen, J., Rosenbluth, M., Coppi, B.: Stabilization by shear and negative V''. In: Plasma Physics and Controlled Nuclear Fusion Research 1966, vol. I, p. 103 (1966)

14. Goldston, R., Rutherford, P.: Introduction to Plasma Physics. CRC Press, Boca Raton (1995). https://books.google.com/books?id=7kM7yEFUGnAC

15. Gruber, R., Troyon, F., Berger, D., Bernard, L., Rousset, S., Schreiber, R., Kerner, W., Schneider, W., Roberts, K.: ERATO stability code. Comput. Phys. Commun. **21**(3), 323–371 (1981). http://dx.doi.org/10.1016/0010-4655(81)90013-8; http://www.sciencedirect.com/science/article/pii/0010465581900138

16. Haye, R.J.L.: Neoclassical tearing modes and their control. Phys. Plasmas **13**(5), 055501 (2006). http://dx.doi.org/10.1063/1.2180747

17. Jeon, Y.M., Park, J.K., Yoon, S.W., Ko, W.H., Lee, S.G., Lee, K.D., Yun, G.S., Nam, Y.U., Kim, W.C., Kwak, J.G., Lee, K.S., Kim, H.K., Yang, H.L.: Suppression of edge localized modes in high-confinement KSTAR plasmas by nonaxisymmetric magnetic perturbations. Phys. Rev. Lett. **109**, 035004 (2012). http://link.aps.org/doi/10.1103/PhysRevLett.109.035004

18. King, J.D., Haye, R.J.L., Petty, C.C., Osborne, T.H., Lasnier, C.J., Groebner, R.J., Volpe, F.A., Lanctot, M.J., Makowski, M.A., Holcomb, C.T., Solomon, W.M., Allen, S.L., Luce, T.C., Austin, M.E., Meyer, W.H., Morse, E.C.: Hybrid-like 2/1 flux-pumping and magnetic island evolution due to edge localized mode-neoclassical tearing mode coupling in diii-d. Phys. Plasmas 19(2), 022503 (2012). http://dx.doi.org/10.1063/1.3684648

19. Kruskal, M., Schwarzschild, M.: Some instabilities of a completely ionized plasma. Proc. R. Soc. Lond. A: Math. Phys. Eng. Sci. 223(1154), 348–360 (1954). https://doi.org/10.1098/rspa. 1954.0120; http://rspa.royalsocietypublishing.org/content/223/1154/348

20. Loureiro, N.F., Uzdensky, D.A.: Magnetic reconnection: from the Sweet–Parker model to stochastic plasmoid chains. Plasma Phys. Controll. Fusion 58(1), 014021 (2016). http://stacks. iop.org/0741-3335/58/i=1/a=014021

21. Mueller, D., Bell, M., Boody, F., Bush, C., Cecchi, J.L., Davis, S., Dylla, H.F., Efthimion, P.C., Hawryluk, R.J., Hill, K.W., Kilpatrick, S., LaMarche, P.H., Manos, D., McCune, D., Medley, S.S., Milora, S., Murakami, M., Owens, D.K., Schivell, J., Schmidt, G., Sesnic, S., Stratton, B., Tait, G., Ulrickson, M., Wong, K.L., Woolley, R.D., Zarnstorff, M.C.: Discharge control and evolution in TFTR. In: Knoepfel, H. (ed.) Tokamak Start-Up: Problems and Scenarios Related to the Transient Phases of a Thermonuclear Fusion Reactor, pp. 143–157. Springer US, Boston (1986). http://dx.doi.org/10.1007/978-1-4757-1889-8_8

22. Murakami, M., Callen, J., Berry, L.: Some observations on maximum densities in tokamak experiments. Nucl. Fusion 16(2), 347 (1976). http://stacks.iop.org/0029-5515/16/i=2/a=020

23. Nave, M., Wesson, J.: Mode locking in tokamaks. Nucl. Fusion 30(12), 2575 (1990). http:// stacks.iop.org/0029-5515/30/i=12/a=011

24. Parker, E.N.: The solar-flare phenomenon and the theory of reconnection and annihiliation of magnetic fields. Astrophys. J. Suppl. 8, 177 (1963). https://doi.org/10.1086/190087

25. Rutherford, P.H.: Nonlinear growth of the tearing mode. Phys. Fluids 16(11), 1903–1908 (1973). http://aip.scitation.org/doi/abs/10.1063/1.1694232

26. Sweet, P.A.: The neutral point theory of solar flares. In: Proceedings of the International Astronomical Union Symposium on Electromagnetic Phenomena in Cosmical Physics, vol. 6, p. 123 (1956)

27. von Goeler, S., Stodiek, W., Sauthoff, N.: Studies of internal disruptions and $m = 1$ oscillations in tokamak discharges with soft X-ray techniques. Phys. Rev. Lett. 33, 1201–1203 (1974). http://link.aps.org/doi/10.1103/PhysRevLett.33.1201

28. Wagner, F., Becker, G., Behringer, K., Campbell, D., Eberhagen, A., Engelhardt, W., Fussmann, G., Gehre, O., Gernhardt, J., Gierke, G.v., Haas, G., Huang, M., Karger, F., Keilhacker, M., Klüber, O., Kornherr, M., Lackner, K., Lisitano, G., Lister, G.G., Mayer, H.M., Meisel, D., Müller, E.R., Murmann, H., Niedermeyer, H., Poschenrieder, W., Rapp, H., Röhr, H., Schneider, F., Siller, G., Speth, E., Stäbler, A., Steuer, K.H., Venus, G., Vollmer, O., Yü, Z.: Regime of improved confinement and high beta in neutral-beam-heated divertor discharges of the ASDEX tokamak. Phys. Rev. Lett. 49, 1408–1412 (1982). http://link.aps.org/doi/10.1103/ PhysRevLett.49.1408

29. Watson, G.: A Treatise on the Theory of Bessel Functions. Cambridge Mathematical Library. Cambridge University Press, Cambridge (1995). https://books.google.com/books?id= Mlk3FrNoEVoC

30. Wesson, J.: Hydromagnetic stability of tokamaks. Nucl. Fusion 18(1), 87 (1978). http://stacks. iop.org/0029-5515/18/i=1/a=010

31. Wesson, J., Campbell, D.: Tokamaks. International Series of Monographs on Physics. OUP, Oxford (2011). https://books.google.com/books?id=BH9vx-iDI74C

32. White, R.: The Theory of Toroidally Confined Plasmas. Imperial College Press, London (2001). https://books.google.com/books?id=BWpfQgAACAAJ

33. White, R.: The Theory of Toroidally Confined Plasmas. World Scientific, Singapore (2013). https://books.google.com/books?id=8OI7DQAAQBAJ

34. Zohm, H.: Edge localized modes (ELMs). Plasma Phys. Controll. Fusion 38(2), 105 (1996). http://stacks.iop.org/0741-3335/38/i=2/a=001

Chapter 7
Transport

7.1 Moments of the Fokker-Planck Equation

The Fokker-Planck equation, as shown in Sect. 3.3, is the starting point in the discussion of the impact of collisions on a plasma which is not in complete thermodynamic equilibrium, i.e. a plasma with gradients in density, temperature, or asymmetries in the pitch angle distribution of the particles. Here we will explore how the moments of the Fokker-Planck equation can be used to derive the basic fluid-like equations useful for plasma transport calculations.

In its simplest form, the Fokker-Planck equation for species α is written as:

$$\mathscr{D} f_\alpha = \mathscr{C}_\alpha f_\alpha \tag{7.1}$$

with the left-hand side given by the left-hand side of Eq. (3.67) and the right-hand side given by Eqs. (3.74) and (3.78). Moments of these equations are integral expressions in the notional form:

$$\int \mathscr{D} f_\alpha P_n(\boldsymbol{v}) d\boldsymbol{v} = \int \mathscr{C} f_\alpha P_n(\boldsymbol{v}) d\boldsymbol{v}. \tag{7.2}$$

In this equation, the polynomial $P_n(\boldsymbol{v})$ can mean either scalar or tensor quantities such as v^2 or $\boldsymbol{vv}$ in second order, and quantities such as $v^2 \boldsymbol{v}$ in third order. Also, constants can be added or subtracted, so that, for example, if the average velocity $\boldsymbol{u} =< \boldsymbol{v} >$ is calculated, then a moment such as $< (\boldsymbol{v} - \boldsymbol{u}) \, |\boldsymbol{v} - \boldsymbol{u}|^2 >$ may be of physical interest. We also note that the collision operator $\mathscr{C}_\alpha$ can be composed of components representing collisions with all species β, including species α itself:

$$\mathscr{C}_\alpha = \sum_\beta \mathscr{C}_{\alpha/\beta} \tag{7.3}$$

© Springer Nature Switzerland AG 2018
E. Morse, *Nuclear Fusion*, Graduate Texts in Physics,
https://doi.org/10.1007/978-3-319-98171-0_7

The zero-th order moment of f_α is just the density n_α:

$$\int f_\alpha(\mathbf{v})d^3\mathbf{v} = n_\alpha \tag{7.4}$$

We can then derive the zeroth-order moment of the Fokker-Planck equation. We note first that

$$\int \mathscr{C} f_\alpha d^3\mathbf{v} = 0, \tag{7.5}$$

since a particle cannot instantaneously move from one physical position to another instantaneously (no "teleportation" allowed). However we may replace $\mathscr{C} f_\alpha$ with a source term S representing a change of species through charge exchange or fusion reactions, for example. We also note that the Vlasov term $\mathbf{v} \cdot \nabla f$, integrated over $\mathbf{v}$, can have the gradient outside of the integral:

$$\int \mathbf{v} \cdot \nabla f d\mathbf{v} = \nabla \cdot \int f \mathbf{v} d\mathbf{v},$$

since x and v are treated as independent variables. We then define a new quantity $\boldsymbol{u}$ as above, but noting the density n_α:

$$n_\alpha \boldsymbol{u}_\alpha \equiv \int f_\alpha \boldsymbol{v}_\alpha d\boldsymbol{v}_\alpha.$$

We also note that the Vlasov moment term $\int (q/m)(\partial f/\partial \boldsymbol{v}) \cdot (\boldsymbol{E} + \boldsymbol{v} \times \boldsymbol{B})d\boldsymbol{v} = 0$ because, $\int \partial f/\partial \boldsymbol{v} d\boldsymbol{v} = f(+\infty) - f(-\infty) = 0$ and integrating by parts $\int \partial f/\partial \boldsymbol{v} \cdot \boldsymbol{v} \times \boldsymbol{B} d\boldsymbol{v} = -\int f(\partial/\partial \boldsymbol{v} \cdot (\boldsymbol{v} \times \boldsymbol{B})d\boldsymbol{v}$ because of the orthogonality of the components in the cross product. We then obtain the zeroth-order equation for species α as

$$\frac{\partial n_\alpha}{\partial t} + \nabla \cdot (n_\alpha \boldsymbol{u}_\alpha) = S_\alpha. \tag{7.6}$$

The "Eulerian" derivative $\partial/\partial t$ in this equation can be replaced with the comoving, or "Lagrangian" derivative $d/dt = \partial/\partial t + \boldsymbol{u} \cdot \nabla$, obtaining

$$\frac{dn_\alpha}{dt} + n_\alpha \nabla \cdot \boldsymbol{u}_\alpha = S_\alpha. \tag{7.7}$$

This equation represents conservation of particles, or equivalently mass density if every term is multiplied by m_α. The first-order moment of the Fokker-planck equation is written with mass multiplied by m_α as

$$\int (\mathscr{D}(f_\alpha)m_\alpha \boldsymbol{v}_\alpha) - \mathscr{C}(f_\alpha)m_\alpha \boldsymbol{v}_\alpha)\, d^3\boldsymbol{v} = \dot{\boldsymbol{P}}_{ext}. \tag{7.8}$$

Here are some steps leading up to the resulting transport equation. First look at the first two terms in the expression for $\mathscr{D}$ on the left-hand side. Split the velocity term v_α into a random part $\mathbf{w}_\alpha$ and an averaged part $\mathbf{u}$ as described above. Note that $< \mathbf{w} > = 0$ and $< \mathbf{u}\mathbf{w} > = 0$. Then

$$\int (\frac{\partial}{\partial t} + \mathbf{v}_\alpha \cdot \nabla)(m_\alpha v_\alpha f_\alpha) d\mathbf{v}_\alpha = \frac{\partial}{\partial t}(n_\alpha m_\alpha \mathbf{u}_\alpha) + \nabla \cdot \int f_\alpha m_\alpha ((\mathbf{u}+\mathbf{w})(\mathbf{u}+\mathbf{w})) d\mathbf{v}$$

$$= \frac{\partial}{\partial t}(n_\alpha m_\alpha \mathbf{u}_\alpha) + \nabla \cdot (n_\alpha m_\alpha \mathbf{u}_\alpha \mathbf{u}_\alpha + \mathbf{p})$$

(7.9)

Here $\mathbf{p}$ is the pressure tensor such that $p_{ij} = nm < w_i w_j > +p\mathbf{I} \equiv \pi_{ij} + p\delta_{ij}$. Thus $\mathbf{p} = \pi + p\mathbf{I}$, with $\mathbf{I}$ being the 3×3 identity matrix. The scalar pressure p is defined in the usual way from an equation of state

$$p_\alpha = n_\alpha T_\alpha = \frac{1}{3} n_\alpha m_\alpha < w_\alpha^2 > .$$

(7.10)

The term $\nabla \cdot \mathbf{p}$ collapses to a simple gradient of a scalar ∇p in the case of an isotropic plasma. Also note that $\nabla \cdot \mathbf{uu} = \mathbf{u}\nabla \cdot \mathbf{u} + \mathbf{u} \cdot \nabla \mathbf{u}$. We can then subtract $m_\alpha \mathbf{u}_\alpha$ times Eq. (7.6) from Eq. (7.9) to obtain

$$\int \left(\frac{\partial}{\partial t} + \mathbf{v}_\alpha \cdot \nabla\right)(m_\alpha v_\alpha f_\alpha) d\mathbf{v}_\alpha = n_\alpha m_\alpha \frac{d\mathbf{u}_\alpha}{dt} + n_\alpha m_\alpha \mathbf{u}_\alpha \cdot \nabla \mathbf{u}_\alpha + \nabla p + \nabla \cdot \pi_\alpha.$$

(7.11)

The next term in the Vlasov part of the first moment involves the electromagnetic term. As in the zeroth order case, we integrate by parts, again noting that $\partial/\partial \mathbf{v}(\mathbf{v} \times \mathbf{B}) = 0$. We then have

$$\int m_\alpha \mathbf{v}_\alpha \frac{q_\alpha}{m_\alpha}(\mathbf{E} + \mathbf{v}_\alpha \times \mathbf{B}) \cdot \frac{\partial f_\alpha}{\partial \mathbf{v}_\alpha} d\mathbf{v}_\alpha = -n_\alpha q_\alpha (\mathbf{E} + \mathbf{u}_\alpha \times \mathbf{B}).$$

(7.12)

Now we look at the term involving $\int m_\alpha \mathbf{v}_\alpha \mathscr{C} f_\alpha d\mathbf{v}_\alpha$. First note that

$$\int m_\alpha \mathbf{v}_\alpha \mathscr{C}_{\alpha/\alpha} f_\alpha d\mathbf{v}_\alpha = 0$$

(7.13)

since collisions within one species cannot affect the total momentum of that species. This leaves evaluation of the collision term

$$\mathbf{R}_{e/i} = -\mathbf{R}_{i/e} = \int m_e \mathbf{v}_e \mathscr{C}_{e/i} d\mathbf{v}_e,$$

(7.14)

where conservation of total momentum from Newton's law has been assumed. The problem here is what to use for f_e and f_i. If we choose maxwellian distributions with no average velocity for f_e and f_i, this term vanishes. We will re-visit this point later. First, we complete the first moment, which is the momentum equation, without deriving the expression for $\mathbf{R}$ at the moment. The result is:

$$n_\alpha m_\alpha \frac{d\mathbf{u}_\alpha}{dt} + n_\alpha m_\alpha \mathbf{u}_\alpha \cdot \nabla \mathbf{u}_\alpha = -\nabla p_\alpha - \nabla \cdot \boldsymbol{\pi}_\alpha + n_\alpha q_\alpha \left(\mathbf{E} + \mathbf{u}_\alpha \times \mathbf{B} \right) + \mathbf{R}_\alpha.$$

(7.15)

A persistent feature of the moment equations is that a new physical quantity must be defined with each higher moment of the Boltzmann equation. In the zeroth moment equation, the new quantity was $\mathbf{u}$, the average velocity; in the first moment, the new quantities p, T, and $\boldsymbol{\pi}$ were defined. Another feature is that subtractions of lower moment equations times some physical quantity are typically done to produce more meaningful physical relationships. We will now look at the second-order moment equation.

The second moment equation arises from the integral relation,

$$\int (\mathscr{D} f - \mathscr{C} f) \frac{m w^2}{2} d\mathbf{v} = 0.$$

(7.16)

The first term in $\mathscr{D} f m w^2 / 2$ is simplified by subtracting $m w^2 / 2$ times the continuity equation, Eq. (7.7). Also, the $\partial f / \partial v$ term does not contribute because upon integrating by parts, it has a term multiplied by $< \mathbf{w} >$, which is zero. Then the first term gives

$$\int \frac{m_\alpha w_\alpha^2}{2} \mathscr{D} f_\alpha d\mathbf{v} = \frac{3}{2} n_\alpha \frac{dT_\alpha}{dt} + p_\alpha \nabla \cdot \boldsymbol{u} + \boldsymbol{\pi}_\alpha : \nabla \mathbf{u}_\alpha + \nabla \cdot \mathbf{q}_\alpha.$$

(7.17)

Here the new quantity to be defined is the heat flow vector $\mathbf{q}$, defined as

$$\mathbf{q}_\alpha = \int \mathbf{w} \frac{m w^2}{2} d\mathbf{v}.$$

The collisional term $\int (m_\alpha w_\alpha^2 / 2) \mathscr{C}_\alpha f_\alpha$ contains only one term such as $\mathscr{C}_{i/e}$ or $\mathscr{C}_{e/i}$. We label this term as Q_α, the interspecies heat transfer. If there is only one ionic species, the two are related by the frictional heating:

$$Q_e = -Q_i - \mathbf{R} \cdot \mathbf{u}.$$

(7.18)

The overall second moment equation, representing the flow of heat, is then given by

$$\frac{3}{2} n_\alpha \frac{dT_\alpha}{dt} + p_\alpha \nabla \cdot \boldsymbol{u} = -\nabla \cdot \mathbf{q}_\alpha - \boldsymbol{P}_\alpha : \nabla \mathbf{u}_\alpha + Q_\alpha.$$

(7.19)

We note that the term $\nabla \cdot \mathbf{q}$ represents the change of the heat energy in a control volume by conduction, $\boldsymbol{P} : \nabla \mathbf{u}$ represents the viscous heating (with viscous stress tensor $\boldsymbol{P}_\alpha$ resulting from collisions), and with Q_α as defined above as the interspecies heat transfer, this equation represents the first law of thermodynamics:

$$dU = -pdV + dQ. \tag{7.20}$$

Here the entropy-producing terms dQ are on the right-hand side of Eq. (7.19).

7.2 Braginskii Transport Coefficients

We will now set about solving for the various transport terms arising from collisions. As stated earlier, if f_e and f_i are chosen to be maxwellian distributions, then $\mathbf{R} = 0$ and there is no interspecies momentum exchange. The viscous terms $\boldsymbol{\pi}_\alpha$ will also vanish, as will the heat flows $\mathbf{q}_\alpha$. In general, some perturbation must be added to the maxwellian zeroth-order distribution function that represents a solution to the Vlasov equation, i.e. it has no sources nor sinks in velocity space or physical space. A method to do this was first done in the 1910–1920 time frame by Chapman and Enskog to treat transport in ordinary gases [4]. The method was adapted to plasma transport in the 1950s. Here we follow the method first used by Braginskii in 1957 [2]. The perturbed distributions can be written as [3]:

$$f_e = f_{e0}(1 + \Phi), \tag{7.21}$$

where the expression for Φ can be written in terms of a vector $\boldsymbol{\Phi}$ and a tensor form ϕ_{ij} with

$$\Phi(v) = \mathbf{v} \cdot \boldsymbol{\Phi}\left(v^2\right) + \Phi_{ij}\left(v^2\right)\left(v_i v_j - \frac{v^2}{3}\delta_{ij}\right). \tag{7.22}$$

Only the vector part is necessary to solve for the drag and heat conduction terms, and only the tensor part is necessary for the viscous terms. (Here the Einstein summation convention is employed on the tensor products.) Note that the tensor part is a product with the fluid shear tensor $\mathbf{uu} - (u^2/3)\mathbf{I}$, and this form insures that no viscous stresses are induced by pure rotations or pure compressions ($\mathbf{u} \propto \mathbf{r}$). The vector quantity $\boldsymbol{\Phi}$ for the electrons can be written in terms of the three components of ∇T_e:

$$\boldsymbol{\Phi}\left(v^2\right) = A\nabla_\parallel \ln T_e + A'\nabla_\perp \ln T_e + A''\boldsymbol{\Omega}_{ce} \times \nabla \ln T_e \tag{7.23}$$

where A coefficients can be written as a sum of Laguerre polynomials (see [1]). The larger the number of polynomials chosen, the greater the exactness of the transport treatment. Braginskii found that the first two even polynomials and the first two odd polynomials are coupled together, so that only two independent polynomials per A-coefficient are required. (Treatments since that time have included as many as fifty terms, resulting in expressions only a few percent different than Braginskii's in most cases.)

We start by finding the expression for the drag coefficient $\mathbf{R}$. Forming an algebraic relationship with the Laguerre polynomial parameters and using the form of Eq. (7.23) outlined above, Braginskii found that, for a $Z = 1$ plasma,

$$\mathbf{R}_{e/i} = \mathbf{R}_{i/e} \equiv \mathbf{R} = \mathbf{R}_u + \mathbf{R}_{th} \tag{7.24}$$

with

$$\mathbf{R}_u = ne \left(\frac{\mathbf{j}_{\|}}{\sigma_{\|}} + \frac{\mathbf{j}_{\perp}}{\sigma_{\perp}} \right). \tag{7.25}$$

Here the "conductivities" are given by

$$\sigma_{\|} = 1.96\sigma_{\perp} = 1.96\frac{n_e^2 \tau_e}{m_e} \tag{7.26}$$

with the basic electron collision time τ_e given by

$$n\tau_e = \frac{3 \cdot [4\pi \epsilon_0]^2 \sqrt{m_e} T_e^{3/2}}{4\sqrt{2\pi} \ln \Lambda e^4} = \frac{3.44 \times 10^{11} T_e(\text{eV})^{3/2}}{\ln \Lambda}. \tag{7.27}$$

Note the MKS units used for Eq. (7.27). (The first formula in cgs units is the same without the $[4\pi \epsilon_0]^2$ term; Eq. (7.26) is unchanged.) Here $\log \Lambda$ is the Coulomb logarithm as defined in Chap. 3. The thermal force term $\mathbf{R}_{th}$ is given by

$$\mathbf{R}_{th} = -0.71 n \nabla_{\|} T_e - \frac{3n}{2\omega_{ce}\tau_e} \hat{\mathbf{b}} \times \nabla_{\perp} T_e. \tag{7.28}$$

The thermal force reflects the imbalance in Coulomb collisions from particles approaching from one side versus another, similar to the formation of a directed flow in a diffusive equilibrium leading to a Fick's law-type behavior.

A word of caution is appropriate when discussing Braginskii's "conductivities." Whereas the force equilibrium in the direction parallel to the magnetic field can give rise to a situation where in steady state the momentum balance essentially gives $J_{\|} = \sigma_{\|} E_{\|}$, this is not the situation for fields transverse to $\mathbf{B}$. In this case, the first-order effect is an $\mathbf{E} \times \mathbf{B}$ motion for both species, and the current flowing in the direction of an applied electric field must be solved for with both perpendicular directions in play. Thus the quotes around "conductivity": these are really the reciprocals of drag coefficients, with the units of conductivity.

Next we look at the Braginskii heat conduction coefficients. These are derived from the self-collisions within each species. The characteristic time for electron collisions is given above as Eq. (7.27). The corresponding ion–ion collision rate is given by

$$n\tau_i = \frac{3 \cdot [4\pi\epsilon_0]^2 \sqrt{m_i} T_i^{3/2}}{4\sqrt{\pi} \ln \Lambda e^4} = \frac{2.09 \times 10^{13} (m/m_p)^{1/2} T_i(\text{eV})^{3/2}}{\ln \Lambda}. \qquad (7.29)$$

Notice that besides the change from electron mass and temperature to ion mass and temperature, there is also a change of $\sqrt{2}$ in the collisional rates. This is because electron–electron collisions occur at roughly the same rate as electron–ion collisions, whereas ion–ion collisions dominate over ion–electron collisions (for an equivalent $90°$ scattering time) by a factor of roughly $\sqrt{m_i/m_e}$.

The electron and ion heat fluxes are given by

$$\begin{aligned}
\mathbf{q}_e &= -\kappa_\parallel^e \nabla_\parallel T_e - \kappa_\perp^e \nabla_\perp T_e + \kappa_\wedge^e \mathbf{b} \times \nabla_\perp T_e \\
\mathbf{q}_i &= -\kappa_\parallel^i \nabla_\parallel T_i - \kappa_\perp^i \nabla_\perp T_i + \kappa_\wedge^i \mathbf{b} \times \nabla_\perp T_i,
\end{aligned} \qquad (7.30)$$

and the thermal conductivities are given by

$$\begin{aligned}
\kappa_\parallel^e &= 3.2 \frac{n T_e \tau_e}{m_e} & \kappa_\perp^e &= 4.7 \frac{n T_e \tau_e}{m_e (\omega_{ce} \tau_e)^2} & \kappa_\wedge^e &= \frac{5}{2} \frac{n T_e \tau_e}{m_e (\omega_{ce} \tau_e)} \\
\kappa_\parallel^i &= 3.9 \frac{n T_i \tau_i}{m_i} & \kappa_\perp^i &= 4.2 \frac{n T_i \tau_i}{m_i (\omega_{ci} \tau_i)^2} & \kappa_\wedge^i &= \frac{5}{2} \frac{n T_i \tau_e}{m_i (\omega_{ci} \tau_i)}.
\end{aligned} \qquad (7.31)$$

Note that these expressions are written with the dimensionless quantity $\omega_\alpha \tau_\alpha$ held together. The reason for this is to illustrate the vastly disparate size of these coefficients depending on the directions involved. Numerical examples are given in the following box.

Examples of $\omega_{ce}\tau_e$ and $\omega_{ci}\tau_i$ for Magnetically Confined Plasma

Take a plasma with a 10^{20} m^{-3} electron density at $T_e = 10$ keV and $B = 4.0$. Using Eq. (7.27), $\tau_e = 196\,\mu$s. The electron cyclotron frequency $\omega_{ce} = 7.03 \times 10^{11}$ s^{-1}. Then $\boxed{\omega_{ce}\tau_e = 1.38 \times 10^8.}$

Similarly, for ions with $m = 2.5 m_p$ (average mass of DT fuel ions), Eq. (7.29) gives $\tau_i = 18.8$ ms. Then $\omega_{ci} = 1.535 \times 10^8$ s^{-1} and $\boxed{\omega_{ci}\tau_i = 2.88 \times 10^6.}$

The large size of the dimensionless quantity $\omega_\alpha \tau_\alpha$ shows that for most magnetically confined plasma example, $\kappa_\parallel \gg \kappa_\wedge \gg \kappa_\perp$ for both species. But it is also interesting to look at the mass dependence of these heat transfer coefficients. We note that, while the two coefficients for $\kappa_\wedge$ are the same ($\kappa_\wedge = (5/2)nT/(qB)$), the parallel electron heat conduction coefficient is roughly a factor of $(m_i/m_e)^{1/2}$ larger than the parallel ion heat transfer coefficient. The reverse is true for perpendicular ion heat conduction: $\kappa_\perp^i/\kappa_\perp^e \approx (m_i/m_e)^{1/2}$. This is because the dimensionless quantity $\omega_\alpha \tau_\alpha \propto m^{-1/2}$. Thus we would expect, for equal temperature gradients, that the ions would conduct heat approximately $100\times$ faster than the electrons out of the plasma, since the radial direction is perpendicular to the magnetic field. This is the reverse of the experimental finding, which is that most of the heat conduction in tokamaks is through the electron channel. This is because this model of heat transport by binary collisions in the electron channel is inaccurate because of the presence of other physics such as magnetic turbulence and microinstabilities.

Next we should discuss the strange "wedge term" in the heat conduction, $\kappa_\wedge$. We notice, as shown in the previous paragraph, $\kappa_\wedge = (5/2)nT/(qB)$ for both species. We note that the collision rate τ does not appear, nor does the mass of the particles. In fact this term is not really due to collisions at all, but related to the magnetization drift discussed in Chap. 3. It can also be thought of as a type of gyroscopic precession, where a force on a gyrating particle, which can be thought of as a spinning object, precesses in a direction perpendicular to both the force vector applied and the spin vector. If we think of a cylindrical column of plasma with the confining field along the z-axis and the temperature gradient in the radial direction, the wedge thermal conductivity causes the heat to circulate azimuthally, "chasing its own tail." Since there is no azimuthal temperature gradient in this example, there is no entropy generated and this does not violate the second law of thermodynamics. Sometimes this term is referred to as the "Bohm term," since an ad-hoc model for the anomalous loss of heat from magnetic confinement devices, attributed to David Bohm, had a similar form for the *perpendicular* heat loss $\kappa_\perp \propto (T_e/B)$. But perpendicular heat conduction at this rate, actually transporting heat out of the plasma, is far from harmless.

We now look at parallel electron heat conduction. An example is shown in the box below.

Example: Parallel Electron Heat Conduction

Suppose that a plasma at $n_e = 10^{20}\,\text{m}^{-3}$ has a 1 eV per meter temperature gradient at a temperature of 15 keV; i.e., it is $T_e = 15{,}001\,\text{eV}$ at one point and $T_e = 15{,}000\,\text{eV}$ one meter away. What is the electron heat conduction along **B**?

$$\tau_e = \frac{3.44 \times 10^{11} T_e[\text{eV}]^{3/2}}{\ln \Lambda n_e[\text{m}^{-3}]}$$

$$= \frac{2.0 \times 10^{10} T_e[\text{eV}]^{3/2}}{n_e[\text{m}^{-3}]}$$

$$= \frac{2 \times 10^{10} \cdot (15000)^{3/2}}{10^{20}}$$

$$= 3.67 \times 10^{-4}\,\text{s} \tag{7.32}$$

$$\kappa_\parallel^e = 3.2 \frac{nT_e\tau_e}{m_e}$$

$$= 3.2 \frac{10^{20}\cdot(15000\times1.6\times10^{-19})\cdot3.67\times10^{-4}}{9.11\times10^{-31}}$$

$$= 3.10\times10^{32}\,\mathrm{m^{-1}\,s^{-1}}$$

$$\nabla T_e = 1.6\times10^{-19}\,\mathrm{J\,m^{-1}}$$

$$q_e = \kappa_\parallel^e\nabla T_e = 3.10\times10^{32}\cdot1.6\times10^{-19} = 4.96\times10^{13}\,\mathrm{W\,m^{-2}}$$

A big number! Note that this calculation is just on the range of validity with this shallow temperature gradient. The characteristic length L is

$$L = \left|\frac{T_e}{\nabla T_e}\right| = 15{,}000\,\mathrm{m} \tag{7.33}$$

and the mean free path $\lambda = v_{the}\tau_e = \sqrt{T_e/m_e}\tau_e = 5.13\times10^7\cdot3.67\times10^{-4} \approx 18{,}000\,\mathrm{m}$

Here we see that even the shallowest of electron temperature gradients along **B** can cause huge heat fluxes. The general conclusion is that electron temperatures must be roughly constant along magnetic field lines.

Cross-field heat conduction rates are rather more modest in the Braginskii theory. An example for cross-field ($\kappa_\perp$) heat conduction is given in the box below.

Example: Cross-Field Ion Heat Conduction
Suppose that the ion temperature near the edge in a D+ plasma is 1.0 keV and the temperature gradient there is -5.0 keV per meter. The magnetic field is 4.0 T. What is the ion heat conduction there?

$$\kappa_\perp^i = 2.0 \frac{nT_i\tau_i}{m_i(\omega_{ci}\tau_i)^2}$$

$$\tau_i = \frac{2.09\times10^{13}\sqrt{m/m_p}T_i[\mathrm{eV}]^{3/2}}{\ln\Lambda n_i[\mathrm{m^{-3}}]}$$

$$= \frac{2.09\times10^{13}\sqrt{2}\cdot(1000)^{3/2}}{16\cdot10^{20}}$$

$$= 5.84\times10^{-4}\,\mathrm{s}$$

$$\omega_{ci} = \frac{1.6 \times 10^{-19} \cdot 4}{2 \cdot 1.67 \times 10^{-27}} \tag{7.34}$$

$$= 1.916 \times 10^8 \, \text{s}^{-1}$$

$$\omega_{ci}\tau_i = 1.916 \times 10^8 \cdot 5.84 \times 10^{-4}$$

$$= 1.12 \times 10^5$$

$$\kappa_\perp^i = 2.0 \frac{10^{20} \cdot 1000 \times 1.6 \times 10^{-19} \cdot 5.84 \times 10^{-4}}{2 \times 1.67 \times 10^{-27} \cdot \left(1.12 \times 10^5\right)^2}$$

$$= 4.47 \times 10^{17} \, \text{m}^{-1} \, \text{s}^{-1}$$

$$\nabla T_i = 5 \cdot 1.6 \times 10^{-16} = 8.0 \times 10^{-16} \, \text{J} \, \text{m}^{-1}$$

$$(q_i)_\perp = -\kappa_\perp^i \nabla T_i = 4.47 \times 10^{17} \cdot 8.0 \times 10^{-16} = 357. \, \text{W} \, \text{m}^{-2}$$

Now we look at the inter-species heating term Q_{ie}. Braginskii gives a particularly simple form for this term:

$$Q_{ie} = \frac{3m_e}{m_i} \frac{n(T_e - T_i)}{\tau_e}. \tag{7.35}$$

A numerical example is given in the following box.

Example: Inter-Species Heating
Suppose that a D+ plasma at $n_e = 10^{20} \, \text{m}^{-3}$ has an electron temperature of 15 keV and an ion temperature of 20 keV. What is the volumetric rate of electron heating?
Note that $\tau_e = 3.67 \times 10^{-4}$ s as before. Then

$$Q_{ie} = 3\frac{m_e}{m_i} \frac{n\,(T_i - T_e)}{\tau_e}$$

$$= 3\left(\frac{9.11 \times 10^{-31}}{2 \cdot 1.67 \times 10^{-27}}\right) \cdot 10^{20} \cdot \left(\frac{(20,000 - 15,000) \cdot 1.6 \times 10^{-19}}{3.67 \times 10^{-4}}\right)$$

$$= 178 \, \text{kW} \, \text{m}^{-3}$$

The electron interspecies heating rate is the reverse of the ion inter-species heating rate, but also including the generation of heat by collisional friction:

$$Q_e = -Q_i - \mathbf{R} \cdot \mathbf{u}. \tag{7.36}$$

Notice the characteristic time for the inter-species heating. Ignoring the frictional heating, which is essentially the Ohmic or Joule heating of the plasma ηj^2, consider a situation where the electrons are initially at a temperature $T_0 + \Delta T/2$ and the ions are at a temperature $T_0 - \Delta T/2$. Then the differential equations for the electron and ion temperatures are given by

$$\frac{3}{2}\frac{dT_e}{dt} = -\frac{3(T_e - T_i)(m_e/m_i)}{\tau_e}$$
$$\frac{3}{2}\frac{dT_i}{dt} = \frac{3(T_e - T_i)(m_e/m_i)}{\tau_e} \tag{7.37}$$

Subtracting the second equation from the first and letting $T_e - T_i = \Delta T$ gives

$$\frac{d\Delta T}{dt} = \frac{4(m_e/m_i)}{\tau_e}\Delta T. \tag{7.38}$$

This has solution $\Delta T(t) = \Delta T(0)\exp(-t/\tau_{EQ})$, with

$$\tau_{EQ} = \frac{m_i}{4m_e}\tau_e.$$

So the equipartition time τ_{EQ} is on a rather long timescale.

Example: Equipartition Time at Fusion Temperatures
Consider a DT plasma with a 25 keV electron temperature. What is the characteristic time for electron–ion equipartition?
   Note that $\tau_e = 3.67 \times 10^{-4}$ s as before. Here the mass ratio $= 2.5m_p/m_e = 2.5 \times 1836 = 4590$. Then

$$\tau_{EQ} = \frac{3.67 \times 10^{-4} \cdot 4590}{4} = 0.42\,\text{s}$$

We do not expect to see electron–ion equipartition before plasma energy confinement is at a level meeting the $Q = 1$ level of confinement, i.e. $n\tau_E \geq 5 \times 10^{19}\,\text{m}^{-3}\,\text{s}$.

Most laboratory plasmas have lower energy confinement times than the electron–ion equipartition time, so that situations where $T_e \neq T_i$ are fairly common. A classic example of a situation where T_e and T_i are radically different is a fluorescent light bulb. Here $T_e \approx 5\,\text{eV}$ and is clamped by the line radiation from the mercury vapor plasma. The Hg ions themselves are in equilibrium with the walls of the tube, and so are typically at $T_i \approx 0.025 \rightarrow 0.05\,\text{eV}$.

Next we look at the viscosity coefficients. The stress tensor $\boldsymbol{\pi}$ is symmetric and has the following on-diagonal structure:

$$\boldsymbol{P}_{xx} = -\frac{\eta_0}{2}\left(W_{xx} + W_{yy}\right) - \frac{\eta_1}{2}\left(W_{xx} - W_{yy}\right) - \eta_3 W_{xy},$$

$$\boldsymbol{P}_{yy} = -\frac{\eta_0}{2}\left(W_{xx} + W_{yy}\right) + \frac{\eta_1}{2}\left(W_{xx} - W_{yy}\right) + \eta_3 W_{xy}, \qquad (7.39)$$

$$\boldsymbol{P}_{zz} = -\eta_0 W_{zz},$$

and off-diagonal terms

$$\boldsymbol{P}_{xy} = \boldsymbol{P}_{yx} = -\eta_1 W_{xy} + \frac{\eta_3}{2}\left(W_{xx} - W_{yy}\right),$$

$$\boldsymbol{P}_{xz} = \boldsymbol{P}_{zx} = -\eta_2 W_{xz} - \eta_4 W_{yz}, \qquad (7.40)$$

$$\boldsymbol{P}_{yz} = \boldsymbol{P}_{zy} = -\eta_2 W_{yz} + \eta_4 W_{xz}.$$

Here the rate-of-strain tensor is given by

$$W_{jk} = \frac{\partial u_j}{\partial x_k} + \frac{\partial u_k}{\partial x_j} - \frac{2}{3}\delta_{jk}\boldsymbol{\nabla}\cdot\boldsymbol{u}. \qquad (7.41)$$

The five viscosity coefficients for the electrons are

$$\eta_0^e = 0.73 n T_e \tau_e \qquad \eta_1^e = 0.51\frac{n T_e \tau_e}{(\omega_{ce}\tau_e)^2} \qquad \eta_2^e = 2.0\frac{n T_e \tau_e}{(\omega_{ce}\tau_e)^2},$$

$$\eta_3^e = -\frac{n T_e \tau_e}{2\omega_{ce}\tau_e}, \qquad \eta_4^e = -\frac{n T_e \tau_e}{\omega_{ce}\tau_e}, \qquad\qquad (7.42)$$

where the dimensionless parameter $\omega_c\tau$ has been left intact to show the ordering. The viscosity coefficients for the ions are

$$\eta_0^i = 0.96 n T_i \tau_i \qquad \eta_1^i = \frac{3}{10}\frac{n T_i \tau_i}{(\omega_{ci}\tau_i)^2}, \qquad \eta_2^i = \frac{6}{5}\frac{n T_i \tau_i}{(\omega_{ci}\tau_i)^2},$$

$$\eta_3^i = \frac{n T_e \tau_i}{2\omega_{ci}\tau_i}, \qquad \eta_4^i = \frac{n T_i \tau_i}{\omega_{ci}\tau_i}. \qquad\qquad (7.43)$$

As is the case with the thermal conductivity, the viscosity coefficients appear with differing orders in $\omega_c\tau$, which leads to quite disparate sizes for the transport coefficients when this parameter is large. Of special interest is the first term η_0 for the ions: on the surface this term appears to be vastly larger than any other viscous effect in the plasma, with an effective volumetric force of

$$-\boldsymbol{\nabla}\cdot\boldsymbol{P}_z^i \sim -\frac{\partial}{\partial z}\left(\eta_0^i\frac{\partial u_z}{\partial z}\right) \sim n_i T_i \tau_i\frac{\partial^2 u_z^i}{\partial z^2}, \qquad (7.44)$$

which can be approximated by taking $\tau_i = \lambda_i/v_{thi}$, and taking $\partial^2/(\partial z)^2 \to L_u$, with L_u being some velocity scale length:

$$-\nabla \cdot \boldsymbol{P}^i{}_z \sim (n_i T_i/L)(\lambda/L)(u/v_{thi}). \tag{7.45}$$

However, the problem here is the scale length for λ, the mean free path, for ions (as well as electrons). See the box below.

Example: Ion and Electron Mean Free Paths in Magnetic Fusion
Find $\lambda = v_{thi}\tau_i$ for a DT fuel ion at $T_i = 15\,\mathrm{keV}$, $n_i = 10^{20}\,\mathrm{m}^{-3}$:

$$\begin{aligned}
\tau_i &= \frac{2.09 \times 10^{13}\sqrt{m/m_p}\,T_i[\mathrm{eV}]^{3/2}}{\ln \Lambda n_i[\mathrm{m}^{-3}]} \\[2mm]
&= \frac{2.09 \times 10^{13}\sqrt{2.5} \cdot (15000)^{3/2}}{17.5 \cdot 10^{20}} \\[2mm]
&= 34.5 \times 10^{-3}\,\mathrm{s} \\[2mm]
v_{thi} &= \sqrt{T_i/m_i} \\[2mm]
&= \left(\frac{15000 \cdot 1.6 \times 10^{-19}}{2.5 \cdot 1.67 \times 10^{-27}}\right)^{1/2} \\[2mm]
&= 7.58 \times 10^5\,\mathrm{ms}^{-1} \\[2mm]
\lambda_i &= v_{thi}\tau_i = 2.62 \times 10^4\,\mathrm{m}
\end{aligned}$$

Since $\tau_e/\tau_i = (1/\sqrt{2})(m_e/m_i)^{1/2}$ and $v_{the}/v_{thi} = (m_i/m_e)^{1/2}$ for $T_e = T_i$, the electron mean free path is shorter by a factor of $\sqrt{2}$, or

$$\lambda_e = 1.8 \times 10^4\,\mathrm{m}$$

as given earlier. Thus electron and ion mean free paths are on roughly the same length scale.

The problem here is that if we took the velocity scale length L_u to be some reasonable size (say, 1 m), then the parallel electron viscous force is much too large and would exceed the pressure gradient by a factor $\sim 1/M(L/\lambda)$, where $M = u/v_{th}$ is the Mach number, typically on the order of a few percent in laboratory plasmas. It would be impossible for this level of force to be balanced by any amount of pressure anisotropy. The answer to this conundrum is that a basic assumption of the Braginskii transport hierarchy is that the mean free path is much smaller than the gradient length: $\lambda \ll L$. This obviously is not the case for the examples given here. For the perpendicular components, one can show that because of the additional factor of $(\omega_c\tau)^2$ contained in the expressions for η_i and η_2, the restriction on the perpendicular gradient length $L_\perp$ is much less restrictive: $L_\perp \gg \sqrt{\lambda r_e}$, where r_e is the electron gyroradius. for the example given here, $r_e = m v t h \perp e/eB = 1.03 \times 10^{-4}$ m, and this gives $\sqrt{\lambda_e r_r} = 1.3$ m for the minimum scale length for the application of Braginskii theory, making the applicability marginal at best. The next section explores the transport coefficients using a hierarchy somewhat more suited to magnetic confinement fusion problems.

7.3 Gyro-Kinetic Transport

As seen in the last section, the classical MHD ordering for scale lengths and timescales is often not appropriate for transport calculations relevant to magnetically confined fusion devices. In this section we examine alternative orderings of the parameters involved, and will attempt to show a set of transport relationships which might be a bit more realistic for the intended applications.

Let us first examine the small parameters assumed in the MHD limit [23] (see Appendix 1). We can identify four dimensionless ratios. The first is the ratio of ion gyroradius to the transport scale length, $\delta \equiv \rho_i/L$. The smallness of this parameter is essential for any fluid-theory formulation; otherwise, the plasma response is nonlocal. Second, we look at the allowed inverse timescale ω for quantities to change and compare this to the ion gyrofrequency Ω_{ci}, hence $\epsilon \equiv \omega/\Omega_{ci}$. Next we look at the inverse Mach number of the flows $\xi \equiv u/v_{thi}$. Finally we have the normalized pressure $\beta \equiv (v_{thi}/v_A)^2$, with $v_{thi} = \sqrt{2T_i/m_i}$ and the Alfvén speed $v_A = B/\sqrt{\mu_0 n_i m_i}$. (It should be noted that these four dimensionless numbers are not independent, as $\beta = \delta/\epsilon$.)

In the MHD scheme, the Ohm's law can be written as

$$\mathbf{J} = \sigma(\mathbf{E} + \mathbf{u} \times \mathbf{B})$$

where the velocity $\mathbf{u}$ is the mass-weighted average of the electron and ion velocities, i.e. $\mathbf{u}(m_e + m_i) = m_e \mathbf{u}_e + m_i \mathbf{u}_i$, which is strongly weighted by the large ion mass, so that $\mathbf{u} \approx \mathbf{u}_i$. In the MHD approximation, the ordering $\mathbf{u}/v_{thi} = \mathcal{O}(1)$, i.e. Mach numbers can be of order unity. The characteristic frequencies are such that $\epsilon = \omega/\Omega_{ci} = \mathcal{O}(\delta)$ and the normalized pressure $\mathcal{O}(\delta)$. As mentioned before, the mean free path $\lambda \ll L$, which is inaccurate for hot plasmas in the direction along the magnetic field.

In another ordering scheme, called "Hall MHD," one orders the pressure-gradient term in the momentum equation, Eq. (7.15), as second order in δ and we retain some of the intrinsically two-fluid aspects of the momentum equation. We then have $\mathbf{E} + \mathbf{u}_e \times \mathbf{B} = 0$ by ignoring electron viscosity and pressure (with $\beta = \mathcal{O}(\delta^2)$ and arrive at (noting that $\mathbf{J} = ne(\mathbf{u}_i - \mathbf{u}_e)$):

$$\mathbf{E} + \mathbf{u} \times \mathbf{B} = \frac{1}{ne}\mathbf{J} \times \mathbf{B} + \eta\mathbf{J} \tag{7.46}$$

In the Hall MHD ordering, the frequencies $\omega = \mathcal{O}(\Omega_{ci})$ and the ion flows can be large compared to the sound speed: $\mathbf{u} = \mathcal{O}v_{thi}/\delta$. With the plasma pressure ordering $\beta = \mathcal{O}(\delta^2)$ given above, the $\mathbf{J} \times \mathbf{B}$ force is countered largely by the ion momentum: $\mathbf{J} \times \mathbf{B} \approx m_i n_i (d\mathbf{u}_i/dt)$. With this ordering, the ordinary Alfvén waves can exist, but also a fast, dispersive wave known as a "whistler" wave, which can complicate the numerical simulation of plasmas with Hall MHD. Because of the high frequencies allowed in Hall MHD, it is useful to describe systems with fast evolution, such as shocks, z-pinches, and gas-filled switch tubes.

Another ordering, called the "Drift ordering," is probably the most realistic ordering for hot plasmas with relatively slow dynamics but arbitrarily large mean free paths. In this ordering, we take the frequency range $\epsilon \equiv \omega/\Omega_{ci} = \mathcal{O}(\delta^2)$. Flow velocities must also be slow: $\mathbf{u} = \mathcal{O}(\delta v_{thi})$. However the dimensionless pressure $\beta = \mathcal{O}(1)$. The equation of motion with this ordering becomes

$$ -\nabla p + \mathbf{J} \times \mathbf{B} = m_i \frac{d\mathbf{u}_i}{dt} + \nabla \cdot \boldsymbol{\pi}_i, \tag{7.47} $$

where the ion viscosity term can be further divided up into separate contributions from anisotropy in velocity space (generating a so-called "gyro-viscous" term), from collisional effects arising from trapped particles (the "neoclassical" effects, which will be described in the next section), and sometimes an artificial viscosity term $\propto \mu_A \nabla^2 \mathbf{u}_i$, required in some numerical simulation schemes to prevent numerical instabilities. The drift model does not allow whistler waves like those allowed in the Hall MHD model, but it does allow the presence of "kinetic Alfvén waves," which are a type of magnetosonic wave allowed at finite β.

A summary of the features of these three ordering schemes is shown as Table 7.1.

In the drift model, the treatment of the viscous terms coming directly out of the Boltzmann equation from a moment method of some kind is generally very difficult. Instead, these coefficients are usually obtained from a reduced form of the Vlasov-Boltzmann equation called the "drift-kinetic equation" [7, 9]:

$$ \frac{\partial f}{\partial t} + (\mathbf{v}_\| + \mathbf{v}_D) \cdot \nabla f + \left\{ \frac{\mu}{mv} \frac{\partial B}{\partial t} + \frac{e}{mv} (\mathbf{v}_\| + \mathbf{v}_D) \cdot \mathbf{E} \right\} \frac{\partial f}{\partial v} = \mathscr{C}(f). \tag{7.48} $$

In this equation, v is the scalar velocity of the particle, $\mathbf{v}_\|$ and $\mathbf{v}_D$ are the parallel and cross-field drift components of the guiding center motion, and μ is the magnetic moment. This equation is the result of averaging over the phase angles of the particles as they orbit around the guiding-center magnetic field, resulting in a more tractable number of independent variables.

Table 7.1 Parameters of three transport ordering schemes

Model	$\mathbf{u}_i$	ω	β	$\mathbf{J} \times \mathbf{B}$	Whistlers	KAW
Hall MHD	v_{thi}/δ	Ω_{ci}	$\mathcal{O}(\delta^2)$	$m_i n(d\mathbf{u}/dt) + \mathcal{O}(\delta)$	No	Yes
Ideal MHD	v_{thi}	$\delta\Omega_{ci}$	$\mathcal{O}(\delta)$	$\mathcal{O}(\delta)$	No	No
Drift	δv_{thi}	$\delta^2 v_{thi}$	$\mathcal{O}(1)$	$\nabla p + \mathcal{O}(\delta^2)$	No	Yes

From [22]

7.4   Neoclassical Transport

Neoclassical transport refers to transport in toroidal geometry that is predicated on a binary collision model, like classical transport, but containing a more detailed geometrical model, i.e. retaining toroidal effects, and also accounting for more sophisticated models for the ion and electron distribution functions due to the departure from isotropy from the nonuniform fields.

The lowest-order effect of the toroidicity correction can be seen from expressions for the perpendicular current and perpendicular heat flux in a torus. (Here "perpendicular" means the direction perpendicular to both the magnetic field **b** and the radial vector **r**, i.e. a direction lying in a flux surface and perpendicular to the magnetic field.) Consider the perpendicular component of the current from the force balance $\mathbf{J} \times \mathbf{B} = \nabla p$:

$$\mathbf{J}_\perp = \frac{\mathbf{B} \times \nabla p}{B^2} \tag{7.49}$$

We can decompose the toroidal and poloidal magnetic fields using Eqs. (5.35) and (5.37) given earlier. These give

$$\frac{\mathbf{B} \times \nabla \Psi}{B^2} = \frac{F}{B}\mathbf{b} - R\hat{\phi} \tag{7.50}$$

Then, using $\nabla p = (dp/d\Psi)\nabla\Psi$, we have

$$\mathbf{J}_\perp = \frac{\mathbf{B} \times \nabla p}{B^2} = \left(\frac{F\mathbf{B}}{B^2} - R\hat{\phi} \right) \frac{dp}{d\Psi}. \tag{7.51}$$

However, the overall current $\nabla \cdot \mathbf{J} = 0$ in equilibrium due to conservation of charge, but $\nabla \cdot \mathbf{J}_\perp \neq 0$ for the representation of $\mathbf{J}_\perp$ given in Eq. (7.51) above. This necessitates a parallel current $\mathbf{J}_\parallel$ in order to make the overall $\nabla \cdot \mathbf{J} = 0$. Writing out the expression for $\nabla \cdot \mathbf{J} = 0$ then gives, using the vector identity $\nabla \times (A\mathbf{B}) = \mathbf{B} \cdot \nabla A$ since $\nabla \cdot \mathbf{B} = 0$:

$$0 = \nabla \cdot \left(\frac{J_\parallel \mathbf{B}}{B} + \mathbf{J}_\perp \right) = \mathbf{B} \cdot \nabla \left(\frac{J_\parallel}{B} + \frac{F}{B^2}\frac{dp}{d\Psi} \right). \tag{7.52}$$

We can solve this first-order differential equation by setting the contributions of the two terms on the right-hand side of Eq. (7.52) to zero, and then adding a homogeneous term, noting that $\mathbf{B} \cdot \nabla K(\psi) = 0$ for any function K:

$$J_\parallel = -\frac{F}{B}\frac{dp}{d\Psi} + K(\psi)B. \tag{7.53}$$

We then make flux-surface averages of the quantities in Eq. (7.53), using the definition that

$$< A > \equiv \frac{\int A d\ell / B}{\int d\ell / B}. \tag{7.54}$$

Since K is constant on a flux surface, we find that upon averaging:

$$J_{\parallel} = \frac{B}{< B^2 >} \langle J_{\parallel} B \rangle - \frac{F(\psi)}{B} \frac{dp}{d\psi} \left(1 - \frac{B^2}{< B^2 >} \right). \tag{7.55}$$

Here the $J_{\parallel}$ term represents the usual current flow along the magnetic field lines, induced by an induction electric field in ohmically heated tokamaks (or by some other momentum source in general) and the second term is a new term, vanishing in the case of a straight cylinder but present in toroidal geometry. The second term is called the Pfirsch-Schlüter current, first reported in 1962 [19]. Summing the components of $\mathbf{J} = \mathbf{b} J_{\parallel} + \mathbf{J}_{\perp}$ gives

$$\mathbf{J} = -\frac{dp}{d\psi} R\hat{\phi} + K(\psi)\mathbf{B}. \tag{7.56}$$

Now we examine the diamagnetic flow. The perpendicular component of the flow comes from the momentum equation, Eq. (7.9), given to lowest order (ignoring acceleration, tensor pressure, and friction) by:

$$n_\alpha \mathbf{u}_{\alpha\perp} = n_\alpha(\psi)\omega_\alpha(\psi) \left(R\hat{\psi} - \frac{F}{B}\mathbf{b} \right) \tag{7.57}$$

with

$$\omega_\alpha(\psi) \equiv -\frac{d\Phi}{d\psi} - \frac{1}{n_\alpha q_\alpha} \frac{dp}{d\psi}. \tag{7.58}$$

We note that, in a similar manner to the argument taken for a parallel contribution to the current $J_{\parallel}$ from $\nabla \cdot \mathbf{J} = 0$, we require that $\nabla \cdot n\mathbf{u}_\alpha = 0$. This gives

$$n_\alpha u_{\alpha\parallel} = \frac{F(\psi)n_\alpha(\psi)\omega_\alpha(\psi)}{B} + K_\alpha(\psi)B. \tag{7.59}$$

This gives for the total flow:

$$n_\alpha \mathbf{u}_\alpha = n_\alpha(\psi)\omega_\alpha(\psi)R\hat{\phi} + K_\alpha(\psi)\mathbf{B}. \tag{7.60}$$

Thus the two allowed flow terms are a toroidal rigid rotation of the volume element surrounding each flux surface (but with a rotation rate allowed to vary from one flux surface to the next) and a flow parallel to the magnetic field, also allowed to vary from one flux surface to the next. Summing the charge-weighted components of the flow by species, one recovers Eq. (7.56), with $K(\Psi) = \sum_\alpha q_\alpha K_\alpha(\Psi)$.

The heat conduction rates follow in exactly the same way. Starting from the Braginskii heat conduction rates given in Eqs. (7.30) and (7.31), one notices that the radial and perpendicular heat flows have very disparate rates when $\Omega\tau \gg 1$ for each species, with the "wedge" term dominating. For both ion and electron species the form of the wedge heat conduction is the same, differing by the appropriate species index, so that we can write

$$\mathbf{q}_{\perp\alpha} = \frac{5p_\alpha}{2m_\alpha\Omega_\alpha}\mathbf{b}\times\nabla T_\alpha, \tag{7.61}$$

and requiring that $\nabla\cdot\mathbf{q}_\alpha = 0$ for the total (parallel+perpendicular) heat flux again gives

$$q_{\alpha\parallel} = -\frac{5n_\alpha T_\alpha F(\Psi)}{2q_\alpha B}\frac{dT_\alpha}{d\Psi} + L_\alpha(\Psi)B \tag{7.62}$$

and then the total heat conduction by species α becomes

$$\mathbf{q}_\alpha = -\frac{5n_\alpha T_\alpha}{2q_\alpha B}\frac{dT_\alpha}{d\Psi}R\hat{\phi} + L_\alpha(\Psi)\mathbf{B}. \tag{7.63}$$

To proceed from here, one needs to find a form for the second term in Eq. (7.63), i.e. the heat flow along $\mathbf{B}$. However, as we have seen earlier, the Braginskii estimate for $\mathbf{q}_\parallel$ does not apply if the mean free path for collisions exceeds the transport scale lengths involved, and one has to use the drift-kinetic equation to find an appropriate expression for $\mathbf{q}_\parallel$. The structure of the heat conduction thus depends on the collisionality of the plasma. This is the subject of the next section.

7.4.1   Orbits and Collisionality

The toroidal geometry of tokamaks and stellarators leads to more complex orbital motion that exists in straight cylindrical geometry. This is because both the toroidal and poloidal magnetic fields have a (roughly) $1/R$ dependence, and thus are stronger on the inside part of a flux surface than the outside. This causes the drift motion due to the grad-B and curvature effects shown in Sects. 3.4.4 and 3.4.5 to cause particles to migrate from one flux surface to another as they gyrate in the magnetic field. For particles with a sufficiently large pitch angle in velocity with respect to the magnetic field, a mirroring of the particle can occur, reversing the velocity component along $\mathbf{B}$, as outlined in Sect. 3.4.6. Figure 7.1a shows a poloidal projection of the guiding center of a particle orbit in a tokamak magnetic field similar to ITER. Note that the guiding center trajectory eventually returns to the same place in R, Z as it started from, making closed trajectories. However, the drift motion causes the orbit to trace a path through two different radial positions on the midplane, leading to a

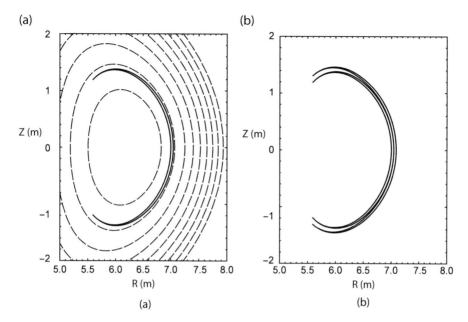

Fig. 7.1 Banana orbits in ITER-like magnetic fields. (**a**) Orbit shown for a deuteron with $W_\perp = 16$ keV and $W_{tot} = 20$ keV with an initial point at $R = 7.0$ m on the midplane, (**b**) The same orbit shown with the orbit after a $180°$ collision on the outer midplane point of its orbit

banana-shaped path. These particles are trapped in a manner similar to the particles in a mirror-machine type of fusion device, or in dipole fields such as the earth's magnetic field.

One should also note that a qualitatively different orbital path is found for particles near the magnetic axis. Figure 7.2 shows the orbit of a 3.5 MeV alpha particle born near the magnetic axis with mostly perpendicular energy. This "potato" orbit shows a substantial radial width. For particles with a large gyroradius, the model of the particle being treated as having a guiding center that is drifting through the flux surfaces in accordance with the grad-B and curvature drifts is no longer accurate, and orbits must be computed from a direct solution from the exact equation of motion. Hence Fig. 7.2 is a direct orbital integration, and the black band is actually a plot of the drifting gyroorbit with hundreds of gyroperiods.

As shown in the discussion of mirror-trapped particles in Sect. 3.4.6, only those particles with a pitch angle α (at the outer midplane) such that $\sin^2 \alpha \geq B_{min}/B_{max}$ are trapped. If we define $\epsilon \equiv r/R_0$ as the inverse aspect ratio for a circular tokamak with nominal major radius R_0, and the fraction f_T of particles that are trapped on a given flux surface, the trapped particle fraction is given in the limit of large aspect ratio ($\epsilon \ll 1$) and circular plasma cross section by [5, 15, 21]:

$$f_T = 1 - \frac{3}{4} < h^2 > \int_0^1 \frac{\lambda d\lambda}{\langle (1 - \lambda h)^{1/2} \rangle} \approx 1.46\sqrt{\epsilon} + \mathcal{O}(\epsilon) \qquad (7.64)$$

Fig. 7.2 Potato-type orbit of a 3.5 Mev α in an ITER-like magnetic field

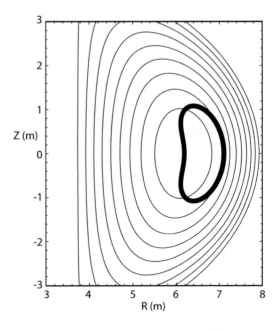

Here $<\ >$ denotes the average of a quantity on a flux surface and $h = 1 + \epsilon \cos\theta$, with θ being the poloidal angle. Additionally Ref. [15] gives an upper and lower bound on the trapped particle fraction for arbitrary poloidal plasma shape:

$$f_T = (1.35 \to 1.5)\sqrt{\epsilon} + \mathcal{O}(\epsilon) \qquad (7.65)$$

Figure 7.1b shows the impact of collisions on the banana orbits. For the extreme case of a 180° collision at the outer midplane point, one can see that the particles orbit becomes another banana orbit with the inside of the new orbit joining the old orbit at the midplane. The width Δ of the banana orbit at the midplane can be obtained from the bounce time and the drift velocity, $\Delta \approx v_d \tau_b \approx \rho_\theta \sqrt{\epsilon}$, with ρ_θ being the "poloidal gyroradius," $\rho_\theta = \rho B / B_\theta$. Thus if collisions are infrequent enough such that the trapped particles orbit a few times before a collision, the transport is enhanced over the classical values because the appropriate stepsize used in transport estimates switches from being the gyroradius to the banana width. However, only the fraction of the particles that are trapped is involved, and the angular deflection required in a collision to remove or insert a particle from or to a banana orbit must be taken into consideration to decide whether or not the banana orbits exist on the timescale for the trapped particles to survive a complete banana trajectory. Since the characteristic length for travel around a banana orbit $L \approx qR$, this gives a characteristic bounce time $\tau_B \approx qR/v_\parallel$. The trapped particles occupy a fraction of velocity space $\propto \sqrt{\epsilon}$, so that we can estimate $v_\parallel/v \approx \epsilon^{1/2}$. Thus

$$\tau_B \approx \frac{qR}{\epsilon^{1/2} v_{th}}. \qquad (7.66)$$

Similarly, the rate of collisions in or out of a banana orbit is determined by the square of the angular deflection required, $\Delta\theta^2 \approx (\sqrt{\epsilon})^2 = \epsilon$. The collision rate ν_{eff} to scatter by an angle $\Delta\theta$ is approximately $\nu_{eff} = \nu_{90}/(\Delta\theta)^2$, where ν is the characteristic 90° scattering frequency. So the condition where the particle executes many banana orbits before being scattered is given by the normalized collision frequency ν_*:

$$\nu_* = \nu_{eff}\tau_B \approx \frac{qR\nu_{90}}{\epsilon^{3/2}v_{th}} \ll 1. \qquad (7.67)$$

Plasmas (or places within a given plasma) meeting this criterion are said to be in the "Banana" transport regime. In the opposite limit, if there are many collisions per bounce period, the concept of trapped vs. untrapped particles becomes meaningless, and collisions between all pitch angles are frequent enough that anisotropies in velocity space cannot develop. For this case we consider the "whole" collision process which is based on the 90° scattering time, and compare this to the mean free path $L \approx qR$ for all of the particles:

$$\tau_B\nu_{90} \gg 1, \qquad (7.68)$$

which in units of the normalized collisionality in neoclassical theory@collisionality in neoclassical theory ν_* collisionality ν_* gives

$$\nu_* \gg \epsilon^{-3/2}. \qquad (7.69)$$

Plasmas in this collisionality regime are said to be in the "Pfirsch-Schlüter regime. Notice, however, that these regimes do not overlap if $\epsilon \ll 1$. The intermediate regime has

$$1 < \nu_* < \epsilon^{-3/2}. \qquad (7.70)$$

Plasmas in this state of collisionality are said to be in the "Banana-Plateau" or simply "Plateau" regime. We will now look at the calculation of the transport in these three regimes.

7.4.2 Pfirsch-Schlüter Transport

For the case of high collisionality, one can assume that the pressure is isotropic, but the temperature T can vary along the flux surface. We can expand the temperature in a series

$$T_i = T_{i0} + T_{i1} + \ldots \qquad (7.71)$$

with $T_{i0} =$ constant. Then requiring that the in-surface heat flux has $\nabla \cdot \mathbf{q} = 0$ gives

$$\nabla \cdot \left(\kappa_\parallel^i \nabla_\parallel T_{i1} - \kappa_\wedge \mathbf{b} \times \nabla T_{i0} \right) = 0. \tag{7.72}$$

Because of axisymmetry, $\partial/\partial\phi = 0$ and then we can rewrite Eq. (7.72) as

$$\frac{\partial}{\partial\theta} \left(\frac{\kappa_\parallel^i}{B} \nabla_\parallel T_{i1} - \frac{F(\Psi)\kappa_\wedge}{B} \frac{dT_{i0}}{d\Psi} \right) = 0. \tag{7.73}$$

This equation can be integrated over θ to give

$$\nabla_\parallel T_{i1} = \frac{F(\Psi)\kappa_\wedge^i}{\kappa_\parallel^i} \frac{dT_{i0}}{d\Psi} + L_i(\Psi)B \tag{7.74}$$

with the constant of integration $L_i(\Psi)$ determined such that the flux average $\langle B\nabla_\parallel T_i \rangle = 0$, which follows from the definition of flux average (this evaluates to $\int \nabla_\parallel T_{i1} d\ell = 0$). This gives

$$\nabla_\parallel T_{i1} = \frac{F(\Psi)\kappa_\wedge^i}{\kappa_\parallel^i} \frac{dT_{i0}}{d\Psi} \left(1 - \frac{B^2}{\langle B^2 \rangle} \right). \tag{7.75}$$

Finally, we can look at the radial heat transport. The classical radial ($\nabla\Psi$ direction) is just $-\kappa_\perp \nabla_\perp T_{i0}$ in keeping with Braginskii's equations for planar geometry. The neoclassical addition comes from Eq. (7.75) using the wedge heat transfer. This gives

$$
\begin{aligned}
\mathbf{q}_i^{PS} \cdot \nabla\Psi &= -\left(\kappa_\perp^i \nabla_\perp T_{i0} + \kappa_\wedge^i \mathbf{b} \times \nabla T_{i1} \right) \cdot \nabla\Psi \\
&= -\left[\kappa_\perp^i |\nabla\Psi|^2 + \frac{(F(\Psi)\kappa_\wedge^i)^2}{\kappa_\parallel^i} \left(1 - \frac{B^2}{\langle B^2 \rangle} \right) \right] \frac{dT_{i0}}{d\Psi} \\
&= \left(\mathbf{q}_i^{cl} + \mathbf{q}_i^{PS} \right) \cdot \nabla\Psi.
\end{aligned}
\tag{7.76}
$$

Here "cl" stands for "classical and "PS"stands for "Pfirsch-Schlüter." We can evaluate the flux surface average of the heat flux given here. If we assume that $B = B_0/(1 + \epsilon \cos\theta)$ (the result for a large aspect ratio, circular bore tokamak), then the average (to lowest order in ϵ)

$$\left\langle 1 - \frac{B^2}{\langle B^2 \rangle} \right\rangle \approx \frac{2\epsilon^2}{B_0^2} \tag{7.77}$$

which gives

$$\langle \mathbf{q}_i^{PS} \cdot \nabla \Psi \rangle \approx -\frac{2\left(\kappa_\wedge^i r B\right)^2}{\kappa_\parallel^i} \frac{dT_{i0}}{d\Psi}. \tag{7.78}$$

For ions with $Z = 1$, Braginskii's value for $\kappa_\wedge^2/(\kappa_\parallel \kappa_\perp) \approx 0.8$ so that we can write, using the relation that $q = (r/R)B_0/B_\theta$:

$$\langle q_{ir} \rangle = \frac{\langle \mathbf{q}_i \cdot \nabla \Psi \rangle}{|\nabla \Psi|} = -\kappa_\perp^i \left(1 + 1.6q^2\right) \frac{dT_i}{dr}. \tag{7.79}$$

Since the safety factor is ≈ 3 in the outer parts of the plasma, we conclude that the enhancement of transport due to the toroidal effects is a factor of $10 \rightarrow 20$ over the classical (cylindrical approximation) value. This factor remains in place in the other transport regimes, and other trapped-particle effects come into play as well, generally degrading confinement further.

7.4.3 Plateau Transport

For the regime with collisionality just under the Pfirsch-Schlüter regime, with $1 < \nu^* < \epsilon^{-3/2}$, we have a situation where the passing orbits can be treated as essentially collisionless, but where the trapped particles suffer collisions with a rate above the characteristic bounce frequency. In this case the passing particles dominate the transport process, but the transport rate must be calculated from the drift-kinetic equation rather than Braginskii's equations. We can write the drift-kinetic equation as

$$\dot{\theta}\frac{\partial f_1}{\partial \theta} + \dot{v}_\parallel \frac{\partial f_1}{\partial v_\parallel} + \dot{v}\frac{\partial f_1}{\partial v} + \mathbf{v}_d \cdot \nabla f_0 = C\left(f_1\right). \tag{7.80}$$

The two velocity variables $v_\parallel$ and v are slowly varying along the particle trajectories:

$$\dot{v}_\parallel = -\frac{\mu \nabla_\parallel B}{m} = -v_\perp^2 \frac{\epsilon \sin \theta}{2qR}$$

$$\dot{v} = \frac{qv_\parallel E_\parallel}{mv}. \tag{7.81}$$

Here f_0 is taken to be a maxwellian

$$f_0 = \left(\frac{m}{2\pi T}\right)^{3/2} \exp\left(-\frac{E - q\Phi}{T}\right).$$

We replace the term $\mathbf{v}_d \cdot \nabla f_0$ with $v_d(\partial f_0/\partial r) \sin\theta$, reflecting the primarily z-directed drift direction, and we use the notation $\xi = v_\parallel/v$ as the cosine of the pitch angle. Then Eq. (7.80) becomes

$$\frac{v}{qR}\left(\xi\frac{\partial f_1}{\partial\theta} - \epsilon\sin\theta\frac{1-\xi^2}{2}\frac{\partial f_1}{\partial\xi}\right) - C\left(f_1\right) = -v_d\frac{\partial f_0}{\partial r}\sin\theta. \tag{7.82}$$

We now model the perturbed distribution function f_1 as a drifting maxwellian with a component that is localized around nearly perpendicular velocities $\xi \ll 1$. representing the second term as a perturbation h, we have

$$f_1 = \frac{mv_\parallel u_\parallel}{T}f_0 + h. \tag{7.83}$$

We then look at the collision term $C(f_1) \approx C(h)$. Using the collision operator in the Lorentz form

$$C\left(f_1\right) \approx v\mathscr{L}f_1 = v\left[\frac{1}{2}\frac{\partial}{\partial\xi}\left(1-\xi^2\right)\frac{\partial}{\partial\xi}\right]f_1.$$

For $\xi \ll 1$, we can simplify $\mathscr{L} \approx (1/2)\partial^2/\partial\xi^2$, and then the drift-kinetic equation becomes

$$\xi\frac{\partial h}{\partial\theta} - \frac{\epsilon\sin\theta}{2}\frac{\partial h}{\partial\xi} - \frac{vqR}{2v}\frac{\partial^2 h}{\partial\xi^2} = s\sin\theta, \tag{7.84}$$

where the source term s is given by

$$s = -\frac{v_d qR}{v}\frac{\partial f_0}{\partial r} + \frac{mvu_\parallel\epsilon}{2T}f_0.$$

We now introduce a normalized pitch angle variable $\eta = \xi\hat{v}^{1/3}$, with $\hat{v}$ being the normalized collision frequency $\hat{v} = v/(2v/(qR))$. Then Eq. (7.84) becomes

$$\eta\frac{\partial h}{\partial\theta} - \frac{\epsilon\sin\theta}{\hat{v}^{2/3}}\frac{\partial h}{\partial\eta} - \frac{\partial^2 h}{\partial\eta^2} = \frac{s}{\hat{v}^{1/3}}\sin\theta \tag{7.85}$$

The second term on the left-hand side of Eq. (7.85) can be ignored for cases where $\epsilon^{3/2} \ll \hat{v}$, and then this resultant equation is similar to an Airy equation. The solution of this equation is

$$h = \frac{s}{\hat{v}^{1/3}}\int_0^\infty \exp{-\tau^3/3}\sin(\theta - \eta\tau)d\tau. \tag{7.86}$$

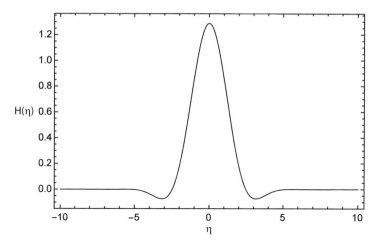

Fig. 7.3 The function $H(\eta)$ for neoclassical plateau mode

Working only with the part of this solution which is even in η gives

$$h^{even} = \frac{s\sin\theta}{\hat{v}^{1/3}} \int_0^\infty \exp(-\tau^3/3)\cos(\eta\tau)d\tau = \frac{s\sin\theta}{\hat{v}^{1/3}} H(\eta), \qquad (7.87)$$

where $H(\eta)$ can be written in terms of Kelvin functions and hypergeometric functions as

$$H(\eta) = \frac{\begin{array}{c} 4\sqrt{3}\pi\sqrt[4]{\eta^2}\left(\mathrm{ber}_{-\frac{1}{3}}\left(\frac{2}{3}\left(\eta^2\right)^{3/4}\right) - \mathrm{ber}_{\frac{1}{3}}\left(\frac{2}{3}\left(\eta^2\right)^{3/4}\right)\right. \\[2mm] \left. -\mathrm{bei}_{-\frac{1}{3}}\left(\frac{2}{3}\left(\eta^2\right)^{3/4}\right) + \mathrm{bei}_{\frac{1}{3}}\left(\frac{2}{3}\left(\eta^2\right)^{3/4}\right)\right) \end{array}}{18\sqrt{2}} \\[4mm] -\frac{1}{2}\eta^2 {}_1F_4\left(1; \frac{2}{3}, \frac{5}{6}, \frac{7}{6}, \frac{4}{3}; -\frac{\eta^6}{1296}\right). \qquad (7.88)$$

A plot of this function is shown as Fig. 7.3. Note that

$$\int_{-\infty}^\infty H(\eta)d\eta = \pi.$$

Returning to the original pitch-angle variable ξ then gives the function $h(\xi)$ approaching a delta function as $\hat{v} \to 0$:

$$\lim_{\hat{v}\to 0} h^{even}(\xi) = \pi s\delta(\xi)\sin\theta. \qquad (7.89)$$

The ion heat flux is then given by

$$\langle \mathbf{q} \cdot \nabla r \rangle = \left\langle \int f_1 \left(\frac{mv^2}{2} - \frac{5}{2} \right) v_d \sin \theta d^3 v \right\rangle = -\frac{v_{th}^3}{q R \Omega_{ci}^2} n_i \frac{3\pi^{1/2} q^2}{4} \frac{dT_i}{dr}.$$

(7.90)

The remarkable part of Eq. (7.90) is that the ion heat conduction does not explicitly depend on the collision frequency in the plateau regime. The thermal diffusivity, which can always be written in the form $\chi \approx (\text{stepsize})^2 \times$ collision frequency, takes a stepsize of an ion gyroradius and the ion bounce frequency $\hat{\omega}_i$, with a Pfirsch-Schlüter-like multiplier on the order of the safety factor squared. Thus Eq. (7.90) can be written as

$$\langle \mathbf{q} \cdot \nabla r \rangle = \hat{\omega} \rho_i^2 n_i \frac{3\pi^{1/2} q^2}{4} \frac{dT_i}{dr}.$$

(7.91)

The electron heat flux derivation falls along similar lines. It is not discussed here because it is not seen in the experiments. Electron heat conduction in tokamaks is anomalously large compared to the neoclassical estimates.

7.4.4 Banana Regime

When the effective collision frequency for the trapped particles $\nu_{eff} = \nu/\epsilon$ is much less than the bounce frequency $\omega_b \approx \epsilon^{1/2} v_{th}/(qR)$, then the trapped particles execute many orbits in the plasma before having a collision. This causes the velocity space effects for the trapped particles to become important. The drift kinetic equation, including the parallel electric field caused by induction $E_\parallel^{(A)} = -\partial A_\parallel/\partial t$, is given by

$$v_\parallel \nabla_\parallel f_1 + \mathbf{v}_d \cdot \nabla f_0 - \frac{q_e v_\parallel E_\parallel^{(A)}}{T} f_0 = C(f_1).$$

(7.92)

(Here q_e is the electric charge of the species, to differentiate from the safety factor q.) Note that the expansion $f = f_0 + f_1 + \ldots$ is in powers of the gyroradius-to-length ratio ρ/L, which is taken as small. The drift term can be written as

$$\mathbf{v}_d \cdot \nabla f_0 = F(\Psi) v_\parallel \nabla_\parallel \left(\frac{v_\parallel}{\Omega} \right) \frac{\partial f_0}{\partial \Psi}.$$

(7.93)

Since we are going to focus on the ion transport, we note that the perturbation to the electrons and ions caused by the parallel electric field is the so-called "Spitzer problem" and was treated by Spitzer and Härm in 1953 [24]. The perturbed distribution functions resulting are the solution to

$$C_\alpha (f_{\alpha S}) = -\frac{q_{e\alpha} v_\parallel E_\parallel^{(A)}}{T_\alpha} f_{\alpha 0}. \tag{7.94}$$

This perturbation to the distribution function is not important to the discussion of ion heat transport and the $E_\parallel$ term will be dropped. We then write down a simpler version of Eq. (7.93) as

$$v_\parallel \nabla_\parallel (f_1 - F) = C (f_1) \tag{7.95}$$

with

$$F_i \equiv -\frac{F(\Psi) v_\parallel}{\Omega_i} \frac{\partial f_{i0}}{\partial \Psi} = -\left[\frac{d \ln n_i}{d \Psi} + \frac{q_i}{T_i} \frac{d \Phi}{d \Psi} + \left(\frac{mv^2}{2T_i} - \frac{3}{2} \right) \frac{d \ln T_i}{d \Psi} \right] f_{i0}. \tag{7.96}$$

Here Φ is the electric scalar potential, $F(\Psi) = RB_\phi$ from the Grad-Shafranov equation, and $\Omega_i = q_i B / m_i$. We then take the perturbed distribution function f_1 and part it out into two terms:

$$f_{i1} = g_i + F_i$$

and then g_i is taken as a constant along the poloidal angle θ. Then g_i contains all of the information concerning the velocity-space perturbations due to trapped particles. Taking a flux surface average of Eq. (7.95) then shows that since g_i is essentially a constant of integration, we must have $g_i = 0$ for the trapped particles. It remains to evaluate g_i for the passing particles.

We can also simplify the full Fokker-Planck operator C with a simpler version which includes a simpler pitch-angle scattering term called the Lorentz term and a simplified first-order term representing the drag forces. This is called the "model collision operator." The Lorentz scattering term

$$\mathscr{L} = \frac{2hv_\parallel}{v^2} \frac{\partial}{\partial \lambda} \lambda v_\parallel \frac{\partial}{\partial \lambda}, \tag{7.97}$$

with $\lambda = (v_\perp / v)^2 (B_0 / B)$ representing a pitch angle variable and $h = [1 + (r/R) \cos \theta]^{-1} = B_0 / B$, with $B_0^2 \equiv \langle B^2 \rangle$. The friction term C_D for the ions in the model collision operator is given as a function of $\mathbf{u}$, which is a collision-weighted average velocity.

$$C_D^{ii} = v_D^{ii}(v) \frac{m_i \mathbf{v} \cdot \mathbf{u}}{T_i} f_{i0} \tag{7.98}$$

with the collision frequency v_D^{ii} given by

$$v_D^{ii}(v) = \left(\frac{n_i q_i^4 \ln \Lambda}{4\pi \epsilon_0^2 m_i^2 v_{thi}^3} \right) \frac{\phi(x) - G(x)}{x^3} \tag{7.99}$$

with

$$\phi(x) = \mathrm{erf}(x) = \frac{2}{\sqrt{\pi}} \int_0^x \exp(-y^2)\, dy, \quad G(x) = \frac{\phi(x) - x\phi'(x)}{2x^2}, \quad x = \frac{v}{v_{thi}}$$

and $v_{thi} = (2T_i/m_i)^{1/2}$. We can then assemble the model collision operator as

$$C_{ii}(f_i) = v_D^{ii}(v)\left(\mathscr{L}(f_i) + \frac{m_i \mathbf{v}\cdot\mathbf{u}}{T_i} f_{i0}\right). \tag{7.100}$$

Keeping only the component of $\mathbf{v}\cdot\mathbf{u}$ parallel to the magnetic field gives $\mathbf{v}\cdot\mathbf{u} \approx v_\parallel u$. After taking flux-surface averages, a relatively simple equation for g_i emerges:

$$\frac{\partial}{\partial\lambda}\lambda\langle v_\parallel\rangle\frac{\partial g_i}{\partial\lambda} = -\frac{v^2}{2}s_i(v,\Psi)f_{i0}, \text{ with}$$

$$s_i(v,\Psi) \equiv \frac{F(\Psi)}{\hat{\Omega}_i}\frac{\partial\ln f_{i0}}{\partial\Psi} + \left\langle\frac{u_i}{h}\right\rangle\frac{m_i}{T_i}. \tag{7.101}$$

We solve this equation with the assumption that $g_i \to 0$ at the bounday between the trapped and passing particles in velocity space, i.e. where $\lambda = \lambda_c = B_0/B_{max}$. Recalling that the independent variables for g_i are v, Ψ, and λ, this equation is an ordinary differential equation in λ for g_i, and then

$$g_i(\lambda, v, \Psi) = H(\lambda_c - \lambda)V_\parallel(\lambda, v, \Psi)s_i(v, \Psi)f_{i0} \tag{7.102}$$

where

$$V_\parallel(\lambda, v, \Psi)s_i(v, \Psi) = \frac{v^2}{2}\int_\lambda^{\lambda_c}\frac{d\lambda'}{\langle v_\parallel(\lambda')\rangle} = \frac{\sigma v}{2}\int_\lambda^{\lambda_c}\frac{d\lambda'}{\langle\sqrt{1 - \lambda'/h(\theta)}\rangle} \tag{7.103}$$

with $H(x)$ being the Heavyside step function and s being the sign function of $v_\parallel$, i.e. $s = 1$ for $v_\parallel > 0$ and $s = -1$ for $v_\parallel < 0$. $V_\parallel$ is defined in such a way that as the inverse aspect ratio becomes small, $V_\parallel \to v_\parallel$.

Now using $f_{i1} = g_i + F_i$ with the above result gives an expression for the perturbed ion distribution function:

$$f_{i1} = -\frac{F(\Psi)}{\hat{\Omega}_i}\left(hv_\parallel - HV_\parallel\right)\frac{\partial f_{i0}}{\partial\Psi} + \frac{m_i HV_\parallel}{T_i}\left\langle\frac{u_i}{h}\right\rangle f_{i0} \tag{7.104}$$

Equation (7.104) now reveals the nature of why it is the trapped particle population, rather than the passing particle population, that dominates the transport: the term $hv_\parallel - HV_\parallel$ is small for the passing particles (where $H = 1$ and $h \approx 1$ for a large aspect ratio), so that the trapped particles contribute mostly to this term. The evaluation of the quantity $< u_i/h >$ remains to be done. The flow velocity u_i in the model collision operator must be chosen to conserve momentum, and this results in a somewhat subtle form for the effective mass flow velocity **u**, which is a collision-weighted average:

$$\mathbf{u}_i = \int \mathbf{v} v_D^{ii}(v) f_i d^3 v \Big/ \int v_D^{ii} \frac{m_i v^2}{3T_i} f_i d^3 v \qquad (7.105)$$

which then gives

$$\{v_d^{ii}\}\left\langle \frac{u_i}{h} \right\rangle = \left\langle \frac{1}{hn_i} \int v_d^{ii} v_\| f_{i1} d^3 v \right\rangle \qquad (7.106)$$

with the velocity-space average defined by

$$\{A(v)\} = \int A \frac{m v_\|^2}{nT} f_0 d^3 v$$

and then, performing the velocity-space integration and using the results of Eq. (7.104) gives

$$f_{i1} = -\frac{F(\Psi)}{\Omega_i} \frac{\partial f_{i0}}{\partial \Psi} + \frac{F(\Psi) H(\lambda_c - \lambda)}{\hat{\Omega}_i} \left(\frac{m v^2}{2T_i} - 1.33 \right) \frac{\partial \ln T_i}{\partial \Psi} f_{i0}. \qquad (7.107)$$

We then can then obtain the expression for the ion heat flux. Using

$$\langle \mathbf{q}_i \cdot \nabla \Psi \rangle = -\frac{F(\Psi)}{q_i B_0} \left\langle h \int \frac{m v^2}{2} m_i v_\| v_D^{ii} \left(-f_{i1} + \frac{m_i v_\| u_i}{T_i} f_{i0} \right) d^3 v \right\rangle \qquad (7.108)$$

If we evaluate this in the limit of large aspect ratio, $\epsilon \to 0$, we find

$$\langle \mathbf{q}_i \cdot \nabla \Psi \rangle = -0.92 f_t \frac{n_i F(\Phi)^2 T_i}{m_i \Omega_{i\theta}^2 \tau_i} \frac{dT_i}{d\Psi}. \qquad (7.109)$$

Here the collision frequency has been converted into the Braginskii collision time $\tau_i = \tau_{ii}\sqrt{2}$. We then use the trapped particle fraction f_t for a large aspect ratio, circular cross section tokamak, $f_t = 1.46\sqrt{\epsilon}$, and noting that $d\Psi = RBdr$ gives an expression for the average ion heat flux:

$$\frac{dq_i}{dr} = -1.35\epsilon^{1/2} \frac{nT_i}{m\Omega_{i\theta}^2 \tau_i} \frac{dT_i}{dr} \qquad (7.110)$$

Note that to show a parallel to the Pfirsch-Schlüter heat conduction, we can write $\Omega_{i\theta} = \epsilon \Omega_i q_s$, where q_s is the safety factor $\epsilon B / B_\theta$. Then the expression becomes

$$\frac{dq_i}{dr} = -1.35 q^2 \epsilon^{-3/2} \frac{nT_i \tau_i}{m(\Omega_i \tau_i)^2} \frac{dT_i}{dr}, \qquad (7.111)$$

which shows that the neoclassical ion heat flux has increased by a factor of $\approx \epsilon^{-3/2}$ from the Pfirsch-Schlüter value. Since ϵ ranges from $\approx 0.1 \rightarrow 1/3$ in the regions with strong temperature gradients, this is a fairly substantial increase.

As the aspect ratio decreases, the approximation that $\epsilon = r/R$ is a small number is no longer accurate. At unity aspect ratio, the trapped particle fraction becomes one, and a more careful transport analysis must be done. Reference [8] gives details and an exact result for $\epsilon = 1$. The result is

$$\langle \mathbf{q}_i \cdot \nabla \Psi \rangle = -\frac{2 n_i m_i T_i F(\Psi)^2}{q_i^2 \tau_i} \left\langle B^{-2} \right\rangle \frac{dT_i}{d\Psi}. \tag{7.112}$$

7.5 Turbulence and Transport

Estimates of transport of heat by electrons in toroidal fusion devices using neoclassical transport do not match experimental experience. The predicted neoclassical electron heat conduction is roughly $(m_e/m_i)^{1/2} \approx 1/60$ times the neoclassical ion heat conduction, whereas the measured electron heat conduction is typically ten times larger than the ion heat conduction. Various theories have been put forward to explain the measured rates, but they have not been rigorously demonstrated to be solid enough to predict the performance of next-step fusion devices. An attempt to outline the various modes of thought is given here. The review articles by Wootton et al. [25] and by Liewer [14] are also excellent compendia of directions in turbulent plasma transport theory.

The general formulation of transport due to fluctuations is given in [25]. Denoting j for the species number (and not including impurities) gives some general results for the transport of particles and energy from perturbed electric and magnetic fields $\tilde{E}_\theta$ and $\tilde{b}_r$ and perturbed densities and temperatures $\tilde{n}_j$ and $\tilde{T}_j$ for each species. We can write the contribution to particle transport from the $\tilde{E}$-like and $\tilde{b}$-like fluctuation terms as:

$$\Gamma_j^f = \Gamma_j^{f,E} + \Gamma_j^{f,b} \tag{7.113}$$

with

$$\Gamma^{f,E} = < \tilde{E}\tilde{n}_j > /B_\phi \tag{7.114}$$

and

$$\Gamma_j^{f,b} = -\frac{< \tilde{j}_\parallel \tilde{b}_r >}{q_e B_\phi} = g_1\left(\frac{\tilde{b}_r}{B_\phi}\right). \tag{7.115}$$

Similarly we can write the fluctuation-induced heat flux terms as

$$Q_j^f = Q_j^{f,E} + Q_j^{f,b} \tag{7.116}$$

with

$$Q^{f,E} = \frac{3}{2} \frac{n_j < \tilde{E}_\theta \tilde{T}_j >}{B_\phi} + \frac{3}{2} \frac{T_j < \tilde{E}_\theta \tilde{n}_j >}{B_\phi} \tag{7.117}$$

and

$$Q_j^{f,b} = g_{2j} \left(\frac{\tilde{b}_r}{B_\phi} \right) \nabla T_j. \tag{7.118}$$

Here g_1 and g_2 are functions which can be arrived at from a theoretical side as well as an experimental side. Experimentally, it is found that contributions to the magnetic fluctuation-driven particle flux, associated with the function(s) g_1 are small and usually ignored. Function g_2 is related to the growth of magnetic stochasticity through magnetic perturbations. A model by Rechester and Rosenbluth [20] gives an estimate of g_2 for the case of weak turbulence when the quasi-linear model holds. This model takes on different forms depending on the collisionality in the plasma, i.e. whether to use the parallel thermal conductivity $\chi_{\parallel}$ or to assume that the electron heat conduction is limited by free streaming. In these two regimes, the function g_{2e} for electrons is given by

$$g_{2e} = \begin{cases} \pi R v_{the} \left(\dfrac{\tilde{b}_r}{B_\phi} \right)^2 & \text{collisionless} \\[4ex] \chi_e^{\parallel} \left(\dfrac{\tilde{b}_r}{B_\phi} \right)^2 & \text{collisional.} \end{cases} \tag{7.119}$$

In the case of strong turbulence, g_{2e} is given by

$$g_{2e} = \delta_\perp v_{the} \left[\left(\frac{\tilde{b}_r}{B_\phi} \right)^2 \right]^{1/2}. \tag{7.120}$$

Here $\delta_\perp$ is a characteristic length and is taken as $1/k_\perp$, where $k_\perp$ is the perpendicular wavenumber for the dominant mode.

Data fitting the strong turbulence model given in Eq. (7.119) was shown in a study on JET [16] where the edge value of $|\tilde{b}|$ multiplied by the edge safety factor $q(a)$ is plotted against the reciprocal of the energy confinement time, as shown in Fig. 7.4. What was not measurable at the time was the structure of

Fig. 7.4 Reciprocal energy
confinement time τ_E^{-1} vs.
edge safety factor-edge
magnetic fluctuation level
$|\tilde{b}|q(a)$ measured on JET for
ohmically heated (OH) and
neutral beam injected (NBI)
plasmas. From [16]. Used by
permission, IAEA

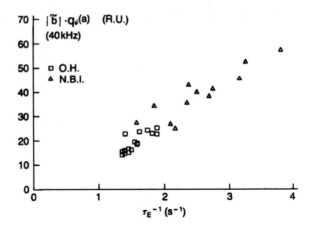

Fig. 7.4 Reciprocal energy confinement time τ_E^{-1} vs. edge safety factor-edge magnetic fluctuation level $|\tilde{b}|q(a)$ measured on JET for ohmically heated (OH) and neutral beam injected (NBI) plasmas. From [16]. Used by permission, IAEA

the perturbed magnetic fields inside the plasma. However, measurements of the
runaway electron fraction, which is known to correlate with the formation of
magnetic islands on resonant surfaces inside the plasma, also show a correlation
with energy confinement time. Newer diagnostics such as "fast MSE" (MSE means
motional Stark effect) have been deployed at D-III-D and other devices and allow
for internal measurement of magnetic activity [12]. These have been used for energy
confinement correlations with MHD features such as neoclassical tearing mode
growth [13].

Other studies have concentrated on the presence of electrostatic modes to explain
the anomalous transport in tokamaks. Plasma models frequently show evidence
of unstable electrostatic mode growth as a result of modes driven by gradients in
the electron and ion temperature and density. The so-called electron temperature
gradient (ETG) instability and ion temperature gradient (ITG) instability are usually
found in computer simulations of confined toroidal plasma using gyrokinetic (GK)
simulations. GK codes simplify the coupled Vlasov and Maxwell equations by using
gyro-phase averages of the ion motion and representing the electron population as
a fluid having density variations due to Boltzmann's law $\tilde{n}_e = e\tilde{\phi}/T_e$. Some codes
allow for high-fidelity simulations including trapped-particle effects. These codes
typically show the presence of trapped particle modes, called TEM modes (not
to be confused with vacuum electromagnetic waves with the same name!), which
typically show a frequency band of instability related to the bounce frequency of
the electrons.

Experimental studies have confirmed the presence of ITG, ETG, and TEM modes
in tokamak plasmas [6, 10]. Figure 7.5 shows a measurement on Tore Supra using
a CO_2 laser scattering system. The consistency of the spectrum as a function of
the dimensionless wavenumber $k\rho_i$ points to a turbulent cascade which supports a
transport scaling law consistent with "gyro-Bohm" transport [17], that is:

$$\chi \propto \left(\frac{\rho_i}{a}\right)\frac{T}{B} \propto \frac{T^{3/2}}{aB^2} \tag{7.121}$$

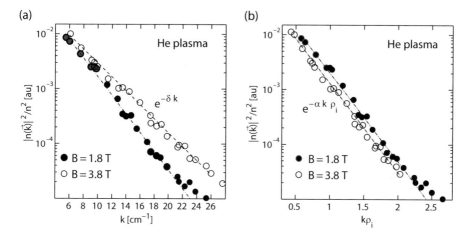

Fig. 7.5 Spectrum of density fluctuations associated with electrostatic turbulence in Tore Supra using a CO_2 laser scattering system. (**a**) shows the change in the k spectrum with magnetic field; (**b**) shows the relatively constant spectrum as a function of $k\rho_i$. From [10]

Fig. 7.6 Evidence of ETG turbulence using microwave scattering at a frequency near the upper hybrid resonance. Sidebands are consistent with one mode in the hot plasma and two modes farther out in minor radius. From [6]

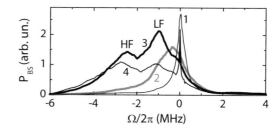

Another interesting study was carried out on the FT-2 tokamak at the Ioffe Institute in Russia [6]. Here a millimeter-wave microwave source with a directional launching horn was aimed at the tokamak plasma and scattering was observed from the sidebands generated. The findings were consistent with an ETG mode in the plasma. Figure 7.6 shows some data from this experiment.

7.6 Empirical Scaling Laws

The wide gamut of theoretical models for heat transport in tokamaks has not given comfort in trusting in the predictive power of these models to extrapolate to reactor-grade burning plasmas. However, a large database of observed confinement times exists, drawing from eleven relatively large devices worldwide. There have been many different empirical models put forward which attempt to fit the observed

energy confinement times with known plasma parameters such as toroidal magnetic
field, average plasma density, major and minor radii, and geometrical factors,
all taken to some power. The resulting "wind-tunnel scaling," referring to the
success (and limitations) of these models for aerodynamic situations, has then been
optimized for the minimum rms error between measured and predicted performance
over the broadest possible range of parameter space. Of paramount importance in
conducting these correlation models is consistency in measurement techniques. For
example, in order to obtain a number for the experimentally observed confinement
time, one must get an accurate measurement of the radiation loss and the rate of
change of the total plasma thermal energy, since these are subtracted out from the
total injected power and alpha heating power. Only data with all of these quantities
rigorously determined are kept in the database. Since plasmas with auxiliary heating
generally see a reduction in energy confinement time, dependence on injected
power is allowed, even though this creates some circular logic, since by definition
$\tau_E = W_{th}/P_L$.

For the 1997 database, the following criteria were used (from [11]): (1) H-mode
data only, with no restriction on heating scheme, (2) all essential data available, (3)
pellet discharges are excluded, (4) limits on dW/dt were in place, (5) limits on total
radiation were in place, (6) limits on q_{95}, the safety factor at 95% flux were in place,
(7) limits on fast ion energy content were used, (8) limits on β were in place, (9) hot
ion H-mode data are excluded, and (10) 1987 JET data are excluded.

A correlation that has withstood the test of time better than some others is the
"ELMy H-mode" correlation [11] $\tau_{th,98y2}$, given by:

$$\tau_{th,98y2} = 0.0562 \, I_p^{0.93} B_t^{0.15} n_{19}^{0.41} P_L^{-0.69} R^{1.97} \epsilon^{0.58} \kappa_a^{0.78} M^{0.19}, \qquad (7.122)$$

with the plasma current I_p in MA, toroidal magnetic field B_t in T, density n_{19} in
units of $10^{19} \, \text{m}^{-3}$, loss power $P_L = P_{heat} - dW/dt - P_{rad}^{core}$ in MW, major radius
R in m, inverse aspect ratio $\epsilon = a/R$, elongation $\kappa_a = V/(2\pi^2 R a^2)$ with plasma
volume V, and average ion mass M in units of proton masses. The quality of this fit
is shown in Fig. 7.7.

More recently, other scaling laws have come about using more data, especially
including the data from lower aspect ratio devices such as NSTX and MAST. Also,
increased faith in gyro-Bohm scaling has resulted in correlations where exact gyro-
Bohm scaling was forced on the model. An example of such a scaling law is given
in [18]:

$$\tau_{th}^{EGB} = 0.028 \, I_p^{0.83} B_T^{0.07} n_{19}^{0.49} P_L^{-0.55} R^{1.81} a^{0.3} \kappa_a^{1.75} M^{0.14} \qquad (7.123)$$

This model is very similar to the $\tau_{th,98y2}$ except for the large change in the exponent
for plasma elongation factor κ, reflecting the high performance of the spherical
tokamaks.

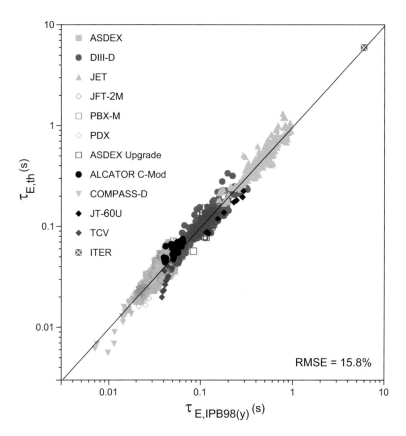

Fig. 7.7 Comparison of experimentally determined energy confinement time with the scaling law $\tau_{th,98y2}$. Note that the point labeled "ITER" is for an earlier, larger design: the current design fits the same line but with a 2.3 s confinement time with a major radius of 6.2 m instead of the original design with $R = 8.14$ m. From [11]

Problems

7.1 A tokamak ($B_\phi = 5.2$ T, $n_i = 10^{20}$ m^{-3}, R=3.1 m, $A = 3$, $q(0.9a) = 3$, 50–50 D-T, circular cross section) has an ion temperature profile given by

$$T_i(r) = T_{i0}\left(1 - \left(\frac{r}{a}\right)^2\right)$$

with $T_{i0} = 15$ keV. Find the total heat conduction through the $r = 0.9a$ surface from perpendicular Braginskii ion heat conduction.

7.2 Do the same for Pfirsch-Schlüter heat conduction.

7.3 Do the same assuming that the ions are in the banana regime. Justify whether or not this assumption is valid by a calculation of ν^*.

7.4 Suppose that a tokamak with the same geometrical factors as ITER was to be built with the same magnetic field, safety factor, density, temperature, Z_{eff}, and β_N. Find the major radius R which, using the $\tau_{th,98y2}$ scaling law, would yield a device which could achieve ignition.

References

1. Abramowitz, M.: Handbook of Mathematical Functions, With Formulas, Graphs, and Mathematical Tables. Dover Publications, New York (1974)
2. Braginskii, S.I.: Transport phenomena in a completely ionized two-temperature plasma. Zh. Eksp. Teor. Fiz. **33**, 645 (1957). English translation: Sov. Phys. JETP **6**, 494 (1958)
3. Braginskii, S.I.: Transport in a plasma. In: Leontovich, M.A. (ed.) Reviews of Plasma Physics, vol. 1, pp. 205–311. Consultants Bureau, New York (1965)
4. Chapman, S., Cowling, T.: The Mathematical Theory of Non-uniform Gases: An Account of the Kinetic Theory of Viscosity, Thermal Conduction, and Diffusion in Gases. University Press, Cambridge (1953). https://books.google.com/books?id=22XNjwEACAAJ
5. Galeev, A.A., Sagdeev, R.Z.: Transport phenomena in a collisionless plasma in a toroidal magnetic system. Sov. Phys. JETP **26**(1), 233 (1968)
6. Gurchenko, A., Gusakov, E., Altukhov, A., Stepanov, A., Esipov, L., Kantor, M., Kouprienko, D., Dyachenko, V., Lashkul, S.: Observation of the ETG mode component of tokamak plasma turbulence by the uhr backscattering diagnostics. Nucl. Fusion **47**(4), 245 (2007). http://stacks.iop.org/0029-5515/47/i=4/a=001
7. Hazeltine, R.D.: Recursive derivation of drift-kinetic equation. Plasma Phys. **15**(1), 77 (1973). http://stacks.iop.org/0032-1028/15/i=1/a=009
8. Hazeltine, R.D., Hinton, F.L., Rosenbluth, M.N.: Plasma transport in a torus of arbitrary aspect ratio. Phys. Fluids **16**(10), 1645–1653 (1973). http://aip.scitation.org/doi/abs/10.1063/1.1694191
9. Held, E.D., Callen, J.D., Hegna, C.C.: Conductive electron heat flow along an inhomogeneous magnetic field. Phys. Plasmas **10**(10), 3933–3938 (2003). http://dx.doi.org/10.1063/1.1611883
10. Hennequin, P., Sabot, R., Honoré, C., Hoang, G.T., Garbet, X., Truc, A., Fenzi, C., Quéméneur, A.: Scaling laws of density fluctuations at high-k on tore supra. Plasma Phys. Controll. Fusion **46**(12B), B121 (2004). http://stacks.iop.org/0741-3335/46/i=12B/a=011
11. ITER Physics Expert Group on Confinement and Transport, ITER Physics Expert Group on Confinement Modelling and Database and ITER Physics Basis Editors: Chapter 2: Plasma confinement and transport. Nucl. Fusion **39**(12), 2175 (1999). http://stacks.iop.org/0029-5515/39/i=12/a=302
12. King, J.D., Makowski, M.A., Allen, S.L., Holcomb, C.T., Geer, R., Ellis, R., Morse, E.C.: A fast MSE measurement of MHD magnetic fluctuations on DIII-D. In: APS Meeting Abstracts (2009)
13. King, J.D., Haye, R.J.L., Petty, C.C., Osborne, T.H., Lasnier, C.J., Groebner, R.J., Volpe, F.A., Lanctot, M.J., Makowski, M.A., Holcomb, C.T., Solomon, W.M., Allen, S.L., Luce, T.C., Austin, M.E., Meyer, W.H., Morse, E.C.: Hybrid-like 2/1 flux-pumping and magnetic island evolution due to edge localized mode-neoclassical tearing mode coupling in DIII-D. Phys. Plasmas **19**(2), 022503 (2012). http://dx.doi.org/10.1063/1.3684648
14. Liewer, P.C.: Measurements of microturbulence in tokamaks and comparisons with theories of turbulence and anomalous transport. Nucl. Fusion **25**(5), 543 (1985). http://stacks.iop.org/0029-5515/25/i=5/a=004

15. Lin-Liu, Y.R., Miller, R.L.: Upper and lower bounds of the effective trapped particle fraction in general tokamak equilibria. Phys. Plasmas **2**(5), 1666–1668 (1995). http://dx.doi.org/10.1063/1.871315
16. Malacarne, M., Duperrex, P.: Turbulent fluctuations and confinement in JET. Nucl. Fusion **27**(12), 2113 (1987). http://stacks.iop.org/0029-5515/27/i=12/a=011
17. Petty, C., Luce, T.: Projections of gyroradius scaling experiments to an ignition tokamak. Nucl. Fusion **37**(1), 1 (1997). http://stacks.iop.org/0029-5515/37/i=1/a=I01
18. Petty, C.C., DeBoo, J.C., Haye, R.J.L., Luce, T.C., Politzer, P.A., Wong, C.P.C.: Feasibility study of a compact ignition tokamak based upon gyrobohm scaling physics. Fusion Sci. Technol. **43**(1), 1–17 (2003)
19. Pfirsch, D., Schlüter, A.: Report of the Max Planck Institute. Max-Planck-Institut Report, MPP/PA/7/62 (1962)
20. Rechester, A.B., Rosenbluth, M.N.: Electron heat transport in a tokamak with destroyed magnetic surfaces. Phys. Rev. Lett. **40**, 38–41 (1978). http://link.aps.org/doi/10.1103/PhysRevLett.40.38
21. Rosenbluth, M.N., Hazeltine, R.D., Hinton, F.L.: Plasma transport in toroidal confinement systems. Phys. Fluids **15**(1), 116–140 (1972). http://aip.scitation.org/doi/abs/10.1063/1.1693728
22. Schnack, D.: Ordered fluid equations. Unpublished notes (2003). http://w3.pppl.gov/cemm/Project/schnack2fr2.pdf
23. Schnack, D.: Lectures in Magnetohydrodynamics: With an Appendix on Extended MHD. Lecture Notes in Physics. Springer, Berlin (2009). https://books.google.com/books?id=Ebon7NTbL0EC
24. Spitzer, L., Härm, R.: Transport phenomena in a completely ionized gas. Phys. Rev. **89**, 977–981 (1953). http://link.aps.org/doi/10.1103/PhysRev.89.977
25. Wootton, A.J., Carreras, B.A., Matsumoto, H., McGuire, K., Peebles, W.A., Ritz, C.P., Terry, P.W., Zweben, S.J.: Fluctuations and anomalous transport in tokamaks. Phys. Fluids B: Plasma Phys. **2**(12), 2879–2903 (1990). http://dx.doi.org/10.1063/1.859358

Chapter 8
Stellarators

8.1 Introduction and History

If we start the clock for experimental studies of fusion plasma with Kantrowitz in 1938 at Langley (see Chap. 1) [10], then we also start with one of the earliest theoretical observations regarding magnetically confined plasma: a purely toroidal magnetic field in a symmetrical geometry has no equilibrium. (One can see this from the Grad-Shafranov equation described in Chap. 5 where the basic coordinate is the poloidal flux Ψ, which is zero everywhere in a purely toroidal geometry.) An early explanation by Enrico Fermi, relying on a drift model for the plasma instead of a fluid model, gives a similar result: the plasma has an unchecked radial force (in the cylindrical $\hat{R}$ direction), which will cause it to expand in this direction. The solution for this in a long line of experiments with only ohmic heating such as ZETA in England and the early tokamaks such as T-3 in the former USSR was to add a toroidal current to the plasma, keeping the plasma equilibrium symmetrical, but inherently generating a poloidal magnetic field, which also requires the use of supplemental poloidal fields from external coils to maintain equilibrium. The modern tokamaks are based upon this approach, which has the drawback that if this internal toroidal current is generated inductively, then the discharge is inherently non-steady state as there is no such thing as a "DC transformer." Long-pulse devices today rely on radiofrequency-driven current drive or neutral beam injection to lengthen the pulse time, but these add complexity and expense to the concept.

Another way to think about the existence of a poloidal field in a symmetrical system such as a tokamak is that it represents a "rotational transform," i.e. a magnetic line follows an orbit around the poloidal direction as it advances toroidally, and thus the magnetic lines experience the high-field side towards the geometric axis and the low-field side at some maximum radial extent. An alternative way to obtain magnetic fields with a rotational transform is to create a magnetic field with a twisted topology in a non-symmetrical, fully three-dimensional equilibrium. This was the original vision of Lyman Spitzer at Princeton. The original stellarator built

© Springer Nature Switzerland AG 2018
E. Morse, *Nuclear Fusion*, Graduate Texts in Physics,
https://doi.org/10.1007/978-3-319-98171-0_8

Fig. 8.1 Figure-eight
stellarator configuration

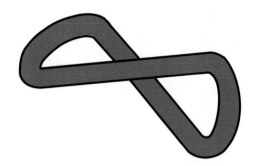

Fig. 8.2 Lyman Spitzer next
to the Model A stellarator,
built in 1953. The photo was
taken in 1983, when the
machine was transported to
the Smithsonian. Princeton
Plasma Physics Laboratory

at Princeton achieved this rotational transform by twisting the torus into a figure-eight type of shape as shown in Fig. 8.1. The actual device constructed at Princeton in 1953 is shown in Fig. 8.2.

Following the figure-eight stellarator at Princeton, a number of alternative methods of making a toroidal plasma free of net toroidal current with a rotational transform have emerged. Three of these are shown in Fig. 8.3. In the first of these, Fig. 8.3a, the currents in adjacent helical windings flow in opposite directions and the currents in the TF coils are all the same. This configuration was used for the Wendelstein 7-A machine built at IPP Garching in Germany, as well as some earlier experiments. (Historical note: the original name for the classified fusion program in the US was Project Matterhorn, in homage to Lyman Spitzer's love of mountain climbing. Wendelstein is the name of a large mountain peak about 70 km from Garching.) Figure 8.3b shows a torsatron/heliotron design. Here the helical currents flow in the same direction, and a set of poloidal field coils typically provide "return currents"—that is, they carry current in the opposite direction to the toroidal component of the current in the helical coils. The Large Helical Device, or LHD, located in Toki, Japan, uses this configuration. This machine uses superconducting magnets. A view of the mechanical components is shown in Fig. 8.4. A third stellarator concept is shown in Fig. 8.3c. This design, called the

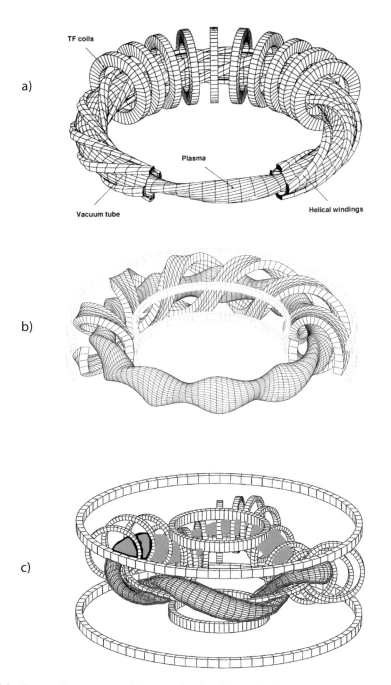

Fig. 8.3 Three stellarator types: (**a**) conventional stellarator similar to Wendelstein 7A, (**b**) a torsatron/heliotron similar to the LHD machine , and (**c**) a heliac design. From [20]. Used with permission, Springer

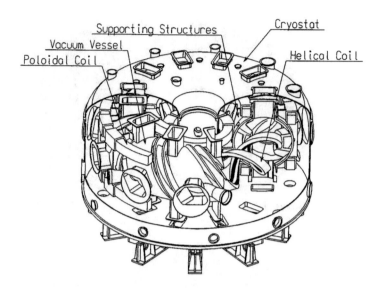

Fig. 8.4 Heliotron/torsatron LHD device. From [6]. Used with permission, Springer

heliac configuration, has a central ring coil that the plasma wraps around. A device of this kind called TJ-II is operating in Madrid, Spain. While the reactor possibilities for this design are limited because of the difficulty of shielding the central conductor from neutrons, the machine has a large gamut of operating conditions which have been useful for physics studies.

All of the previously mentioned stellarator configurations involve some continuously wound coils–that is, some coils going around the machine toroidally. The problem is especially seen in the LHD machine, where the superconducting helical coils had to be wound in place. (This does have the advantage, however, that there is no splicing together of current leads in the coil, eliminating the risk of ohmic connections.) Another type of stellarator design has come about where the coils are discrete and thus modular. In this case rotational transform is provided by designing the coils with a non-planar shape, giving them a sort of "wobbly" look. This concept is sometimes called a "helias" configuration. The Wendelstein 7-AS machine tested the concept on a medium scale at Garching, and recently the Wendelstein 7-X machine has come online in Greifswald, Germany. A diagram showing the fifty superconducting nonplanar coils is shown as Fig. 8.5. In normal operation these are the only coils in use; however, the actual machine has an additional twenty planar coils for use in some advanced experiments. The machine took over a decade to build at a cost of about 1.06 billion euro. It made its first plasma in December 2015.

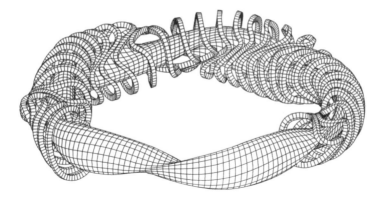

Fig. 8.5 Coil set and plasma for W7-X stellarator. (A set of twenty additional planar coils, used for advanced modes of operation, are not shown.) From [2]. Used with permission. Copyright 1992 American Nuclear Society, Lagrange Park, Illinois USA

8.2 MHD Equilibrium

A general non-axisymmetric equilibrium for stellarators requires a trip back to first principles. We start by observing that the total energy in the plasma is given by (using units where $\mu_0 = 1$)[1]:

$$W_P = \int dV \left[\frac{B^2}{2} + \frac{p}{\gamma - 1}, \right] \tag{8.1}$$

and the energy in a vacuum region surrounding the plasma is given by

$$W_V = \int dV \frac{B^2}{2} e. \tag{8.2}$$

A stable equilibrium is obtained when the hamiltonian $H = W_P - W_V$ is stationary, that is, when small perturbations added to the equilibrium increase H.

We also need to construct fields which satisfy $\nabla \cdot B = 0$, and this can be done by giving the magnetic fields in "Clebsch form":

$$B = \nabla s \times \nabla \psi. \tag{8.3}$$

We presume that the magnetic lines lie in surfaces which form nested tori (in the topological sense: these surfaces are not circular). Then we take s to be a coordinate which is related to the enclosed magnetic flux. The other generalized coordinate ψ contains the periodic structure. Then we can write

$$\psi = -u + \mu v + f(s, u, v). \tag{8.4}$$

Here v is proportional to the angle (in cylindrical coordinates) about the $R = 0$ axis and u is a generalized poloidal angle. The function μ is a function of s only and it is the rotational transform, i.e. the number of poloidal orbits that a field line makes through a travel of one orbit toroidally.

In order to generate equilibria for steady-state stellarators, it is usual to prescribe that the total toroidal current $I(s)$ inside a flux surface ($s = $ constant) is zero. This can be thought of as a boundary condition for the energy minimization. Also, the vacuum field in the region between the plasma and the wall must satisfy $\nabla \cdot \mathbf{B} = \nabla \times \mathbf{B} = 0$, and this means that the vacuum field can be written as $\mathbf{B} = \nabla \phi$ for some potential ϕ satisfying Laplace's equation $\nabla^2 \phi = 0$. If there is a tight-fitting conducting wall around the plasma, then the energy is minimized by the usual MHD equilibrium equation $\nabla p = \mathbf{J} \times \mathbf{B}$, with $\mathbf{J} = \nabla \times \mathbf{B}$. If a vacuum region is present around the plasma, then a jump condition is added:

$$\frac{1}{2} B_P^2 + p = \frac{1}{2} B_V^2. \tag{8.5}$$

Determination of the equilibrium is essentially a minimization procedure carried out on the hamiltonian $H = W_P - W_V$ with some prescribed plasma boundary. An early version of this technique is described in the BETA code [1]. The plasma boundary is adjusted to allow for a convergent solution for the flux function ψ given in 8.4. As an example of an optimized stellarator design (similar to W7-X in its "standard configuration"), Ref. [15] gives:

$$R(u, v) = A + R_{0,1} \cos v + (1 - \Delta_{1,0} - \Delta_0 \cos v) \cos u$$
$$+ \Delta_{2,0} \cos(2u) - \Delta_{1,-1} \cos(u - v)$$
$$+ \Delta_{2,-1} \cos(2u - v) + \Delta_{2,-2} \cos(2u - 2v) \tag{8.6}$$
$$Z(u, v) = Z_{0,1} \sin v + (1 + \Delta_{1,0} - \Delta_0 \cos v) \sin u$$
$$+ \Delta_{2,0} \sin(2u) + \Delta_{1,-1} \sin(u - v)$$
$$+ \Delta_{2,-1} \sin(2u - v) - \Delta_{2,-2} \sin(2u - 2v).$$

An example of a Helias-type equilibrium computed with the BETA code is shown in Fig. 8.6. The variational method has been extended to include equilibria with a vacuum region surrounding the plasma, thus necessitating the solution of a vacuum field problem with boundary conditions of both plasma-induced and coil-induced magnetic fields. An example of a rapidly convergent equilibrium solver with this capability is VMEC [12] (Variational Moments Equilibrium Code), similar to BETA described above, with NESTOR (NEumann Solver for TOroidal Regions) which solves the elliptic equation with Neumann boundary conditions for the vacuum region.

Stellarator equilibrium codes have evolved over time. The earliest codes used the assumption that the flux surface are nested tori, i.e. there are no stochastic magnetic lines. This criterion has been relaxed in some newer efforts such as the PIES code

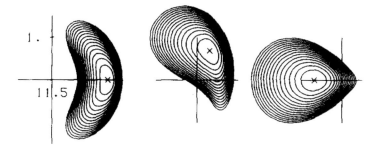

Fig. 8.6 Flux contours (computed using the BETA code) at $v = 0, \ \pi/2,$ and π for the equilibrium with external boundaries given by (8.6) with the parameters given in the text. From [15]. Used with permission, Elsevier

Fig. 8.7 Poincare plots of magnetic fields in the W7-X stellarator. From[9]. Used with permission, IOP Publishing

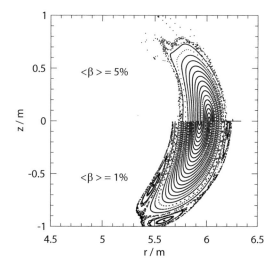

(for "Princeton iterative equilibrium solver") [5], which allows for more general magnetic topology. An example of a model of the W7-X stellarator equilibria with Poincaré plots of the magnetic field using the PIES code is shown in Fig. 8.7.

Figure 8.6 shows a computed flux surface model using the boundary given by Eq. (8.6) with parameters $A = 11.5$, $R_{0,1} = 0.8$, $Z_{0,1} = 0.4$, $\Delta_{1,0} = 0.1$, $\Delta_0 = 0.07$, $\Delta_{2,0} = 0.05$, $\Delta_{1,-1} = 0.29$, $\Delta_{2,-1} = 0.24$, and $\Delta_{2,-2} = 0.07$. The pressure model used was $p(s) = p_0(1-s)$, with p_0 chosen so that the average beta $<\beta>= 0.05$.

8.3 Stability

MHD stability to low-order modes (such as the $m = 2$, $n = 1$ mode) using the computational methods outlined above often manifests itself in the failure to converge, i.e. the initial iterative scheme driven by a reduction in plasma energy seeks an entirely different equilibrium. The inherent three-dimensional character of stellarators makes this possible, since the magnetic perturbations can appear to be adjustments to the equilibrium. Since tokamak equilibria are computed from axisymmetric solutions, this "voluntary" MHD dynamics does not arrive. Some physical insight is required to determine if such an instability is a physical one or a numerical artifact. In practice this can be determined by decreasing the mesh size h (in the s direction) and finding the extrapolated growth rate as $h \to 0$. This can represent either a physical instability or a numerical instability with the code. One can expect the propensity for this behavior to stop for highly localized modes, where the algorithms tend to be diffusive. For higher values of n, m, MHD stability studies of stellarators have started by generating an equilibrium as generated with the variational approach as outlined above and then analyzing these equilibria for stability using the Mercier criterion as outlined in Chap. 6 for tokamaks. In stellarator geometry, the Mercier criterion takes the form [1]:

$$\Omega = \Omega_s + \Omega_w + \Omega_d \geq 0 \text{ stable.} \tag{8.7}$$

Here

$$\frac{\pi^2 \mu^2}{s} \Omega_s = \frac{(\mu')^2}{4} - \mu' \int D \, du \, dv \frac{(\mathbf{J} - I'\mathbf{B}) \cdot \mathbf{B}}{(\nabla s)^2} \tag{8.8}$$

and

$$\frac{\pi^2 \mu^2}{s} \Omega_w = p' \left[V'' - p' \int \frac{D \, du \, dv}{B^2} \right] \int D \, du \, dv \frac{B^2}{(\nabla s)^2} \tag{8.9}$$

and

$$\frac{\pi^2 \mu^2}{s} \Omega_d = \left[\int D \, du \, dv \frac{\mathbf{J} \cdot \mathbf{B}}{(\nabla s)^2} \right]^2 - \int D \, du \, dv \frac{B^2}{(\nabla s)^2} \int D \, du \, dv \frac{(\mathbf{J} \cdot \mathbf{B})^2}{B^2 (\nabla s)^2}. \tag{8.10}$$

Note that the volume element $D \, du \, dv \, ds$ is the incremental toroidal volume element $V'(s)ds$, with D representing a jacobian, and that primes are taken to mean differentiation by the flux surface parameter s. The physical meaning of these terms are as follows. Ω_s is a shear-driven term. The term I' inside the integrand of Eq. (8.7) vanishes for the standard stellarator case where there is no averaged toroidal current enclosed by a flux surface. The term Ω_w contains the effect of a magnetic well V''. The last term Ω_d is related to the energy of formation of magnetic islands in the plasma and is more or less proportional to the island width d. This term is always negative.

An advantage of the Mercier criterion is that it only involves equilibrium quantities, i.e. no variational principle is required. A disadvantage is that the term in $\mathbf{J} \cdot \mathbf{B}$ in Eqs. (8.7) and (8.10) may become numerically inaccurate at places where the shear $\mu = n/m$, i.e. at rational surfaces, because the expansion of $\mathbf{J} \cdot \mathbf{B}/B^2$ results in a resonant denominator $\propto 1/(n - \mu m)$. This can be mitigated to some extent by adding a small flattening of the pressure profile near the resonant surface.

A more direct approach to full 3-D MHD stability problems has appeared with the advances in computational speed. This again involves minimization of an energy functional similar to the equilibrium codes discussed earlier, but now solving for the perturbed energy in the plasma with fluid displacements in the curvilinear coordinate system described for the equilibrium case [17]. We can write down the perturbed energy as:

$$ W_P = \frac{1}{2} \int dV \left[|\mathbf{C}|^2 - \mathscr{A}(\boldsymbol{\xi} \cdot \nabla s)^2 + \gamma p (\nabla \cdot \boldsymbol{\xi})^2 \right]. \tag{8.11} $$

Here

$$ \mathbf{C} = \nabla \times (\boldsymbol{\xi} \times \mathbf{B}) + \frac{\mathbf{j} \times \nabla s}{|\nabla s|^2} \boldsymbol{\xi} \cdot \nabla s. \tag{8.12} $$

The destabilizing term in this equation is the only non-positive-definite term and that is $\mathscr{A}$, which is given by:

$$ \mathscr{A} = 2|\nabla s|^{-4} (\mathbf{j} \times \nabla s) \cdot (\mathbf{B} \cdot \nabla) \nabla s. \tag{8.13} $$

This term represents the destabilization brought on by current in the plasma. The variational problem resulting from the perturbed potential energy given by Eq. (8.11) is given by

$$ \lambda W_k - W_P = \frac{\lambda}{2} \int dV \rho |\xi|^2 - W_P = \text{minimum}. \tag{8.14} $$

The selection of a coordinate system local to the field line structure is important in being able to write down a concise expression for W_P. If we choose

$$ \mathbf{e}^1 = \frac{\nabla s}{|\nabla s|}, \quad \mathbf{e}^2 = \frac{\nabla s \times \mathbf{B}}{|\nabla s|B}, \quad \mathbf{e}^3 = \frac{\mathbf{B}}{B}, $$

we can then eliminate one component of $\boldsymbol{\xi}$ by invoking an incompressibility argument $\nabla \cdot \boldsymbol{\xi} = 0$ (since the operator $\mathbf{B} \cdot \nabla$ is nonsingular) and then writing down the term $\xi_\parallel = \mathbf{e}^3 \cdot \boldsymbol{\xi}$ in terms of two remaining two displacement components. We then have $\mathbf{C} = C^1 \mathbf{e}^1 + C^2 \mathbf{e}^2 + C^3 \mathbf{e}^3$. After some algebra, one obtains:

$$
C^1 = \frac{1}{|\nabla s|} \mathbf{B} \cdot \nabla \xi^s,
$$

$$
C^2 = -\frac{|\nabla s|}{B\sqrt{g}} \left(\sqrt{g} \mathbf{B} \cdot \nabla \eta - \iota' F_T'^2 \xi^s + \frac{\mathbf{j} \cdot \mathbf{B}}{|\nabla s|^2} \sqrt{g} \xi^s \right.
$$

$$
\left. + \frac{\tilde{\sigma} B}{|\nabla s|^2} \sqrt{g} \mathbf{B} \cdot \nabla \xi^s \right),
\tag{8.15}
$$

$$
C^3 = \frac{1}{B\sqrt{g}} \left[I_{tor} \frac{\partial \eta}{\partial \phi} - I_{pol} \frac{\partial \eta}{\partial \theta} + \left(F_T' I_{pol} + F_P' I_{tor} \right) \frac{\partial \xi^s}{\partial s} \right.
$$

$$
\left. + \left(I_{tor} F_P'' + I_{pol} F_T'' \right) \xi^s - p' \sqrt{g} \xi^s + \tilde{\beta} \mathbf{B} \cdot \nabla \xi^s \right].
$$

Here $\sqrt{g}$ is the jacobian for the transformation from cartesian coordinates (x, y, z) and the curvilinear coordinates (s, θ, ϕ), $\eta = \mathbf{e}^1 \cdot (\xi \times \mathbf{B})$, ι (iota) is the winding number (reciprocal safety factor), F_T and F_P are the toroidal and poloidal flux functions, and two auxiliary functions $\tilde{\beta}$ and $\tilde{\sigma}$ are defined by a field-line integration

$$
\sqrt{g} \mathbf{B} \cdot \nabla \tilde{\beta} = - \left(F_T' \frac{\partial}{\partial \phi} + F_P' \frac{\partial}{\partial \theta} \right) \tilde{\beta} = p' \left(\sqrt{g} - V' \right)
\tag{8.16}
$$

with $\int V'(s)ds$ being the volume enclosed within one field period. Then the shear function $\tilde{\sigma}$ is defined as

$$
\tilde{\sigma} = -\frac{1}{F_T' B} \left(B^2 g_{s\theta} - I_{tor} \tilde{\beta} \right).
\tag{8.17}
$$

Finally we need an expression for $\mathscr{A}$, the potentially destabilizing coefficient appearing in the energy equation Eq. (8.11). This is given by

$$
\sqrt{g} \mathscr{A} = F_T'' I_{pol}' + F_P'' I_{tor}' - F_T'' \frac{\partial \tilde{\beta}}{\partial \phi} - F_P'' \frac{\partial \tilde{\beta}}{\partial \theta}
$$

$$
+ \frac{|\mathbf{j}|^2 \sqrt{g}}{|\nabla s|^2} - p' \frac{\partial \sqrt{g}}{\partial s}
\tag{8.18}
$$

$$
+ \sqrt{g} \mathbf{B} \cdot \nabla \left(\frac{\sqrt{g} j^\phi g^{\theta s} - \sqrt{g} j^\theta g^{\phi s}}{|\nabla s|^2} \right).
$$

The energy functional is given by inserting Eqs. (8.15) and (8.18) into Eq. (8.11), with the corresponding trial functions ξ^s and η decomposed into Fourier harmonics in θ and ϕ and finite elements (hat functions) in s. The quadratic terms present in the energy principle tend to couple the mode numbers together, unlike the case in axisymmetric systems. However, only certain modes within a "family" are coupled

Fig. 8.8 Perturbed pressure
contours at the nonplanar
$\phi = 0$ surface for an unstable
perturbation which is
predominantly $m = 4$. The
minimum perturbed pressure
is near the dotted line on the
outboard side, and the
minimum perturbed pressure
occurs near the two dashed
lines. From [17]. Used by
permission, American
Institute of Physics

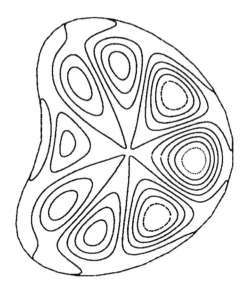

together, and thus the active components for any particular mode of instability can
be found through a set of selection rules involving the toroidal mode spacing and
the Fourier components in the equilibrium. Thus a mode space with potentially
hundreds of thousands of modes (and thus, naively, say, $10^5 \times 10^5$ matrices) can
be reduced dramatically by judicious choice of mode selection. A starting point
would be to choose an initial dominant mode, such as one where $m + n/\iota \approx 0$, and
then assign family members based on the selection rules. The energy minimization
is then reduced to the solution of an eigenvalue problem $\mathbf{P}.\mathbf{x} = \lambda \mathbf{K}.\mathbf{x}$, and this
can be solved iteratively using standard methods (block cyclic reduction and LU
decomposition).

Results of one such stability analysis are shown as Fig. 8.8. Here a low-shear
helias-type equilibrium with four toroidal periods, aspect ratio $A = 8$, $< \beta > =$
0.0205, and $0.733 \leq \iota \leq 0.762$ is analyzed for stability with starting poloidal
and toroidal mode numbers $(m, n) = (4, -3)$. One can see that the minimum and
maximum perturbed pressures are located on the outboard side of the plasma at zero
toroidal angle, where the plasma is bean-shaped.

8.4 Transport

8.4.1 Collisionless Transport, Omnigeneity, and Quasi-Symmetry

Stellarators can have two unfavorable transport mechanisms that symmetrical
devices do not have. These are caused by the essentially non-axisymmetric magnetic

fields present. The first of these effects is that some particle orbits possess drift characteristics such that particles may drift out of the device without collisions. This is especially important for alpha particle confinement because of their large orbits and rapid motion. A second effect is that like-particle collisions can result in particles being swept into a loss cone, reminiscent of mirror-machine transport. The design of stellarator magnetic fields must take these two possibilities into account, along with MHD equilibrium and stability.

Collisionless loss can be minimized by trying to design the system for "omnigeneity," defined as having all of the particles on a single magnetic field line possessing the same drift surface [8]. This also means that the time-averaged radial drift motion for the particles is zero [16]. A discussion of the particle orbits in complex three-dimensional magnetic fields is simplified by the use of a special coordinate system known as " Boozer coordinates" [4]. Boozer coordinates are a more natural form for an expression of the magnetic fields, although they result in more complexity in mapping this coordinate system back to the "real world" of, say, cylindrical coordinates $\{R, \theta, Z\}$, necessary for magnet design, diagnostics, limiter heat loads, and so on. The Boozer form uses three coordinate directions $\{\psi, \chi, \zeta\}$ and results in two equivalent forms for the magnetic field **B**:

$$\mathbf{B} = \nabla \times \boldsymbol{\psi} + q(\psi)\nabla\psi \times \nabla\chi$$

$$\mathbf{B} = i_t(\psi)\nabla\chi + i_p(\psi)\nabla\zeta + \tilde{i}(\psi, \chi, \phi)\nabla\psi. \tag{8.19}$$

Here i_t and i_p are the normalized toroidal and poloidal currents with $i_t = \mu_0/(2\pi)I_{tor}(\psi)$ and $i_p = \mu_0/(2\pi)I_{pol}(\psi)$, and $q(\psi)$ is the safety factor. The function $\tilde{i}$ is arbitrary except that it must have an average value of zero over a full period in either χ or ζ.

An advantage of this dual representation is that the jacobian from the coordinate system $\{\psi, \chi, \zeta\}$ to a cartesian system can be obtained by using the dot product of the two forms for **B** given in Eq. (8.19). The jacobian $\mathscr{J}$ is given by

$$\mathscr{J} = \nabla\psi \cdot (\nabla\chi \times \nabla\zeta) = \frac{B^2}{i_t + qi_p} = f(\psi)B^2. \tag{8.20}$$

We then turn our attention to the particle motion resulting from the transit of a particle (or more accurately, its guiding center, where we assume that the gyroradius is small compared to the characteristic magnetic length). We appeal to an argument from classical mechanics that the quantity $J = \int \mathbf{p} \cdot d\mathbf{l}$ is an adiabatic invariant for every particle in the system, whether it is trapped or not by the changing magnetic field as it moves through the system. If we assume that the magnetic moment $\mu = mv_\perp^2/(2B)$ is also conserved, we have

$$J = 2v \oint_{l_1}^{l_2} \sqrt{1 - \lambda B}.$$

Here $\lambda = 1/B_{max} = \mu/E$, with B_{max} as the magnetic field at which the particle would bounce (which may or may not exist in the plasma). If a particle is to drift freely on a flux surface without a radial drift, there must be a contiguous field line on the flux surface with the property that the action integral J is conserved. The coordinate direction perpendicular to both ψ and b is given by an angle α such that

$$\alpha = \chi - \iota \zeta$$

in Boozer coordinates. Then the property of omnigeneity is given by

$$\frac{\partial}{\partial \alpha} J(\psi, \alpha, l) = 0. \tag{8.21}$$

A simple example of an omnigeneous field is one where every line on a given flux surface has the same maximum and minimum magnetic field, and only one of each. This, however, is too restrictive, and less restrictive cases with several maxima and minima have been shown to exist [16].

Another more design-oriented approach to minimizing collisionless loss is to create a magnetic equilibrium where the general three-dimensional scalar function $B = B(\psi, \chi, \eta)$ can be reduced (approximately) to a two-dimensional form such as $B \approx B(\psi, \zeta)$ ("quasi-poloidal") or $B \approx B(\psi, \chi)$("quasi-toroidal"). Devices possessing helical symmetry also meet the criterion of minimizing collisionless particle loss. These equilibria, taken together, are said to be "quasi-symmetrical" arrangements. Some compromise is often required, however, to obtain a physically realizable magnet and manageable heat loads. The W7-X design, with its good-but-not-perfect quasi-poloidal symmetry, has already demonstrated fair ion and electron confinement, but it will not produce alpha particles, which are a more stringent test of collisionless loss.

8.4.2 Neoclassical Transport

Neoclassical transport in stellarators closely follows the treatment for neoclassical transport as described in Chap. 7. One major difference is the handling of the radial electric field. In tokamaks, the toroidal symmetry means that there is a conserved generalized angular momentum P_ϕ which forms a constraint of the motion of the particles, and the particles in both passing and banana-like orbits do not drift across flux surfaces. Collisions are required for there to be radial flux, and the momentum balance involved in electron–ion collisions results in neoclassical transport becoming "automatically ambipolar." The lack of toroidal asymmetry in stellarators causes some of the trapped particles to drift across flux surfaces. Figure 8.9 shows the variation of the magnetic field $|\mathbf{B}|$ along a magnetic line, and one can see that particles with a velocity vector almost perpendicular to $\mathbf{B}$ will be

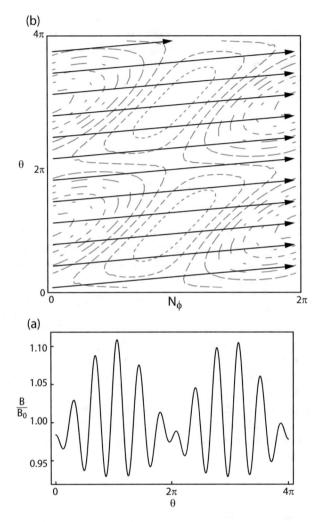

Fig. 8.9 (b) Contour plot of |**B**| at a normalized radius $r = 0.5$. Magnetic field line (straight in Boozer coordinates) shown. (a) Plot of |**B**| along a single field line at $r = 0.5$. Both are for the W7-X standard configuration. Four terms are kept in the Boozer expansion for the field. Data from [3]

trapped into a small region of poloidal angle. It is this class of particles that exhibit most of the radial drift. Generally the loss rates for collisionless loss will be different for ions than for electrons, and this in turn will be balanced by a radial electric field. Consequently, the particle and energy fluxes become tied together through the radial electric field [3, 14].

As in the case of tokamak neoclassical transport, we seek a solution to the drift-kinetic equation to solve for a perturbed distribution function f_1 and proceed to take moments of this to arrive at the transport coefficients. Following [3], we write

$$f_1 = -\frac{q\,R_0\,B_0\langle \mathbf{E}\cdot\mathbf{B}\rangle}{T\langle B^2\rangle}\,f_M\,\widehat{f_I}$$

$$+\frac{v_d\,R_0}{v}\left(\frac{1}{n}\frac{dn}{dr}-\frac{q\,E_r}{T}+\left(K-\frac{3}{2}\right)\frac{1}{T}\frac{dT}{dr}\right)f_M\,\widehat{f_{II}}. \tag{8.22}$$

Here $K = mv^2/(2T)$ is the normalized energy, $v_d = mv^2/(2q\,R_0\,B_0)$ is the characteristic ∇B drift velocity, f_M is the maxwellian distribution, and f_I and f_{II} are dimensionless quantities specified by the equations

$$\frac{R_0}{v}\mathscr{V}(\widehat{f_I}) - \frac{R_0 v}{v}\mathscr{L}(\widehat{f_I}) = -p\frac{B}{B_0},$$

$$\frac{R_0}{v}\mathscr{V}(\widehat{f_{II}}) - \frac{R_0 v}{v}\mathscr{L}(\widehat{f_{II}}) = -\frac{1}{v_d}\frac{dr}{dt}. \tag{8.23}$$

Here the operator $\mathscr{V}$ is given by

$$\mathscr{V}(f_1) = \left(pv\frac{\mathbf{B}}{B}+\frac{E_r\nabla r\times\mathbf{B}}{\langle B^2\rangle}\right)\cdot\nabla f_1$$

$$-\frac{v(1-p^2)}{2B^2}\mathbf{B}\cdot\nabla B\frac{\partial f_1}{\partial p} \tag{8.24}$$

and the Lorentz pitch-angle scattering operator $\mathscr{L}$ given by

$$\mathscr{L}(f_1) = \frac{1}{2}\frac{\partial}{\partial p}\left((1-p^2)\frac{\partial f_1}{\partial p}\right) \tag{8.25}$$

with the pitch angle variable $p = v_{\parallel}/v$. The radial drift velocity is given by

$$\frac{dr}{dt} = \frac{mv^2(1+p^2)}{2q\,B^3}(\mathbf{B}\times\nabla B)\cdot\nabla r. \tag{8.26}$$

Moments of the perturbed distribution function are thus obtained, resulting in a matrix of transport coefficients L_{ij} defined so that

$$I_i = -n\sum_{j=1}^{3}L_{ij}\,A_j. \tag{8.27}$$

Here the fluxes I_n are defined as the radial particle flux

$$I_1 = \langle\boldsymbol{\Gamma}\cdot\nabla r\rangle = \left\langle\int d^3v\,\frac{dr}{dt}\,f_1\right\rangle, \tag{8.28}$$

the energy flux

$$I_2 = \left\langle \frac{\mathbf{Q}}{T} \cdot \nabla r \right\rangle = \left\langle \int \mathrm{d}^3 v \; K \frac{\mathrm{d}r}{\mathrm{d}t} f_1 \right\rangle, \tag{8.29}$$

and the parallel current

$$I_3 = \frac{\langle \mathbf{J} \cdot \mathbf{B} \rangle}{q \, B_0} = \left\langle \int \mathrm{d}^3 v \; p v \frac{B}{B_0} f_1 \right\rangle. \tag{8.30}$$

The thermodynamic forces A_j are defined as

$$A_1 = \frac{1}{n} \frac{\mathrm{d}n}{\mathrm{d}r} - \frac{q E_r}{T} - \frac{3}{2} \frac{1}{T} \frac{\mathrm{d}T}{\mathrm{d}r},$$

$$A_2 = \frac{1}{T} \frac{\mathrm{d}T}{\mathrm{d}r}, \tag{8.31}$$

$$A_3 = -\frac{q B_0 \langle \mathbf{E} \cdot \mathbf{B} \rangle}{T \langle B^2 \rangle}.$$

The transport coefficients L_{ij} are then given by

$$L_{ij} = \frac{2}{\sqrt{\pi}} \int_0^\infty \mathrm{d}K \; \sqrt{K} \, \mathrm{e}^{-K} \, D_{ij}(K) \, h_i h_j, \tag{8.32}$$

with $h_1 = h_3 = 1$ and $h_2 = K$. The "mono-energetic transport coefficients" D_{ij} are then the flux-averaged moments of f_1:

$$D_{11} = D_{12} = D_{21} = D_{22} = -\frac{v_d^2 R_0}{2v} \left\langle \left(\int_{-1}^1 \mathrm{d}p \; \frac{1}{v_d} \frac{\mathrm{d}r}{\mathrm{d}t} \widehat{f_{\mathrm{II}}} \right) \right\rangle,$$

$$D_{13} = D_{23} = -\frac{v_d R_0}{2} \left\langle \left(\int_{-1}^1 \mathrm{d}p \; \frac{1}{v_d} \frac{\mathrm{d}r}{\mathrm{d}t} \widehat{f_{\mathrm{I}}} \right) \right\rangle,$$

$$D_{31} = D_{32} = -\frac{v_d R_0}{2} \left\langle \left(\int_{-1}^1 \mathrm{d}p \; p \frac{B}{B_0} \widehat{f_{\mathrm{II}}} \right) \right\rangle, \tag{8.33}$$

$$D_{33} = -\frac{v R_0}{2} \left\langle \left(\int_{-1}^1 \mathrm{d}p \; p \frac{B}{B_0} \widehat{f_{\mathrm{I}}} \right) \right\rangle.$$

Now we will discuss qualitatively what the impact of locally trapped particles have on the transport. The nature of trapped particles is such that they spend a large amount of time near the turning points in their bouncing motion. If collisions are infrequent, these particles can experience a strong radial drift which will not average to zero because the particles are localized to a small poloidal region. The symmetrical part of the perturbed distribution function $\widehat{f_{\mathrm{II}}}$ then has a part which

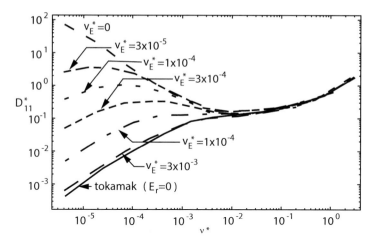

Fig. 8.10 Transport coefficient D_{11} as a function of collisionality ν^* and normalized $E \times B$ drift. Data from [3]

is inversely proportional to the mean free path $\lambda^* = 1/\nu^*$. The overall transport coefficients D_{11} and D_{22} then scale as $(\text{stepsize})^2 \times$ collision frequency, with a stepsize determined by a number $\propto 1/\nu$. This creates an overall transport scaling $\propto v_d^2 \nu^{-1} \propto T^{7/2}$. This "$1/\nu$ regime" is very unfavorable to the operation of stellarators at high temperatures. In particular, if gradient lengths are held constant, then the heat conduction rates scale as $T^{9/2}$ in this low-collisionality (sometimes called the "lmfp" regime for "long mean free path").

As mentioned earlier, however, the presence of a radial electric field may change this situation. If the rotation rate $\omega_{E \times B}$ becomes large enough, then the trapped-particle population becomes effectively de-phased and the plasma transport no longer follows a $1/\nu$ transport law at low collisionality. In fact, an effective drift velocity $v_E^* = E/(B v_{th})$ as small as $\approx 3 \times 10^{-3}$ has been shown to completely eliminate the $1/\nu$ regime. Figure 8.10 shows the result of an international code benchmarking exercise which showed the overall variation of D_{11} parameterized by v_E^*. Early results from W7-X show that radial electric fields are sufficiently high to obtain $1/\nu$ suppression, at least over the interior half of the plasma [13].

8.5 Density Limit, Anomalous Transport, and Scaling Laws

8.5.1 Density Limit

Stellarators have a density limit which is largely determined by impurity radiation at the edge of the plasma and the prospect of radiative collapse when the density exceeds some critical value [19]. The lack of overall toroidal current precludes using a simple model such as the Greenwald density limit used in tokamaks. However, a simple model for the density limit based on data from W7-AS was given by Giannone et al. [7]:

$$n_c = [1.46 \pm 0.07(P_{abs}/V_p)^{0.48\pm0.03} B_0^{0.54\pm0.05}] \times (10^{20}\,\mathrm{m}^{-3}), \qquad (8.34)$$

with B in T, V in m^{-3} and P in MW. By comparing this data to experimentally observed density limits in the AUG tokamak, it was found that the stellarator could sustain a significantly higher density than the tokamak with similar conditions [18].

8.5.2 Anomalous Transport

Parallel to the tokamak experience, one can also expect to see the impact of plasma turbulence on transport. Magnetic fluctuations have been observed on W7-AS [11]. An L-H transition similar to that observed on tokamaks has also been seen in stellarators, with a reduction in edge plasma magnetic activity seen at the time of the transition. The frequency band where this magnetic activity changes the most is in the range of 50 $\rightarrow$ 200 kHz. It has not been possible to find whether this turbulence is due to drift waves or resistive MHD modes. In addition, edge localized mode (ELM) activity is seen, and a sort of "ELMy H-mode" confinement has been seen.

8.5.3 Scaling Laws

The parallels between the tokamak and the stellarator in the arena of overall energy confinement, with all of its complexity and uncertainty, thus begs the question of whether there are empirical transport scaling laws for the stellarator drawn from a large international database. Several attempts to develop scaling laws have been done. One that has been frequently quoted is the "International Stellarator Database," and the most current one dates to 2004. The "ISS04" scaling law is given by [21]:

$$\tau_E^{\mathrm{ISS04}} = 0.134a^{2.28} R^{0.64} P^{-0.61} \bar{n}_e^{0.54} B^{0.84} \iota_{2/3}^{0.41} \qquad (8.35)$$

Here a and R are in meters, P is in MW, n is in units of $10^{19}\,\mathrm{m}^{-3}$, B is in T, and $\iota_{2/3} = 1/q$ at $r = 2/3a$. (Note that some authors refer to this as ℓ in an analogy to $\hbar$, since earlier usage of ι differed from most current usage by a factor of 2π.)

Note that, similarly to the tokamak scaling law, there is a degradation of confinement time with injected power ($P^{-0.61}$ here vs. $P^{-0.69}$ for $\tau_{th,98y2}$, the ITER design basis). The density scaling is also similar: $n_e^{0.54}$ for ISS04 vs. $n_e^{0.41}$ for $\tau_{th,98y2}$. At first it may appear that the magnetic field scaling is very different between ISS04 and $\tau_{th,98y2}$, but that is mostly due to the separating out a current dependence in $\tau_{th,98y2}$, which can also be written in terms of $B_{pol}a = B_{pol}\epsilon R$. The ι dependence in ISS04 can also be written in terms of B_{pol}/B_{tor}, so that overall, these scaling laws are very similar. The murky field of "wind-tunnel scaling" is as alive in the stellarator world as in the tokamak world.

Problems

8.1 A "low-mirror" alternative to the standard case equilibrium for W7-X has, in Boozer coordinates,

$$B(\theta, \phi) = 1 + b_{10}\cos\theta + b_{11}\cos(\theta - N\phi)$$

with $N = 5$ and with coefficients $b_{11} = -0.04335$, $b_{10} = -0.01794$, and $\iota = 0.8623$ at $r = 0.5a$. Make plots of the magnetic field contours with the straight-line magnetic field trajectories and a plot of the magnetic field vs. θ for this configuration similar to Fig. 8.9 for the standard case.

8.2 For the conditions of Problem 8.1, find the maximum mirror ratio and thus determine the maximum trapped particle fraction f_T.

8.3 Estimate the time for a 1.0 keV H^+ ion with a pitch parameter $p = 0$ to drift to the last closed flux surface with the configuration of Problem 8.1.

8.4 Using the ISS04 scaling law and the density limit of Eq. (8.34) (assuming that P is the alpha heating power), find the predicted size R of a stellarator reactor using W7-X parameters for ι, aspect ratio, B, etc. for an ignition device ($Q = \infty$).

References

1. Bauer, F., Betancourt, O., Garabedian, P.: Magnetohydrodynamic Equilibrium and Stability of Stellarators. Springer, New York (1984). http://www.springer.com/us/book/9781461297536
2. Beidler, G.G.C., Harmeyer, E., Lotz, W., KiβLinger, J., Merkel, P., Nührenberg, J., Rau, F., Strumberger, E., Wobig, H.: Modular stellarator reactors and plans for Wendelstein 7-X. Fusion Sci. Technol. **21**(3P2B), 1767–1778 (1992)
3. Beidler, C., Allmaier, K., Isaev, M., Kasilov, S., Kernbichler, W., Leitold, G., Maaßberg, H., Mikkelsen, D., Murakami, S., Schmidt, M., Spong, D., Tribaldos, V., Wakasa, A.: Benchmarking of the mono-energetic transport coefficients-results from the international collaboration on neoclassical transport in stellarators (ICNTS). Nucl. Fusion **51**(7), 076001 (2011). http://stacks.iop.org/0029-5515/51/i=7/a=076001
4. Boozer, A.H.: Establishment of magnetic coordinates for a given magnetic field. Phys. Fluids **25**(3), 520–521 (1982). http://aip.scitation.org/doi/abs/10.1063/1.863765
5. Drevlak, M., Monticello, D., Reiman, A.: PIES free boundary stellarator equilibria with improved initial conditions. Nucl. Fusion **45**(7), 731 (2005). http://stacks.iop.org/0029-5515/45/i=7/a=022

6. Fujiwara, M., Yamazaki, K., Okamoto, M., Todoroki, J., Amano, T., Watanabe, T., Hayashi, T., Sanuki, H., Nakajima, N., Itoh, K., Sugama, H., Ichiguchi, K., Murakami, S., Motojima, O., Yamamoto, J., Satow, T., Yanagi, N., Imagawa, S., Takahata, K., Tamura, H., Nishimura, A., Komori, A., Inoue, N., Noda, N., Sagara, A., Kubota, Y., Akaishi, N., Satoh, S., Tanahashi, S., Chikaraishi, H., Mito, T., Yamada, S., Yamaguchi, S., Sudo, S., Sato, K.N., Watari, T., Kuroda, T., Kaneko, O., Ohkubo, K., Kitagawa, S., Ando, A., Idei, H., Tsumori, K., Kubo, S., Kumazawa, R., Mutoh, T., Oka, Y., Sato, M., Seki, T., Shimozuma, T., Takeiri, Y., Hamada, Y., Narihara, K., Kawahata, K., Fujisawa, S., Hidekuma, S., Minami, T., Yamada, I., Ejiri, A., Tanaka, K., Sasao, M., Iguchi, H., Watanabe, K.Y., Yamada, H., Ohyabu, N., Suzuki, H., Iiyoshi, A.: Large helical device (LHD) program. J. Fusion Energ. **15**(1), 7–153 (1996). http://dx.doi.org/10.1007/BF02266926

7. Giannone, L., Baldzuhn, J., Burhenn, R., Grigull, P., Stroth, U., Wagner, F., Brakel, R., Fuchs, C., Hartfuss, H.J., McCormick, K., Weller, A., Wendland, C., NBI Team, ECRH Team, W7-AS Team, Itoh, K., Itoh, S.I.: Physics of the density limit in the W7-AS stellarator. Plasma Phys. Controll. Fusion **42**(6), 603 (2000). http://stacks.iop.org/0741-3335/42/i=6/a=301

8. Hall, L.S., McNamara, B.: Three-dimensional equilibrium of the anisotropic, finite-pressure guiding-center plasma: theory of the magnetic plasma. Phys. Fluids **18**(5), 552–565 (1975). http://aip.scitation.org/doi/abs/10.1063/1.861189

9. Helander, P., Beidler, C.D., Bird, T.M., Drevlak, M., Feng, Y., Hatzky, R., Jenko, F., Kleiber, R., Proll, J.H.E., Turkin, Y., Xanthopoulos, P.: Stellarator and tokamak plasmas: a comparison. Plasma Phys. Controll. Fusion **54**(12), 124009 (2012). http://stacks.iop.org/0741-3335/54/i=12/a=124009

10. Heppenheimer, T.A.: The Man-Made Sun: The Quest for Fusion Power. Little, Brown, Boston (1983). See also http://history.nasa.gov/SP-4305/ch2.htm

11. Hirsch, M., Holzhauer, E., Baldzuhn, J., Branas, B., Fiedler, S., Geiger, J., Geist, T., Grigull, P., Hartfuss, H.J., Hofmann, J., Jaenicke, R., Konrad, C., Koponen, J., Kuhner, G., Pernreiter, W., Wagner, F., Weller, A., Wobig, H., W7-AS Team: Edge transport barrier and edge turbulence during h-mode operation in the W7-AS stellarator. In: Proceedings of 16th Interantional Conference on Fusion Energy (1996, Montreal), vol. 2, p. 315 (1997)

12. Hirshman, S., van RIJ, W., Merkel, P.: Three-dimensional free boundary calculations using a spectral Green's function method. Comput. Phys. Commun. **43**(1), 143–155 (1986). http://dx.doi.org/10.1016/0010-4655(86)90058-5; http://www.sciencedirect.com/science/article/pii/0010465586900585

13. Klinger, T., Alonso, A., Bozhenkov, S., Burhenn, R., Dinklage, A., Fuchert, G., Geiger, J., Grulke, O., Langenberg, A., Hirsch, M., Kocsis, G., Knauer, J., Krämer-Flecken, A., Laqua, H., Lazerson, S., Landreman, M., Maaßberg, H., Marsen, S., Otte, M., Pablant, N., Pasch, E., Rahbarnia, K., Stange, T., Szepesi, T., Thomsen, H., Traverso, P., Velasco, J.L., Wauters, T., Weir, G., Windisch, T., Team, T.W.X.: Performance and properties of the first plasmas of Wendelstein 7-X. Plasma Phys. Controll. Fusion **59**(1), 014018 (2017). http://stacks.iop.org/0741-3335/59/i=1/a=014018

14. Mynick, H., Hitchon, W.: Effect of the ambipolar potential on stellarator confinement. Nucl. Fusion **23**(8), 1053 (1983). http://stacks.iop.org/0029-5515/23/i=8/a=006

15. Nührenberg, J., Zille, R.: Stable stellarators with medium β and aspect ratio. Phys. Lett. A **114**(3), 129–132 (1986). http://dx.doi.org/10.1016/0375-9601(86)90539-6; http://www.sciencedirect.com/science/article/pii/0375960186905396

16. Parra, F.I., Calvo, I., Helander, P., Landreman, M.: Less constrained omnigeneous stellarators. Nucl. Fusion **55**(3), 033005 (2015). http://stacks.iop.org/0029-5515/55/i=3/a=033005

17. Schwab, C.: Ideal magnetohydrodynamics: global mode analysis of three-dimensional plasma configurations. Phys. Fluids B Plasma Phys. **5**(9), 3195–3206 (1993). http://dx.doi.org/10.1063/1.860656

18. Stäbler, A., Burhenn, R., Grigull, P., Hofmann, J., , McCormick, K., Müller, E., Neuhauser, J., Niedermeyer, H., Reiter, D., Schneider, R., Steuer, K.H., Weller, A., Würsching, E., Zohm, H., ASDEX Teams, W7-AS Teams: Comparison of density limit physics on the ASDEX tokamak and the Wendelstein 7-AS stellarator. In: Proceedings of 14th International Conference on Plasma Physics and Controlled Nuclear Fusion Research (Würzburg 1992), IAEA-CN-56/C-2-6, p. 523 (1993)

19. Wobig, H.: On radiative density limits and anomalous transport in stellarators. Plasma Phys. Controll. Fusion **42**(9), 931 (2000). http://stacks.iop.org/0741-3335/42/i=9/a=301

20. Wobig, H., Wagner, F.: 7 Magnetic confinement fusion: stellarator. In: Nuclear Energy. Landolt-Börnstein - Group VIII Advanced Materials and Technologies, SpringerMaterials, vol. 3B, Springer, Berlin (2005). Part of SpringerMaterials, http://dx.doi.org/10.1007/10857629_17; http://materials.springer.com/lb/docs/sm_lbs_978-3-540-31712-8_17

21. Yamada, H., Harris, J., Dinklage, A., Ascasibar, E., Sano, F., Okamura, S., Talmadge, J., Stroth, U., Kus, A., Murakami, S., Yokoyama, M., Beidler, C., Tribaldos, V., Watanabe, K., Suzuki, Y.: Characterization of energy confinement in net-current free plasmas using the extended international stellarator database. Nucl. Fusion **45**(12), 1684 (2005). http://stacks.iop.org/0029-5515/45/i=12/a=024

Chapter 9
Plasma Heating in Magnetic Fusion Devices

9.1 Introduction and History

9.1.1 Survey of Heating Methods

Historically, four major approaches to heating plasmas in magnetically confined fusion experiments have been explored. These are:

- Compressional/shock heating
- Ohmic heating
- Neutral beam injection
- Radiofrequency wave heating

The first two methods, compressional/shock heating and ohmic heating, are not expected to be relied upon for use in the most prominent magnetic fusion devices such as tokamaks and stellarators. But both have a rich history in the development of all magnetic fusion concepts. Both of these schemes involve inducing energy into the plasma by changing magnetic fields. Since it is the goal of all magnetic confinement schemes to provide the longest burn time per pulse in order to reduce transient thermal stress in materials and to provide steady conditions for heat transfer in blankets and thermal conversion cycles, transient magnetic fields in fusion power reactors are usually considered something worth avoiding. While the current tokamak designs still employ an ohmic heating (OH) coil which creates plasma current by induction, the extension of the inductive current drive from the OH coil has been supplemented by non-inductive current drive to extend pulse lengths far beyond that achievable by inductive current drive alone. So first we explore how we arrived at where we are today.

The earliest fusion schemes involved pinches, as outlined in Chap. 1. In the early z-pinch concepts the plasma was expected to collapse radially, increasing both the density and temperature as it "pinched down." A shock-heated z-pinch plasma with good diagnostics is described in [88]. In these plasmas, collisionless shocks were

© Springer Nature Switzerland AG 2018
E. Morse, *Nuclear Fusion*, Graduate Texts in Physics,
https://doi.org/10.1007/978-3-319-98171-0_9

induced into the plasma by the radial compression of the plasma column as current was applied along the z-axis. Typical Alfvénic Mach numbers (the ratio of the radial velocity to the Alfvén speed) in the range from two to five could be generated, resulting in plasma temperatures in the $80 \rightarrow 100\,\text{eV}$ range. A comprehensive theory of collisionless shock heating was given by Manheimer and Boris [78]. Here the electrostatic turbulence was treated by solving a dispersion relation treating the ion-acoustic wave spectrum. The results of this analysis were fairly consistent with the experiment described in [88]. A more recent comprehensive review of shock heating physics, as applied to z-pinches, is given in [92].

While no large-scale z-pinch experiments are currently being planned for fusion energy applications, the potentially unstable nature of z-pinch plasmas has generated quite a bit of interest within the weapons-effects community as a way of simulating the high-intensity transient photon and neutron fields generated by nuclear weapons. Use of fast z-pinch drivers with various target structures for the production of these intense fields is also reviewed in [92].

Ohmic heating is a natural concomitant of plasma generation in magnetic fusion devices, since plasma pressure generated in the magnetic equilibria must be balanced by a $J \times B$ force. The resistivity η then creates a thermal energy production term in the plasma given by ηJ^2. In practical units, the plasma resistivity is given by

$$\eta = 2.8 \times 10^{-8} Z T_e (\text{keV})^{-3/2} \; \Omega - \text{m}. \tag{9.1}$$

The important part of this expression is the $T_e^{-3/2}$ dependence. This shows that at the temperatures required to sustain a plasma with substantial fusion alpha heating, the ohmic heating power is very small. If one considers just the loss due to bremsstrahlung without any transport losses, there is a range in temperature where neither fusion alpha heating nor ohmic heating is very large, and the net balance of power gain vs. loss cannot be sustained in the range roughly between 2.0 and 5.0 keV (see Fig. 9.1). While the early success of the tokamak experiments with only ohmic heating such as the famous T-3 experiment in the former Soviet Union, which obtained kilovolt temperatures, Fig. 9.1 clearly shows how this is possible, in spite of the extremely small amount of ohmic heating power available at higher temperatures.

While the classical predictions for ohmic heating paint a fairly dismal picture for the success of taking an ohmically heated plasma up in temperature to the point where it will have thermonuclear gain, effort has been expended to find "work-arounds," where anomalously high resistivity can be induced into the plasma through turbulence or other means. However, most turbulence studies have shown that magnetic turbulence induces lower resistivity, not higher, at constant temperature: [16] is a good example. Trapped-particle effects with turbulence have also been analyzed from the standpoint of anomalous resistivity, but again, generally trapped-particle studies have shown a negative effect on resistivity, see, e.g., [23]. The majority of the work on anomalous resistivity has been devoted to explaining the higher rates of magnetic reconnection seen in laboratory and astrophysical settings; this would imply higher resistivity, but these ideas have not translated into a practical way of raising the ohmic heating power in toroidal magnetic fusion devices.

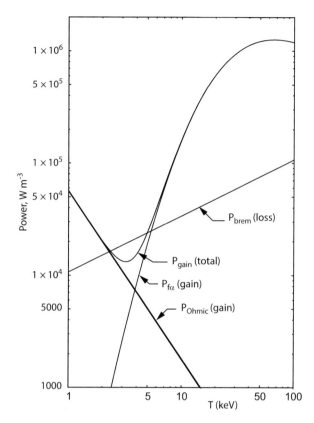

Fig. 9.1 Power gain and loss terms for an ohmically heated magnetic confinement plasma. Shown are bremsstrahlung losses, fusion alpha heating, and ohmic heating, for $Z = 2$, current density $J = 10^6 \, \mathrm{Am}^{-2}$, and density $n = 10^{20} \, \mathrm{m}^{-3}$, 50-50 D-T plasma. Note power balance deficit for temperatures $2 < T < 5 \, \mathrm{keV}$

9.2 Neutral Beam Injection

A great deal of the success of the magnetic fusion program can be attributed to the development of powerful, high energy neutral beams. The general scheme for creating these beams is to generate ions of hydrogen isotopes, typically either deuterium or tritium, accelerating these to high energies using conventional electrostatic acceleration, and then passing these ions through a gas cell containing neutral molecules of the same species. Charge exchange processes take place in the neutralizer cell, and then the beam may be sent through a magnetic analyzer to remove the unwanted remaining charged ions. A schematic of a system of this kind (this is the TFTR neutral beam injection system, designed at Berkeley and deployed at Princeton) is shown as Fig. 9.2.

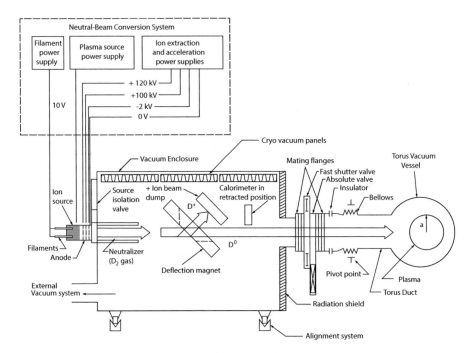

Fig. 9.2 Schematic of the TFTR neutral beam injection system. From PPPL-1475 (1978), Princeton Plasma Physics Laboratory

There are several facets of neutral beam injection system design. We start with the basic physics involved in turning an ion beam into a neutral beam.

9.2.1 Neutralization

Suppose that we have an accelerator which produces only D^+ ions. We have a neutralizer cell which can convert these D^+ ions to fast D^0 atoms by charge exchange. The fast D^0 atoms, however, can have another collision with the background neutralizer gas molecules (D_2 in this case), causing a charge exchange reaction that turns the fast D^0 back into a D^+. We use F_1 and F_0 to denote the fractions of D^+ and D^0, respectively, and denote the charge exchange reactions of the fast particles on the background neutralizer gas as σ_{10} and σ_{01} for the reactions $\underline{D^+} + D_2 \rightarrow \underline{D^0} + D_2^+$ and $\underline{D^0} + D_2 \rightarrow \underline{D^+} + D^- + D^0$, where the underline represents the fast particle and the non-underline species are particles at the background velocities. (There are additional final states for the slow particles leaving the charge exchange process, and these are included in the cross sections.) Then we can write down a set of simple differential equations for the particle fractions as a function of the target line density $\pi = \int n_{D_2} dx$, in molecules per m^2:

$$\frac{dF_0}{d\pi} = F_1\sigma_{10} - F_0\sigma_{01}$$

$$\frac{dF_1}{d\pi} = F_0\sigma_{01} - F_1\sigma_{10}, \tag{9.2}$$

with initial condition $F_1(0) = 1$.

Notice that this is really just one first-order ODE, since the forcing terms on the right are not linearly independent; in fact, $F_1 + F_0 = 1$ everywhere. The solution to this set of equations is given by

$$F_0(\pi) = \frac{\sigma_{10}}{\sigma_{10} + \sigma_{01}} \{1 - \exp[-(\sigma_{10} + \sigma_{01})\pi]\}. \tag{9.3}$$

From Eq. (9.3) it is clear that in the thick-target limit $\pi \to \infty$, the efficiency, defined as charged particle beam power in / neutral particle beam power out, which in this case is simply $F_0(\infty)$, is given by

$$\eta_\infty = \frac{\sigma_{10}}{\sigma_{10} + \sigma_{01}}. \tag{9.4}$$

The cross sections for a full gamut of charge exchange reactions relevant to neutral beam injection systems is given in Table 9.1. Notice that the table gives the cross sections for deuterons. The cross section data can be used for tritons traveling at the same speed as a deuteron, i.e. multiplying the projectile energies by 1.5 allows this data to be used for T^+ ions. This is because all that matters here is the projectile velocity, since the atomic physics only involves the interaction of the electrons, and the nuclear mass is unimportant. (In fact, most of the data given in Table 9.1 was obtained using hydrogen, and then the energies for the equivalent deuteron were obtained by multiplying the actual energies in the experiments by a factor of two.)

Sample Neutral Beam Efficiency Calculation

Suppose that we have a 100 keV D^+ beam, what is the neutralization efficiency for a thick-target D_2 gas cell ?

Table 9.1 gives the following values for the charge exchange reactions:

$$\sigma_{10} = 17 \times 10^{-21}\ m^2$$

$$\sigma_{01} = 14 \times 10^{-21}\ m^2$$

so then the thick-target neutralization efficiency is given by

$$\eta_\infty = \frac{\sigma_{10}}{\sigma_{10} + \sigma_{01}} = \frac{17}{17 + 14} = \boxed{54.8\%}$$

> Now suppose that we wanted to switch to a tritium beam. At what energy
> for a T^+ ion would we have the same neutralization efficiency?
>   Since the cross section data is derived from the velocity of the fast particle
> involved, a triton with the same speed as the deuteron would have the same
> charge exchange cross sections, and since energy $= (1/2)mv^2$, this represents a
> $\boxed{150\,\text{keV}}$ triton.

One sees from the table that there are also reactions of the type that give $\underline{D^0} \rightarrow$
$\underline{D^-}$ and $\underline{D^+} \rightarrow \underline{D^-}$ reactions as well as their inverses. These were ignored in the
previous calculations, and we see from Table 9.1 that these cross sections are much
smaller than σ_{01} and σ_{10}, and make a small correction to the calculation given above.
In general we may write rate equations for all three charge states $\{-1,0,1\}$ as [13]:

$$\frac{dF_i}{d\pi} = \sum_{j \neq i} F_j \sigma_{j,i} - F_i \sum_{j \neq i} \sigma_{i,j} \qquad i,\, j = D^+, D^0, D^-. \tag{9.5}$$

The three-state rate equations have forcing terms which are linearly independent,
unlike the previous case, and one result of this is that there can be a maximum
efficiency of conversion to neutrals at some intermediate value of target thickness
π. The more modern neutral beam injection systems will be based upon D^-, rather
than D^+ ion sources, mostly because higher efficiencies can be obtained at high
energies, but another benefit in D^- sources is the rather low value of gas loading
(π) for optimum performance. High gas loads in the neutralizer are difficult to deal
with, as a great deal of vacuum pumping power must be available for the neutralizer
gas pressures (typically 0.1 Pa or 10^{-3} torr) to not affect the vacuum pressure in the
plasma chamber (typically 10^{-6} Pa or 10^{-8} torr). Figure 9.4 shows the "optimum"
gas loading for all five possible deuterium atomic and molecular ions. (For those
cases where the optimum gas load is infinity, the optimum is defined as that value
of π where the neutralization efficiency is 95% of the ultimate value, reflecting the
compromise between efficiency and gas loading.)

For positive-ion based sources, a pure D^+ beam is generally not possible. Typical
positive ion source results are ion components which are 75% D^+, 15% D_2^+, and
10% D_3^+ [13]. If mixed-species ions are sent through a common accelerator, then
the fast ions emerge with the D_2^+ and D_3^+ ions having one-half and one-third of the
D+ energy. These species at lower energy are less desirable because they will tend to
have faster ionization and charge exchange processes in the plasma and thus deposit
more of their energy near the plasma's edge, which is not only less desirable as a
heating strategy but can also lead to instabilities. The overall set of rate equations
for all four atomic and molecular ions (D^+, D^-, D_2^+, and D_3^+) and the two neutrals
produced (D^0 and D_2^0) are given by [13]:

Table 9.1 Cross section data for neutralizer atomic physics reactions of interest for the production of D^0 and D_2^0 neutral atoms/molecules. Values in parenthese are inferred values (not directly measured)

Energy	D^+	D^0			D^-		D_2^+			D_2^0			D_3^+			
keV/deut	σ_{10}	σ_{1-1}	σ_{01}	σ_{0-1}	σ_{-10}	σ_{-11}	$\sigma_{D_2^0}$	σ_{D^0}	σ_{D^+}	$\sigma_{D_2^+}$	σ_{D^0}	σ_{D^+}	$\sigma_{D_2^0}$	$\sigma_{D_2^+}$	σ_{D^0}	σ_{D^+}
10	83	0.13	8.0	1.6	100	8.5	46	74	22	7.7	25	3.3	35	11	72	11
20	80	0.45	9.3	2.5	108	9.0	36	83	22	12	(17.0)	4.9	41	12.5	91	16
50	47	0.80	13	1.6	85	8.5	16	63	24	19	(9.5)	7.0	30	10.8	83	21
100	17	0.10	14	0.75	65	8.0	4.3	35	24	19	(7.3)	6.9	10	8.2	50	24
200	2.5	0.01	10.8	(0.25)	47	(5.3)	0.7	13	19	15	5.0	5.8	3.5	5.6	25	22.3
500	0.046	(0.0)	6.0	(0.06)	25	(2.2)	(0.04)	3.7	10	7.0	3.2	3.8	1.1	3.0	13	15
1000	0.0012	(0.0)	3.3	(0.02)	15	(0.75)	(0.006)	2.0	5.6	4.2	2.2	2.4	0.63	1.8	6.5	7

Units are 10^{-21} m^2 per D_2 molecule. From [13], with data from [7, 9, 70, 80–82, 89, 108, 119], and [12]

$$\frac{dF_{D_3^+}}{d\pi} = -F_{D_3^+} \left(\frac{1}{3} \sum_i \sigma_{D_3^+,i} + \frac{2}{3} \sum_k \sigma_{D_3^+,k} \right)$$

$$\frac{dF_k}{d\pi} = F_\ell \sigma_{\ell,k} - F_k \left(\sigma_{k,\ell} + \frac{1}{2} \sum_i \sigma_{k,i} \right) + F_{D_3^+} \sigma_{D_3^+,k} \qquad (9.6)$$

$$\frac{dF_i}{d\pi} = \sum_{j \neq i} F_j \sigma_{j,i} - F_i \sum_{j \neq i} \sigma_{i,j} + \sum_k F_k \sigma_{k,i} + F_{D_3^+} \sigma_{D_3^+,i},$$

where

$$k, \ell = D_2^+, D_2^0$$

$$i, j = D^+, D^0, D^-. \qquad (9.7)$$

The factors of 1/2 and 1/3 in Eq. (9.6) are there because the cross sections are defined in terms of output particles rather than input particles to the reaction; it is relatively easy to show that this set of equations conserves deuterons ($3F_{D_3^+} + 2F_{D_2^+} + 2F_{D_2^0} + F_{D^+} + F_{D^0} + F_{D^-} = 1$).

We can then explore some applications of Eq. (9.6). First, Fig. 9.3 shows the neutralization efficiency vs. gas loading for D^-, D^+, D_2^+, and D_3^+ at energies of 200 keV per nucleon. Notice that D^-, D_2^+, and D_3^+ all show an optimum neutralization efficiency at some intermediate value of π under 10^{20} m^{-2}. For the case of D^-, the design of an optimized neutralizer is straightforward; for D_2^+ and D_3^+, it is likely that these were produced in a source where the intention was to maximize D^+ production, and the D_2^+ and D_3^+ species are "along for the ride." Figure 9.4 shows the optimum target thickness for a large range of deuteron energies. In the case of D^+, where there is never an optimum target thickness, the optimum is chosen to be where the neutralization efficiency is at 95 % of the infinite-thickness value. One can see that at 1.0 MeV energy, the optimum D^+ target thickness has climbed to around 10^{21} m^{-2}, but this is a moot point, because the neutralization efficiency for D^+ at 1.0 MeV is almost zero.

Next we would like to calculate the overall efficiency for the production of neutral beam power from an input ion beam. Figure 9.5 shows the results of such a calculation from the data given in Table 9.1. For D^+ and D^- input beams, these efficiencies are given by $F_{D^0}(\pi)$, where π is taken as infinity for the D^+ case and $\pi = \pi_{opt}$ for the D^- case. There is a subtlety associated with the D_2^+ and D_3^+ cases, however, because here both D_2^0 and D^0 are present after the neutralization. For D_2^+ injection into the neutralizer, D_2^0 emerges with the full beam energy and D^0 emerges with half the beam energy. For D_3^+, D_2^0 emerges at two-thirds of the beam energy and D^0 emerges with one-third of the D_3^+ beam energy. Therefore we can write

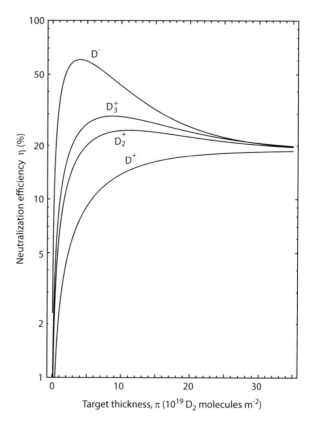

Fig. 9.3 Neutralization efficiency vs. gas load π for deuterium ions D^-, D^+ at 200 keV, D^{2+} at 400 keV, and D^{3+} at 600 keV. Data from [13]

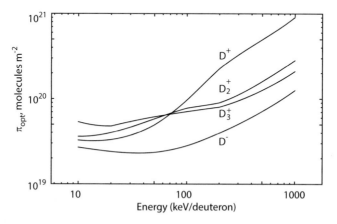

Fig. 9.4 Optimum gas thickness π vs. energy per deuteron for deuterium ions D^-, D^+, D^{2+}, and D^{3+}. (For reactions with a peak efficiency, this represents the peak value; for other cases, it represents the point where the efficiency is 95% of maximum.) Data from [13]

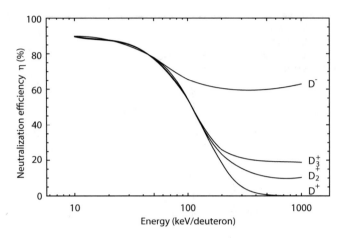

Fig. 9.5 Neutralization efficiency for deuterium species D^-, D^+, D^{2+}, and D^{3+} with either infinite thickness or optimum thickness. Data from [13]

$$\eta = F_{D_2^0}(\pi) + \frac{1}{2}F_{D^0}(\pi) \qquad\qquad D_2^+ \text{ injection}$$

$$= \frac{2}{3}F_{D_2^0}(\pi) + \frac{1}{3}F_{D^0}(\pi) \qquad\qquad D_3^+ \text{ injection.}$$

(9.8)

Curves of the neutralization efficiency defined in this way are given in Fig. 9.5.

9.2.2   Penetration into Plasma

We can now address the subject of what the penetration length of the neutral beam into the plasma is. In this case, we have a different set of atomic physics processes to consider from those in the neutralizer. With a fast neutral beam entering the plasma, one component of the ionization is electron impact ionization:

$$e^- + \underline{D^0} \rightarrow \underline{D^+} + e^- + e^-$$

A useful formula for the electron impact ionization has been given by [76]:

$$\sigma_{e-I} = 10^{-18}\frac{a_1\log\left(\dfrac{E}{P_1}\right)\left(1 - b_1\exp\left(-c_1\left(\dfrac{E}{P_1} - 1\right)\right)\right)}{E\,P_1}\,\text{m}^2.$$

(9.9)

Here $a_1 = 4.0$, $b_1 = 0.60$, and $c_1 = 0.56$ (all dimensionless), and $P_1 = 13.6\,\mathrm{eV}$ (the ionization potential of hydrogen), and electron energy E in eV. The formula can only be used for $E > P_1$.

In order to use this cross section data for the case of a fast neutral atom interacting with electrons, we note that for electron temperatures above $\approx 5 \rightarrow 10\,\mathrm{eV}$, the neutral atoms can be considered stationary against a maxwellian spectrum of electrons with $v_{the} \gg v_0$. We thus must take a maxwellian average of the electron background, obtaining an equivalent cross section

$$< \sigma_{e-I} >= \frac{< \sigma_{e-I} v_e >}{v_0}.$$

The effective cross section for electron ionization is thus temperature dependent because of this maxwellian averaging. For very cold electrons ($T_e < 5\,\mathrm{eV}$) we make the opposite assumption, i.e. that the electrons are stationary and the cross section is given by the Lotz cross section (Eq. (9.9)) with the appropriate conversion of neutral atom energy into electron energy for the same relative velocity. (A cold electron region such as this may represent the plasma outside the closed flux surfaces in a tokamak or stellarator, for example.)

Forms for the proton impact and charge exchange cross sections have been given by Riviere [91]. The proton impact ionization cross section is:

$$\sigma_{p-I} = \begin{cases} 10^{\left(-0.8712 \left(\log_{10} E\right)^2 + 8.156 \log_{10} E - 38.833\right)} & E < 150\,\mathrm{keV} \\ 3.6 \times 10^8 E^{-1} \log_{10}(0.1666E) & E > 150\,\mathrm{keV} \end{cases} \mathrm{m}^2. \tag{9.10}$$

Here E is in electron volts. Riviere [91] also gives an expression for the charge exchange cross section for the reaction $\underline{H^0} + H^+ \rightarrow H^+ + \underline{H^0}$:

$$\sigma_{cx} = \frac{0.6937 \times 10^{-18} \left(1 - 0.155 \log_{10} E\right)^2}{1 + 0.1112 \times 10^{-14} E^{3.3}} \,\mathrm{m}^2 \tag{9.11}$$

with E in electron volts.

Adding up the contributions to ionization from electron impact, ion impact, and charge exchange then results in a total cross section for ionization of the neutral beam in the plasma. Figure 9.6 shows the contributions from all three processes. We can invert the total cross section to get a mean free path $n\lambda$ for the interaction of the neutral beam with the plasma. This is shown in Fig. 9.7. (Note that the energy scale is doubled for D^0 atoms and tripled for T^0 atoms.) Since we would like a quantity much less than $1/e$ of the beam energy leaving the plasma, we would like the plasma

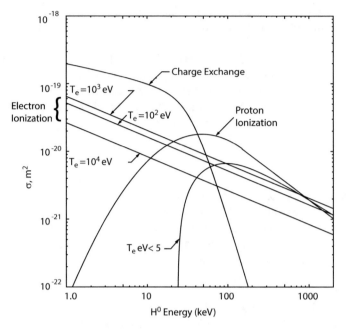

Fig. 9.6 Cross sections of relevance for the ionization of neutral beams in target plasma. Data from [13], with original data from [109] and [91]

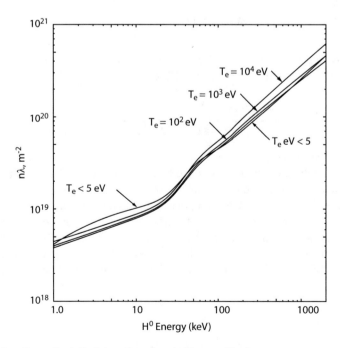

Fig. 9.7 Mean free path $n\lambda$ for interaction of neutral beam with plasma

to represent several mean free paths to the beam traversal. Typically, the beams are configured for tangential injection, and then the key length parameter for the plasma target is the mean chord length of a torus, which is $\overline{L} = 2(2Ra)^{1/2}$ [109].

9.2.3 Accelerator Physics

Space Charge and the Child-Langmuir Law

The vacuum electric fields in an accelerator structure are modified by the presence of ions or electrons in the space between the electrodes. If one considers Poisson's law

$$\nabla^2 \Phi = -\frac{\rho}{\varepsilon_0},$$

then the vacuum solution to $\nabla^2 \Phi = 0$ is modified by the presence of charge carriers in the region. As a simple example, consider an accelerator consisting of two infinite planar grids spaced a distance d apart. Let the plane at $x = 0$ have a potential of zero and the plane at $x = d$ have a potential V_0. Suppose that particles at zero energy are injected at $x = 0$ and accelerate towards the planar grid at $x = d$. At any point x the energy of the particle is $(1/2)mv^2 = -q\Phi(x)$. In steady state, there is no charge buildup in the gap, and therefore the current J =constant across the domain from $x = 0$ to $x = d$. Then the space charge is given by

$$\rho = \frac{J}{v} = \frac{J}{\left(\dfrac{-2q\Phi}{m}\right)^{1/2}}. \tag{9.12}$$

The Poisson equation then has the form

$$\frac{d^2\Phi}{dx^2} = \frac{K}{\Phi^{1/2}} \tag{9.13}$$

with $K = (-J/\varepsilon_0)(-m/(2q))^{1/2}$. We can change to dimensionless coordinates where $x' = x/d$ and $\Phi' = \Phi/V_0$. With no loss of generality we can let q be negative and V_0 be positive, and then

$$\frac{d^2\Phi'}{dx'^2} = \frac{K'}{\Phi'^{1/2}} \tag{9.14}$$

with $K' = Kd^2V_0^{-3/2}$. Then the boundary conditions for Eq. (9.14) are $\Phi(0) = 0$ and $\Phi(1) = 1$.

This nonlinear, second-order ODE has a surprisingly simple solution. By multiplying both sides by $2d\Phi/dx$, one has

$$2\frac{d\Phi'}{dx'}\frac{d^2\Phi'}{dx'^2} = 2\frac{d\Phi'}{dx'}\frac{K'}{\Phi'^{1/2}} \tag{9.15}$$

which can then be integrated to give

$$\left(\frac{d\Phi'}{dx'}\right)^2 = 4K'\Phi'^{1/2} + K_1. \tag{9.16}$$

Several special cases exist. For $K' = 0$, $d\Phi'/dx' = $ constant $=1$ will fit the boundary conditions. This is the vacuum solution to Poisson's equation, which then is $\Phi'(x') = x'$. Another solution exists where $d\Phi'/dx' = 0$ at $x' = 0$. This is the marginal solution where the electric field at $x = 0$ is reduced to zero, and particles have lost their initial electric field "push" allowing them to enter the accelerator region. For this marginal solution, $K_1 = 0$ and taking the square root of both sides of Eq. (9.16) gives

$$\frac{d\Phi'}{dx'} = 2K'^{1/2}\Phi'^{1/4} \tag{9.17}$$

and this has an exact integral

$$(4/3)\Phi'^{3/4} = 2K'^{1/2}x \tag{9.18}$$

after applying the boundary condition $\Phi'(0) = 0$. Since $\Phi'(1) = 1$, we then know that $\phi' = x^{4/3}$ which then requires that $K' = 4/9$. Plugging in the physical constants, this gives the celebrated Child-Langmuir law:

$$J_{CL} = \frac{4}{9}\varepsilon_0 \left(\frac{2q}{m}\right)^{1/2}\frac{V_0^{3/2}}{d^2}. \tag{9.19}$$

It is also interesting to explore other cases. For an accelerator carrying half the Child-Langmuir critical current, we have $K' = 2/9$ and $K_1 = 4/9$; the algebraic form for Φ' is quite complicated but it is easily shown to be between the two special cases. Also, if particles are given some initial velocity as they enter the acceleration region, then an equilibrium current above the Child-Langmuir value can be sustained. In this case Eq. (9.14) is modified to read

$$\frac{d^2\Phi'}{dx'^2} = \frac{K'}{(\Phi' + w_0)^{1/2}} \tag{9.20}$$

where w_0 is the particle's input energy, normalized to the accelerating potential. Figure 9.8 shows the potential distribution for all four cases. For the cases with $J >$

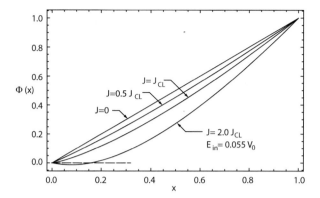

Fig. 9.8 Potential distribution $\Phi(x)$ in a planar accelerator showing the effects of space charge. The vacuum field case ($J = 0$ and cases with one-half, one, and two times the Child-Langmuir limiting current are shown. For the case with $J = 2.0 J_{CL}$, an initial energy of 0.055 times the accelerator's potential energy is required to launch the particles into the acceleration zone

J_{CL}, some initial energy must be provided to the particles. These higher currents and thus space charges may result in the potential going negative towards the beginning of the acceleration zone; when the particles are electrons, this negative region is referred to as a "virtual cathode."

The Child-Langmuir law in its abstract form gives the limit on the available current density for a planar accelerator with transparent electrodes of infinitesimally small spacing. The "real-world" implementations have somewhat lower performance. Another important observation about the Child-Langmuir law is that it suggests that larger currents can be handled with shorter distances d involved in the primary acceleration stage, as well as higher voltages V_0, since $J_{CL} \propto V_0^{3/2}/d^2$. However, both of these design incentives are directly opposite that required to avoid breakdowns on the electrodes. A good design rule is to have V_0/d be no more than $100\,\text{kV}$ per cm. This may not prevent all arcs but coupled with fast shutdown circuitry and good arc detection instrumentation will result in a fairly robust system.

As current levels increase, and often long before the Child-Langmuir limit is hit, beam quality issues emerge, because of space charge modification to the vacuum electric fields. This has resulted in extensive efforts to shape the electrodes in a certain way to provide optimum potential distributions, and thus beam quality, at the design current density.

9.2.4   ITER Design

The overall design for the source and acceleration region of the ITER heating neutral beam (HNB) injector is shown as Fig. 9.9. The HNB accelerator uses a multi-aperture, multi-grid concept (MAMuG) and is designed to produce a negative ion

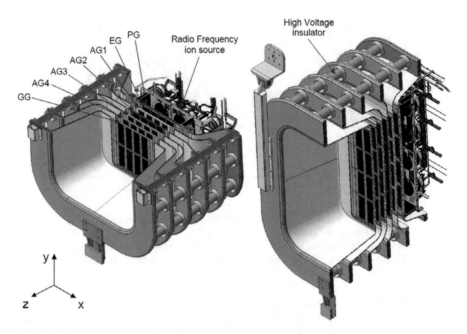

Fig. 9.9 Design of ITER neutral beam injector, showing source, plasma grid (PG), extraction grid (EG), acceleration grids (AG1-AG4), and final grounded grid (GG). From [26]. Used with permission, International Atomic Energy Agency

beam composed of 1280 beamlets. The accelerator is designed to produce 40 A of 1.0 MeV D^- beams or 46 A of 870 keV H^- beams to the neutralizer. Each of the two beamlines is designed to deliver 16.5 MW of either D^0 or H^0 beam to the plasma. The design is a collaborative effort between laboratories in France, Italy, Japan, and the ITER organization itself.

The design for the ITER neutral beam injection system makes use of a suite of numerical simulation codes which calculate electric field profiles with the space charge effects added self-consistently. A detailed description of the codes used is found in [86]. The ITER NBI design uses many (1280 per injector) small "beamlets" with circular accelerator electrodes and as such can be treated well with two-dimensional (r, z) modeling. The SLACCAD code, developed at Cadarache in the late 1980s, is based on a code developed at the Stanford Linear Accelerator Laboratory (SLAC) by Hermannsfeldt in the late 1970s [54], which has been adopted for many accelerator design efforts since. Figure 9.10 shows the results of integrated computer-aided design which couples the charge density, electric field profiles, and ion optics together. Figure 9.11 shows the SLACCAD simulation of the current design, showing the high quality overall optics for the acceleration system (with a few strays).

There is a tendency for the beamlets being processed through the acceleration system to diverge from each other due to electrostatic forces. To counter this, a

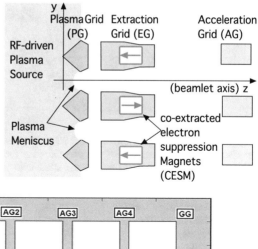

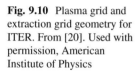

Fig. 9.10 Plasma grid and extraction grid geometry for ITER. From [20]. Used with permission, American Institute of Physics

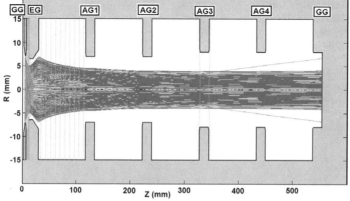

Fig. 9.11 SLACCAD simulation of a single beamlet through the HNB accelerator, assuming $T_i = 0.5\,\text{eV}$. From [26]. Used with permission, International Atomic Energy Agency

set of "kerbs," an extra rounded part, are added downstream to the intermediate acceleration grids (AG1 through AG4, as labeled Fig. 9.9). These shaping electrodes create an electrostatic lens which deflects the outermost beamlets back towards the center of the bundle. Figure 9.12 shows the kerbs added to the basic accelerator grid structure.

The high voltage ($<1.0\,\text{MV}$) and negative polarity in the ITER design makes suppression of electrons in the accelerator of paramount importance, since if unchecked, electrons would create a huge energy loss and large heat deposition in random places in the beamline. The current design now includes an array of permanent magnets embedded in the extraction grid (EG), which trap and deflect any secondary electrons produced before they are accelerated. Design features of the magnetic array for the EG are found in [20]. The design concept for the magnetic array is the combination of a system of magnets called the "Asymmetric Deflection Compensation Magnets" (ADCM) with another system called the "Co-Extracted Electron Suppression Magnets" (CESM). The latter is a set of alternating magnets

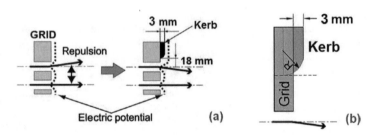

Fig. 9.12 "Kerbs" (also known as field shaping plates) deployed in the design for ITER ion accelerator. (**a**) Demonstration of the cancellation of the electrostatic self-forces by field shaping, (**b**) close-up of kerb design. From [26]. Used with permission, International Atomic Energy Agency

Fig. 9.13 Permanent magnet array built into the EG for electron suppression. From [20]. Used with permission, American Institute of Physics

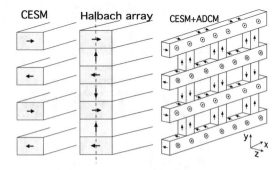

laid out in the **Z** direction which achieves electron suppression to first order, but causes some drift of the ions as they accelerate. The former is the addition of a smooth, alternately varying magnetic field imposed by the addition of a Halbach array [44], an elegant design for a periodic magnetic field found by Halbach at Berkeley in the late 1970s. The addition of the Halbach field virtually eliminates the effects of ion drift due to the electron suppression magnets. Figure 9.13 shows the integration of the permanent magnets comprising the ADCM+CESM structure now forming the ITER EG design.

Now we examine the ion source for ITER. We start the discussion by showing the departure from earlier positive-ion designs. In earlier designs, D^+ ions, along with other species, were extracted from a plasma consisting of D^+, D_2^+, and D_3^+ species. Generally plasmas with the greatest D^+ species were the most desirable, and these could be obtained by careful control of the density and temperature of the source plasma. Typically these plasmas were created using a set of filaments heated to the point of thermionic emission, and the resulting electrons were used to create an arc plasma with typically up to 1000 A of discharge current at voltages on the order of 100 V. Later positive ion sources have attempted to use RF power to replace some of the arc power, but this method was never brought around to the large-scale NBI systems for the large fusion experiments.

When the need for negative-ion based sources became clear, many different configurations were tried. There were experiments based on RF-driven sources with magnetic filtering to extract the D^- ions [74], as well as attempts to create a two-region plasma where one high-temperature plasma component could be used to create rotationally and vibrationally excited D_2^{**} molecules, which could then drift into a low-temperature plasma region where cold electrons could interact with the excited D_2^{**} dimers creating D^- ions [55]. (The need to avoid a mid-temperature plasma region around 1 eV temperature is because to the large de-excitation cross section for the excited dimers by D_0 atoms, which are generated by heated filaments, for example.) No solutions were found with high extracted currents of D^-.

The abandonment of the "volumetric" approaches to D^- production was due to the relatively high yield associated with the production of negative hydrogenic ions with a reactive metal such as cesium:

$$D_0 + Cs + e^- \rightarrow Cs + D^-$$

$$D^+ + Cs + e^- + e^- \rightarrow Cs + D^-.$$

In this case the Cs can be thought of as a catalyst, i.e. it has no change of state during the reaction, but facilitates the electron transfer due to its low work function. The cesium catalytic reaction has been known for a long time; it has also been known that cesium absorbed on molybdenum (another low work function material) performs better than tungsten or other materials. An example of a large scale Cs-based NBI system was demonstrated by the JAERI group [60]. Experimental and theoretical efforts leading up to the current ITER RF-driven NBI source are described in [43, 97], and [83].

The overall design for the ITER negative ion NBI source is described in [49]. A view of a single source cell is shown in Fig. 9.14. Here an RF coil is wound tangentially on an alumina cylinder. Inside the source, a Faraday shield is placed with Mo-coated Cu components against the back of the cylinder and around the side wall of the cylinder. The side-wall covering has slots perpendicular to the RF coil, allowing penetration of the RF but keeping the plasma and Cs vapor from covering the alumina insulator. The Faraday shield components have internal water coolant channels for heat dissipation, which can be substantial with over 100 kW of RF drive per cell. A Cs oven is attached to the system, allowing constant replenishment of the Cs surface. The individual ion sources are part of a block of eight as shown in Fig. 9.15. The backplane of this system includes the capacitors and buswork for the RF drive, which are required to be local to the RF coils because of the inherent mis-match in impedances between the drive coils and the transmission lines from the RF drive hardware, tens of meters away.

Fig. 9.14 Close-up view of one cell of the ITER RF-driven negative ion source. From [79]. Used by permission, American Institute of Physics

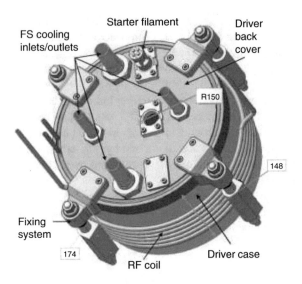

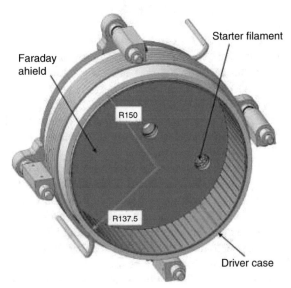

9.3   Radiofrequency Wave Heating

9.3.1   Introduction and History

The earliest attempt at terrestrial nuclear fusion, as referenced in Chap. 1 was done by Kantrowitz in 1938 at Langley [53]. This experiment consisted of a torus with a supplied toroidal field and a lab-built 150 W RF oscillator. Thus this experiment represents not only the first thermonuclear fusion experiment, but it also marks the

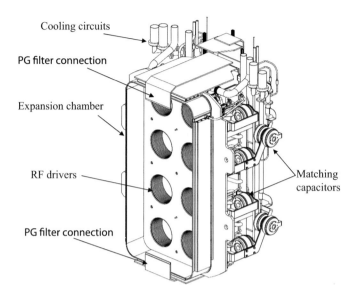

Cooling circuits

PG filter connection

Expansion chamber

RF drivers

Matching
capacitors

PG filter connection

Fig. 9.15 View of the ITER RF-driven negative ion source subassembly. From [49]. Used with permission, International Atomic Energy Agency

beginning of the study of RF heating in plasmas. By 1960, the model B65 stellarator at Princeton was fitted up with a 11.5 MHz, 200 kW RF source which could be tuned to slightly below the ion cyclotron resonance frequency (ICRF), and neutrons were observed when the resonant condition was satisfied [102]. The driving force for this development was Thomas Stix, who developed an antenna wrapped around a toroidal section with coil sections of alternating polarity, thus fixing the toroidal wavenumber being launched. This early phased-array technique, the "Stix antenna," is the model for all ICRF antenna structures to this day. Figure 9.16 shows Stix's seminal ICRF experiment on the B65 stellarator.

Many of the early stellarator experiments had some sort of low-frequency (ion cyclotron) heating system. In laboratories such as Princeton, these sources were available for tokamak experiments, although they were not as powerful as the ohmic heating present in the early tokamak experiments and played little role. A remarkable development in the Soviet Union, however, was the invention of the gyrotron. Gyrotrons were capable of producing large amounts of power (tens of kilowatts and beyond) at frequencies relevant to electron cyclotron heating (ECH) (30–100s of GHz). The first ECH heating experiment was carried out on a small Soviet tokamak, Tuman-3 [40]. Later, the Soviet tokamak TM-3 used one of these sources, and produced clear evidence of electron heating as a result [2–4].

Lower hybrid (LH) heating, operating at roughly the geometric mean in frequencies between ICRF and ECH frequencies ($0.5 \rightarrow 5.0$ GHz), was of interest early on because it showed the propensity for current drive as well as heating [36], thus

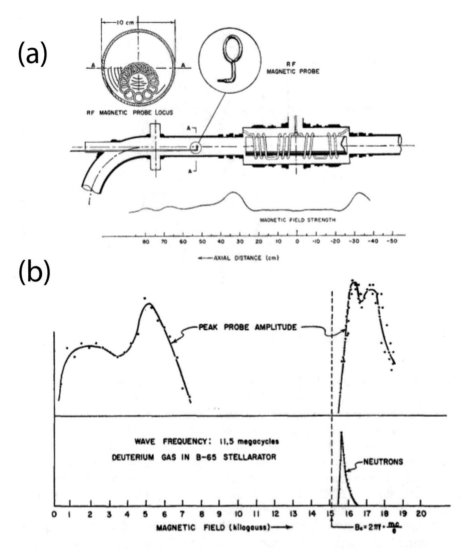

Fig. 9.16 First large-scale attempt at heating fusion plasma with radiofrequency wave heating. (**a**) Schematic of antenna and magnetic field geometry. (Note that the rotational fields were turned off; this is essentially a mirror field.) (**b**) Magnetic field fluctuations and neutron signal, showing resonance at $\omega = \Omega_{ci} - \delta\omega$. From [102]. Used with permission, American Institute of Physics

potentially removing an obstacle to steady-state operation of tokamaks. A proof-of-principle experiment was done at Princeton in 1980 [121], which clearly showed evidence of LH current drive. This led to full-scale deployment of LH heating and current drive on PLT [14], where LH power of 0.5 MW was available, and LH drive currents to 420 kA were produced.

Alfvén wave heating was first demonstrated on a small Soviet stellarator RT-0
and reported on in 1968 [27]. Other early experiments on small stellarators RT-02,
Uragan-2, Proto-Cleo, and Heliotron-D are described in [28, 41, 65, 85, 113], and
[31]. Experiments on tokamaks using Alfvén wave heating were first reported on by
de Chambrier from the Swiss tokamak TCA [25]. Subsequent Alfvén wave heating
experiments were performed on TEXTOR [11] and Tokapole [120].

9.3.2 RF Heating Basics: Propagation and Absorption of RF Energy in Plasma

Cold Plasma Waves

To understand waves in plasmas, we start with Maxwell's equations for the curl of
E and **B**, using an electric displacement **D**:

$$\nabla \times \mathbf{E} = -\frac{\partial \mathbf{B}}{\partial t}$$
$$\nabla \times \mathbf{B} = \mu_0 \frac{\partial \mathbf{D}}{\partial t},$$
$$(9.21)$$

and we define

$$\frac{\partial \mathbf{D}}{\partial t} = \varepsilon_0 \frac{\partial \mathbf{E}}{\partial t} + \mathbf{J}. \qquad (9.22)$$

(Note that in plasmas we do not allow for a permeability function, so we take $\mathbf{B} = \mu_0 \mathbf{H}$.) We can then define a dielectric tensor $\boldsymbol{\varepsilon}$ such that

$$\mathbf{D} = \varepsilon_0 \boldsymbol{\varepsilon} \cdot \mathbf{E}.$$

We assume that solutions to the Maxwell equations with a wave-like nature in a
uniform medium and small amplitudes can be written in a form

$$(\mathbf{E}(\mathbf{x}, t), \mathbf{B}(\mathbf{x}, t)) = e^{i(\mathbf{k} \cdot \mathbf{x} - \omega t)} (\mathbf{E}(\omega, \mathbf{k}), \mathbf{B}(\omega, \mathbf{k})).$$

We can further separate the free-space dielectric current from the plasma contribution to the current (through **J**) by writing

$$\boldsymbol{\varepsilon}(\omega, \mathbf{k}) = \mathbf{I} + \sum_s \boldsymbol{\chi}_s(\omega, \mathbf{k}).$$

Here **I** is the 3×3 identity matrix and the sum is over all species components s.

To get the contribution to the dielectric tensor from the plasma currents, we look at the equations of motion for the electrons and ions in a straight magnetic field $\mathbf{B} = B_0 \hat{\mathbf{z}}$ and assume that B_0 is much larger than the perturbation $\mathbf{B}$ field represented by the wave. (We also make the assumption here that the electron and ions have no thermal motion, hence initial conditions for the particles are dropped, and there is no drift motion for the particles, meaning that they experience electric and magnetic fields at the point of their origin. This is the essence of the"cold-plasma" assumption.) These are

$$- i\omega\, m_s v_{sx} = +q_s v_y B_0 + q_s E_x$$
$$- i\omega\, m_s v_{sy} = -q_s v_x B_0 + q_s E_y$$
$$- i\omega\, m_s v_{sz} = q_s E_z. \tag{9.23}$$

Solving for v_{sx}, v_{sy}, and v_{sz}, and substituting $B = m_s \Omega_s / q_s$ to introduce the cyclotron frequency Ω_s, we obtain

$$v_{sx} = \frac{q_s \left(-E_y \Omega_s + i E_x \omega \right)}{m_s \left(\omega^2 - \Omega_s^2 \right)}$$

$$v_{sy} = \frac{q_s \left(E_x \Omega_s + i E_y \omega \right)}{m_s \left(\omega^2 - \Omega_s^2 \right)} \tag{9.24}$$

$$v_{sz} = -\frac{i E_z q_s}{\omega m_s}.$$

Notice here that Ω_s is an "algebraic" quantity, i.e. it is negative for electrons and positive for ions.

We can use the forms for the velocities v_s to construct the dielectric tensor $\boldsymbol{\varepsilon}$ by using $\mathbf{J} = n_s q_s \mathbf{v}_s$. Then we have $\boldsymbol{\chi} \cdot \mathbf{E} = (1/(-i\omega\varepsilon_0))n_s q_s \mathbf{v}_s$. It is useful to use some auxiliary expressions (R, L, S, D, P) to construct the dielectric tensor:

$$R = 1 - \sum_s \frac{\omega_{ps}^2}{\omega(\omega + \Omega_s)}$$

$$L = 1 - \sum_s \frac{\omega_{ps}^2}{\omega(\omega - \Omega_s)}$$

$$S = \frac{1}{2}(R + L) \tag{9.25}$$

$$D = \frac{1}{2}(R - L)$$

$$P = 1 - \sum_s \frac{\omega_{ps}^2}{\omega^2}.$$

Here $\omega_{ps}^2 = n_s^2 q_s^2 / (m_s \varepsilon_0)$ is the square of the plasma frequency for species s. Then the dielectric tensor takes the rather simple form:

$$\varepsilon = \begin{pmatrix} S & -iD & 0 \\ iD & S & 0 \\ 0 & 0 & P \end{pmatrix}. \tag{9.26}$$

Note that the R function becomes infinite at the electron cyclotron frequency (recall that Ω_e is negative), and the L function becomes infinite at each of the ion cyclotron frequencies. This is because electrons spiral in a right-handed manner on $\mathbf{B}$, whereas the ions spiral in a left-handed sense.

We can complete the wave equation by taking another curl of the first equation in Eq. (9.21). We note that $\nabla \times \rightarrow i\mathbf{k} \times$ and we define $\mathbf{n} = c\mathbf{k}/\omega$ and use $\mu_0 = 1/(\varepsilon_0 c^2)$. Without loss of generality we take $k_y = 0$ since the z-axis is an axis of symmetry for the system. We take θ to be the angle between the $\mathbf{k}$ vector and $\mathbf{B}$. Then the $\mathbf{n} \times \mathbf{n} \times$ operator becomes

$$\mathbf{n} \times \mathbf{n} \times \mathbf{E} = \begin{pmatrix} -n^2 \cos^2 \theta & 0 & n^2 \cos \theta \sin \theta \\ 0 & -n^2 & 0 \\ n^2 \cos \theta \sin \theta & 0 & -n^2 \sin^2 \theta \end{pmatrix} \begin{pmatrix} E_x \\ E_y \\ E_z \end{pmatrix}.$$

and the wave equation

$$\mathbf{n} \times (\mathbf{n} \times \mathbf{E}) + \varepsilon \cdot \mathbf{E} = 0$$

becomes

$$\begin{pmatrix} S - n^2 \cos^2 \theta & -iD & n^2 \cos \theta \sin \theta \\ iD & S - n^2 & 0 \\ n^2 \cos \theta \sin \theta & 0 & P - n^2 \sin^2 \theta \end{pmatrix} \begin{pmatrix} E_x \\ E_y \\ E_z \end{pmatrix} = 0. \tag{9.27}$$

General solutions to this matrix equation without restrictions on the amplitudes of $\mathbf{E}$ require that the determinant of the matrix operating on $\mathbf{E}$ be zero. Despite the apparent sixth-order nature of the determinant, one can show that the determinant is divisible by n^2 and thus a quadratic equation in n^2 can be obtained:

$$An^4 - Bn^2 + C = 0, \qquad \text{where}$$
$$A = S\sin^2\theta + P\cos^2\theta,$$
$$B = RL\sin^2\theta + PS\left(1 + \cos^2\theta\right) \qquad (9.28)$$
$$C = PRL.$$

Manipulating the trigonometric terms yields an expression:

$$\tan^2\theta = -\frac{\left(n^2 - R\right)\left(n^2 - L\right)}{\left(Sn^2 - RL\right)\left(n^2 - P\right)}. \qquad (9.29)$$

From this expression, one can immediately determine the possible branches of the dispersion relation for propagation parallel to the magnetic field ($\tan\theta = 0$), and for perpendicular propagation ($\tan\theta = \infty$). For parallel propagation,

$$P = 0, \qquad n^2 = R, \qquad n^2 = L$$

and for perpendicular propagation

$$n^2 = \frac{RL}{S}, \qquad n^2 = P.$$

For the parallel case, the $P = 0$ solution represents the plasma oscillation (see Chap. 3), and the other two cases are called the "R" wave and the "L" wave, respectively, and the right- and left- circular polarization is consistent with these names. In fact, reading the second row of the matrix equation for $\mathbf{E}$ in Eq. (9.27) gives

$$\frac{iE_x}{E_y} = \frac{n^2 - S}{D}$$

and for $n^2 = R$ we have $iE_x/E_y = +1$ and for $n^2 = L$ we have $iE_x/E_y = -1$. For the perpendicular case, the first solution is called the "extraordinary" wave (or X-mode and the second is the "ordinary" wave or O-mode. (Note that as $B \to 0$, the only solutions in all directions are $P = 0$ and $n^2 = P$, and this is the plasma oscillation and cold unmagnetized plasma wave, as described in Chap. 3.)

The observation that two circularly polarized solutions exist for $\theta = 0$ brings up the notion of re-defining the electric field components to be

$$E^\pm = \frac{1}{2}\left(E_x \pm iE_y\right) \qquad (9.30)$$

and then the susceptibility tensor can also be written in terms of the right- and left-handed components as

$$\chi_s^\pm = -\frac{\omega_{ps}^2}{\omega\left(\omega \mp \Omega_s\right)}$$

$$\chi_{s,zz} = -\frac{\omega_{ps}^2}{\omega^2}.$$

(9.31)

We also define

$$v_s^\pm = \frac{1}{2}\left(v_x \pm i v_y\right)$$

and then the currents are

$$J_s^\pm = n_s q_s v_s^\pm \qquad = -i\omega\varepsilon_0 \chi_s^\pm E^\pm$$

$$J_{s,z} = n_s q_s v_{s,z} \quad = -i\omega\varepsilon_0 \chi_{s,zz} E_z.$$

(9.32)

Using these definitions and the properties of the susceptibility components $\chi_s^\pm$ then gives

$$\frac{i v_{s,x}}{v_{s,y}} = \frac{\left(\chi_s^+ + \chi_s^-\right)\left(i E_x/E_y\right) - \left(\chi_s^+ - \chi_s^-\right)}{\left(\chi_s^+ - \chi_s^-\right)\left(i E_x/E_y\right) - \left(\chi_s^+ + \chi_s^-\right)}$$

$$= -\frac{\left(\omega + \Omega_s\right)\left(n^2 - R\right) + \left(\omega - \Omega_s\right)\left(n^2 - L\right)}{\left(\omega + \Omega_s\right)\left(n^2 - R\right) - \left(\omega - \Omega_s\right)\left(n^2 - L\right)}.$$

(9.33)

Notice that the electron and ion motion perpendicular to **B** are also perfectly circular for purely R or L modes of propagation. Note, however, from Eq. (9.31), that the perpendicular velocities (and thus currents) go to infinity for left-handed waves for ions at $\omega = +\Omega_i$ and for electrons on right-handed waves at $\omega = -\Omega_e$ (a positive frequency). Thus the resonant nature of these waves is revealed here.

9.3.3   Cutoffs and Resonances

Allis [5, 6] suggested that the special cases where $n^2 \to 0$ be called "cutoffs" and $n^2 \to \infty$ be called "resonances." While it is straightforward to show that as a wave approaches cutoff, the wave energy is reflected back, the situation with resonances does not necessarily lead to absorption of the wave. In both cases, however, one has the notion that the characteristics of the medium are slowly changing over the distance of a reciprocal wavenumber (the so-called WKB approximation), and this condition does not strictly apply at either a cutoff or a resonance. Furthermore, the dielectric tensor must have a non-hermitian part for there to be coupling of wave energy into thermal energy in the plasma. This cannot happen for a perfectly hermitian dielectric tensor such as given in Eq. (9.26), so that one must develop a more sophisticated model in order to capture the essence of resonant energy coupling. Nevertheless, applying the canonical definitions of cutoffs and resonances to the cold-plasma dispersion relations gives some results providing insight.

For propagation parallel to **B**, Eq. (9.28) can be written, assuming that there is only one ion species and after dropping some terms of order Zm_e/m_i smaller than other terms:

$$\omega^2 = \omega_{pe}^2$$

$$\frac{k_\parallel^2 c^2}{\omega^2} = \frac{\omega^2 \pm \omega\Omega_e + \Omega_e\Omega_i - \omega_{pe}^2}{(\omega \pm \Omega_i)(\omega \pm \Omega_e)} \tag{9.34}$$

The cutoffs occur where the numerator is zero. We obtain a simple quadratic equation:

$$\omega^2 \pm \omega\Omega_e + \Omega_e\Omega_i - \omega_{pe}^2 = 0 \tag{9.35}$$

The upper sign is for the R wave and the lower sign is for the L wave. Solving for the cutoff frequencies (using only the positive roots of the quadratic) gives

$$\omega_R = +\frac{|\Omega_e|}{2} + \frac{1}{2}\sqrt{\Omega_e^2 - 4\Omega_e\Omega_i + 4\omega_{pe}^2}$$

$$\omega_L = -\frac{|\Omega_e|}{2} + \frac{1}{2}\sqrt{\Omega_e^2 - 4\Omega_e\Omega_i + 4\omega_{pe}^2} \tag{9.36}$$

Here the absolute values have been used to avoid confusion, since Ω_e is negative. We can assume that $\Omega_i|\Omega_e| \ll \omega_{pe}^2$ and drop this term for approximate work. Note that in this limit $\omega_R\omega_L = \omega_{pe}^2$, i.e. the two cutoff frequencies are on either side of ω_{pe}, with $|\Omega_e|$ as the difference in these frequencies. Also, for $|\Omega_e| \ll \omega_{pe}$, the two cutoff frequencies approach ω_{pe}, as one expects that the unmagnetized cutoff should reappear in the no-field limit.

We should also note that at frequencies below the R cutoff and also below the cyclotron frequency $\omega < |\Omega_e|$, the R wave re-emerges since both the numerator and denominator of the right-hand side of Eq. (9.34) have changed sign. Thus an R wave is possible everywhere $\omega < |\Omega_e|$. Similarly, the L wave re-emerges at a frequency $\omega < \Omega_i$, so that at low frequencies below Ω_i, there are both R waves and L waves.

The resonances for parallel propagation should be fairly obvious from Eq. (9.34): recalling that Ω_e is negative, choosing the $+$ sign (right-hand polarization) produces a resonant denominator at $\omega = -\Omega_e$, with the $-$ sign giving a resonant denominator at $\omega = +\Omega_i$.

For perpendicular propagation, we have two branches of the dispersion relation: one representing $n_\perp^2 = P$ and the other representing $n_\perp^2 = RL/S$. These are given, respectively, by:

$$k_\perp^2 c^2 = \omega^2 - \omega_{pe}^2$$

$$\frac{k_\perp^2 c^2}{\omega^2} = \frac{\left(\omega^2 + \omega\Omega_e + \Omega_e\Omega_i - \omega_{pe}^2\right)\left(\omega^2 - \omega\Omega_e + \Omega_e\Omega_i - \omega_{pe}^2\right)}{\left(\omega^2 - \omega_{LH}^2\right)\left(\omega^2 - \omega_{UH}^2\right)}$$

(9.37)

Here two new frequencies are defined: these are the lower hybrid resonance frequency ω_{LH}:

$$\frac{1}{\omega_{LH}^2} = \frac{1}{\Omega_i^2 + \omega_{pi}^2} + \frac{1}{|\Omega_i\Omega_e|}$$

(9.38)

and an upper hybrid resonance frequency ω_{UH}:

$$\omega_{UH}^2 = \Omega_e^2 + \omega_{pe}^2$$

(9.39)

Again terms of ordering Zm_e/m_i compared to one have been dropped.

From Eq. (9.37), it is apparent that the numerator has the form of the numerator(s) of Eq. (9.34), with each choice of sign, multiplied together. Therefore the cutoffs for perpendicular propagation remain $\omega = \omega_R$ and $\omega = \omega_L$ as in the case of parallel propagation. The resonances are at the lower hybrid and upper hybrid resonance frequencies as defined above.

Some simple examples are given here to illustrate the general nature of cutoffs and resonances as they apply to waves launched some distance away from the cutoff or resonance at arbitrary angles to magnetic field and gradients in the magnetic field and density. For the first example, shown in Fig. 9.17, we illustrate the propagation of an RF wave in an unmagnetized plasma with a density which is a linear function of x. We ignore the effects of the ions and then the dispersion relation reduces to:

$$\omega^2 = \omega_{pe}^2(x) + k^2 c^2 = \omega_{pe}^2(x) + (k_x^2 + k_z^2)c^2$$

(9.40)

with $\omega_{pe}^2(x)/\omega^2 = x$, i.e. the cutoff $\omega = \omega_{pe}$ occurs at $x = 1$. The wave is launched at $x = 0$ with various initial angles set by the values of k_x and k_z. If the initial value of k_z is very small, then the wave advances almost to the cutoff surface at $x = 1$ and then bounces back. As k_z increases, the maximum x-position attained is farther away from the cutoff. This is because the dispersion characteristics are independent of z, which in turn means that $k_z =$constant. Thus the cutoff density (and thus ω_{pe}^2) in the x-direction is lowered because of the portion of the squared wavenumber k^2 which is "locked up" in the k_z component. In this case the true definition of cutoff is the point where $k_x^2 = 0$, not where $k^2 = 0$.

Fig. 9.17 Ray tracing
applied to the case of an
electromagnetic wave
launched into an
unmagnetized cold plasma at
various launch angles with
density linearly increasing
from $x = 0$ to the right, with
$\omega_{pe} = \omega$ at $x = 1$

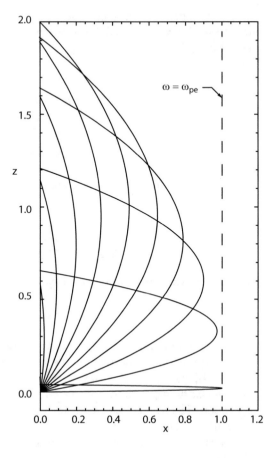

(As a historical note, the reflection of RF waves at grazing incidence was observed with the emergence of television after WWII. Television signals at a frequency of $40 \rightarrow 50$ MHz broadcast in England were sometimes viewable in the US, thousands of miles away. Even though the maximum plasma frequency in the ionosphere is about 6 MHz, a cutoff and reflection can occur at a place where $\omega \gg \omega_{pe}$, similar to Fig. 9.17 (in spherical geometry, of course!))

For the next example, we look at a plasma with a constant density but with a field in the z-direction which is decreasing uniformly for $z = 0$ such that

$$|\Omega_e(z)|/\omega = 2 - z.$$

(Although this magnetic field has $\nabla \cdot \mathbf{B} \neq 0$, ignoring Maxwell's equation for the static magnetic field does no harm here.) The plasma frequency was constant and chosen so that $\omega_{pe}^2/\omega^2 = 0.5$. We launch RF waves at $x = z = 0$ with a variety

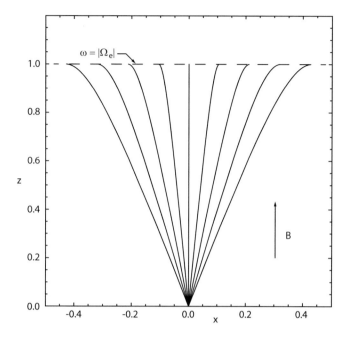

Fig. 9.18 Cold plasma ray tracing for a plasma with constant density and $\omega > \omega_{pe}$, and with magnetic field increasing linearly upward with $\omega = |\Omega_e|$ at $z = 1$

of wavenumbers k_x and k_z which satisfy the R branch of the dispersion relation, i.e. with the larger of the two values of n allowed by the quadratic equation in n^2 given by Eq. (9.28). Figure 9.18 shows the ray trajectories at launch and beyond. One can see that the ray trajectories have spread out somewhat as they have propagated, but all of the rays launched eventually encounter the resonance at $\omega = |\Omega_e|$. (The ray calculations were run for a time $t = 5L/c$, adequate for the waves to get very close to the resonance.)

In this example it is also important to look at the index of refraction in the z-direction $n_z = ck_z/\omega$. This is shown in Fig. 9.19 for a representative wave. Here one notices that the index of refraction n_z goes to infinity as the ray approaches the resonance. One might ask why all of the waves launch encounter the same resonance condition, even though the resonances given by Eqs. (9.34) and (9.37) show that the resonances clearly are functions of the angle of incidence. But note that in this example, there are no x-derivatives in the dispersion relation (thanks in part to ignoring Maxwell for the equilibrium!) and therefore k_x is a constant. Thus as the wave approaches resonance, $k_z \gg k_x$, and all of the launched rays have $\mathbf{k} \parallel \mathbf{B}$ as they approach the resonant surface. (Note that this also shows that the wave trajectory is determined by the group velocity $d\omega/d\mathbf{k}$ and not the phase velocity $(\omega/|\mathbf{k}|)\,\hat{\mathbf{k}}$.)

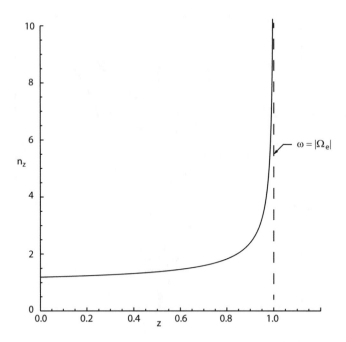

Fig. 9.19 Parallel index of refraction $n_\parallel$ vs. z for the case of a wave launch along **B** as shown in Fig. 9.18

9.3.4   Wave-Normal Surfaces

The quadratic equation in n^2 given by Eq. (9.28) can be re-written in terms of the inverse of the index of refraction, $u = 1/n$, yielding

$$Cu^4 - Bu^2 + A = 0, \tag{9.41}$$

and the vector $\mathbf{u} = +u \sin\theta \hat{\mathbf{x}} + u \cos\theta \hat{\mathbf{z}}$ is a dimensionless phase velocity for the wave. We can also use a cartesian representation for $\mathbf{k}$, so that

$$(u_x, u_z) = k_x(\omega/k^2), k_z(\omega/k^2).$$

Since the only trigonometric terms in Eq. (9.41) are terms in $\sin^2\theta$ and $\cos^2\theta$, one can see that the plots of $1/n$ vs. angle will have fourfold symmetry. Figure 9.20 shows wave-normal surfaces for some representative cases.

Figure 9.20a shows the case for very low frequency propagation where $\omega \ll \Omega_i$ (in this case, $|\Omega_e|/\omega_{pe} = 1.5$ the mass ratio $m_i/m_e = 3600$, and $\omega = \Omega_i/100$). We note that in this case, for propagation along **B**, there are two modes of propagation with identical phase velocities. This phase velocity is much less than the speed of light and is in fact the familiar Alfvén speed $v_A = B/\sqrt{\mu_0 \rho}$ used earlier in the

Fig. 9.20 Wave-normal
surfaces at three different
frequencies

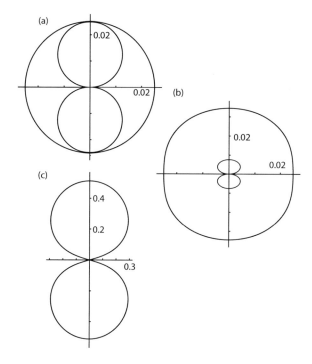

description of MHD waves, so that $n_\parallel = v_A/c$, typically much less than one. As the angle of propagation increases, one of the waves remains at the same speed and the other drops to zero: these are called the compressional Alfvén wave and the shear Alfvén wave, respectively. Figure 9.20b shows how the wave-normal surfaces change as the frequency is raised: this chart is generated with the same conditions as part (a) but with a frequency just under the ion cyclotron frequency, $\omega = \Omega_i/1.1$. While the two wave-normal surfaces have the same topological form, there is now a clear separation of the modes by their phase velocities, which is about a four-to-one ratio for parallel propagation. For these conditions, these waves are called the "fast wave" and the "slow wave." Waves of this type play a role in ion cyclotron heating, as we shall see later. Figure 9.20c shows another example with $|\Omega_e|/\omega_{pe} = 1.5$, mass ratio $m_i/m_e = 3600$, but now with $\omega = 400\Omega_i = |\Omega_e|/9$. Here there is only one solution to the dispersion relation, and the wave only exists for propagation with the **k** vector within some angle with **B**. Waves in this region of parameter space are called "whistler" waves. (The name has to do with early experience from WWI, where military personnel would attempt to pick up enemy telephone conversations by listening for "crosstalk" on open telephone lines adjacent to lines in use. While not seen from the wave-normal surface drawing, there is a strong variation in the group velocity vs. frequency in propagation parallel to **B**. Lightning in the upper atmosphere would excite waves in the magnetosphere, which would be heard on the telephone lines as a "whistle" with descending pitch for a second or so.)

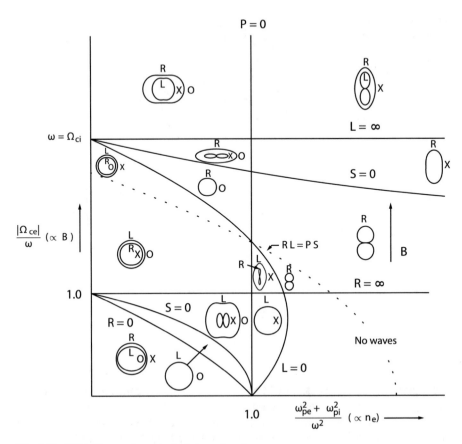

Fig. 9.21 CMA diagram for a single-ion species plasma

9.3.5   Accessibility and the CMA Diagram

An important question to be answered for any type of RF heating design is whether or not the wave launched at the edge of the plasma has access to the intended resonance. Although the roots of the dispersion relation given in Eq. (9.28) may appear straightforward, the combination of the three fundamental dimensionless parameters $(\omega_{pe}^2 + \omega_{pi}^2)/\omega^2$, $|\Omega_e|/\omega$, and $|\Omega_i|/\omega$, and the angle of propagation θ make for a rather complex system of propagation. A useful graphical tool to describe the modes of propagation in different parts of parameter space is shown in Fig. 9.21. This chart was introduced by P. C. Clemmow and R. F. Mullaly in 1955 [21], and in a modified form by Allis in 1959 [5]. The parameter space is described by choosing $|\Omega_e|/\omega$ as the y-axis and $(\omega_{pe}^2 + \omega_{pi}^2)/\omega^2$ as the x-axis. Because the ratio of ion cyclotron resonance to electron cyclotron resonance is a very large number in actuality (1:3600 in deuterium), an artificially low ion-to-electron mass ratio is typically used (for the chart shown here, a mass ratio of 1:2.5 was used). It can be shown that the change of mass ratio does not affect the organization of the regions on the CMA diagram.

The basic resonances at $\omega = |\Omega_e|$ and $\omega = \omega_i$ appear as straight horizontal lines on the CMA diagram and are labeled $R = \infty$ and $L = \infty$, respectively. The cutoff at $(\omega_{pe}^2 + \omega_{pi}^2) = \omega^2$ also appears as a vertical line on the diagram and is labeled $P = 0$. The R cutoff and L cutoff equations are simple algebraic expressions in x and y, as is the $S = 0$ curve, representing the lower hybrid and upper hybrid resonances. The curve labeled $RL = PS$ shows that the R and L branches for parallel propagation become re-associated with the X and O branches of perpendicular propagation about this curve.

In a real plasma experiment, the wave is launched from a place where the density is almost zero, and the trajectory of a wave is to generally progress in the x-direction on the CMA diagram. The magnetic field may increase or decrease as the wave travels inside the plasma. Figure 9.22 shows two possible trajectories for an attempt at RF heating at the electron cyclotron resonance frequency at the center of the plasma. Two cases are shown: the lower dashed line represents a wave launched in the X mode perpendicular to the magnetic field from the outboard side of a tokamak, and the upper dashed line shows a launch from the inboard side. Since the magnetic field in a tokamak is proportional to $1/R$, the field is lower on the outboard side of the plasma than at the center. The wave will encounter the R cutoff before it can get to the electron cyclotron resonance at $R = \infty$. Thus there is no accessibility to the electron cyclotron resonance for an outboard-side launch. The situation is different, however, for an inboard-side X-mode launch: here B is greater at the launch point than it is inside, and the upper dashed line in Fig. 9.22 shows the unencumbered path for the wave in CMA space to make it to the resonance.

One also notices that an O-mode launched from the outboard side encounters the electron cyclotron resonance without experiencing the R cutoff, provided that $\omega_{pe} < |\Omega_e|$ at all points in the trajectory. This is in fact the base case for ECRH and ECCD (electron cyclotron current drive) on ITER, which has launchers for 170 GHz microwaves on the low field side [50] and the ratio of $\omega_{pe}/|\Omega_e|$ is around 0.9:1.7. However, the cold plasma dispersion relation does not indicate that the O mode, which is connected to the L wave for perpendicular propagation, has a resonance at the electron cyclotron resonance frequency. Again, however, more sophisticated physics shows that other effects, such as relativity, trapping, and effects due to finite electron temperature, assure fairly adequate coupling of the RF energy at the electron cyclotron resonance. Similarly, plasmas are known to absorb energy at the second harmonic of the electron cyclotron resonance, $\omega = 2|\Omega_e|$. This extra resonance, not on the CMA diagram because it is a warm-plasma effect, will have access from an outboard-side launch with either very high-frequency sources (340 GHz for ITER) or at reduced magnetic field configurations (somewhere around half the nominal toroidal field, say, 2.6 T).

Fig. 9.22 CMA diagram
showing trajectory of waves
launched from the edge of a
plasma, where $n = 0$, to the
electron cyclotron resonance,
depending on the increase or
decrease of the magnetic field
from its value at the edge to
the resonance, where
$B =_r= m_e|\Omega_e/q_e|$

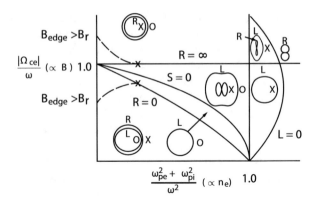

9.3.6 Ray Tracing

The notion of a "wave packet" is an important component to the understanding
of the wave solution for a free particle in quantum mechanics, although the original
concept came from work by W. R. Hamilton in 1828 [45]. The germ of the concept is
the conservation of "wave action" for a wave in slowly varying conditions (the WKB
approximation). The modern interpretation of Hamilton's result is from Landau and
Lifshitz [73] and Weinberg's article in 1962 [117]. The general idea is that the local
wave solution can be written as $A \exp(i(\mathbf{k} \cdot \mathbf{x} - \omega t)$, but with A and $\mathbf{k}$ as slowly
varying. The frequency ω may vary as well, but only in a medium that is explicitly
varying in time. We assume that there is a conserved quantity $g(\mathbf{k}, \mathbf{x}, \omega, t)$ that
is exactly zero along the path of the ray. (An example of such a function is the
dispersion determinant D given earlier, but sometimes it is simpler to work with
simpler forms if we are only concerned with a particular branch or approximate
form to the dispersion relation.). Then the condition for g to be conserved along a
ray path is that

$$\delta g = \frac{\partial g}{\partial \mathbf{k}} \cdot \delta \mathbf{k} + \frac{\partial g}{\partial \mathbf{x}} \cdot \delta \mathbf{x} + \frac{\partial g}{\partial \omega} \delta \omega + \frac{\partial g}{\partial t} \delta t = 0. \tag{9.42}$$

If we consider τ to be the distance along the ray, then the following set of
differential relations for $(\mathbf{k}, \mathbf{x}, \omega, t)$ can be constructed:

$$\frac{d\mathbf{x}}{d\tau} = \frac{\partial g}{\partial \mathbf{k}},$$

$$\frac{d\mathbf{k}}{d\tau} = -\frac{\partial g}{\partial \mathbf{x}},$$

$$\frac{dt}{d\tau} = -\frac{\partial g}{\partial \omega}, \tag{9.43}$$

$$\frac{d\omega}{d\tau} = \frac{\partial g}{\partial t}.$$

This satisfies Eq. (9.42) exactly. Now, using the ordinary differential form $d\tau = -dt/(\partial g/\partial \omega)^{-1}$ from the third line of Eq. (9.43), we have:

$$\frac{d\mathbf{x}}{dt} = -\frac{\partial g}{\partial \mathbf{k}}\left(\frac{\partial g}{\partial \omega}\right)^{-1},$$

$$\frac{d\mathbf{k}}{dt} = +\frac{\partial g}{\partial \mathbf{x}}\left(\frac{\partial g}{\partial \omega}\right)^{-1}, \qquad (9.44)$$

$$\frac{d\omega}{dt} = \frac{\partial \omega}{\partial t}.$$

If we choose a simple form for g such that $g = \omega - \hat{\omega}(\mathbf{k}, \mathbf{x}, t)$, then the ray tracing equations simplify to

$$\frac{d\mathbf{x}}{dt} = \frac{\partial \hat{\omega}}{\partial \mathbf{k}},$$

$$\frac{d\mathbf{k}}{dt} = -\frac{\partial \hat{\omega}}{\partial \mathbf{x}}, \qquad (9.45)$$

$$\frac{d\omega}{dt} = \frac{\partial \hat{\omega}}{\partial t},$$

which recovers the simple expression for the group velocity $\mathbf{v}_g = \partial \hat{\omega}/\partial \mathbf{k}$ from simple wave-packet analysis.

One notices that the equations presented in (9.45) have the exact form of Hamilton's equations from classical dynamics, with $\hat{\omega}$ standing in for the Hamiltonian $\mathscr{H}$ and the wave vector $\mathbf{k}$ standing in for the canonical momentum $\mathbf{p}$. Thus we can borrow many important concepts from Hamiltonian mechanics. For one thing, in cartesian geometry (x, y, z), if there is asymmetry in one coordinate direction, such as $\partial D/\partial z = 0$, then the corresponding canonical "momentum" (k_z in this case) is a constant. This property already appeared in the dispersion relations involved in Figs. 9.17, 9.18, and 9.19 shown earlier. Furthermore, in a system with azimuthal symmetry, such as a tokamak with $\partial/\partial \phi = 0$, with ϕ being the toroidal coordinate, then the quantity Rk_ϕ is conserved for the same reason that Rp_ϕ would be conserved in particle motion (without a toroidal vector potential: there is no such analog in the ray tracing equations). Looking at the phase dependence along the ray path in the toroidal direction, this translates into $\exp(ik_\phi \cdot (R\phi)) = \exp(in\phi)$, where the "quantum number" $n = Rk_\phi$ is conserved.

Another result, which has a rather lengthy classical derivation, concerns the transport of energy in a plasma by a bundle of waves which can be thought of as a "ray bundle" of nearby wavelets, each satisfying the ray-tracing equations given above. We can treat the energy of each wavelet as $W_{wp} = N_{wp}\hbar\omega$, where N_{wp} is the number of photons in an electromagnetic wave packet or the number of plasmons in an electrostatic wave packet. The conservation of photons is then determined by equation similar to the particle conservation law in fluid mechanics, which then gives

$$\frac{\partial N_{wp}}{\partial t} + \nabla \cdot \left(\mathbf{v}_g N_{wp} \right) = 2\gamma N_{wp}. \tag{9.46}$$

Here γ is the growth rate (or damping rate if negative) for the wave and is derived in the classical limit by

$$\gamma(\omega, \mathbf{k}, \mathbf{x}, t) = -\frac{\varepsilon_0}{2} \frac{\omega(\mathbf{k}, \mathbf{x}, t)}{W(\omega, \mathbf{k}, \mathbf{x}, t)} \mathbf{E}(\omega(-\mathbf{k}), -\mathbf{k}, \mathbf{x}, t) \cdot \boldsymbol{\varepsilon}_a \cdot \mathbf{E}(\omega, \mathbf{k}, \mathbf{x}, t). \tag{9.47}$$

Here $\boldsymbol{\varepsilon}_a$ is the anti-hermitian part of the dielectric tensor and accounts for all energy sources and sinks at this place in $(\omega, \mathbf{k}, \mathbf{x})$ space at time t. For a non-time-varying system where ω is constant, then Eq. (9.46) can be written as a transport equation for the wave-packet energy density as

$$\frac{\partial W_{wp}}{\partial t} + \nabla \cdot \left(\mathbf{v}_g W_{wp} \right) = 2\gamma W_{wp}. \tag{9.48}$$

Ray tracing has been a useful tool for determining the accessibility of RF energy to resonance in systems which have proven to be too complicated for predictions based upon simplified geometry and a cold plasma approximation could not be used because effects such as toroidicity and the effect of warm-plasma modifications to the dielectric tensor have been important. An example of an early application of the ray-tracing method to the problem of lower hybrid accessibility in tokamaks was given in Ref. [17]. Here a ray-tracing model with a warm-plasma contribution to the dispersion determinant D was used (a term in n^6 was added to the dispersion relation) with a model toroidal geometry. Even with the simple model used, a host of effects, including stochasticity of the rays involved and mode conversion, were seen that did not show up in slab-model calculations.

9.3.7 Bernstein Waves

The cold plasma dispersion relation is derived with the notion that the electron and ion dynamics of the particles in the plasma being exposed to the electric and magnetic fields due to the wave are experiencing those fields at a point, and the radius of gyration of the particles around an equilibrium field line is negligible compared to the characteristic lengths of the wave itself, i.e. $k_\perp \rho_e$ and $k_\perp \rho_i \ll 1$. For the case of motion parallel to the magnetic field, we must assume that $k v_{th} \ll \omega$ for this to hold as well.

The dielectric tensor for the plasma is modified when this condition is dropped. The perturbed current due to the wave must then be calculated by an ensemble average of the particle's motion due to the wave field. In other words, we must solve the perturbed Vlasov equation for the system. Starting with the Vlasov equation, and linearizing the equation and keeping only the first-order terms ($f = f_0 + \delta f$), gives an expression for the perturbed distribution function

$$\frac{d\delta f_\alpha}{dt} = -q_\alpha \left(\mathbf{E}_1 + \mathbf{v} \times \mathbf{B}_1\right) \cdot \frac{\partial f_{\alpha 0}}{\partial \mathbf{p}}, \tag{9.49}$$

which can be integrated along the path of the particle to give

$$d\delta f_\alpha(\mathbf{x}, \mathbf{p}, t) = -q_\alpha \int_{-\infty}^{t} dt' \left(\mathbf{E}_1\left(\mathbf{x}', \mathbf{p}', t'\right) + \mathbf{v}' \times \mathbf{B}_1\left(\mathbf{x}', \mathbf{p}', t'\right)\right) \cdot \frac{\partial f_{\alpha 0}\left(\mathbf{p}'\right)}{\partial \mathbf{p}'}. \tag{9.50}$$

(In both Eqs. (9.49) and (9.50) we have chosen $\mathbf{p} = \gamma m_0 \mathbf{v}$ instead of $\mathbf{v}$ as the velocity-like dependent variable for f, which allows for a relativistically consistent approach, if necessary. Note that the relativistic factor $\gamma = 1 + \mathbf{p} \cdot \mathbf{p}/(m_0 c)^2$.) We let the perturbed fields $\mathbf{E}_1\left(\mathbf{x}', t\right)$ and $\mathbf{B}_1\left(\mathbf{x}', t\right)$ have a time dependence $\exp\left(i\mathbf{k} \cdot \mathbf{x}' - i\omega t'\right)$. We can replace the $\mathbf{v}' \times \mathbf{B}'$ term in Eq. (9.50) by noting from Maxwell's equation that $\mathbf{B}_1 = (kc/\omega) \times \mathbf{E}_1$, and then we have:

$$d\delta f_\alpha(\mathbf{x}, \mathbf{p}, t) = -q_\alpha \int_{-\infty}^{t} dt' \exp\left(i\mathbf{k} \cdot \mathbf{x}' - i\omega t'\right)$$

$$\times \mathbf{E}_1 \cdot \left[\mathbf{I}\left(1 - \frac{\mathbf{v}' \cdot \mathbf{k}}{\omega}\right) + \frac{\mathbf{v}'\mathbf{k}}{\omega}\right] \cdot \frac{\partial f_{\alpha 0}\left(\mathbf{p}'\right)}{\partial \mathbf{p}'}. \tag{9.51}$$

To model the particle dynamics, we keep the assumption that the magnetic field is constant as in the cold-plasma case and thus $\mathbf{B} = B_0 \hat{\mathbf{z}}$. The equation of motion for the (unperturbed) orbit is then

$$\frac{d\mathbf{p}'}{dt} = \Omega \mathbf{p}' \times \hat{\mathbf{z}} \tag{9.52}$$

where $\Omega = q B_0/(\gamma m_0)$. Note that the particle energy is constant as well as v, Ω, and γ.

We then have a solution for a particle at position x, y at time t and velocity $v_x = v_\perp \cos\phi$ and $v_y = v_\perp \sin\phi$ at earlier times t' by using a time-difference variable $\tau = t - t'$ by writing, for the velocities:

$$v'_x = v_\perp \cos(\phi + \Omega\tau)$$

$$v'_y = v_\perp \sin(\phi + \Omega\tau) \tag{9.53}$$

$$v'_z = v_\parallel$$

and for the positions x', y', z':

$$x' = x - \frac{v_\perp}{\Omega} [\sin(\phi + \Omega\tau) - \sin\phi]$$

$$y' = y + \frac{v_\perp}{\Omega} [\cos(\phi + \Omega\tau) - \cos\phi] \qquad (9.54)$$

$$z' = z - v_\parallel \tau.$$

We define an angle θ such that $k_x = k_\perp \cos\theta$ and $k_y = k_\perp \sin\theta$. We can then examine the phase factor in Eq. (9.51) as

$$\exp\left(i\mathbf{k} \cdot \mathbf{x}' - i\omega t'\right) = \exp\left(i\mathbf{k} \cdot \mathbf{x} - i\omega t + i\beta\right) \qquad (9.55)$$

with

$$\beta = -\frac{k_\perp v_\perp}{\Omega} [\sin(\phi - \theta + \Omega\tau) - \sin(\phi - \theta)] + \left(\omega - k_\parallel z\right)\tau \qquad (9.56)$$

and some derivatives related to $\partial f_0/\partial\mathbf{p}$ appearing in Eq. (9.51) are needed:

$$U = \frac{\partial f_0}{\partial p_\perp} + \frac{k_\parallel}{\omega}\left(v_\perp \frac{\partial f_0}{\partial p_\parallel} - v_\parallel \frac{\partial f_0}{\partial p_\perp}\right)$$

$$V = \frac{k_\perp}{\omega}\left(v_\perp \frac{\partial f_0}{\partial p_\parallel} - v_\parallel \frac{\partial f_0}{\partial p_\perp}\right) \qquad (9.57)$$

$$W = \left(1 - \frac{n\Omega}{\omega}\right)\frac{\partial f_0}{\partial p_\parallel} + \frac{n\Omega p_\parallel}{\omega p_\perp}\frac{\partial f_0}{\partial p_\perp}.$$

We can then use these quantities U, V, and W to find a compact form for δf:

$$\delta f(\mathbf{x}, \mathbf{p}, t) = -q \exp\left(i\mathbf{k} \cdot \mathbf{x} - i\omega t\right) \int_0^\infty d\tau\, e^{i\beta}\Big\{ E_x U \cos(\phi + \Omega\tau)$$

$$+ E_y U \sin(\phi + \Omega\tau) + E_z \left[\frac{\partial f_0}{\partial p_\parallel} - V \cos(\phi - \theta + \Omega\tau)\right]\Big\}. \qquad (9.58)$$

Then the dielectric tensor $\boldsymbol{\varepsilon}(\omega, \mathbf{k})$ can be written, as before:

$$\boldsymbol{\varepsilon}(\omega, \mathbf{k}) = \mathbf{I} + \sum_s \boldsymbol{\chi}_s(\omega, \mathbf{k}), \qquad (9.59)$$

where the susceptibilities $\boldsymbol{\chi}_s$ are defined by

$$\mathbf{J} = -i\omega\varepsilon_0 \sum_s \boldsymbol{\chi}_s \cdot \mathbf{E}. \qquad (9.60)$$

Without loss of generality we can set $\theta = 0$, i.e. the only components of $\mathbf{k}$ are k_x and k_z, as was done in the cold-plasma wave case. We next can deal with the phase term $\exp(i\beta)$ appearing in Eq. (9.58), which has terms of form $\exp(iz\sin(\phi + \Omega\tau))\exp(iz\sin\phi)$, where $z \equiv k_{\perp}v_{\perp}/\Omega$. Using Bessel's identity

$$e^{iz\sin Q} = \sum_{n=-\infty}^{\infty} e^{inQ} J_n(z), \qquad (9.61)$$

with Q representing ϕ and $-(\phi - \Omega\tau)$ for the two exponentials and J_n being the usual Bessel function of order n, along with similar expressions found by taking derivatives in ϕ and z. Substituting in the dependence on τ from the Bessel identities back into Eq. (9.58) results in an infinite integral over τ with an exponential of a purely imaginary argument. We let ω have a small imaginary part, i.e. we assume that the wave has a slowly growing part. The the τ-integration converges, and we have integrals of the form

$$-q \int_0^{\infty} d\tau \exp\left[i\left(\omega - k_{\parallel}v_{\parallel} - n\Omega\right)\tau\right] = \frac{-iq}{\omega - k_{\parallel} - n\Omega}. \qquad (9.62)$$

These substitutions result in an expression for the susceptibility χ in terms of products of Bessel functions and their derivatives and integration of momenta and the operators U, V and W over the velocity space. This expression is

$$\chi_s = \frac{\omega_{p0s}^2}{\omega\Omega_{0s}} \int_0^{\infty} 2\pi p_{\perp} dp_{\perp} \int_{-\infty}^{\infty} dp_{\parallel} \left[\hat{\mathbf{e}}_{\parallel}\hat{\mathbf{e}}_{\parallel} \frac{\Omega}{\omega} \left(\frac{1}{p_{\parallel}} \frac{\partial f_0}{\partial p_{\parallel}} - \frac{1}{p_{\perp}} \frac{\partial f_0}{\partial p_{\perp}} \right) p_{\parallel}^2 \right.$$

$$\left. + \sum_{n=-\infty}^{\infty} \left(\frac{\Omega p_{\perp} U}{\omega - k_{\parallel}v_{\parallel} - n\Omega} \mathbf{T}_n \right) \right] \qquad (9.63)$$

where the tensor $\mathbf{T}_n$ is given by:

$$\mathbf{T}_n = \begin{pmatrix} \dfrac{n^2 J_n^2}{z^2} & \dfrac{in J_n J_n'}{z} & \dfrac{n J_n^2 p_{\parallel}}{z p_{\perp}} \\[2ex] -\dfrac{in J_n J_n'}{z} & (J_n')^2 & -i\dfrac{J_n J_n' p_{\parallel}}{p_{\perp}} \\[2ex] \dfrac{n J_n^2 p_{\parallel}}{z p_{\perp}} & i\dfrac{J_n J_n' p_{\parallel}}{p_{\perp}} & \dfrac{J_n^2 p_{\parallel}^2}{p_{\perp}^2} \end{pmatrix}. \qquad (9.64)$$

Here J_n and J_n' have an argument $z = k_{\perp}v_{\perp}/\Omega$. The integration must be taken over a "Landau contour," meaning that the integration over $p_{\parallel}$ should be done along the real axis in complex $p_{\parallel}$ space, with a small semicircular indentation around the poles in the expression where $\omega - k_\parallel v_\parallel - n\Omega = 0$. Note also that the density has been moved into the numerator in the leading term ω_{p0s}^2, so that the normalization taken for the distribution function is $\int 2\pi p_\perp dp_\perp \int p_\parallel f_0 = 1$.

For the special case of a nonrelativistic maxwellian (in $v_\perp$) plasma with a distribution function given by:

$$f_0(v_\perp, v_\parallel) = h(v_\parallel)\frac{1}{\pi w_\perp^2} \exp\left(-\frac{v_\perp^2}{w_\perp^2}\right),$$

$$v_\perp^2 = v_x^2 + v_y^2, \qquad\qquad w_\perp^2 = \frac{2T_\perp}{m}, \qquad\qquad (9.65)$$

$$\int_{-\infty}^{\infty} dv_\parallel h(v_\parallel) = 1.$$

We can make use of a Bessel identity [116]:

$$\int_0^\infty t\,dt\, J_\nu(at) J_\nu(bt) \exp(-p^2 t^2) = \frac{1}{2p^2} \exp\left(-\frac{a^2 + b^2}{4p^2}\right) I_\nu\left(\frac{ab}{2p^2}\right)$$

$$(\text{Re } \nu > -1, |\arg p| < \pi/4).$$
$$(9.66)$$

Here I_n is the modified Bessel function, $I_n(\lambda) = i^{-n} J_n(i\lambda)$. Then the dielectric tensor becomes

$$\boldsymbol{\chi}_s = \left[\hat{\mathbf{e}}_\parallel \hat{\mathbf{e}}_\parallel \frac{2\omega_p^2}{\omega k_\parallel w_\perp^2} <v_\parallel> + \frac{\omega_p^2}{\omega} \sum_{-\infty}^{\infty} \exp(-\lambda)\mathbf{Y}_n(\lambda)\right]_s. \qquad (9.67)$$

Here $\lambda \equiv k_\perp^2 \rho_i^2/2 = k_\perp^2 T/(m\Omega^2)$, and

$$\mathbf{Y}_n(\lambda) = \begin{pmatrix} \dfrac{n^2 I_n}{\lambda} A_n & -in(I_n - I_n')A_n & \dfrac{k_\perp}{\Omega}\dfrac{nI_n}{\lambda} B_n \\[3mm] in(I_n - I_n')A_n & \left(\dfrac{n^2}{\lambda}I_n + 2\lambda I_n - 2\lambda I_n'\right)A_n & \dfrac{ik_\perp}{\Omega}\left(I_n - I_n'\right)B_n \\[3mm] \dfrac{k_\perp}{\Omega}\dfrac{nI_n}{\lambda} B_n & -\dfrac{ik_\perp}{\Omega}\left(I_n - I_n'\right)B_n & \dfrac{2(\omega - n\Omega)}{k_\parallel w_\perp^2}I_n B_n \end{pmatrix}. $$
$$(9.68)$$

Here the coefficients A_n and B_n are moments of the parallel distribution function $h(v_\parallel)$, given by:

$$A_n = \int_{-\infty}^{\infty} dv_{\parallel} \frac{H(v_{\parallel})}{\omega - k_{\parallel} - n\Omega}$$

$$B_n = \int_{-\infty}^{\infty} dv_{\parallel} \frac{v_{\parallel} H(v_{\parallel})}{\omega - k_{\parallel} - n\Omega}$$

$$(9.69)$$

where $H(v_{\parallel})$ is defined as

$$H(v_{\parallel}) = -\left(1 - \frac{k_{\parallel} v_{\parallel}}{\omega}\right) h(v_{\parallel}) + \frac{k_{\parallel} w_{\perp}^2}{2\omega} h'(v_{\parallel}). \tag{9.70}$$

Here $w_{\parallel} = (2T_{\parallel}/m)^{1/2}$. The case where there is no average drift motion along $\mathbf{B}$ ($V = 0$) and the parallel velocity distribution is also maxwellian with the same temperature as the perpendicular temperature, $T_{\parallel} = T_{\perp}$, has a particularly simple form for the A_n and B_n coefficients:

$$A_n = \frac{1}{k_{\parallel} w_{\parallel}} Z_0(\zeta_n)$$

$$B_n = \frac{1}{k_{\parallel}} [1 + \zeta_n Z_0(\zeta_n)] = -\frac{1}{2k_{\parallel}} \frac{dZ_0(\zeta_n)}{d\zeta_n} \tag{9.71}$$

$$\zeta_n = \frac{\omega - k_{\parallel} V - n\Omega}{k_{\parallel} w_{\parallel}}$$

where $Z_0(\zeta)$ is a variation on the Fried-Conte plasma dispersion function $Z(\zeta)$ given by [37]:

$$Z(\zeta) = \frac{1}{\sqrt{\pi}} \int_{-\infty}^{\infty} dz \frac{\exp(-z^2)}{z - \zeta}, \qquad \mathrm{Im}\, \zeta > 0 \tag{9.72}$$

or equivalently $Z(\zeta) = i\sqrt{\pi} \exp(-\zeta^2)[1 + \mathrm{erf}(i\zeta)]$. The modified function Z_0 used here is defined so that

$$Z_0(\zeta) = i \int_0^{\infty\,\mathrm{sgn}\,k_{\parallel}} dz \exp\left(i\zeta z - \frac{z^2}{4}\right)$$

$$= Z(\zeta) \qquad \text{for } k_{\parallel} > 0$$

$$= -Z(-\zeta) \qquad \text{for } k_{\parallel} < 0. \tag{9.73}$$

The general form for χ given by Eq. (9.68), even with the special case of an isotropic maxwellian plasma with only one species, still yields terms with infinite series with terms like

$$\chi_{xx} = \frac{\omega_p^2}{\omega}\exp(-k_\perp^2 T_\perp/(m\Omega^2))\sum_{-\infty}^{\infty}\frac{n^2 I_n(k_\perp^2 T_\perp/(m\Omega^2))}{k_\perp T_\perp/(m\Omega^2)}\frac{Z_0((\omega - n\Omega)/(k_\| w_\|))}{k_\| w_\|},$$

$$(9.74)$$

and if one considers the Z_0 function to be an infinite series, it looks like a daunting task to represent the warm-plasma dielectric susceptibilities in terms of elementary functions. However, a number of special cases and approximations of this dielectric tensor contain a great deal of the physics of plasma waves. These cases are itemized here.

1. **Cold plasma waves.** First we look at the limit where $T \to 0$. We note that the series expansion of the modified Bessel function is given by

$$I_n(\lambda) = \frac{1}{n!}\left(\frac{\lambda}{2}\right)^n\left[1 + \frac{(\lambda/2)^2}{1(n+1)} + \frac{(\lambda/2)^4}{1\cdot 2(n+1)(n+2)}\right.$$

$$\left. + \frac{(\lambda/2)^6}{1\cdot 2\cdot 3(n+1)(n+2)(n+3)} + \cdots\right].$$

$$(9.75)$$

For the term Y_{xx}, note that the contribution from $n = 0$ is zero and the lowest-order terms are for $n = \pm 1$. Then $I_1(\lambda) \approx \lambda/2$. Then $A_{\pm 1} = 1/(k_\| w_\|)Z_0((\omega \mp \Omega)/(k_\| w_\|))$. Since the parallel thermal velocity $w_\| = (2T/m)^{1/2} \to 0$, we can use the asymptotic form for the Z_0 function as $\zeta \to \infty$, which is

$$Z_0(\zeta) = -\frac{1}{\zeta} - \frac{1}{2\zeta^3} - \cdots + i\sigma\sqrt{\pi}\,\mathrm{sgn}(k_\|)\left(\exp(-\zeta^2)\right).$$

$$(9.76)$$

Here $\sigma = 0$ for $\mathrm{sgn}(k_\|)\mathrm{Im}(\zeta) = \mathrm{sgn}(\mathrm{Im}(\omega)) > 0$, $\sigma = 2$ for $\mathrm{sgn}(k_\|)\mathrm{Im}(\zeta) = \mathrm{sgn}(\mathrm{Im}(\omega)) < 0$, except that $\sigma = 1$ if $|\mathrm{Re}\zeta| \gg 1$ and $|\mathrm{Re}\zeta||\mathrm{Im}(\zeta)| \le \pi/4$. Then we have, keeping only the term in $1/\zeta$:

$$\chi_{xx} = -\frac{1}{2}\frac{\omega_p^2}{\omega}\left[\frac{1}{\omega(\omega - \Omega)} + \frac{1}{\omega(\omega + \Omega)}\right],$$

$$(9.77)$$

and then, using $\boldsymbol{\varepsilon} = \mathbf{I} + \boldsymbol{\chi}$,

$$\varepsilon_{xx} = 1 - \frac{1}{2}\frac{\omega_p^2}{\omega}\left[\frac{1}{\omega(\omega - \Omega)} + \frac{1}{\omega(\omega + \Omega)}\right]$$

$$= \frac{R + L}{2}$$

$$= S$$

$$(9.78)$$

where we have used the cold-plasma definitions of R, L, and S from Eq. (9.24). The other terms proceed similarly, with the exception that the lowest term in χ_{zz} is for $n = 0$. The overall asymptotic form for the hot-plasma dielectric tensor as

$T \to 0$ is the same as the cold-plasma dielectric tensor given in Eq. (9.26):

$$\lim_{T \to 0} \boldsymbol{\varepsilon}^h = \begin{pmatrix} S & -iD & 0 \\ \\ iD & S & 0 \\ \\ 0 & 0 & P \end{pmatrix}. \tag{9.79}$$

2. **Electrostatic waves in unmagnetized plasma.** We now look at the case where the magnetic field is small, and we look at electrostatic waves, where $\delta \mathbf{B} \propto \mathbf{k} \times \mathbf{E} \approx 0$. We can set $\lambda = 0$ by assuming that $k_\perp \to 0$. Then keeping only the $n = 0$ term in χ_{zz} and keeping ζ finite, we arrive at an expression

$$\varepsilon_{zz} = 1 - \sum_s \frac{\omega_{ps}^2}{k_\parallel^2 w_\perp^2} Z_0'(\zeta_s) = 0. \tag{9.80}$$

For high frequency waves, we can drop the ion susceptibility term, so that:

$$\frac{\omega^2}{\omega_{pe}^2} = \zeta^2 Z_0(\zeta). \tag{9.81}$$

Then using Eq. (9.76) and assuming that $|\mathrm{Re}\zeta| |\mathrm{Im}\zeta| \leq \pi/4$, we have

$$Z'(\zeta) = +1/\zeta^2 + 3/(2\zeta^4) + \cdots - 2i\sqrt{\pi}\zeta \exp(-\zeta^2). \tag{9.82}$$

Then, noting that $\zeta = \omega/(kw_\parallel) = \omega/(\sqrt{2}kv_{the})$, with $v_{the} \equiv (T_e/m_e)^{1/2}$, and dropping the small imaginary term:

$$\omega^2 = \omega_{pe}^2 + 3kv_{the}^2. \tag{9.83}$$

This is the famous Langmuir wave dispersion relation, also called the Bohm-Gross plasma dispersion relation. (Plasma waves following this dispersion relation are sometimes called "plasmons.") This wave is purely electrostatic, since it was derived using only the ε_{zz} term in the plasma dielectric tensor.

We now turn our attention to the imaginary term, which is small when $|\mathrm{Re}\zeta| \gg 1$:

$$\mathrm{Im}\, Z_0'(\zeta) \approx -2\sqrt{\pi}\zeta \exp(-\zeta^2) \tag{9.84}$$

Suppose that we excite a wave with a real frequency ω, i.e. a plasma excitation from a constant-amplitude frequency source. Then we can deduce that there must be an imaginary part to the wavenumber k, since the imaginary part of the dispersion relation is just

$$\text{Im}\left[\zeta^2 Z_0'(\zeta)\right] = 0. \tag{9.85}$$

Since the leading term in the $\zeta \gg 1$ expansion of $Z_0'(\zeta)$ is $1/\zeta^2$, the first term in the expansion of $\zeta^2 Z_0'(\zeta)$ has no imaginary part, and we must turn to the $1/\zeta^4$ term in order to find an imaginary part to balance the equation. Taking a Taylor expansion on the $1/\zeta^4$ term then yields:

$$\frac{\text{Im}\,k}{k} \approx \frac{2\sqrt{\pi}\zeta^5}{3}\exp(-\zeta^2). \tag{9.86}$$

This expression shows that there is some attenuation of the wave amplitude $\propto \exp(-[\text{Im}\,k]x)$ as the wave propagates in the $+\hat{x}$ direction. This wave damping phenomenon is called Landau damping, first examined by L. D. Landau in 1946 [72]. At first this result was rather controversial, because it was not believed that a plasma could exhibit wave damping without collisions present. However, the effect is real, and experimental measurements have confirmed the effect, as shown in Figs. 9.23 and 9.24. The explanation for this collisionless damping process is that the wave stores energy in a non-maxwellian distortion in the electron distribution function. A flattening of the distribution function in the neighborhood of the phase velocity $v \approx \omega/k$ starts to appear as the wave amplitude is increased. In fact, since we then have a situation where the average velocity $< v > \propto \int d^3 v f(v)v$ is nonzero, this phenomenon can be used to drive currents in the plasma.

One should also notice that the complex roots to the warm-electron dispersion relation $\omega^2/\omega_{pe}^2 = \zeta^2 Z_0'(\zeta)$ deviate from Eqs. (9.83) and (9.85) at fairly low values of ω/ω_{pe} above one, so that these equations are not reliable at $\omega/\omega_{pe} > 1.1$. The actual roots to the Vlasov dispersion relation are shown in Fig. 9.24 for values of $1 \leq \omega/\omega_{pe} \leq 2.5$, and there is hardly any semblance to the Bohm-Gross dispersion relation over this range. (This is important to keep in mind when one is studying stimulated Raman scattering (SRS) in laser fusion applications, where the nonlinearly produced sideband is a warm plasma wave of this type.)

3. **Ion acoustic waves.** We now introduce the distinction between "warm" and "hot." The above treatment for the electrostatic electron plasma wave leading to the Bohm-Gross dispersion relation uses an asymptotic form for the plasma dispersion function $Z_0(\zeta_n)$, and ζ for the $n = 0$ term is $\zeta_0 = \omega/(\sqrt{2}kv_{the})$. We assume that for the derivation of the electrostatic plasma wave that $\text{Re}(\zeta_0) \gg 1$. Enough expansion terms (two) are kept in order to show the departure from the cold-plasma limit. This is the "warm" plasma response, but this expansion breaks down as ζ is decreased. If we take $\text{Re}\zeta_0 \ll 1$, then the electron thermal velocity is much larger than the characteristic velocity ω/k, and we call this a "hot" plasma case. Then a Taylor expansion of $Z_0(\zeta)$ in ζ rather than $1/\zeta$ becomes appropriate.

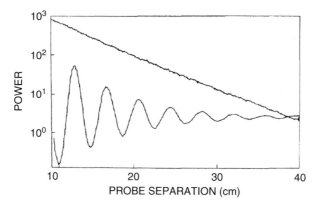

Fig. 9.23 Experimental observation of Landau damping. Upper curve is received power (scale on left); lower curve is interferometer output of density (arbitrary linear units). From [77]. Used with permission, American Physical Society

Fig. 9.24 Experimental confirmation of Landau damping. Curves for Re k and Im k are from Vlasov theory. The dash-dot curve is the Bohm-Gross dispersion relation $\omega^2 = \omega_{pe}^2 + 3k^2 v_{the}^2$. The asymptotic line $\omega = \sqrt{3}k v_{the}$ is also shown. From [29]. Used with permission, American Institute of Physics

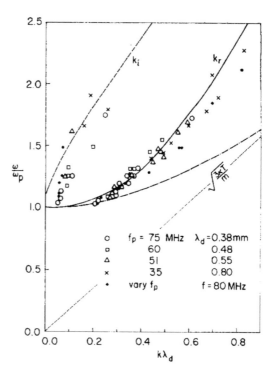

The Taylor expansion for $Z_0(\zeta)$ is

$$Z_0(\zeta) = -2\zeta + \frac{2 \cdot 2\zeta^3}{3 \cdot 1} - \frac{2 \cdot 2 \cdot 2\zeta^5}{5 \cdot 3 \cdot 1} + \cdots + i\sigma\sqrt{\pi}\,\text{sgn}(k_\parallel)\left(\exp(-\zeta^2)\right).$$

$$(9.87)$$

Ignoring the imaginary term and using only the first term in the power series yields $Z_0'(\zeta) = -2$ and then the electron component to the dielectric susceptibility χ_{zz} is given by:

$$\chi_{zz}^{(e)} = \frac{1}{k^2 \lambda_{De}^2}. \tag{9.88}$$

Now we let the ions be "warm," i.e. $|\mathrm{Re}\zeta_0^i| \gg 1$, so that

$$\chi_{zz}^i = -\frac{\omega_{pi}^2}{\omega^2}\left(1 + \frac{\text{``3''}k v_{thi}^2}{\omega^2}\right), \tag{9.89}$$

similar to the warm electron response calculated earlier. Here we have put the "3" in quotes because this is the ion response for a one-dimensional adiabatic equation of state for the ions. If the ions are considered to be in a three-dimensional isotropic regime, i.e. having some scalar pressure, then the "3" would be replaced with γ, the ratio of specific heats, which is 5/3 for a monatomic gas. The combined electron and ion susceptibilities then gives a dispersion relation of form

$$\varepsilon_{zz}^i = 1 - \frac{\omega_{pi}^2}{\omega^2}\left(1 + \frac{\gamma k v_{thi}^2}{\omega^2}\right) + \frac{\omega_{pe}^2}{\omega^2} = 0. \tag{9.90}$$

If we approximate $1 + 3k^2 v_{thi}^2 \approx 1/(1 - 3k^2 v_{thi}^2)$ and set $k^2 \lambda_{De}^2 \to 0$, we obtain a solution. This has an approximate solution (remembering that $\omega_{pi}^2 = n_e Z^2 e^2/(\varepsilon_0 m_i)$ for the ions):

$$\omega^2 \approx k^2 \left(\frac{Z T_e + \gamma T_i}{m_i}\right). \tag{9.91}$$

This is the ion-acoustic wave dispersion relation. (In the physics of laser fusion, the factor of Z is important, as this wave represents a sideband involved in stimulated Brillouin scattering (SBS), which happens in the hohlraum, where the average charge state is $< Z > \approx 30 \to 40$.)

Notice the somewhat odd-looking appearance of the characteristic phase velocity $(Z T_e/m_i)^{1/2}$ when $T_e \gg T_i$. This illustrates the nature of this as an acoustic wave, and is the analogy of the wave velocity $c_s = (\gamma p/\rho)^{1/2}$ for acoustic waves in gases: the pressure is dominated by the electron pressure, but the mass density almost entirely due to the ion mass.

4. **Unmagnetized plasma ions and electrons both "hot."**
   Using the definition of the Debye length $\lambda_{Ds}^{-2} = n_s q_s^2/(\varepsilon_0 T_s)$ and summing over species s:

$$k^2 = \frac{1}{2}\sum_s \frac{1}{\lambda_{Ds}^2} Z_0''(\zeta_s). \tag{9.92}$$

If we assume that both the ions and electrons are hot and we evaluate the function $Z_0'(\zeta_s) \approx -2$ for each species, we find that there are no waves, but there is the Debye shielding explored in Chap. 3:

$$k^2 \approx -\sum_s \frac{1}{\lambda_{Ds}^2}.$$ (9.93)

Thus electric fields associated with charges in the plasma tend to fall off exponentially, $\phi \propto \exp(-r/\lambda_{Dtot})$, where $\lambda_{Dtot}^{-2} = \lambda_{De}^{-2} + \lambda_{Di}^{-2}$.

5. **Perpendicular propagation of electrostatic waves in warm, magnetized plasma.** We now turn our attention to the case where we have propagation of an electrostatic wave in the $\hat{x}$ direction with an electric field also in the $\hat{x}$ direction, and a nonzero magnetic field in the $\hat{z}$ direction. For high frequency waves where we can ignore the ion motion, the dispersion relation is given by

$$\varepsilon_{xx} = 0 = 1 - \frac{\omega_{pe}^2}{\omega} \sum_{n=-\infty}^{\infty} \frac{e^{-\lambda} n^2 I_n(\lambda)}{\lambda} \frac{Z_n(\zeta_n)}{k_\parallel w_\parallel}.$$ (9.94)

Here $\lambda = k^2 w_\perp^2/(2\Omega_e^2) = k^2 v_{the}^2/\Omega_e^2 = k^2 \rho_e^2$, and $\zeta_n = (\omega - n\Omega_e)/(k_\parallel w_\parallel) = (\omega - n\Omega_e)/(\sqrt{2}k v_{the})$. We then make the assumption that $|\zeta_n|$ is large (excluding frequencies exactly at harmonics of the cyclotron frequency) so that $Z_0(\zeta_n \approx -1/\zeta_n = -k_\parallel w_\parallel/(\omega - n\Omega_e)$. Then the dispersion relation becomes, with $q = \omega/\Omega_e$ [51, 94]:

$$\varepsilon_{xx} = 0 = 1 - \frac{\omega_{pe}^2}{\Omega_e^2} \sum_{n=-\infty}^{\infty} \frac{e^{-\lambda} I_n(\lambda)}{\lambda} \frac{n^2}{q^2 - n^2}.$$ (9.95)

This dispersion relation thus has an infinite series of terms, each term having a resonant denominator at $\omega = \pm n\Omega_e$. At first glance this might appear somewhat difficult to calculate. Remarkably, however, a one-line analytic expression is available for this dispersion relation. It was derived in [94] and written in a more usable form in [51]. It is:

$$\varepsilon(\omega, k) = 1 + \frac{1 - {}_2F_2\left(1, \frac{1}{2}; 1 + \omega/\Omega_{ce}, 1 - \omega/\Omega_{ce}; -2\lambda_e\right)}{(k\lambda_{De})^2} = 0.$$ (9.96)

Here $\lambda_e = k^2 T_e/(m_e \Omega_{ce}^2) = (k\rho_e)^2$. The hypergeometric function ${}_2F_2$ is available as `HypergeometricPFQ` in *Mathematica*. A *Mathematica* script is given here for evaluation of the dielectric constant ε_{xx}:

```
epsilon[k_] =
1 + 1/(k lambdaDe)^2
```

```
(1 - HypergeometricPFQ[{1, 1/2}, {1 + x, 1 - x}, -2
lambdae])
```

Here $x = \omega/|\Omega_e|$. There are an infinite set of solutions to the dispersion relation $\varepsilon_{xx} = 0$ at each value of k, which tend to be near harmonics of the electron cyclotron frequency as k becomes large. The waves that these represent are called the electron Bernstein wave, after I. Bernstein who elucidated their properties in 1958 [15].

Shown in Fig. 9.25 is a plot of the dispersion relation for an electrostatic Bernstein electron wave in the vicinity of the third harmonic. Also shown in Fig. 9.26 are plots of the first four roots for the electron Bernstein wave, with the harmonic of the cyclotron frequency subtracted and the difference multiplied by n^2. This shows that the wavenumber for the frequency maximum is $\propto (n - 1)$, and the higher roots are only slightly separated from $n|\Omega_e|$. Also, note that the asymptotic value of the first root as $k \to 0$ is the upper hybrid resonance

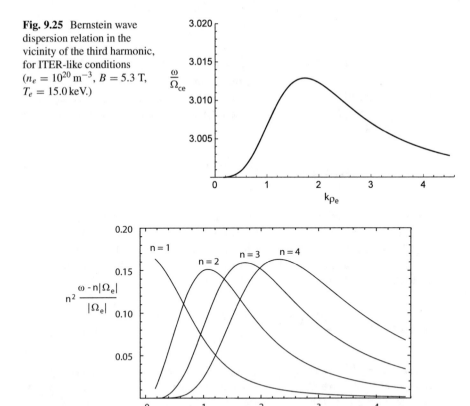

Fig. 9.25 Bernstein wave dispersion relation in the vicinity of the third harmonic, for ITER-like conditions ($n_e = 10^{20}\,\mathrm{m}^{-3}$, $B = 5.3$ T, $T_e = 15.0\,\mathrm{keV}$.)

Fig. 9.26 Plots of $n^2(\omega - n|\Omega_e|)/|\Omega_e|$ for the first four roots of the electrostatic electron Bernstein wave for ITER-like conditions

frequency, $\omega_{UH} = \sqrt{\omega_{pe}^2 + \Omega_e^2}$ if this value is $\ll 2|\Omega_e|$, and approaches $|2\Omega_e|$ as the density ($\propto \omega_{pe}^2$) is increased.

Low-frequency Bernstein waves associated with ion motion also exist. In this case, the electron dielectic response $\chi_{xx}^{(e)}$ is nonzero but takes on a hot-plasma limit, $\chi_{xx}^{(e)} \approx \omega_{pe}^2/\Omega_e^2$ and then the dispersion equation is similar to (9.96), but with the warm dielectric response for the ions substituted for the electron response. This gives

$$\varepsilon_{xx}(\omega, k) = 1 + \frac{\omega_{pe}^2}{\Omega_e^2} + \frac{1 - 2F_2\left(1, \tfrac{1}{2}; 1 + \omega/\Omega_i, 1 - \omega/\Omega_i; -2\lambda_i\right)}{(k\lambda_{Di})^2} = 0.$$
(9.97)

Now $x = \omega/\Omega_i$. A plot of ε_{xx} vs. frequency for a fixed $k_x\rho_i$ is shown as Fig. 9.27. A dispersion relation for the lowest-frequency branch ($\Omega_i \leq \omega \leq 2\Omega_i$) is shown as Fig. 9.28. In addition to the roots to $\varepsilon_{xx} = 0$ at intervals between the ion cyclotron resonance frequencies, there is a zero near the lower hybrid resonance frequency $\omega_{LH} \approx 30\Omega_i$ for the ITER-like conditions used here. This zero at the lower hybrid resonance frequency also gives rise to a warm plasma wave with a dispersion relation given by

$$\omega^2 = \omega_{LH}^2 + \gamma_{LH}k_x^2 T_i/m_i,$$
(9.98)

and a graph of this dispersion relation is shown as Fig. 9.29.

Also, the real reactor-grade plasma will have multiple ion species, so that the actual dispersion characteristics will be more complex. For example, $\Omega_T/\Omega_D = 2/3$ for a D-T plasma, and so the harmonic structure is richer for that case.

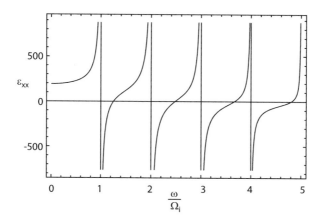

Fig. 9.27 Ion Bernstein wave dielectric constant ε_{xx} for a fixed $k\rho_i = 2.0$. ITER-like conditions ($n_e = 10^{20}\,\mathrm{m}^{-3}$, $B = 5.3\,\mathrm{T}$, $T_i = 15.0\,\mathrm{keV.}$), but with D$^+$ ions only, are assumed

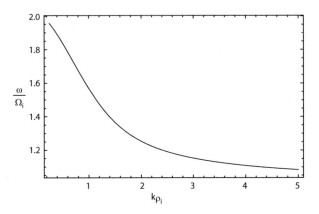

Fig. 9.28 ω vs. k_{xx} for the lowest branch of the ion Bernstein dispersion relation ($\Omega_i \leq \omega \leq 2\Omega_i$). Same conditions as Fig. 9.27

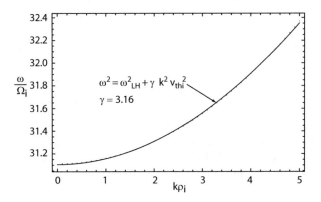

Fig. 9.29 Warm-plasma ion plasma wave near the lower hybrid frequency. Same conditions as Fig. 9.27

9.3.8 Mode Conversion and Tunneling

In most cases of radiofrequency wave propagation in plasmas, the dissipation of energy as a wave passes through the medium is small. However, the cold-plasma wave theory gives a hint regarding conditions where energy transfer might take place in an efficient manner between the wave, which can be regarded as a coherent motion of the ions and electrons, into an incoherent motion of these particles, i.e. the generation of heat. These regions are typically where the cold-plasma wave theory shows that the local index of refraction goes to infinity. We note, however, that the Wentzel-Kramers-Brillouin (WKB), which assumes that k is a slowly varying function of position, breaks down at these resonances. A more refined approach might be to turn the wave dispersion relation of form $b(x)k^2 + c(x) = 0$ back into a differential equation:

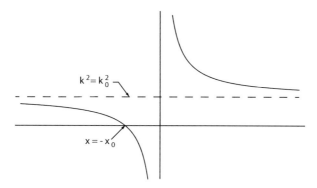

Fig. 9.30 Model dispersion relation suggested by Budden [19]

$$\frac{d^2 E}{dx^2} + k(x)^2 E = 0. \tag{9.99}$$

A simple model which contains both a cutoff and a resonance was proposed by Budden [19, 107]:

$$k^2(x) = k_0^2 \left(1 + \frac{x_0}{x}\right). \tag{9.100}$$

This simple dispersion relation is shown as Fig. 9.30. By adding a small imaginary part to the denominator x in Eq. (9.100), a singularity at $x = 0$ can be avoided, and the solution can be written in terms of Airy functions, i.e. Bessel functions of order $\pm 1/3$. Reflection and transmission coefficients can be assigned for waves moving towards $x = 0$ from either the right or the left. For a wave inbound from the right, one finds that the transmission coefficient $T = \exp(-\eta)$, with $\eta = k_0\pi/2$, which is the number of quarter wavelengths that the wave is in cutoff. This "tunneling" process is reminiscent of a quantum-mechanical tunneling problem. The reflection coefficient $R = 0$ for the wave inbound from the right, as it encounters the resonance before the cutoff. Thus the total energy in the reflected and transmitted wave $R^2 + T^2 = \exp -2\eta$. This is less than one. For a wave inbound from the left, $R = 1 - \exp(-\eta)$ and $T = \exp(-\eta)$ and the total scattered energy is $R^2 + T^2 = 1 - \exp(-\eta) + \exp(-2\eta)$, which is also less than one. Thus the Budden model fails to conserve energy, which is an unsatisfying result. Furthermore, the reflection and transmission of energy are functions only of the tunneling layer's thickness, and not on any other physical properties. While certain wave propagation studies can be done using the Budden-type model, attention has turned to a more sophisticated dispersion model which avoids these shortcomings.

The more modern approach was pioneered by Stix in 1965 [100]. This adds a fourth-order term to the dispersion relation, modeling the warm-plasma effects:

$$ak_x^4 + (b_r + ib_i)k_x^2 + c = 0. \tag{9.101}$$

Fig. 9.31 Plot of
perpendicular index of
refraction n_x^2 vs. position
(with density increasing from
right to left) for waves at a
frequency of the lower hybrid
resonance at the center of the
drawing. Dotted lines show
dispersion relations outside
their regions of validity. From
[100]. Used by permission,
American Physical Society

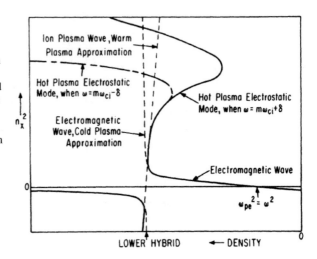

Here we have also added an imaginary part to the b coefficient, which can model
dissipative processes such as resistivity, collisions, and Landau damping. If we
convert this to a representative differential equation, and ignore the imaginary part
of the b coefficient, we have an equation of type

$$y^{iv} + \lambda^2(xy'' + \beta y) = 0. \tag{9.102}$$

This equation, being a fourth-order ODE with non-constant coefficients, was
studied by Wasow in 1950 [115] and is sometimes referred to as the Wasow
equation. It is also sometimes called the Standard Equation [101]. The fourth-order
nature of this equation implies that there are four linearly independent solutions.
These solutions can be found using an integral transform method, where the choice
of a contour in the complex plane determines the branch of the dispersion relation.
While the mathematical method is somewhat complex, the physical nature of the
waves involved can be gleaned from the underlying waves found from the simplified
dispersion relations given earlier. For example, Fig. 9.31, from Stix's 1965 paper,
shows the situation near the lower hybrid resonance for a wave incoming from
the right. The plasma density is increasing towards the left, and the cold plasma
electromagnetic wave takes on the characteristics of a warm-ion plasma wave. This
phenomenon is called "mode conversion." The converted wave will typically have
much greater absorption locally than the starting electromagnetic wave, so that
energy is transferred into the plasma.

9.3.9 RF Heating: Resonant Frequencies in Plasma

We now survey the possible frequency ranges for RF heating of fusion plasma. They are:

1. **Ion cyclotron resonance**

   The ion cyclotron resonance frequency takes a very simple form:

$$\Omega_{ci} = \frac{qB}{m_i}.$$

   Typical frequency parameters are shown in the box following.

A D^+ ion in a 5.3 T magnetic field (ITER nominal magnetic field) has a cyclotron resonance frequency:

$$f_{ci} = \frac{\Omega_{ci}}{2\pi} = \frac{1.6 \times 10^{-19} \cdot 4.0}{2\pi \cdot 2 \times 1.67 \times 10^{-27}} = 40.4\,\text{MHz}$$

This is just below the VHF television band and very high-power sources are commercially available in this frequency range.

Notice that a triton in the same field will have a gyrofrequency $2/3\times$ that of a deuteron, or about 26.9 MHz in this case. An H^+ ion will have $F_H = 2f_D$, or around 80.8 MHz.

At first glance it might seem that a strong candidate for plasma heating of the ions might be shining radiofrequency waves into the plasma at one of the ion cyclotron resonance frequencies. However, nature has conspired to make this a less attractive option than one might think, especially in a plasma where a single ionic species is dominant. The reason has to do with the wave polarization near a majority ion cyclotron frequency. For a wave launch from the low-field side in a tokamak, the CMA diagram shows that at frequencies around Ω_i, with $\Omega_i < \omega$, there is only one propagating mode and this is called the "fast wave," or the compressional Alfvén wave. (Calling this wave the fast wave is sort of a misnomer because its phase velocity is near the Alfvén speed, typically one fiftieth of the speed of light.)

In the low-frequency limit the cold-plasma dielectric tensor produces a result for the ratio of the left-handed to right-handed electric field components for a single-ion-species plasma as:

$$\frac{E_x + i E_y}{E_x - i E_y} = \frac{-R - n_x^2}{L - n_z^2} = \frac{\dfrac{\Omega_i}{\omega + \Omega_i} - n_{\parallel}^2}{\dfrac{\Omega_i}{\omega - \Omega_i} - n_{\parallel}^2}, \tag{9.103}$$

which, since $n_\parallel^2 > 0$, gives

$$\frac{E_x + iE_y}{E_x - iE_y} \leq \frac{\omega - \Omega_i}{\omega + \Omega_i}. \tag{9.104}$$

Thus the electric field polarization for the fast wave is almost entirely right-handed when the wave is close to the ion cyclotron resonance and therefore little power absorption can occur. If the wave was launched from the high-field side, there is another wave available, called the ion cyclotron wave or the "slow wave," also called the shear Alfvén wave. While this has a more favorable polarization for coupling energy to the ions, this wave tends to be involved in mode conversion to an electrostatic wave, which will be dissipated by the electrons at places not necessarily near the center of the plasma.

There can be more than one ionic species in the plasma. If we take the case of two ionic species, we have a situation where a new resonance occurs, called the Buchsbaum two-ion hybrid resonance [18].

Looking again at propagation perpendicular to the magnetic field, we find the dispersion relation in the cold-plasma model is given by

$$An_\perp^4 - Bn_\perp^2 + C = 0 \tag{9.105}$$

with

$$A = S = 1 + \sum_s \frac{\omega_{ps}^2}{\omega^2 - \Omega_s^2}$$

$$B = RL + PS \tag{9.106}$$

$$C = PRL$$

with S, R, L, and P defined in Eq. (9.28). Note that the solution to this bi-quadratic equation is

$$n_\perp^2 = \frac{B \pm F}{2A}, \tag{9.107}$$

with $F = RL - PS$. We then note that a resonance ($n_\perp \to \infty$) occurs when $S \to 0$. If we look at low frequencies where we can ignore the contribution to S by the electrons (equivalent to taking $|\Omega_e| \to \infty$), and ignoring the "1" in the equation for S (equivalent to assuming that $\omega_{pi} \gg \Omega_i \approx \omega$) and then setting the sum of the two ionic terms in S to zero gives a simple expression for the two-ion hybrid resonance:

$$\omega^2 = \frac{\Omega_1{}^2\omega_{p2}{}^2 + \Omega_2{}^2\omega_{p1}{}^2}{\omega_{p1}{}^2 + \omega_{p2}{}^2} \tag{9.108}$$

which can be re-written as

$$\omega^2 = \Omega_1 \Omega_2 \frac{\alpha_1 \Omega_2 + \alpha_2 \Omega_1}{\alpha_1 \Omega_1 + \alpha_2 \Omega_2}. \tag{9.109}$$

Here $\alpha_1 = n_1 Z_1 / (n_1 Z_1 + n_2 Z_2)$, i.e. it is the charge-weighted fraction of the ions in species 1. Note a peculiarity of the Buchsbaum two-ion resonance: as $\alpha_1 \to 0$, the resonance frequency approaches Ω_1, the "ghost" ion cyclotron frequency!

For small fractions of the ions in one species, i.e. a minority species, Eq. (9.104) does not apply and rather high amounts of absorbed energy can happen at the minority ion cyclotron resonance. Many successful RF heating experiments have been carried out on JET with hydrogen and deuterium minority ions (4–5% each) in a tritium main plasma [52].

Another possibility exists for heating provided through the warm plasma dielectric response. Unlike the cold-plasma approximation, the warm-plasma dielectric response includes resonances at harmonic multiples of the fundamental cyclotrons frequencies. As shown earlier, the perturbed plasma current can be represented as an infinite sum of Bessel functions of argument $k_\perp \rho_i$, i.e. terms like $J_n(k_\perp \rho_i)$ appear in the expansion. While the Bessel functions for small argument have magnitude $J_n(x) \propto x^n$ and therefore become small when $k_\perp \rho_i \ll 1$, the low-order resonances $\omega = (n \pm 1)\Omega_i$, especially at the second harmonic ($n = 1$) can have a substantial response if $k_\perp$ is large enough.

There is an additional bonus with cyclotron harmonic heating and that comes about because the ion charge and mass ratios are very close to perfect integer ratios. So, for example, the second harmonic cyclotron resonance frequency for deuterium is almost exactly at the first harmonic for hydrogen. So at this frequency, there can be second-harmonic heating of deuterium and minority heating of hydrogen at the same place in the plasma (or at least at the same value of **B**). Once other ions are considered, such as T, ^{3}He, ^{4}He, (and even Be and C ions from wall impurities), an entire "zoo" of resonances is possible.

We now consider the cold-plasma accessibility condition for the ion cyclotron range of frequencies. Keeping only the largest terms in the cold-plasma dispersion equation (the ones multiplying $\varepsilon_\parallel$), we have

$$n_\perp^2 = \frac{\left(\varepsilon_\perp + i\varepsilon_{xy} - n_\parallel^2\right)\left(\varepsilon_\perp - i\varepsilon_{xy} - n_\parallel^2\right)}{\varepsilon_\perp - n_\parallel^2}. \tag{9.110}$$

Since the quantity $\omega_{pi}^2 / \Omega_i^2 = c^2 / v_A^2$, low-frequency ICH waves have a wave velocity near the Alfvén speed:

$$v_A = \frac{B}{\sqrt{\mu_0 \rho}}.$$

Thus the wavelength along **B** is of the order

$$\lambda = \frac{v_A}{f} = 0.201 \text{ m.}$$

A more precise expression, good for both the fast wave and the ion cyclotron wave, is given by Stix [101]:

$$N_\perp^2 + N_\parallel^2 = \frac{A \left(1 - N_\parallel^2\right)}{A - N_\parallel^2}. \tag{9.111}$$

Here $N_\parallel$ and $N_\perp$ are the indices of refraction in units of the Alfvén speed, i.e. $N_\parallel = (v_A/c)n_\parallel$ and $N_\perp = (v_A/c)n_\perp$, and

$$A = \frac{\Omega_i^2}{\Omega_i^2 - \omega^2}.$$

A graph of $N_\perp$ vs. $N_\parallel$ is shown as Fig. 9.32.

We can see that a launch with $N_\perp^2$ and $N_\parallel^2$ both positive from the low-field side requires that $N_\parallel$ should be between zero and one. We can fix the size of $N_\parallel$ by using an antenna structure with an acceptable spectrum of wavenumbers along B. This is typically done by using a phased-array antenna structure such as shown in Fig. 9.33. Here current-carrying straps are located close to the plasma's edge, and the parallel wavenumber spectrum is determined by the phasing of the elements. (Not shown in this schematic is a metallic structure with a set of slots, cut perpendicular to the straps and between the straps and the plasma to minimize the exposure of the straps to the plasma, which can cause impurity evolution into the plasma and impose a large heat load on the active elements.) The phasing used can be a $0, \pi, 0, \pi$ as shown here, or it can be a $0, 0, \pi, \pi$ which gives about half the value of $N_\parallel$ as the other case.

The actual antenna system for ITER is quite complex in comparison with the simple cartoon shown in Fig. 9.33. Figure 9.34 shows the actual design of the system. One can see that this 45-ton structure contains the essential current straps and the covering Faraday screen, and also the transmission lines, which must pass from the atmospheric side into vacuum, along with various tuning elements, diagnostic pickoff lines, and a shimming system to fine-tune the distance from the antenna structure to the plasma edge. Each antenna module is designed to handle up to 40 MW of radiofrequency power over a fairly large frequency range ($40 \rightarrow 55$ MHz), to accommodate minority ion resonance, current drive, and second harmonic heating scenarios [34].

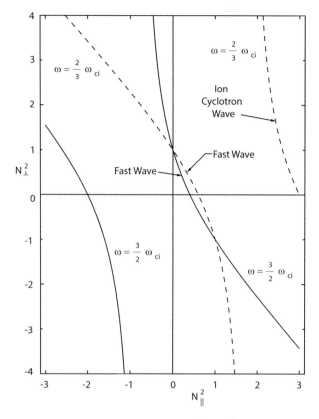

Fig. 9.32 Plot of the normalized indices of refraction $N_\perp^2$ vs. $N_\parallel^2$ for waves at frequencies above and below the ion cyclotron resonance frequency. After [101]

Fig. 9.33 Schematic of phased array antenna structure

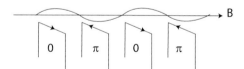

Engineering of ICH&CD Systems

Transmission Lines

(A more extensive derivation of the fields in transmission lines and waveguides can be found in [61].) At low frequencies, coaxial cables are the preferred method of transporting RF energy from the source to the load. Typical cables have an annular cylindrical geometry with an outer radius b and an inner radius a. See Fig. 9.35 for the appropriate geometry. The electric field generated in such a cable is given by

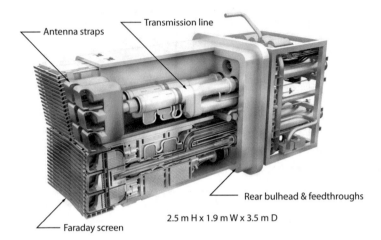

Fig. 9.34 Equatorial antenna for Ion Cyclotron Heating and Current Drive (IC H & CD) for ITER. Assembly is 2.5 mH ×1.8 mW ×3.5 m D., weight 45 tons. Courtesy of Fusion For Energy, http://fusionforenergy.europa.eu/

Fig. 9.35 Coaxial transmission line geometry

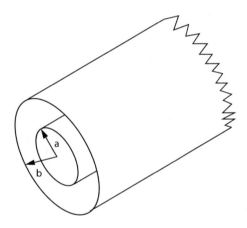

$$\mathbf{E} = \hat{r}\, \frac{E_0 a}{r} \exp(i(kz - \omega t)). \tag{9.112}$$

In a coaxial cable, TEM (transverse electromagnetic) propagation is the available electromagnetic mode. It has no low-frequency cutoff, i.e. the line can work down to DC. For TEM transmission, the magnetic field is given by $|\mathbf{E}| = c|\mathbf{B}|$, with $\mathbf{E} \times \mathbf{B}$ pointing in the direction of propagation. In (r, θ, z) geometry, this means that

$$\mathbf{B} = \hat{\theta} c\, \frac{E_0 a}{r} \exp(i(kz - \omega t)). \tag{9.113}$$

The accumulated voltage $\int E_r dr$ is:

$$V_0 = \int_a^b E_r dr = E_0 a \ln\left(\frac{b}{a}\right) \exp(i(kz - \omega t)). \tag{9.114}$$

The boundary condition at $r = a$ and $r = b$ is that the surface currents $\mathbf{J} = \hat{n} \times \mathbf{H}$, with $\mathbf{H} = \mathbf{B}/\mu_0$, which implies that $J_z = E_0 c/\mu_0$ on the inner conductor and $J_z = -E_0 ca/(b\mu_0)$ on the outer conductor; thus the currents in the conductors are $I = \pm 2\pi ca E_0/\mu_0$. The ratio of voltage to current is thus

$$Z = V_0/I_0 = \frac{1}{2\pi}\sqrt{\frac{\mu_0}{\varepsilon_0}}\ln\left(\frac{b}{a}\right) \approx 60\ln\left(\frac{b}{a}\right). \tag{9.115}$$

There are different possible options for optimizing the conductor radius ratio b/a. If we consider the outer radius b to be fixed, because of cost or physical size constraints, then one possible objective would be to minimize the loss per unit length of the transmission line. The power carried in the cable is given by

$$P = \frac{1}{2}V_0/I_0 = \frac{1}{2}I_0^2 Z, \tag{9.116}$$

and the loss per unit length is given by $(1/2)I_0^2 R'$, where R' is the resistance per unit length in the transmission line, which in turn is given by the conductivity of the material in the transmission line and the skin depth δ of the current in the line. We can calculate the resistivity by treating the resistance per unit length as coming from a uniform current-carrying annulus of thickness δ and of circumference $2\pi a$ and $2\pi b$ for the inner and outer radius, respectively, thus:

$$P' = \frac{1}{2}I_0^2 R' = \frac{1}{2}I_0^2 \frac{1}{2\pi\delta\sigma}\left(\frac{1}{a} + \frac{1}{b}\right), \tag{9.117}$$

with the skin depth δ given by:

$$\delta = \sqrt{\frac{2}{\omega\mu_0\sigma}}. \tag{9.118}$$

Since the skin depth δ is geometry-dependent, we can write the fractional power dissipation per unit length as

$$\frac{P'_{\text{loss}}}{P} \propto \frac{\dfrac{1}{a} + \dfrac{1}{b}}{\ln\left(\dfrac{b}{a}\right)}. \tag{9.119}$$

If we assign a variable $x = b/a$ and keep b fixed, then (9.119) becomes

$$\frac{P'_{loss}}{P} \propto \frac{1+x}{\ln x}, \tag{9.120}$$

and this function has a minimum at $x = 3.59112$, which gives $Z = 76.7\ \Omega$ for lines with a dielectric constant of one (vacuum, air, and He fit this description). For higher dielectric constants, Eq. (9.115) is modified to read $Z = 60/\sqrt{\varepsilon_r}\ln(b/a)$, where ε_r is the relative dielectric constant. (For polyethylene, $\varepsilon_r = 2.2$ which gives $Z = 52\ \Omega$, which explains the ubiquity of 50-ohm cables in laboratories.)

An alternative optimization exists and that is to maximize the power handling in the cable for a specified maximum field E_0 on the center conductor to avoid breakdown. Again holding b fixed, this gives an expression for the power transmitted as

$$P = \frac{1}{2}V_0 I_0 = \frac{1}{2}\left(E_0 a \ln\left(\frac{b}{a}\right)\right)(2\pi a E_0/(\mu_0 c)) \propto \frac{\ln x}{x^2}, \tag{9.121}$$

where again $x = b/a$. This gives an optimum at $x = \exp(1/2) = 1.64872$ and thus $Z = 60 \cdot 1/2 = 30\ \Omega$.

The ITER transmission system is designed with the transmission lines having an impedance of $50\ \Omega$, a compromise between these two design optima [71, 106], with eight lines per antenna with a maximum power handling per line of 2.5 MW. A schematic of the overall transmission line is shown as Fig. 9.36. Note especially the hybrids located near the RF source. The hybrids couple forward

Fig. 9.36 ICH transmission system for ITER. From [106]. Used by permission, American institute of Physics

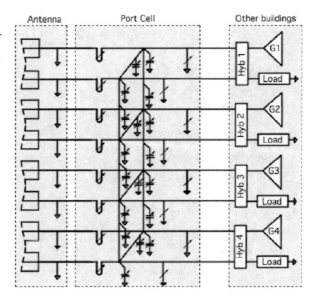

energy from the driving amplifier into the transmission line and reflected power into a dummy load, to prevent damage to the driving amplifier from the reflected power. This is especially important when certain types of ELMs (edge localized modes) create unusually high reflected power.

Ion Cyclotron Heating and Current Drive Sources

The ion cyclotron range of frequencies (30 → 100 MHz) requires heating sources with available output power in the megawatt range. This power and frequency range is adequately covered by vacuum tubes whose parentage goes back to Lee de Forest's first amplifying device, the triode, in 1906. A triode consists of a thermionic cathode surrounded by a wire grid as a control element, and an anode or "plate" as the receiving point for the electron stream launched from the cathode. Modern high-power tubes for this frequency range are tedrodes, i.e. four-element devices, which are similar to triodes but have an additional second grid, called the screen grid, which has a structure similar to the control grid (typically in the "electron shadow" of the control grid) which tends to focus the electron stream in such a way that the space charge of the electrons in the screen-to-anode region can prevent secondary electrons from the anode from reaching the control grid, and providing reduced capacitance between the anode and the control grid. Also, while the original triodes ran with the control grid voltage negative with respect to the cathode, preventing substantial electron current from entering through the control grid, modern high-power tetrodes can run with the control grid going positive during the waveform cycle, making more efficient operation possible but creating heat dissipation in the control grid. Figure 9.37 shows one of the two tube types selected for the ITER IC H&CD drive system. Each generator in Fig. 9.36 consists of two tubes of this type coupled through a quarter-wavelength combiner circuit.

Fig. 9.37 Continental/Eimac 4CM2500K RF power tube, one of two types selected for the ITER IC H&CD system. Tube is 43 cm diameter × 38 cm length and is rated for 1.5 MW output for 3600 s pulses. Courtesy of CPI, Inc

The basic limitation of tubes of this type for application at higher frequencies is the electron time-of-flight between the control grid and anode. When the time-of-flight approaches a quarter cycle of the RF waveform, the oscillatory nature of the anode voltage causes de-phasing of the electron beam, limiting the gain and efficiency of the device. Thus higher-frequency tubes are typically smaller in size to reduce the electron transit time, resulting in diminished power handling. Thus in fusion heating applications, tetrodes are universally used for Ion cyclotron heating and current drive (ICH&CD) applications, but other technologies are required for the higher frequency ranges.

2. **Lower hybrid heating**

The cold plasma dispersion relation gives a lower hybrid resonance frequency, given in Eq. (9.38) as :

$$\frac{1}{\omega_{LH}^2} = \frac{1}{\omega_{pi}^2} + \frac{1}{\Omega_{ce}\Omega_{ci}}.$$

This resonance is caused by the cold-plasma term $S \to 0$ at the LH frequency. An example of a typical LH resonance frequency is shown in the box below.

Consider a 10^{20} m^{-3} density DT plasma in a 5.3 T magnetic field. (Here we take an average ion mass of 2.5 proton masses.) The lower hybrid frequency f_{LH} is given by

$$f_{LH} = \sqrt{\frac{1}{1/f_{pi}^2 + 1/(f_{ce}\,f_{ci})}}.$$

Here we have

$$f_{pi} = \frac{1}{2\pi}\sqrt{\frac{nq^2}{m_i\varepsilon_0}} = \frac{1}{2\pi}\left(\frac{10^{20}\cdot\left(1.6\times10^{-19}\right)^2}{8.85\times10^{-12}\cdot 2.5\times1.67\times10^{-27}}\right)^{1/2}$$

$$= 1.327\times10^9\,\text{Hz}$$

and

$$f_{ce} = \frac{1}{2\pi}\frac{qB}{m_e} = \frac{1}{2\pi}\frac{1.6\times10^{-19}\cdot 5.3}{9.11\times10^{-31}} = 1.48\times10^{11}\,\text{Hz}.$$

and

$$f_{ci} = f_{ce}(m_e/m_i) = f_{ce}/4583 = 2.32\times10^7\,\text{Hz}.$$

Then

$$f_{LH} = \sqrt{\frac{1}{1/f_{pi}^2 + 1/(f_{ce}f_{ci})}} = \boxed{1.13 \times 10^9 \text{ Hz}}.$$

The frequency range for LH resonance is a sort of "Goldilocks" frequency, i.e. not too low so as to require antennae with a large footprint on the plasma, and not so high so as to require rather esoteric driver technology and transmission techniques as we will encounter with electron cyclotron heating and current drive.

Accessibility of Lower Hybrid Wave to Resonance

Here we show a calculation of the accessibility of the lower hybrid resonance to a slow wave launched from the edge of the plasma. The treatment closely follows that of Stix in the original 1962 edition of his book [99] and assumes propagation in slab geometry, i.e. with no toroidal effects.

The cold-plasma dispersion relation gives a quadratic equation in $n_\perp^2$ given by

$$an_\perp^4 - bn_\perp^2 + c = 0, \tag{9.122}$$

with

$$\begin{aligned} a &= S \\ b &= RL + PS - Pn_\parallel^2 - Sn_\parallel^2 \\ c &= P\left(RL - 2Sn_\parallel^2 + n_\parallel^4\right). \end{aligned} \tag{9.123}$$

The root to Eq. (9.122) that is appropriate is the one such that $n_\perp \to \infty$ as $S \to 0$, that is,

$$n_\perp^2 = \frac{b + \left(b^2 - 4ac\right)^{1/2}}{2a}. \tag{9.124}$$

We first look at the propagation characteristics near the plasma edge. Note that the quantity $RL = S^2 - D^2$, and the quantity D is small near the edge since

$$D = \sum_s \frac{\omega_{ps}^2}{\omega^2} \frac{\omega \Omega_s}{\omega^2 - \Omega_s^2} \tag{9.125}$$

and thus $D \propto n$, becoming zero as $n \to 0$. Similarly $S \to 1$ as $n \to 0$, and then Eq. (9.122) can be factored to give

$$\left[n_\perp^2 - P \left(1 - n_\parallel^2 \right) \right] \left[n_\perp^2 - \left(1 - n_\parallel^2 \right) \right] = 0 \tag{9.126}$$

We will show that accessibility to the LH resonance will require that $n_\parallel > 1$ at the edge. In that case, since $P \to 1$ as $n \to 0$, $n_\perp^2 < 0$ at the edge of the plasma and does not become positive until $P < 0$, which is to say, $\omega^2 < \omega_{pi}^2 + \omega_{pe}^2$. This condition happens very close to the edge, however. For the above example, $P = 0$ when the density is about $1.7 \times 10^{16} \, \text{m}^{-3}$, or roughly 10,000 times lower than the central density. Thus the evanescent region is small and little attenuation happens before the LH wave is propagating inward as a slow wave, starting a few millimeters from the antenna.

After the first few millimeters, we will have $|P| \gg 1$ and thus $|P| \gg |S|$. Examination of Eq. (9.122) shows that $n_\perp^2$ will be positive if b, a, and $b^2 - 4ac$ are all positive. We have already shown that S is between zero and one between the edge and the resonance, and b will be positive if

$$n_\parallel^2 > \left| \frac{RL}{P} \right| + |S|, \tag{9.127}$$

but this inequality is less restrictive than the condition that $b^2 - 4ac > 0$. That condition can be expressed using a new variable $x = n_\parallel^2 - S$, which then gives

$$b^2 - 4ac = \left[D^2 + x(P + S) \right]^2 + 4PS \left(D^2 - x^2 \right), \tag{9.128}$$

or equivalently

$$b^2 - 4ac \approx P^2 \left[\left(\frac{D^2}{P} + x \right)^2 + 4 \frac{SD^2}{P} \right]. \tag{9.129}$$

Recalling that $P > 0$ in the propagating region, the quantity in the brackets in Eq. (9.129) will be positive if

$$\frac{D^2}{P} + x > 2 \left| \frac{SD^2}{P} \right|^{1/2}. \tag{9.130}$$

Note that in slab geometry, with density and magnetic field gradients only in the perpendicular direction, $n_\parallel$ is a constant of the propagation. We are interested in the minimum value of $n_\parallel$ which will allow propagation all the way from the edge to the LH resonance. This minimum value for $n_\parallel$ corresponds to the place where the RHS of Eq. (9.130) is a maximum, and this is where

$$S = \left. \frac{P}{P^2 - D^2} \right|_{\text{res}}, \tag{9.131}$$

which generates a necessary and sufficient condition on $n_{\parallel}$:

$$n_{\parallel}^2 > 1 + \left| \frac{D^2}{P} \right|_{res}, \tag{9.132}$$

and at lower hybrid frequencies $D \approx \gamma \Omega_i / \omega$, with γ the low-frequency dielectric constant, $\gamma = \omega_{pi}^2 / \Omega_i^2$, so that Eq. (9.132) can be rewritten as

$$n_{\parallel}^2 > 1 + \left. \frac{\omega_{pe}^2}{\Omega_e^2} \right|_{res}. \tag{9.133}$$

This is the famous Stix-Golant accessibility criterion for lower hybrid accessibility [39, 99]. As a numerical example, for ITER conditions with $n_e = 10^{20}\,\mathrm{m}^{-3}$ and $B = 5.3$ T, this gives $n_{\parallel}(\min) = 1.16$.

The derivation is not exact for toroidal geometry, because $n_{\parallel}$ is not an invariant quantity, but the toroidal wavenumber n_{ϕ} is. Studies of LH accessibility in tokamaks have only been available through ray tracing calculations. References [17] and [59] give examples of such calculations. Figure 9.38 shows two calculations using ray tracing for conditions found in PLT, one with an aspect ratio of 3.37 and another with an aspect ratio of 5.6, and all other parameters kept constant, illustrating the effects of toroidicity on the propagation characteristics. One can see from Fig. 9.38 that one wave coupled to the resonant layer with complete absorption whereas the other did not couple at all. We conclude from this that toroidal effects are important and that the non-conservation of $n_{\parallel}$ in realistic toroidal geometry requires a more careful look.

Lower Hybrid Current Drive

Quasilinear theory

The application of RF energy to a plasma can modify the distribution function of the plasma electrons and ions. Here we concentrate on the distortion of the electron distribution function from maxwellian for sufficiently high electric fields in the plasma. If we look at the Vlasov equation derived earlier, we have

$$\frac{\partial f}{\partial t} + \mathbf{v} \cdot \nabla f + \frac{q}{m} (\mathbf{E} + \mathbf{v} \times \mathbf{B}) \cdot \frac{\partial f}{\partial \mathbf{v}} = 0. \tag{9.134}$$

If we expand f in a series of terms representing the equilibrium f_0 and perturbations $\delta f^{(1)}$, $\delta f^{(2)}$, etc. resulting from an iterative application of the third term in the Vlasov equation, we have for the first-order perturbation $\delta f^{(1)}$ [69]:

$$\frac{d}{dt} \delta f^{(1)} = -\frac{q}{m} (\delta \mathbf{E} + \mathbf{v} \times \delta \mathbf{B}) \cdot \frac{\partial f_0}{\partial \mathbf{v}}. \tag{9.135}$$

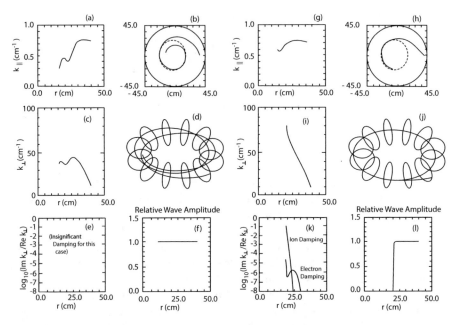

Fig. 9.38 Ray tracing calculation for two PLT-like equilibria with $B_\phi(r = 0) = 2.5$ T, $n_\parallel = 4.5$, $a - 40$ cm, $n_e(0) = 4 \times 10^{19}$ m^{-3}, $\omega/(2\pi) = 800$ MHz, $T_e(0) = 1.5$ keV, $q(a) = 3.7$, $T_i(0) = 0.8$ keV. For (a) → (f), $R_0 = 1.35$ m, and for (g) → (l), $R_0 = 2.25$ m. For (a) and (g), $k_\parallel$ vs. minor radius r, (b) and (h), ray trajectory in the r, θ plane, (c) and (i), $k_\perp$ vs. r, (d) and (j), isometric view of ray path, (e) and (k), $\log_{10}(\mathrm{Im}k_\perp/\mathrm{Re}k_\perp)$, (f) and (l), relative wave amplitude. From [59]. Used with permission, American Institute of Physics

Here d/dt represents the total, or "lagrangian," derivative of the perturbed distribution function. We can integrate over the path of an unperturbed particle representing a member of the f_0 distribution and arrive at an expression for $\delta f^{(1)}$ at time t from the path integral:

$$\delta f^{(1)} = -\frac{q}{m} \int_{-\infty}^{t} dt' \left(\delta \mathbf{E}(\mathbf{x}(t')) + \mathbf{v}((t')) \times \delta \mathbf{B}(\mathbf{x}(t')) \right) \cdot \frac{\partial f_0}{\partial \mathbf{v}}. \tag{9.136}$$

Higher-order corrections to the perturbed distribution function could be found by feeding back $\delta f^{(1)}$ in place of f_0 in Eq. (9.136) to obtain an expression for $\delta f^{(2)}$, and so forth, but stopping the procedure after the first-order perturbation is typical and is referred to as the "quasilinear approximation." If we ignore the gyromotion in the path integral and assume a uniform plasma with $\mathbf{E}_0 = 0$ so that $\delta \mathbf{E} = \mathbf{E}$, then the trajectories are straight lines, and the particle experiences a time-dependent electric field, given by a Fourier integral:

$$\mathbf{E} = \frac{1}{(2\pi)^3} \int \mathbf{E}_k \exp(i((k) \cdot \mathbf{x} - \omega t)) d\mathbf{k}. \tag{9.137}$$

Ignoring the gyromotion is equivalent to ignoring the $\mathbf{v} \times \delta\mathbf{B}$ term in the path integral. The integration over time is thus simply

$$\int_{-\infty}^{t} dt' = \frac{i}{\omega - \mathbf{k} \cdot \mathbf{v}}.$$

We now insert $f = f_0 + \delta f^{(1)}$ into the zeroth-order Vlasov equation to obtain a Fourier component of the perturbed distribution function $\delta f_{\mathbf{k}}$ as

$$\delta f_{\mathbf{k}}^{(1)} = -i \frac{q}{m} \frac{\mathbf{E}_{\mathbf{k}}}{\omega - \mathbf{k} \cdot \mathbf{v}} \cdot \frac{\partial f_0}{\partial \mathbf{v}}. \tag{9.138}$$

We substitute a dummy variable $\mathbf{q}$ into (9.138) such that now $\mathbf{E}$ is given by a Fourier integral over $\mathbf{q}$ and then a double integral over $\mathbf{q}$ and $\mathbf{k}$ appears, with the result from the Fourier integral transform that

$$\frac{1}{(2\pi)^3} \int d\mathbf{q} \, \mathbf{E}_{\mathbf{q}} \exp(i\mathbf{q} \cdot \mathbf{x}) \exp(i\mathbf{k} \cdot \mathbf{x}) = \mathbf{E}_{-\mathbf{k}} \exp(-i(\mathbf{k} \cdot \mathbf{x})).$$

We can then integrate over $\mathbf{k}$ to obtain the quasilinear evolution equation for f_0:

$$\frac{\partial f_0}{\partial t} = i \left(\frac{q}{m}\right)^2 \int \mathbf{E}_{-\mathbf{k}} \cdot \frac{\partial}{\partial \mathbf{v}} \left[\frac{1}{\omega - \mathbf{k} \cdot \mathbf{v}} \mathbf{E}_{\mathbf{k}} \cdot \frac{\partial}{\partial \mathbf{v}} f_0 \right] \frac{d\mathbf{k}}{(2\pi)^3}. \tag{9.139}$$

We can simplify this expression by noting that the spectral energy density $\mathscr{E}_{\mathbf{k}}$ is given by

$$\mathscr{E}_{\mathbf{k}} = \frac{1}{(2\pi)^3} \frac{\varepsilon_0}{2} \mathbf{E}_{-\mathbf{k}} \mathbf{E}_{\mathbf{k}}. \tag{9.140}$$

Then (9.139) can be written in the suggestive form

$$\frac{\partial f_0}{\partial t} = \frac{\partial}{\partial \mathbf{v}} \left(\mathscr{D} \frac{\partial f_0}{\partial \mathbf{v}} \right) \tag{9.141}$$

where

$$\mathscr{D} = \frac{2iq^2}{\varepsilon_0 m^2} \int d\mathbf{k} \frac{\mathscr{E}_{\mathbf{k}}}{\omega - \mathbf{k} \cdot \mathbf{v}}. \tag{9.142}$$

QuasilinearTheory of Lower Hybrid Current Drive

The quasilinear (QL) theory outlined above can be applied to the problem of
current drive from lower hybrid waves. We can write the equation for evolution
of the electron distribution function in the presence of strong RF fields and
collisions, keeping only the parallel terms in the QL diffusion operator, as
[36, 63]:

$$\frac{\partial f}{\partial t} = \frac{\partial}{\partial v_\parallel} \mathcal{D}_{rf}\left(v_\parallel\right) \frac{\partial}{\partial v_\parallel} f + \left(\frac{\partial f}{\partial t}\right)_c. \tag{9.143}$$

Examination of Eq. (9.141) shows that the RF-induced velocity space diffusion-
operator $\mathcal{D}_{rf}$ has a resonant denominator in the integrand where $\omega = \mathbf{k} \cdot \mathbf{v} = k_\parallel v_\parallel$
here. Thus waves with a phase velocity $\omega/k_\parallel = c/n_\parallel$ will be resonant with
electrons with this velocity, and perturbations to the electron distribution function
will be maximized around this velocity. We expect that after some initial transient
time, a steady-state nonmaxwellian distribution function will be reached. We can
write Eq. (9.143) in terms of variables w and τ such that $w = v_\parallel/v_{the}$ and
$\tau = v_0 t$, where $v_{the} = (T_e/m_e)^{1/2}$ is the one-dimensional electron thermal
velocity and v_0 is the characteristic electron relaxation time (see, e.g., [111]). We
also have a normalized velocity-space diffusion operator $\mathcal{D} = \mathcal{D}_{rf}/\left(v_{the}^2/v_0\right)$.
Then Eq. (9.143) becomes

$$\frac{\partial f}{\partial \tau} = \frac{\partial}{\partial w} \mathcal{D}(w) \frac{\partial}{\partial w} f + \left(\frac{\partial f}{\partial \tau}\right)_c. \tag{9.144}$$

The overall form of the operator $\mathcal{D}$, rigorously derived, will end up being
a complicated function of the electric field spectrum and the warm-plasma
dielectric response with its hypergeometric functions, etc., but we can assume
that in some vicinity of $v_\parallel = \omega/k_\parallel$ that the QL process saturates, and that over
that range $\mathcal{D}$ is roughly constant. If we assume a simple box function for $\mathcal{D}$
given by:

$$\mathcal{D}(w) = \begin{cases} D, & w_1 < w < w_2 \\ \\ 0 & \text{else,} \end{cases} \tag{9.145}$$

then we can simplify the collision operator by assuming that the perpendicular
velocity distribution is maxwellian, and we integrate over the perpendicular
components $v_\perp$ to obtain a one-dimensional form for Eq. (9.144) in terms of the
integrated distribution function $F(w) = \int f d\mu$, where $\mu = v_\parallel/v$:

$$\frac{\partial}{\partial \tau} F(w) = \frac{\partial}{\partial w} \mathcal{D}(w) \frac{\partial}{\partial w} F(w) + \frac{2+Z_i}{2} \frac{\partial}{\partial w} \left(\frac{1}{w^3} \frac{\partial}{\partial w} + \frac{1}{w^3}\right) F(w). \tag{9.146}$$

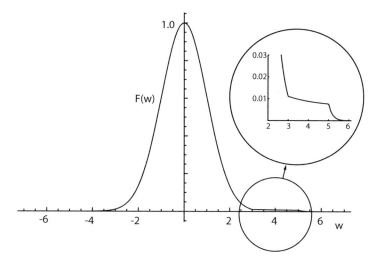

Fig. 9.39 Steady state distribution with quasilinear model for LH current drive. Parameters are $\mathscr{D} = 1/2$, $Z_i = 1$, $w_1 = 3$, $w_2 = 5$. From [63]

Here Z_i is the atomic number of the ions, i.e. $Z_i = 1$ for D-T ions. In steady state, we have $\partial F(w)/\partial \tau = 0$ and then Eq. (9.146) can be integrated to give

$$F(w) = C \, \exp \left(\int^w \frac{-w'dw'}{1 + 2w'^2 \mathscr{D}(w')/(2Z_i + 1)} \right). \tag{9.147}$$

This integral is relatively straightforward to evaluate with the simple form for $\mathscr{D}$ given in Eq. (9.145) and a representative example is shown as Figs. 9.39 and 9.40. (Here the perpendicular velocity distribution is taken as a maxwellian with the same temperature as for the parallel distribution, and the contour plot is truncated to emphasize the perturbed part of the distribution.)

Applications of LH H& CD to ITER

The first phase of ITER operations will not include lower hybrid heating and current drive, but a design for subsequent operations is in place. Some features of the ITER LH H&CD design are worth examining [56, 57]. The conditions for LH accessibility in the ITER plasma for a design frequency of 5.0 GHz are shown in Fig. 9.41. Simulations of the performance of this system, using ray tracing codes and quasilinear models for current drive as discussed above, are shown in Fig. 9.42.

Fig. 9.40 Parallel
distribution function $F(w)$
for the same parameters as
Fig. 9.39. From [63]

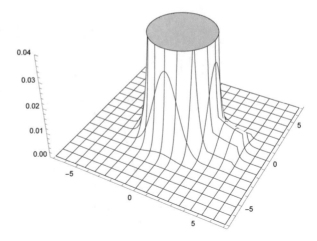

Fig. 9.41 Parameter space
for a 5.0 GHz lower hybrid
system for ITER. From [56]

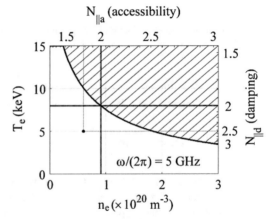

Sources for LH H& CD

The tube technology used for low frequency heating and current drive such as
the ion cyclotron range discussed earlier cannot be carried over to the higher
frequency range needed for lower hybrid heating. As outlined earlier, tetrode-
type tubes cannot be used when the electron transit time in the tube is on
the order of the period of the RF frequency being amplified, which leads to
smaller-format tubes, which then lack the available power, or a new type of
technology. The newer technology came about in 1937 with the invention of the
klystron by Russell and Sigurd Varian. In the klystron, a beam of electrons flows
between the cathode and the anode, and this beam experiences an excitation by a
longitudinal electric field in a cavity excited by an external low-power source as
shown in Fig. 9.43. The acceleration and retardation of the beam in turn causes
compression and rarefaction of the electrons in the beam in a sort of "clumping"
mechanism, similar to the bunching that occurs to cars on a freeway. Additional

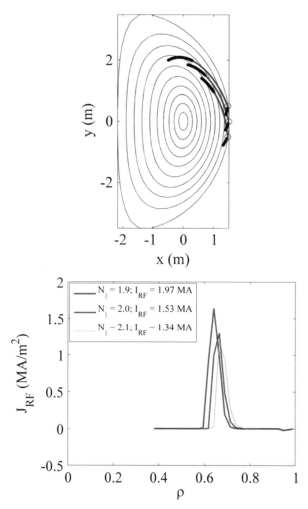

Fig. 9.42 Ray tracing and quasilinear current drive calculations for a 5.0 GHz LH H&CD design for ITER. From [56]

cavities along the beam path are then excited by the progressively more deeply modulated density profile, and this creates more intense electric fields at each resonant cavity. The final cavity has a coupling structure for the RF output, typically delivering the output power to a waveguide. Multi-cavity klystrons can have overall gain of more than 60 dB (a factor of a million in power), allowing the tube to be driven by solid-state amplifiers with only tens of watts required.

As an example of a device under consideration for future ITER lower hybrid heating and current drive, the Thales Electron devices TH 2103C klystron (see Fig. 9.44) amplifier can deliver output power of more than 700 kW CW for 1000 s. It operates at 3.7 GHz with a typical gain of 50 dB. Typically two such tubes are used together with a hybrid coupler, coupling to a common waveguide for transmission of the RF power to the launcher.

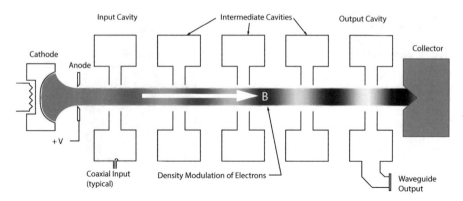

Fig. 9.43   Schematic of klystron

Fig. 9.44   Thales TH2103C
klystron

Waveguides and Launchers for LHH&CD

Power at low frequencies can be carried in coaxial transmission lines, as outlined
above for ion cyclotron heating systems. However, just as a different technology
is required for the sources for the lower hybrid frequency range, so must the
transmission system be changed to accommodate the higher frequencies. The
TEM-type electric and magnetic field distribution, as outlined above for ICH
technology, becomes difficult to implement when the frequency is high enough

so that non-TEM modes can exist in the cable. (This condition is called "multi-moding" and results in enhanced power loss in the line.) Using the nomenclature of Fig. 9.35, the minimum frequency where non-TEM modes can propagate in a coaxial cable occurs when

$$\lambda \approx 2\pi \frac{(a+b)}{2},$$

i.e., when the wavelength is the average circumference of the interelectrode radii. As an example, a six-inch outer diameter coaxial line with a $30\,\Omega$ impedance, with $\varepsilon_r = 1$, will have a high-frequency limit to multi-moding of 780 MHz, too low for most lower hybrid applications. Smaller-diameter lines will, of course, have a higher frequency cutoff, but at the expense of lower power. (The power handling scales as the square of the cable diameter, for a constant electric field limit at the inner conductor.)

The alternative transmission technology with the best performance at LH frequencies is waveguides. Waveguides are simply a duct-type structure with no center conductor. As such they have a low-frequency limit that is related to the relationship between the size of the waveguide and the free-space wavelength of the RF frequency being used. While waveguides come in many different formats, including rectangular, H-shaped, round, and elliptical cross sections and either of constant cross section in the direction of propagation or corrugated in some way, the most common type is of rectangular cross section with smooth-non-corrugated walls. A typical rectangular waveguide is shown as Fig. 9.45.

Fig. 9.45 Rectangular waveguide dimensions

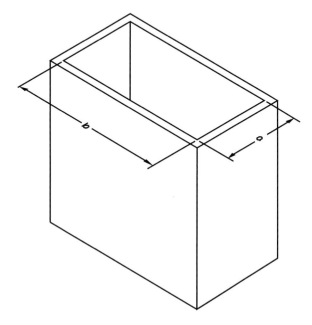

Waveguides can support two different modes of propagation. Since no pure TEM modes exist in a waveguide, modes are identified by having either a purely transverse electric field (called Transverse electric (TE) modes) or a purely transverse magnetic field (called Transverse magnetic (TM) modes). For a rectangular waveguide with $b > a$, the lowest propagating frequency is given by $\lambda_0(\text{max}) = 2b$, i.e. one-half of the free-space wavelength λ_0 must "fit" in the guide. (This mode is called the TE_{10} mode as it has one maximum for the electric field in the x direction (the direction of the b dimension in Fig. 9.45) and is constant in the y direction (the a dimension in Fig. 9.45). The electric and magnetic fields in the guide are then given by [61]:

$$E_y = i \frac{\omega b}{\pi} B_0 \, \sin\left(\frac{\pi x}{b}\right) \exp(i(kz - \omega t)),$$

$$B_x = -i \frac{kb}{\pi} B_0 \, \sin\left(\frac{\pi x}{b}\right) \exp(i(kz - \omega t)), \qquad (9.148)$$

$$B_z = B_0 \, \cos\left(\frac{\pi x}{b}\right) \exp(i(kz - \omega t)).$$

The dispersion relation for the TE_{10} mode is given by

$$\omega^2 = k^2 c^2 + \omega_c^2, \quad \text{with} \quad \omega_c = \pi c / b. \qquad (9.149)$$

The power carried in the waveguide is given by Poynting's theorem:

$$P = \int dA \frac{1}{2} (\mathbf{E} \times \mathbf{H}^*) \cdot \hat{n} = \frac{E_0^2 abc}{4\eta} \sqrt{1 - \left(\frac{\omega}{\omega_c}\right)^2}. \qquad (9.150)$$

Here $\eta = (\mu_0/\varepsilon_0)^{1/2} \approx 377\,\Omega$ and $E_0 = (\omega b/\pi) B_0$ is the peak electric field in the guide. This result has a simple interpretation. If there were a plane wave (TEM wave) with a peak electric field E_0, then the power flux per unit area would be $E_0^2/(2\eta)$, where η is the "impedance of free space," $377\,\Omega$, and the factor of one-half comes from the time average. In the waveguide, the group velocity is $c\sqrt{1 - (\omega/\omega_c)^2}$ (one can think of the RF energy "zig-zagging" through the guide) and the extra factor of one-half comes from averaging the $\sin^2(\pi x/b)$ spatial dependence.

Multipactor Effect

The power handling capability of waveguides and waveguide-based launchers in the LH frequency range is limited not just by thermal limits and breakdown effects due to residual gas but also by a phenomenon called the multipactor effect [67]. The multipactor effect is in essence a situation where an electron near one metallic surface of an RF structure is accelerated by the RF electric field so that it reaches the opposing metallic surface just as the electrical phase of the RF field changes to the opposite sign, whereby secondary electrons produced at the site of the original electron's impact are produced, which then accelerate towards the surface where the original electron was born, and so forth, causing an avalanche of electrons and leading to an arc breakdown.

A simplified treatment of the multipactor effect follows. This assumes that the electron is born exactly when the electric field (considered constant across two parallel plates) is changing sign. The electron is considered to be "born" as a free electron just outside the metallic plate with zero energy. The plates are assumed to be at positions $x = \pm d/2$, with the electron being born at $x = -d/2$. Then the equation of motion for the electron is

$$\ddot{x}(t) = \frac{q E_0}{m} \sin(\omega t). \tag{9.151}$$

Integrating this equation twice, with the boundary conditions that $x(0) = -d/2$ and $\dot{x}(0) = 0$ then gives

$$x(t) = -\frac{q E_0}{m\omega^2} \sin(\omega t) + \frac{q E_0}{m\omega} t - \frac{d}{2}. \tag{9.152}$$

We require that the electron appears at position $x = +d/2$ just as the electric field is changing sign, that is, when $t_{1/2} = N\pi/\omega$, where N is an odd positive integer. Using this time to solve for E_0 gives an equation

$$x(t_{1/2}) = -\frac{q E_0}{m\omega^2} \sin(\omega t_{1/2}) + \frac{q E_0}{m\omega} t_{1/2} - \frac{d}{2}, \tag{9.153}$$

or equivalently

$$\frac{d}{2} = -\frac{q E_0}{m\omega^2} \sin(\omega \frac{N\pi}{\omega}) + \frac{q E_0}{m\omega} \frac{N\pi}{\omega} - \frac{d}{2}. \tag{9.154}$$

Note that the term in $\sin(\omega t_{1/2})$ is zero for all N. Then, simplifying and substituting $f = \omega/(2\pi)$ and $E_0 = V_0/d$ gives a relation for the frequency-distance product fd:

$$fd = \frac{N}{2\sqrt{\pi}} \sqrt{\frac{q V_0}{m}}. \tag{9.155}$$

This gives a relationship between fd and the voltage V which scales as $V \propto (fd)^2$, which turns out to not fit the experimental data very well. Experimentally, a relationship closer to $V \propto fd$ is seen. An early explanation for this behavior was given by Hatch and Williams [48]. Here the energy of the secondary electron produced was taken into account, which gives an upper and lower limit to the energy of the primary electron producing the secondaries: if the energy of the

incident electron is too low, say, below 50 eV, the work function of the metal
and other details of the metallic structure result in limited secondary electron
production (<1 secondary per primary), and if the incident electron energy is
too high, say above 5 keV, then the range of the primary electron in the metal
is too long to produce secondaries appearing back at the surface of the metal.
Hatch and Williams used a model where the velocity of the secondary was taken
to be a fraction $1/k$ of the primary electron velocity, independent of electron
energy inside of some energy range. This so-called "constant-k" theory has no
basis in the actual secondary emission properties of metals but seems to represent
the physical observations better than the simple model given above, and gives
the correct experimentally observed $V \propto fd$ scaling. The results of the Hatch
and Williams model are shown as Fig. 9.46. Since the Hatch study in 1958,
other refinements on the theory of multipactor discharges have been undertaken,
which have included more realistic models of the secondary emission process,
accounting for the angular and velocity dependence of the secondary electrons,
and including the effects of both the RF and static magnetic fields at the
discharge point. A more modern treatment, including computer simulation of the
coupled electron-electromagnetic field problem using particle-in-cell techniques,

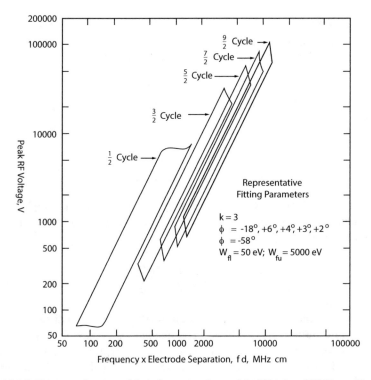

Fig. 9.46 Multipactor voltage vs. fd product using the model of Hatch and Williams. From [48].
Used with permission, American Institute of Physics

is described in [42]. While the overall breakdown voltage has dependence on a larger gamut of factors, a general "rule-of-thumb" for the breakdown voltage can be given as

$$V_b \geq 54 f d \qquad (9.156)$$

with f in GHz and d in mm. This correlation is valid for $f d > 2\,\text{GHz-mm}$ and is conservative by a factor of $30 \rightarrow 50\%$ for most of the range in fd being considered for LH H &CD antenna structures.

LH Launching Structures

Phased array techniques similar to those outlined for the ICH range of frequencies are typically used to launch LH waves into the plasma. Controlling the parallel index of refraction $n_\parallel$ is of paramount importance, for reasons of accessibility and current drive. The power transmission to the antenna is done with waveguides rather than coaxial lines, as outlined above. The multipactor problem as described above is especially limiting at the launching grill, where the path length in the direction of the electric field is typically at its minimum value. To complicate matters even further, the load impedance presented by the plasma can vary with the position of the launcher vis-a-vis the plasma edge, and the plasma and neutral gas outside the closed flux surfaces can contribute to breakdowns and arcing. Designing a load-agile, breakdown-resistant, and low-sputtering LH antenna is as much an art as a science.

An example of the state of the art in LH antenna design is given in [10] and shown in Fig. 9.47. This antenna is designed for use on Tore Supra, with

Fig. 9.47 "ITER-like" PAM launcher for Tore Supra. From [10]. Used by permission, Elsevier

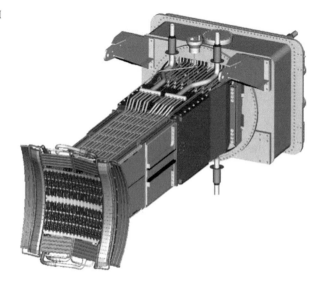

the motivation that the design can be adapted for ITER, should an LH H&CD campaign on ITER be approved. This design optimizes RF coupling into the plasma by the insertion of quarter-wavelength passive waveguide stubs between each active waveguide transmitting RF energy from the source, hence the name PAM, for "passive-active multijunction." The entire eight-tonne structure can be moved by hydraulics through a 20 cm range, to optimize coupling and to pull it back when not in use to reduce metal sputtering into the plasma. The antenna is designed for a power handling capacity of 2.7 MW at a frequency of 3.7 GHz.

3. **ECH&CD**

The theoretical knowledge and experimental experience with electron cyclotron heating and current drive (ECH&CD) is well established [35, 75]. Electron cyclotron resonance is given by the basic formula (for a nonrelativistic electron at a stationary point along the magnetic field):

$$\Omega_e = \frac{q_e B}{m_e}. \tag{9.157}$$

As an example, we use the magnetic field strength on the magnetic axis of ITER of $B = 5.3$ T and then we have $f_{ce} = \Omega_{ce}/(2\pi) = 148.3$ GHz. A more general expression for the resonance frequency takes account for: (1) harmonic multiples ℓ of the fundamental cyclotron frequency, (2) the effect of relativity, which changes the mass of the electron $m_e = \gamma m_{e0}$, and (3) allows for the Doppler effect due to parallel motion of the electron along the magnetic field $v_\parallel$. Then the more general resonance condition is given by

$$\omega = \frac{\ell \Omega_e}{\gamma} + k_\parallel v_\parallel = \frac{\ell \Omega_e}{\gamma} + \frac{n_\parallel v_{e\parallel} \omega}{c}. \tag{9.158}$$

Here we use n, $n_\parallel$, and $n_\perp$ to represent the total, parallel, and perpendicular indices of refraction, respectively. We can evaluate this resonance condition by substituting the dimensionless variable $u = \gamma v/c$ and then the resonance condition can be written as:

$$\frac{\left(u_\parallel - u_{\parallel 0}\right)^2}{\alpha_\parallel^2} + \frac{u_\perp^2}{\alpha_\perp^2} = 1, \tag{9.159}$$

with the quantities $u_{\parallel 0}$, $\alpha_\parallel$, and $\alpha_\perp$ being defined as

Fig. 9.48 Resonance curves for perpendicular ($n_\parallel = 0$) and oblique ($n_\parallel = -0.5$) propagation at ECH&CD frequency. For the oblique propagation case, the dashed curves are for downshifted resonance and the solid curves are for upshifted resonance. From [35]

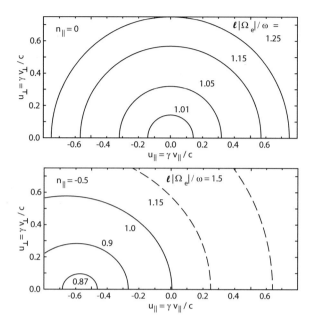

$$u_{\parallel 0} = \frac{n_\parallel \left(\ell|\Omega_e|/\omega\right)}{1 - n_\parallel^2},$$

$$\alpha_\parallel = \frac{\sqrt{n_\parallel^2 + (\ell\Omega_e/\omega)^2 - 1}}{1 - n_\parallel^2},$$

$$(9.160)$$

and

$$\alpha_\perp = \frac{\sqrt{n_\parallel^2 + (\ell\Omega_e/\omega)^2 - 1}}{\sqrt{1 - n_\parallel^2}}.$$

Figure 9.48 shows the locus of resonant particle velocities using Eq. (9.159). As the RF wave moves through the plasma, the magnetic field changes and thus $|\Omega_e|$ changes, resulting in resonance with different parts of velocity space. Note that for perpendicular incidence, however, there is no resonant velocity for $\omega > \ell|\Omega_e|/\gamma$ whereas for oblique incidence both upshifted and downshifted resonant points are possible. The asymmetry of the resonant curves in velocity space for oblique incidence is an essential part of the coupling of ECRH waves leading to current drive.

Accessibility to the electron cyclotron resonance using cold-plasma wave theory was discussed earlier using the CMA diagram and trajectories in the CMA space were shown for inside and outside launch in Fig. 9.22. For an O-mode launch from the outside at the first harmonic, the electron density should be low enough to allow access, which requires that $\omega_{pe} < |\Omega_e|$. (For warm plasma with $\gamma > 1$, the condition is more restrictive: $\omega_{pe} < |\Omega_e|/\gamma$.) For the case of ITER, ECH & CD will be accomplished with waves launched in the O-mode from a number of positions around the perimeter of the plasma. Since the anticipated plasma electron density profile is nearly flat across the central plasma with a density of 1.0×10^{20} m^{-3}, the plasma frequency $f_{pe} \approx 90$ GHz < 170 GHz, so no cutoff is encountered. Figure 9.49 shows ray tracing results for an ECH & CD wave launched from an antenna in the upper poloidal plane on the outside of the $\omega = \Omega_e$ surface. Figure 9.50 shows the energy deposition and current drive profiles for this model.

While cold-plasma theory can be applied using the ray tracing technique to find the dispersion of the ray trajectories and to determine the basic accessibility conditions for the RF wave, more sophisticated methods must be used to assess the coupling of the wave energy to the plasma and to calculate the predicted current drive capability of the RF source. Several different models have been used to determine the RF power absorption profile and RF-driven current density profile.

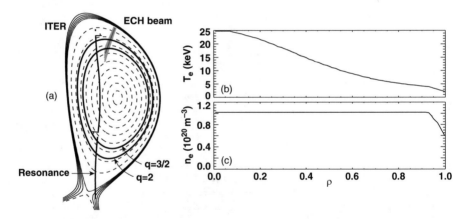

Fig. 9.49 Antenna arrangement for electron cyclotron heating and current drive for ITER. Beam is launched in O-mode at 170 GHz, with the $\omega > \omega_{pe}$ everywhere. (**a**) Calculated ray path, (**b**) assumed electron temperature profile, (**c**) assumed density profile. From [90]. Used with permission, IAEA

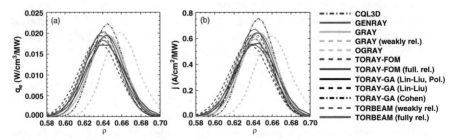

Fig. 9.50 Predicted (**a**) energy deposition profile and (**b**) RF-driven current density for various computer models. From [90]. Used with permission, IAEA

A quasilinear treatment was developed in [8] and applied to the problem of ECH&CD in tokamaks, including relativity and trapped-particle effects, in [22]. The starting point is an expression for the quasilinear heating and current drive given in [8]:

$$\bar{Q}(v) = -\sum_{k\omega\ell} \left\langle \int d^3p \frac{D_{ee}\delta(\omega - k_\parallel v_\parallel - \ell\Omega)}{\gamma} \left[\frac{\partial f}{\partial\varepsilon} + \frac{k_\parallel}{\omega} \frac{\partial f}{\partial p_\parallel} \right] \right\rangle$$

$$\frac{J_\parallel}{B} = -\sum_{k\omega\ell} \left\langle \int d^3p \frac{D_{ee}\delta(\omega - k_\parallel v_\parallel - \ell\Omega)}{\gamma} \left[\frac{\partial f}{\partial\varepsilon} + \frac{k_\parallel}{\omega} \frac{\partial f}{\partial p_\parallel} \right] \right. \tag{9.161}$$

$$\left. \cdot \left[\frac{\partial G}{\partial\varepsilon} + \frac{k_\parallel}{\omega} \frac{\partial G}{\partial p_\parallel} \right] \right\rangle .$$

Here $\langle\rangle$ indicates an average over a flux surface, ε and $p_\parallel$ are the energy and angular momentum of the electrons, and D_{ee} is the quasilinear diffusion in velocity space associated with the wave electric field, given by:

$$D_{ee} = \frac{e^2}{4\varepsilon_0} \gamma \left| E_\parallel v_\parallel J_\ell \left(\frac{k_\perp v_\perp \gamma}{|\Omega_e|} \right) \right.$$

$$\left. + \frac{v_\perp}{\sqrt{2}} \left(E_+ v_\perp J_{\ell-1} \left(\frac{k_\perp v_\perp \gamma}{|\Omega_e|} \right| \right) e^{-i\psi} + E_- v_\perp J_{\ell+1} \left(\frac{k_\perp v_\perp \gamma}{|\Omega_e|} \right| \right) e^{i\psi} \right) \right|^2 . \tag{9.162}$$

Here J_n are the ordinary Bessel functions of order n, and $k = \{k_\perp \cos\psi, k_\perp \sin\psi, k_\parallel\}$. Here the nonrelativistic electron cyclotron frequency $|\Omega_e| = eB/m_{e0}$ is taken as positive and the electric field E is decomposed into left- and right- circularly polarized components such that $E_\pm = (1/\sqrt{2})(E_x \pm iE_y)$.

The function G appearing in the current drive expression Eq. (9.161) remains to be evaluated. This function is related to the modification to the electron distribution function due to trapped particles and in the presence of collisions. We define G such that

$$G = e \langle B_t/R \rangle^{-1} \exp(\varepsilon/T)g. \tag{9.163}$$

Here B_t is the toroidal magnetic field and T is the temperature of the background electrons. The function g satisfies a simplified Fokker-Planck equation

$$v_\parallel \mathbf{b} \cdot \nabla g + \mathscr{C}(g) = v_\parallel \mathbf{b} \cdot \nabla\phi \exp(-\varepsilon/T), \tag{9.164}$$

with ϕ the toroidal coordinate and $\mathscr{C}$ the Fokker-Planck collision operator. The calculation of g is sometimes referred to as the "Spitzer-Härm problem" because of its reference to an early paper by Spitzer and Härm [98] where the conductivity of a uniform plasma was calculated using a distribution function perturbed from maxwellian by the presence of an electric field. Here the solution of Eq. (9.164) is made more difficult by the presence of trapped particles caused by the variation in the magnetic field B along the magnetic field lines. Cohen [22] gives an approximate solution to the bounce-averaged perturbed distribution function with the substitution of variables $h = g \exp(\varepsilon/T)$, with h being defined by

$$h = g \exp(\varepsilon/T) = \langle \mathscr{R}_t/R \rangle v^4 H \left(v_{\perp 0}^2/v^2\right) / \left(4 v v_{th}^2\right). \tag{9.165}$$

here $\mathscr{R}$ is the local value of the mirror ratio B_{tor}/B_0, $\eta = v_{\perp 0}^2/v^2$, the thermal speed $v_{th} = (2T/m_e)^{1/2}$, and the collision frequency v is given by

$$v = \frac{n_e e^4 \ln \Lambda}{4\pi \varepsilon_0^2 m_e^2 v_{th}^3}. \tag{9.166}$$

If we replace the local mirror ratio $\mathscr{R}$ by a square-well approximation, then the function H can be evaluated from the solution to an inhomogeneous Legendre equation, such that

$$H = \frac{-2\lambda}{Z+5} \left[1 - \frac{\lambda_s}{|\lambda|} \frac{P_\alpha(|\lambda|)}{P_\alpha(\lambda_s)} \right]. \tag{9.167}$$

Here $\lambda = \pm(1-\eta)^{1/2} = v_\parallel/v$, λ_s is the value of λ at the bottom of the magnetic well, Z is the average charge of the ions, and P_α is the Legendre function of index α, determined from:

$$\alpha(\alpha+1) = -8/(1+Z). \tag{9.168}$$

Reference [22] also gives the results for a more accurate solution to Eq. (9.164), using careful flux averages of the operators in the Fokker-Planck equation with trapped particles in a tokamak geometry with concentric circular flux surfaces. It was found, however, that the square-well model for the potential was within a few percent of the results from the more exact treatment for all cases considered. The closeness of this approximation gives one some comfort in applying this technique to noncircular tokamaks such as ITER, and provides a benchmark for evaluating newer code results with more sophisticated models.

We can now apply the perturbed distribution function from (9.167) to Eq. (9.161) to find expressions for the RF-derived current and the RF-driven heating effect. It is convenient to replace the actual current drive $J_\parallel$ and heating Q by dimensionless variables J and P_d such that $J = -J_\parallel/(e n_e v_{th})$ and $P_d = Q/(n m v_{th}^2 v)$, with $v_{th} = (2T_e/m_e)^{1/2}$ and v given by Eq. (9.166). We further define a current normalized to the magnetic field such that $J_0 = J(B/B_0)$. For a single mode of interaction, the current-drive-to-heating ratio is given by

$$\frac{J_0}{P_d} = \frac{-m^2c^4 \int d\varepsilon\gamma \, [D_{ee}\mathscr{L}(f)\mathscr{L}(FH)]_{p_{\|a},\theta_a}}{4T \int d\varepsilon\gamma \, [D_{ee}\mathscr{L}(f)]_{p_{\|a},\theta_a}}. \tag{9.169}$$

Here $p_{\|a}$ and θ_a denote the values of $p_\|$ and the poloidal angle θ_a resulting in resonance. After an assumption that the overall distribution function $f \propto \exp(-\varepsilon/T)$, we arrive at a simple expression for J/P_d using Eq. (9.169) as an integral along the ray path with the resonant condition $p_\|^{res} = p_\|^{res}(\gamma, \ell\Omega_0/\omega)$. We find that

$$\frac{J}{P_d} = -\frac{mc^2}{2T Q_\ell} \int d\gamma \alpha_\ell(\gamma) \left(F_\gamma H + \beta F H'\right) \tag{9.170}$$

with $H' = dH/d\eta$ from Eq. (9.169), and α_ℓ and β given by

$$\alpha_\ell = (\gamma - Y)^j \left\{\gamma^2 - 1 - \left[(\gamma - Y)/n_\|\right]^2\right\}^\ell$$
$$\times \exp\left[-mc^2(\gamma - \gamma_0)/T_h\right],$$
$$\beta = \left[2(\gamma - Y)/\left(\mathscr{R}\left(\gamma^2 - 1\right)\right)\right] \tag{9.171}$$
$$\times \left\{\left[\gamma(\gamma - Y)/\left((\gamma^2 - 1)\, n_\|^2\right)\right] - 1\right\}.$$

Here $j = 0$ for the E_- component of the electric field, $j = 1$ for the $E_\|$ part, and $j = 0$ and ℓ is replaced by $\ell + 2$ for the $E_\perp$ part (which is typically a very small contribution). We also use $Y = \ell\Omega_0/\omega$, with Ω_0 as the nonrelativistic cyclotron frequency. The quantity Q_ℓ in Eq. (9.170) is the normalized power deposition given by

$$Q_\ell = \int d\gamma \alpha_\ell, \tag{9.172}$$

which is given in [22, Appendix].

The calculations for J/P_d are dependent on the geometry through the mirror ratio $\mathscr{R}$ and the value of the quantity λ_s, the cosine of the pitch angle for trapping, as it appears in Eq. (9.167). The resonance equation also limits the values of γ vs. Y for which a solution exists. Reference [22] gives some examples for the predicted average $< J > /P_d$ as a function of the inverse aspect ratio r/R for a tokamak with concentric circular flux surfaces. It was found that efficiencies were always higher for the case where the fundamental resonance $\omega = \omega_0$ was at the inside of the flux surface where the magnetic field was the highest along any flux surface, and that the current drive-to-absorbed power ratio dropped as the inverse aspect ratio increased. It was also found that cases where $Y > 1$ ($\omega < \Omega_0$) were less efficient than cases where $\omega > \Omega_0$. Figure 9.51 shows the results of this calculation for the current drive efficiency as a function of electron

Fig. 9.51 Current drive efficiency $< J > / P_d$ from Cohen's model where $r/R = 0.1$, fundamental resonance on the outside of the flux surface, $\varepsilon_0/T = 2$, and $T_h = T$. Data from [22]

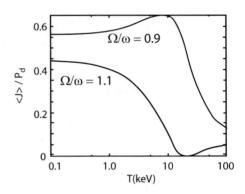

temperature. One can see that fairly good efficiency is obtained at $T_e = 10\,\text{keV}$ and $\Omega/\omega = 0.9$; higher efficiencies would be attainable under these conditions with this frequency relationship holding at the inside of the flux surface instead.

ECH & CD Technology

Waveguides and Antennae

The high frequency of ECH&CD systems results in antenna and transmission systems which are large compared to the free space wavelength $\lambda \approx 2\,\text{mm}$. Thus the engineering of these systems relies upon the concepts used in optics as much as upon conventional RF electromagnetic theory. The high power levels involved, along with the large scale lengths involved, have moved the design of these systems beyond the conventional. At frequencies where the free space wavelength λ is much smaller than the length d of some component in the transmission or delivery system, it is useful to use concepts from optics to describe the electromagnetic propagation. As a starting point, consider the properties of a gaussian electromagnetic wave, the most common type of wave found in ray optics. The solution to Maxwell's equations for such a wave is given by [114]:

$$\mathbf{E}(r, z) = E_0\,\hat{x}\,\frac{w_0}{w(z)}\,\exp\!\left(\frac{-r^2}{w(z)^2}\right)\exp\!\left(-i\left(kz + k\frac{r^2}{2R(z)} - \psi(z)\right)\right).$$

$$(9.173)$$

Here $w(z)$ is the beam radial extent (for the power density to drop to $1/e^2$ of its central value), w_0 is the minimum beam radius, $R(z)$ is the radius of curvature of the beam phase front, $k = 2\pi/\lambda$ is the free space wavenumber, and $\psi(z)$ is an extra phase factor called the Gouy phase (a factor between zero and 2π for the fundamental Gaussian mode, and only affecting the phase and not the amplitude). The beam radius depends on z such that

$$w(z) = w_0\sqrt{1 + \left(\frac{z}{z_R}\right)^2},$$

$$(9.174)$$

with the Rayleigh length z_R given by

$$z_R = \frac{\pi w_0^2}{\lambda}. \tag{9.175}$$

Manipulation of this expression shows that asymptotically $w(z) \approx \theta z$, which describes the half-angle for the divergence of the beam; and thus

$$\theta = \frac{\lambda}{\pi w_0}. \tag{9.176}$$

Using Poynting's theorem then gives an expression for the power transmitted by the beam:

$$P_{beam} = \pi r_0^2 \frac{E_0^2}{2\eta} \tag{9.177}$$

with $\eta = (\mu_0/\varepsilon_0)^{1/2} = 377 \ \Omega$. For $\lambda \ll d$, where d is the diameter of some aperture that the electromagnetic energy is emanating from, the beam has a narrow divergence angle. For typical fusion ECH&CD systems, apertures in the $50 \rightarrow 100$ mm diameter range result in beam spreading angles of a few degrees—not quite the milliradian angles typically found in visible lasers, but still small enough so that the gaussian beam concept has some uses. For these parameters, the concept of "quasioptical" RF transport is used, indicating the reference to the gaussian beam concept in style if not substance.

The quasioptical beam is technically lossless, but transport of the beam for large distances (tens of meters) by purely quasioptical means is impractical without some attempt to focus the beams along their path. The beams of RF energy are launched in a quasioptical mode before they enter the plasma, but typically are transported in "overmoded" waveguides (waveguides with $d \gg \lambda$) up to the final launching assembly. For $\lambda \ll d$, the optimum waveguide configuration appears to be a corrugated waveguide with corrugations having a length and depth on the order of $\lambda/4$, as shown in Fig. 9.52. Waveguides of this sort support an electromagnetic mode known as HE_{11}. The HE_{11} mode has an electric field given by [33]:

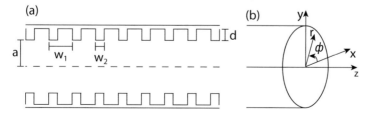

Fig. 9.52 Corrugated waveguide for transport of the HE_{11} mode. (**a**) Cutaway view of waveguide, (**b**) cylindrical geometry for the HE_{11} mode. From [68]. Massachusetts Institute of Technology

Fig. 9.53 Mode patterns for
the HE$_{11}$ mode and HE$_{12}$
mode. (**a**) HE$_{11}$ mode, (**b**)
HE$_{12}$ mode. From [68].
Massachusetts Institute of
Technology

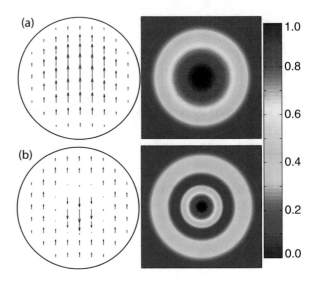

$$E_{\gamma m} = \left(\frac{2P\eta}{\pi P_0}\right)^{1/2} \frac{J_0\,(x_m r/a)}{a\,J_1\,(x_m)}. \tag{9.178}$$

Here x_m is the m-th zero of the Bessel function (i.e., $x_1 = 2.4048$), and R_0 is the
radius of the guide to the top of the corrugations. Note that the structure of the
HE$_{11}$ mode is similar to the quasioptical Gaussian mode except near the edge of
the waveguide, and the losses for this mode are minimized by the relatively small
wall currents needed to support this mode.

Figure 9.53 shows the electric field pattern in the wave guide for the HE$_{11}$
mode and for the HE$_{12}$ mode, the next higher mode. For transitional elements
in the transmission system, such as mirrors and the entrance to the waveguide
system from the source tubes, diffraction results in the generation of higher
modes, starting with the HE$_{12}$ mode. The transmission loss in corrugated
waveguide with $d/\lambda \approx 18$ is very small, on the order of 0.1 dB per 100 m
(note that $-0.1\,\text{dB} = 10^{(-1/100)} = 0.9772$, or $\approx 2.3\%$ loss). Figure 9.54
shows the calculated results for loss in two corrugated waveguide designs,
both with 63.55 mm diameter (the ITER waveguide design dimension). These
show reduced loss compared to smooth waveguide, as shown in the figure. The
corrugated waveguide will start to show significant loss when the period of the
corrugations approaches one-half of the wavelength of the RF signal, as shown
in the figure; this is caused by the appearance of reflections similar to the Bragg-
type diffraction that one sees on an optical grating. Thus the selection of a
higher frequency drive in the future, say for second harmonic heating, will likely
necessitate a redesigned waveguide system.

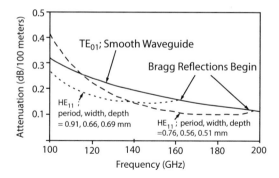

Fig. 9.54 "Comparison of attenuation in smooth-wall and corrugated 31.75-mm aluminum waveguides with bulk resistivity of $4.0 \,\mu$-Ω cm and indicated corrugation geometries. This resistivity and a 30% increase in the attenuation due to surface roughness are assumed in this and successive figures. Bragg reflections occur when the corrugation period exceeds one-half wavelength." From [32]. Used by permission, Taylor and Francis

Additional losses in the transmission system are caused by the need for bends in the waveguide. There are two sources of loss in the bends: one is due to ohmic heating in the metallic reflecting surface, and the other is due to refraction from the reflector due to the finite wavelength. The fractional loss $f_{loss} = P_{loss}/P_{in}$ due to ohmic dissipation in the reflecting wall can be determined from plane-wave theory, using a result given by Stratton in 1941 [103]:

$$
f_{loss} = \begin{cases} 4\left(\dfrac{R}{\eta}\right)\cos\theta, & \text{H plane,} \\[2mm] 4\left(\dfrac{R}{\eta}\right)/\cos\theta, & \text{E plane.} \end{cases} \tag{9.179}
$$

Here R is the effective skin-effect resistance ρ/δ, where ρ is the resistivity of the metal and $\delta = (2\rho/(\omega\mu_0))^{1/2}$ is the skin depth, and $\eta = 377\,\Omega$ is the impedance of free space. "E-plane" and "H-plane" refer to the cases where the electric field E or the magnetic field H lie in the plane of the reflecting surface, and θ is the angle that the normal from the reflecting surface makes with the propagation vector of the RF wave. This loss is very small for the EC H&CD systems of interest here. The diffraction loss is typically the dominant one, and this is estimated by assuming that the loss is similar to that obtained from having a break in the waveguide of length $L = d = 2a$, as shown in Fig. 9.55. The loss from such a miter bend is given by:

$$
f_{loss} = 0.55\left(\frac{\lambda}{d}\right)^{3/2}. \tag{9.180}
$$

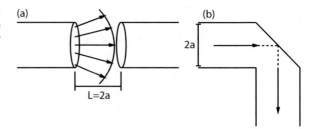

Fig. 9.55 (**a**) Diffraction loss from a radially symmetric gap with length $L = 2a$. (**b**) Miter bend having equivalent loss. From [68]. Massachusetts Institute of Technology

For $d/\lambda = 18$ (the ITER case), this gives a loss on order of 0.03 dB, or a 0.72% loss per bend. While this loss does not occur directly at the reflecting surface, it may point to the need for active cooling on the joint and the neighboring waveguide where the RF power will be absorbed.

Another source of loss in the transmission system are windows which segregate the source end of the system from the launching end, which shares common vacuum conditions with the plasma. The windows provide protection against tritium contamination outside of the torus in case of an off-normal event. Since high-grade diamond grown by chemical vapor deposition has become available at wafer sizes, this has become the material of choice for these windows. CVD diamond windows have been deployed on the DIII-D tokamak [24] and are planned for the ITER tokamak [62, 104]. The superior mechanical strength and high heat conductivity, as well as low RF power loss, are key features for the use of this material for RF windows. The power loss for a thin CVD diamond window is given by [93]:

$$\frac{P_{out}}{P_{in}} = \exp(-\alpha d), \tag{9.181}$$

with

$$\alpha = \frac{2\pi n \tan \delta}{\lambda_0}. \tag{9.182}$$

Here $\tan \delta = \varepsilon''/\varepsilon'$ is the ratio of the imaginary to real parts of the dielectric constant ($\varepsilon' = 5.7$ for CVD diamond), n is the index of refraction ($\approx \sqrt{\varepsilon'} \approx 2.38$ for CVD diamond), and λ_0 is the free space wavelength. For the ITER design, a thickness of $d = 1.11$ mm ($= 1.5\lambda$ in CVD diamond, integer half-wavelengths minimize reflected power loss), this results in a fractional power loss $\alpha d \approx 0.01\%$, which results in an internal heating of the window on order of 200 W maximum.

Gyrotrons

Just as klystrons made a major departure in RF power sources from triodes and other gridded tubes, gyrotrons have supplanted other technologies for the production of high-power millimeter-wavelength sources for ECH & CD. The gyrotron, as its name implies, exploits the removal of energy from an electron beam with a substantial component of its energy in a direction perpendicular to a magnetic field. Implicit in this process is the change in the gyrofrequency of a particle with a change in its energy through the resonance relation used earlier in Eq. (9.158):

$$\omega = \frac{\ell \Omega_e}{\gamma} + k_\| v_{e\|} = \frac{\ell \Omega_e}{\gamma} + \frac{n_\| v_{e\|} \omega}{c},$$

where the relativistic factor $\gamma = 1 + E_k/(m_e c^2)$ is substantially different from one for typical electron beam energies of $60 \rightarrow 100\,\text{keV}$. When the gyrofrequency of the electrons entering a resonance region (such as a cavity resonator or a set of confocal mirrors) is below the resonance, the electrons can be "pulled" into the resonance by a loss of energy, causing γ to decrease and energy to be transferred to electromagnetic fields. This beam-driven instability is called the "cyclotron resonance maser" (CRM) interaction, and was first discovered by Twiss in 1958 [112] in connection with astrophysical radio emissions. Figure 9.56 shows a graph of ω vs. $k_z = n_\| \omega / c$ for the gyrotron interaction.

Critical to the performance of the gyrotron is the generation of a beam with a large component of its energy perpendicular to the magnetic field. The most successful method for doing this is to use an annular cathode embedded in a magnetic field which increases in intensity as the beam is accelerated and then

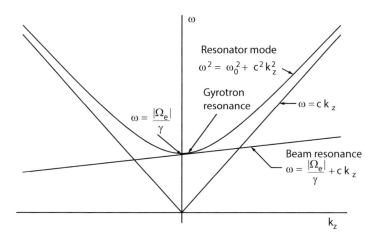

Fig. 9.56 ω vs. k_z for the cyclotron resonance maser interaction used in gyrotrons

Fig. 9.57 Magnetron
Injection Gun (MIG) for
gyrotron

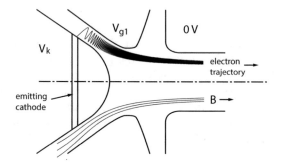

drifts into a resonator. The basic injector structure is called a magnetron injection gun, or MIG, and was first introduced by a Soviet team in 1964 [66]. A schematic of a typical MIG injector is shown as Fig. 9.57. The injection of an annular beam of electrons with $v_\perp / v_\parallel \approx 2$ is ideal for a gyrotron with a circular cavity, since the energy is drawn from the component of velocity perpendicular to the magnetic field and the beam can be configured so that its footprint in the cavity is close to the electric field maximum in the resonator cavity. A modern design, capable of 4 MW of injected electron beam, is described in [38]. Here the distance between the cathode and the first accelerating electrode (typically held at around 30% of the total accelerating voltage) is reduced, which the authors claim calls for a "non-paraxial" design, i.e. the cathode-to-first anode spacing is only a few times the electron gyroradius, requiring a full analysis of the particles and fields with self-consistent electromagnetic codes.

Figure 9.58 shows the overall implementation of the gyrotron for both the conventional format and for quasioptical designs. While quasioptical designs have enjoyed some success [1, 58, 96, 122], they have not replaced conventional gyrotrons, in part because it has been more difficult to match the electron beam footprint to the electric field profile in a Fabry-Perot type confocal resonator. While the quasi-optical design holds the promise of more easily generating a gaussian beam for coupling to the transmission waveguide, gaussian-beam launching structures are now being used in conventional gyrotrons as well, as shown in Fig. 9.59. As conventional gyrotrons are designed for high power at higher frequencies, however, it becomes necessary to use resonator cavities with very high mode numbers, e.g. hundreds of modes above the lowest resonance frequency, with the possibility that neighboring modes might compete for excitation in the device, with possibly worse coupling into the waveguide. The quasi-optical gyrotron can operate with more tolerance towards higher values of d/λ, in the same way that an infrared laser or a visible laser operates with a physical size orders of magnitude larger than the wavelength being produced. Quasioptical gyrotrons thus may have a future for fusion heating applications, especially if operation at multiple harmonics of the electron cyclotron frequency becomes important and interesting.

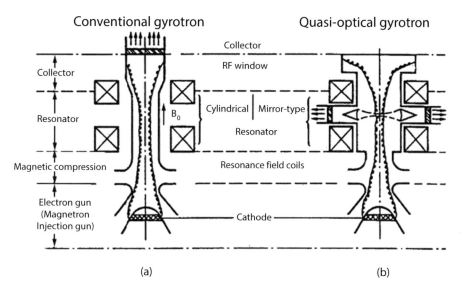

Fig. 9.58 Schematic diagram for (**a**) conventional gyrotron and (**b**) quasi-optical gyrotron. From [64]. Used with permission, Springer

Gyrotrons for the ITER device are expected to produce 1.0 MW in long pulse mode at 170 MHz. Prototypes for the ITER gyrotrons are being developed in India, Russia, Japan, and Europe. As of this writing, the prototypes are still in the testing phase. Figure 9.60 shows the European device being fabricated by Thales Electron Devices.

The ITER ECH & CD System Integration

EC H& CD components, including transmission structures and gyrotron sources, have been described above. Here we will take a comprehensive overview if the ECH & CD system as it has been designed, prototyped, and partially built (as of this writing).

The ECH & CD transmission system for ITER consists of 24 lines of evacuated 63.5 mm diameter corrugated waveguide. Each line is designed to handle 2 MW of RF power at 170 GHz. The design uses corrugated waveguide and will be operated in the HE_{11} mode. Components include 140° and 90° miter bends, waveguide switches, arc detectors, polarization rotators, waveguide pumpouts, and DC breaks, among others. Isolation of the torus plasma from the back end of the EC H& CD system is provided by a set of diamond windows.

Recent experience with large tokamak confinement experiments has highlighted the importance of controlling neoclassical tearing modes (NTMs), as described in Chap. 7. To this end, The ITER design has a set of dedicated ECH antennae specially designed to deliver power to the rational surfaces in the plasma

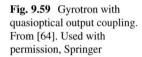
Fig. 9.59 Gyrotron with
quasioptical output coupling.
From [64]. Used with
permission, Springer

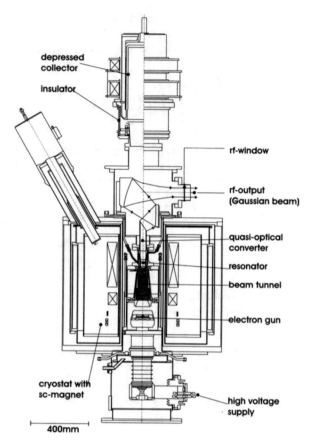

where the most dangerous NTMs are potentially present. Fig. 9.61 shows a
schematic of one of the upper launcher (UL) antenna structures for use on ITER,
which are designed for this purpose. Note the periscope-style mirror arrangement
for the last two sets of mirrors, which are designed to prevent neutrons from
streaming out of the plasma and causing radiation damage in the components
upstream. The UL contains bellows and other mechanical components so that
the launcher can be swept from one position to another to concentrate power on
the NTM zone of interest in a matter of seconds.

Figure 9.62 shows the overall poloidal footprint with port locations, showing
both the upper ports for the UL and also the equatorial ports for ECH & CD along
the midplane.

The gyrotons for the ITER H & CD system are located in a separate room
from the device itself, and thus waveguide runs are tens of meters away, thus
necessitating a low-loss transmission system consisting of overmoded circular
waveguide, quasioptical mirrors, and diamond windows as described above. The
transmission system design was challenged by the overall complexity of the

Fig. 9.60 Prototype 170 GHz gyrotron for ITER from Thales Electron Devices. ©2015 European Joint Undertaking for ITER and the Development of Fusion Energy ("Fusion for Energy")

machine, with competing demands for space from vacuum components, magnets and their charging lines and helium piping, neutral beam injectors, diagnostics, shielding components, and other systems. Figure 9.63 shows the current design. In addition to its current design configuration, the system has the capability for extensions and upgrades as perhaps more powerful high frequency sources will emerge, or other physics demands are placed on the system.

4. **High harmonic fast waves**

   RF heating and current drive in spherical tokamaks such as MAST and NSTX presents a challenge compared to conventional tokamaks due to the relatively low toroidal field in these devices, causing $\omega_{pe}/|\Omega_e| > 1$, making the electron cyclotron resonance inaccessible under cold-plasma wave theory. Ion cyclotron resonance heating is also a problem because the RF antenna size for heating at the fundamental ion cyclotron resonance would result in an antenna size larger than the device would allow. One possible heating scenario that has been explored is high harmonic fast wave (HHFW) heating, where the RF drive is at a frequency many times the fundamental ion cyclotron resonance. This concept has been explored experimentally and theoretically at Princeton using the National Spherical Tokamak Experiment (NSTX) device. In this device, which has a

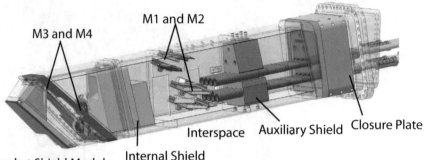

Fig. 9.61 Upper Launcher for ITER, mostly for suppression of neoclassical tearing modes (NTMs). Note mirrors M1/M2 providing a "dogleg" path to prevent neutron streaming, and M3/M4 to provide a variable launch angle. From [105]. Used by permission, Elsevier

Fig. 9.62 ITER ECH&CD ports. RF power is applied through the Upper Launcher (UL) ports and the equatorial ports. Form [104]. Used with permission, Elsevier

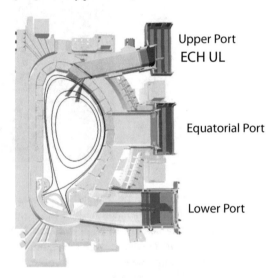

maximum toroidal field of 5.5 kG (0.55 T), RF heating is applied through a phased array antenna similar to that used for ICRF heating on other machines. The frequency used is 30 MHz, which is nominally around the seventh harmonic for the He plasmas currently being generated. The antenna system is shown as Fig. 9.64. The antenna has twelve straps with a pitch (distance between straps) of 21.5 cm, resulting in an antenna spread out over about one-quarter of the beltline, since the major radius in NSTX is 1.0 m. The phasing between the straps can be chosen as $\pi/6$, $\pi/2$, or $5\pi/6$. The current design also includes a center tap for each antenna strap, allowing for programming in a poloidal amplitude and phase for each half independently, resulting in effectively twenty-four antennae.

Some rather significant results have been obtained with the HHFW system after some issues with impurity control have been resolved [110]. The system

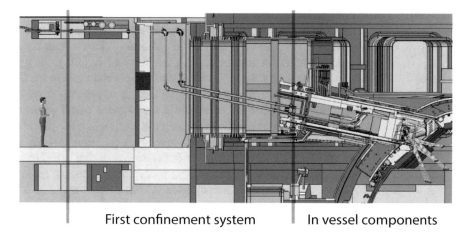

First confinement system   **In vessel components**

Fig. 9.63 Transmission system for ITER ECH&CD. From [105]. Used by permission, Elsevier

Fig. 9.64 High Harmonic fast Wave (HHFW) antenna for NSTX. From [87]. Used with permission, International Atomic Energy Agency

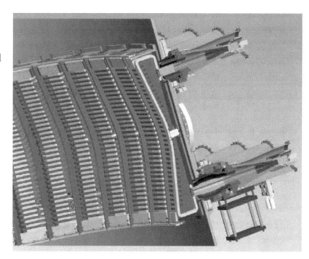

has been particularly effective when combined with a neutral beam injection system providing 90 keV neutral particles to the plasma, with on-axis electron temperatures as high as 6.0 keV in H-mode plasmas. It has been found that roughly 50% of the RF power goes in as heat inside the last closed flux surface (LCFS), and the power is split 2/3 to the electrons and 1/3 into fast ion heating inside the LCFS. Another impact of the HHFW has been the capability of providing a large fraction of the toroidal current through non-inductive RF current drive, with total non-inductive current fractions reaching as high as $f_{CD}^{NI} = 0.65$. This is an especially important milestone for spherical tokamaks, since the area on the inside of the torus is very limited and there is little room for inductive coils.

5. **Electron Bernstein waves**

The issues surrounding RF heating schemes for spherical tokamaks mentioned above, most importantly the plasma frequency-to-electron cyclotron frequency $\omega_{pe}/|\Omega_e| > 1$, have resulted in studies of the possibility of coupling energy into the electron cyclotron resonance by launching a cold-plasma wave in the O mode from the outside of the plasma, and exploiting a mode conversion process from the O mode to the X mode at the point where the O-mode cutoff appears when $\omega = \omega_{pe}$, where the converted X-mode energy is carried into the plasma to a point where $\omega = \omega_{UH}$, the upper hybrid frequency ($\omega_{UH}^2 = \omega_{pe}^2 + \Omega_e^2$). Here the wave energy undergoes a second mode conversion resulting in electron Bernstein waves (EBWs), which then are absorbed by cyclotron damping where $\omega = |\Omega_e|$. The overall path traveled to resonance is thus called the "O-X-B" process and has been well studied for application to high-density tokamaks and to spherical tokamaks [46, 47, 84, 118].

The overall loss in the mode conversion process is dominated by the coupling between the O mode and the X mode at $\omega = \omega_{pe}$. Weitzner [118] gives an expression for the overall efficiency of this process as a function of the frequency ratio $Y = |\Omega_e|/\omega$, the scale length $L = n_e/(dn_e/dr)$ (evaluated at the X-O conversion point where $\omega = \omega_{pe}$), and the parallel and perpendicular wavenumbers $n_\parallel$ and $n_\perp$:

$$\eta = \exp\left\{-\pi k_0 L \sqrt{Y/2}\left[2(1+Y)\left(n_{\parallel\,opt} - n_\parallel\right)^2 + n_\perp^2\right]\right\} \qquad (9.183)$$

Here k_0 is the free-s[ace wavenumber and the optimum $n_\parallel$ for coupling is given by

$$n_{\parallel\,opt}^2 = \frac{Y}{Y+1}. \qquad (9.184)$$

This equation predicts good coupling to the EBW provided that gradient scale lengths are short enough and that the $n_\parallel$ spectrum is localized to the "window" around $n_{\parallel\,opt}$, with modest values of $n_\perp$.

To date no full-scale EBW heating experiments have been performed on spherical tokamaks. There has been a novel first step reported from the Multi-Ampere Spherical Tokamak (MAST) at Culham Laboratory in the U. K. [95]. Here an O-mode wave at 28 GHz was launched for the outside during the startup phase, when $\omega_{pe} < \omega$. The O-X mode conversion layer was therefore non-existent in the plasma, but the RF energy in the O mode penetrated all of the way through the plasma and impinged upon the central rod containing the toroidal field coil. The surface of this rod contained a specially grooved polarizer which converted the O-mode to the X-mode at the inside of the plasma. The reflected X-mode could then proceed back through the electron cyclotron resonant zone, where it did not interact strongly, but then onto the upper hybrid resonance where $\omega > |\Omega_e|$ and the conversion to EBWs with propagation back towards the electron cyclotron resonance could occur. When timed synchronously with the buildup of the poloidal field by programming the vertical field coils with the appropriate ramp-up, currents as high as 30 kA could be driven. The efficacy of this rather complicated "zig-zag" ray path was demonstrated by measurement of high values of the electron temperature (up to 1.0 keV) at the ECH resonant point.

A study of EBW at the NSTX experiment did not actually inject RF power into the device, but instead monitored the electron cyclotron emission from the plasma surviving from the core plasma to the outside [30]. Here the microwave signals in a band from $8 \rightarrow 40$ GHz emanating from the plasma were analyzed, representing the range for $\omega = n|\Omega_e|$ from the center of the plasma, with $n = 1, 2,$ and 3. From the relative signal intensities it was determined that the "B-X-O" conversion process, reciprocal to the O-X-B process that would be used for EBW heating and current drive, represented a $70 \pm 20\%$ conversion efficiency at 15 GHz, the candidate frequency for EBW heating on the machine. The authors concluded that these measurements showed promise for future EBW experiments on spherical tokamaks.

Problems

9.1 Design a positive-ion based deuteron neutral-beam injector using a source which generates 80 % D^+, 10 % D_2^+, and 10% D_3^+, and can deliver 5.0 MW to a DT plasma and delivering a beam of D_0 atoms at 110 keV. Find: (a) the total extracted current from the source, (b) the neutralization efficiency in the neutralizer with 95 % of maximum conversion efficiency, and (c) the surface area required for the source if the accelerating electrode spacing is designed to hold off 60 kV cm^{-1} electric fields, and runs at a current density of $0.15\times$ the Child-Langmuir limiting current density.

9.2 For the NBI design given in Problem 9.1, find the penetration depth of the D_0 ions into a DT plasma with a constant density of $n_e = n_{DT} = 10^{20}$ m^{-3}.

9.3 For the conditions in given in Problem 9.1, find the penetration depth of the D_0 ions into a DT plasma with a constant density of $n_e = n_{DT} = 10^{20}$ m^{-3}.

9.4 Discuss the possible parameter limits on voltage and current density if the ITER negative-ion source were converted to neutral tritium (T_0) production. What other changes would be required for the injection of neutral ^{3}He particles?

9.5 For a deuterium plasma with $n_e = 10^{20}$ m^{-3} and $B = 4.0T$, plot wave normal surfaces (plots of $1/n$ vs. angle θ) at frequencies of 10 MHz, 100 MHz, 1 GHz, 10 GHz, and 100 GHz. Also calculate the electron and ion cyclotron frequencies, the lower hybrid and upper hybrid resonance frequencies, the electron and ion plasma frequencies, and the R and L cutoff frequencies.

9.6 For a 15 keV electron temperature in a 6.2 T field at $n_e = 10^{20}\,\text{m}^{-3}$, plot the dispersion relation for electron Bernstein waves, keeping the first three harmonic modes.

9.7 Assume that the ionosphere is an unmagnetized plasma with an electron density profile

$$n_e = n_0 \left(\frac{r - r_{earth}}{H} \right)$$

with $H = 100$ km and $n_0 = 10^6\,\text{cm}^{-3}$. Assume a spherical earth. Using ray tracing, plot the trajectory of a 40 MHz wave fired at a $2°$ angle above the horizon. (Hint: use a spherical coordinate system so that $\theta = \pi/2$ for both London and New York, with ϕ running between the two cities. Note that $n_\phi = r k_\phi$ is conserved.) Is there an angle for which the wave arrives back at $r = r_{earth}$ at a distance equivalent to the distance between New York and London (5576.47 km)? If so, what is that angle?

9.8 A lower hybrid heating system has a frequency equal to 1.8 times the lower hybrid resonance frequency in a $10^{20}\,\text{m}^{-3}$ D-T plasma at 4.0 T. Find the frequency and the optimum size of rectangular that could be used to carry this power. Estimate the multipactor limit for this size guide and thus estimate the number of guides required for a safety factor of five in RF power for a 10 MW drive system. Use the following chart for standard waveguide sizes.

Waveguide name	Recommended frequency	Cutoff frequency (lowest mode)	Cutoff frequency (next mode)	Dimension b, in (mm)	Dimension a, in (mm)
WR975	0.75–1.15 GHz	0.605 GHz	1.211 GHz	9.75 (247.65)	4.875 (123.825)
WR770	0.97–1.45 GHz	0.766 GHz	1.533 GHz	7.7 (195.58)	3.85 (97.79)
WR650	1.15–1.72 GHz	0.908 GHz	1.816 GHz	6.5 (165.1)	3.25 (82.55)
WR510	1.45–2.20 GHz	1.157 GHz	2.314 GHz	5.1 (129.54)	2.55 (64.77)
WR430	1.72–2.60 GHz	1.372 GHz	2.745 GHz	4.3 (109.22)	2.15 (54.61)
WR340	2.20–3.30 GHz	1.736 GHz	3.471 GHz	3.4 (86.36)	1.7 (43.18)
WR284	2.60–3.95 GHz	2.078 GHz	4.156 GHz	2.84 (72.136)	1.34 (34.036)
WR229	3.30–4.90 GHz	2.577 GHz	5.154 GHz	2.29 (58.166)	1.145 (29.083)
WR187	3.95–5.85 GHz	3.153 GHz	6.305 GHz	1.872 (47.549)	0.872 (22.149)
WR159	4.90–7.05 GHz	3.712 GHz	7.423 GHz	1.59 (40.386)	0.795 (20.193)

9.9 Suppose that powerful, efficient sources at a frequency of 340 GHz became available, i.e. double the current design frequency for the ECH & CD system on ITER. Calculate the losses in deploying this frequency on the current 170 GHz transmission system by consideration of the copper losses and diffraction losses in the total system. Discuss the physics possibilities that such a source deployment could bring.

References

1. Alberti, S., Tran, M.Q., Hogge, J.P., Tran, T.M., Bondeson, A., Muggli, P., Perrenoud, A., Jödicke, B., Mathews, H.G.: Experimental measurements on a 100 GHz frequency tunable quasioptical gyrotron. Phys. Fluids B: Plasma Phys. **2**(7), 1654–1661 (1990). http://dx.doi.org/10.1063/1.859439
2. Alikaev, V., Arsenyev, Y.: High frequency power sources applied for plasma heating in TM-3 Tokamak. Conference report (1977). www.iaea.org/inis/collection/NCLCollectionStore/_Public/08/308/8308213.pdf
3. Alikaev, V., Bobrovskij, G., Ofitserov, M., Poznyak, V., Razumova, K.: Electron-cyclotron heating at the tokamak TM-3. Zh. Eksp. Teor. Fiz. Pis'ma Red. **15**, 33–36 (1972)
4. Alikaev, V.V., et al.: Investigation of the electron energy distribution function and its variation during electron cyclotron resonance heating. In: Plasma Physics and Controlled Nuclear Fusion Research (Proceedings of International Conference, 5th, Tokyo, Japan, 1974), vol. I, p. 241 (1975)
5. Allis, W.P.: Waves in a plasma. Sherwood Coference Controlled Fusion, TID-7582, p. 32 (27–28 April 1959). Also in MIT Research Laboratory of Electronics Quarterly Progress Report **54**, 5 (1959).
6. Allis, W.P., Buchsbaum, S.J., Bers, A.: Waves in Anisotropic Plasmas. MIT Press, Cambridge (1963). ISBN 0-262-51155-X
7. Allison, S.K., Garcia-Munoz, M.: Electron capture and loss at high energies. Pure Appl. Phys. **13**, 721–782 (1962). Atomic and Molecular Processes, http://dx.doi.org/10.1016/B978-0-12-081450-3.50023-0; http://www.sciencedirect.com/science/article/pii/B9780120814503500230
8. Antonsen, T.M., Hui, B.: The generation of current in tokamaks by the absorption of waves in the electron cyclotron frequency range. IEEE Trans. Plasma Sci. **PS-12**(2), 118 (1984)
9. Barnett, C.F., Ray, J.A., Thompson, J.C.: Atomic and molecular collision cross sections of interest in controlled thermonuclear research. Oak Ridge National Laboratory techical report, ORNL-3113(revised), pp. 1–332 (1964)
10. Belo, J., Bibet, P., Missirlian, M., Achard, J., Beaumont, B., Bertrand, B., Chantant, M., Chappuis, P., Doceul, L., Durocher, A., Gargiulo, L., Saille, A., Samaille, F., Villedieu, E.: ITER-like PAM launcher for Tore Supra's LHCD system. Fusion Eng. Des. **74**(1), 283–288 (2005). Proceedings of the 23rd Symposium of Fusion Technology, http://dx.doi.org/10.1016/j.fusengdes.2005.06.173; http://www.sciencedirect.com/science/article/pii/S0920379605003315
11. Bengtson, R., Benesch, J., Chen, G.L., Evans, T., Li, Y.M., Lin, S.H., Mahajan, S., Michie, R., Oakes, M., Ross, D., Valanju, P., Surko, C.: Alfvén wave heating in the PRETEXT tokamak– experiments and theory. In: Gormezano, C., Leotta, G., Sindoni, E. (eds.) Heating in Toroidal Plasmas 1982, pp. 151–160. Pergamon, Oxford (1982). https://doi.org/10.1016/B978-1-4832-8428-6.50023-0; http://www.sciencedirect.com/science/article/pii/B9781483284286500230
12. Berkner, K.H., Morgan, T.J., Pyle, R.V., Stearns, J.W.: Collision cross sections of 400- to 1800-kev H_3^+ ions in collisions with H_2 and N_2 gases and Li Vapor. Phys. Rev. A **8**, 2870–2876 (1973). https://link.aps.org/doi/10.1103/PhysRevA.8.2870
13. Berkner, K., Pyle, R., Stearns, J.: Intense, mixed-energy hydrogen beams for CTR injection. Nucl. Fusion **15**(2), 249 (1975). http://stacks.iop.org/0029-5515/15/i=2/a=009
14. Bernabei, S., Daughney, C., Efthimion, P., Hooke, W., Hosea, J., Jobes, F., Martin, A., Mazzucato, E., Meservey, E., Motley, R., Stevens, J., Goeler, S.V., Wilson, R.: Lower-hybrid current drive in the PLT tokamak. Phys. Rev. Lett. **49**, 1255–1258 (1982). https://link.aps.org/doi/10.1103/PhysRevLett.49.1255
15. Bernstein, I.B.: Waves in a plasma in a magnetic field. Phys. Rev. **109**, 10–21 (1958). https://link.aps.org/doi/10.1103/PhysRev.109.10

16. Biskamp, D.: Anomalous resistivity and viscosity due to small-scale magnetic turbulence. Plasma Phys. Controll. Fusion **26**(1B), 311 (1984). http://stacks.iop.org/0741-3335/26/i=1B/a=004
17. Bonoli, P.T., Ott, E.: Toroidal and scattering effects on lower-hybrid wave propagation. Phys. Fluids **25**(2), 359–375 (1982). http://aip.scitation.org/doi/abs/10.1063/1.863744
18. Buchsbaum, S.J.: Ion resonance in a multicomponent plasma. Phys. Rev. Lett. **5**, 495–497 (1960). https://link.aps.org/doi/10.1103/PhysRevLett.5.495
19. Budden, K.G.: The non-existence of a "Fourth Reflection Coefficient" for radio waves in the ionosphere. In: Physics of the Ionosphere: Report of the Physical Society Conference Cavendish Laboratory, p. 320 (1955)
20. Chitarin, G., Agostinetti, P., Aprile, D., Marconato, N., Veltri, P.: Cancellation of the ion deflection due to electron-suppression magnetic field in a negative-ion accelerator. Rev. Sci. Instrum. **85**(2), 02B317 (2014). http://dx.doi.org/10.1063/1.4826581
21. Clemmmow, P.C., Mullaly, R.F.: Dependence of the refractive index in magneto-ionic theory on the direction of the wave normal. In: Physics of the Ionosphere: Report of the Physical Society Conference Cavendish Laboratory, p. 340 (1955)
22. Cohen, R.H.: Effect of trapped electrons on current drive. Phys. Fluids **30**(8), 2442–2449 (1987). http://aip.scitation.org/doi/abs/10.1063/1.866136
23. Colas, L., Giruzzi, G.: Anomalous resistivity of a toroidal plasma in the presence of magnetic turbulence. Nucl. Fusion **33**(1), 156 (1993). http://stacks.iop.org/0029-5515/33/i=1/a=I16
24. Danilov, I., Heidinger, R., Meier, A., Spaeh, P.: Torus window development for the ITER ECRH upper launcher. In: Twenty Seventh International Conference on Infrared and Millimeter Waves, pp. 161–162. IEEE, Piscataway (2002). https://doi.org/110.1109/ICIMW.2002.1076045
25. de Chambrier, A., Cheetham, A., Heym, A., Hofmann, F., Joye, B., Keller, R., Lietti, A., Lister, J., Pochelon, A., Simm, W., Toninato, J., Tuszel, A.: Alfvén wave absorption studies in TCA . In: Gormezano, C., Leotta, G., Sindoni, E. (eds.) Heating in Toroidal Plasmas 1982, pp. 161–172. Pergamon, Oxford (1982). https://doi.org/10.1016/B978-1-4832-8428-6.50024-2; http://www.sciencedirect.com/science/article/pii/B9781483284286500242
26. de Esch, H., Kashiwagi, M., Taniguchi, M., Inoue, T., Serianni, G., Agostinetti, P., Chitarin, G., Marconato, N., Sartori, E., Sonato, P., Veltri, P., Pilan, N., Aprile, D., Fonnesu, N., Antoni, V., Singh, M., Hemsworth, R., Cavenago, M.: Physics design of the HNB accelerator for ITER. Nucl. Fusion **55**(9), 096001 (2015). http://stacks.iop.org/0029-5515/55/i=9/a=096001
27. Demirkhanov, R.A., Kirov, A.G., Stotland, M.A., Malik, N.I.: Investigations of plasma equilibrium in a torus with high frequency and longitudinal static magnetic fields. In: Lehnert, B. (ed.) Second European Conference on Controlled Fusion and Plasma Physics (1968). Plasma Physics, vol. 10, p. 444 (1968). http://stacks.iop.org/0032-1028/10/i=4/a=308
28. Demirkhanov, R., Kirov, A., Lozovskij, S., Nekrasov, F., Elfimov, A., Il'inskij, S., Onishenko, V.: Plasma heating in a toroidal system by a helical quadrupole RF field with $\omega < \omega_{Bi}$. In: Plasma Physics and Controlled Nuclear Fusion Research 1976, vol. 3, pp. 31–37 (1977)
29. Derfler, H., Simonen, T.C.: Experimental verification of Landau waves in an isotropic electron plasma. J. Appl. Phys. **38**(13), 5014–5020 (1967). http://dx.doi.org/10.1063/1.1709269
30. Diem, S.J., Taylor, G., Caughman, J.B., Bigelow, T., Garstka, G.D., Harvey, R.W., LeBlanc, B.P., Preinhaelter, J., Sabbagh, S.A., Urban, J., Wilgen, J.B.: Electron Bernstein wave research on NSTX and PEGASUS. In: Ryan, P., Rasmussen, D. (eds.) Radio Frequency Power in Plasmas. American Institute of Physics Conference Series, vol. 933, pp. 331–338 (2007). https://doi.org/10.1063/1.2800504
31. Dikij, A., Kalinchenko, S., Kuznetsov, Y., Kurilko, P., Lysojvan, A., Pashney, V., Tarasenko, V., Suprunenko, V., Tolok, V., Shvets, O.: High-frequency heating and equilibrium plasmas in the URAGAN-2 stellarator. In: Plasma Physics and Controlled Nuclear Fusion Research 1976, vol. 2, pp. 129–143 (1977)
32. Doane, J.: Design of circular corrugated waveguides to transmit millimeter waves at ITER. Fusion Sci. Technol. **53**(1), 159–173 (2008). https://doi.org/10.13182/FST08-A1662

33. Doane, J.L., Moeller, C.P.: He$_{11}$ mitre bends and gaps in a circular corrugated waveguide. Int. J. Electron. **77**(4), 489–509 (1994). http://dx.doi.org/10.1080/00207219408926081

34. Durodié, F., Vrancken, M., Bamber, R., Colas, L., Dumortier, P., Hancock, D., Huygen, S., Lockley, D., Louche, F., Maggiora, R., Milanesio, D., Messiaen, A., Nightingale, M.P.S., Shannon, M., Tigwell, P., Schoor, M.V., Wilson, D., and the CYCLE Team, K.W.: Performance assessment of the ITER ICRF antenna. AIP Conf. Proc. **1580**(1), 362–365 (2014). http://aip.scitation.org/doi/abs/10.1063/1.4864563

35. Erckmann, V., Gasparino, U.: Electron cyclotron resonance heating and current drive in toroidal fusion plasmas. Plasma Phys. Controll. Fusion **36**(12), 1869 (1994). http://stacks.iop.org/0741-3335/36/i=12/a=001

36. Fisch, N.J.: Confining a tokamak plasma with RF-driven currents. Phys. Rev. Lett. **41**, 873–876 (1978). https://link.aps.org/doi/10.1103/PhysRevLett.41.873

37. Fried, B.D., Conte, S.D.: The Plasma Dispersion Function. Academic Press, Cambridge (1961). https://doi.org/10.1016/B978-1-4832-2929-4.50009-5; http://www.sciencedirect.com/science/book/9781483229294

38. Glyavin, M.Y., Luchinin, A.G., Manuilov, V.N.: Nonparaxial magnetron injection gun for a high-power pulsed submillimeter-wave gyrotron. Radiophys. Quantum Electron. **52**(2), 150–156 (2009). https://doi.org/10.1007/s11141-009-9114-2

39. Golant, V.E.: Plasma penetration near the lower hybrid frequency. Zh. Tekh. Fiz. **41**, 2492 (1971). English translation: Sov. Phys.-Tech. Phys. **16**, 1980 (1972)

40. Golant, V., Fedorov, V.: RF Plasma Heating in Toroidal Fusion Devices. Plenum Publishing Corp, New York (1989)

41. Golovato, S.N., Shohet, J.L.: Plasma heating by Alfvén wave excitation in the Proto-Cleo stellarator. Phys. Fluids **21**(8), 1421–1427 (1978). http://aip.scitation.org/doi/abs/10.1063/1.862385

42. Goniche, M., Mhari, C.E., Francisquez, M., Anza, S., Belo, J., Hertout, P., Hillairet, J.: Modelling of power limit in RF antenna waveguides operated in the lower hybrid range of frequency. Nucl. Fusion **54**(1), 013003 (2014). http://stacks.iop.org/0029-5515/54/i=1/a=013003

43. Hagelaar, G.J.M., Boeuf, J.P., Simonin, A.: Modeling of an inductive negative ion source for neutral beam injection. AIP Conf. Proc. **993**(1), 55–60 (2008). http://aip.scitation.org/doi/abs/10.1063/1.2909176

44. Halbach, K.: Design of permanent multipole magnets with oriented rare earth cobalt material. Nucl. Instrum. Methods **169**(1), 1–10 (1980). http://dx.doi.org/10.1016/0029-554X(80)90094-4; http://www.sciencedirect.com/science/article/pii/0029554X80900944

45. Hamilton, W.: Theory on systems of waves. Trans. R. Ir. Acad. **15**, 69–174 (1828). https://books.google.com/books?id=TpY_AAAAYAAJ&pg=PA69#v=onepage&q&f=false

46. Hansen, F.R., Lynov, J.P., Michelsen, P.: The O-X-B mode conversion scheme for ecrh of a high-density tokamak plasma. Plasma Phys. Controll. Fusion **27**(10), 1077 (1985). http://stacks.iop.org/0741-3335/27/i=10/a=002

47. Hansen, F.R., Lynov, J.P., Maroli, C., Petrillo, V.: Full-wave calculations of the O-X mode conversion process. J. Plasma Phys. **39**(2), 319–337 (1988). https://doi.org/10.1017/S0022377800013064

48. Hatch, A.J., Williams, H.B.: Multipacting modes of high-frequency gaseous breakdown. Phys. Rev. **112**, 681–685 (1958). https://link.aps.org/doi/10.1103/PhysRev.112.681

49. Hemsworth, R., Decamps, H., Graceffa, J., Schunke, B., Tanaka, M., Dremel, M., Tanga, A., Esch, H.D., Geli, F., Milnes, J., Inoue, T., Marcuzzi, D., Sonato, P., Zaccaria, P.: Status of the ITER heating neutral beam system. Nucl. Fusion **49**(4), 045006 (2009). http://stacks.iop.org/0029-5515/49/i=4/a=045006

50. Henderson, M., Heidinger, R., Strauss, D., Bertizzolo, R., Bruschi, A., Chavan, R., Ciattaglia, E., Cirant, S., Collazos, A., Danilov, I., Dolizy, F., Duron, J., Farina, D., Fischer, U., Gantenbein, G., Hailfinger, G., Kasparek, W., Kleefeldt, K., Landis, J.D., Meier, A., Moro, A., Platania, P., Plaum, B., Poli, E., Ramponi, G., Saibene, G., Sanchez, F., Sauter, O., Serikov, A., Shidara, H., Sozzi, C., Spaeh, P., Udintsev, V., Zohm, H., Zucca, C.: Overview of the ITER EC upper launcher. Nucl. Fusion **48**(5), 054013 (2008). http://stacks.iop.org/0029-5515/48/i=5/a=054013

51. Henning, F.D., Mace, R.L., Pillay, S.R.: Electrostatic Bernstein waves in plasmas whose electrons have a dual kappa distribution: applications to the saturnian magnetosphere. J. Geophys. Res.: Space Phys. **116**(A12), A12203 (2011). http://dx.doi.org/10.1029/2011JA016965

52. Henriksson, H., Conroy, S., Ericsson, G., Gorini, G., Hjalmarsson, A., Källne, J., Tardocchi, M., contributors to the EFDA-JET Workprogramme: Neutron emission from JET DT plasmas with rf heating on minority hydrogen. Plasma Phys. Controll. Fusion **44**(7), 1253 (2002). http://stacks.iop.org/0741-3335/44/i=7/a=314

53. Heppenheimer, T.A.: The Man-Made Sun: The Quest for Fusion Power. Little, Brown, Boston (1983). See also http://history.nasa.gov/SP-4305/ch2.htm

54. Herrmannsfeldt, W.B.: Electron trajectory program. SLAC Report, vol. 226, pp. 1–118 (1979). http://www.slac.stanford.edu/cgi-wrap/getdoc/slac-r-226.pdf

55. Hiskes, J.R., Karo, A.M.: Generation of negative ions in tandem high-density hydrogen discharges. J. Appl. Phys. **56**(7), 1927–1938 (1984). http://dx.doi.org/10.1063/1.334237

56. Hoang, G., Bécoulet, A., Jacquinot, J., Artaud, J., Bae, Y., Beaumont, B., Belo, J., Berger-By, G., Bizarro, J.P., Bonoli, P., Cho, M., Decker, J., Delpech, L., Ekedahl, A., Garcia, J., Giruzzi, G., Goniche, M., Gormezano, C., Guilhem, D., Hillairet, J., Imbeaux, F., Kazarian, F., Kessel, C., Kim, S., Kwak, J., Jeong, J., Lister, J., Litaudon, X., Magne, R., Milora, S., Mirizzi, F., Namkung, W., Noterdaeme, J., Park, S., Parker, R., Peysson, Y., Rasmussen, D., Sharma, P., Schneider, M., Synakowski, E., Tanga, A., Tuccillo, A., Wan, Y.: A lower hybrid current drive system for ITER. Nucl. Fusion **49**(7), 075001 (2009). http://stacks.iop.org/0029-5515/49/i=7/a=075001

57. Hoang, G.T., Delpech, L., Ekedahl, A., Bae, Y.S., Achard, J., Berger-By, G., Cho, M.H., Decker, J., Dumont, R., Do, H., Goletto, C., Goniche, M., Guilhem, D., Hillairet, J., Kim, H., Mollard, P., Namkung, W., Park, S., Park, H., Peysson, Y., Poli, S., Prou, M., Preynas, M., Sharma, P.K., Yang, H.L., Tore Supra Team: Advances in lower hybrid current drive for tokamak long pulse operation: technology and physics. Plasma Fusion Res. **7**, 2502140 (2012). http://www.jspf.or.jp/PFR/PFR_articles/pfr2012S1/pfr2012_07-2502140.html

58. Hogge, J.P., Tran, T.M., Paris, P.J., Tran, M.Q.: Operation of a quasi-optical gyrotron with a gaussian output coupler. Phys. Plasmas **3**(9), 3492–3500 (1996). http://dx.doi.org/10.1063/1.871499

59. Ignat, D.W.: Toroidal effects on propagation, damping, and linear mode conversion of lower hybrid waves. Phys. Fluids **24**(6), 1110–1114 (1981). http://aip.scitation.org/doi/abs/10.1063/1.863500

60. Inoue, T., Tobari, H., Takado, N., Hanada, M., Kashiwagi, M., Hatayama, A., Wada, M., Sakamoto, K.: Negative ion production in cesium seeded high electron temperature plasmas. Rev. Sci. Instrum. **79**(2), 02C112 (2008). http://aip.scitation.org/doi/abs/10.1063/1.2823899

61. Jackson, J.D.: Classical Electrodynamics, 3rd edn. Wiley, New York (1999)

62. Kajiwara, K., Takahashi, K., Kobayashi, N., Kasugai, A., Sakamoto, K.: Design of a high power millimeter wave launcher for EC H&CD system on ITER. Fusion Eng. Des. **84**(1), 72–77 (2009). https://doi.org/10.1016/j.fusengdes.2008.10.003; http://www.sciencedirect.com/science/article/pii/S0920379608003050

63. Karney, C.F.F., Fisch, N.J.: Numerical studies of current generation by radio-frequency traveling waves. Phys. Fluids **22**(9), 1817–1824 (1979). http://aip.scitation.org/doi/abs/10.1063/1.862787

64. Kartikeyan, M.V., Borie, E., Thumm, M.K.A.: Review of Gyro-Devices, pp. 7–24. Springer, Berlin (2004). https://doi.org/10.1007/978-3-662-07637-8_2

65. Kirov, A.G., Rouchko, L., Sukachov, A., Meleta, E., Kadysh, I.: MHD resonant HF heating in the R-OM stellarator. In: 9th European Conference on Controlled Fusion and Plasma Physics, Oxford, 17–21 September 1979, vol. 1, p. 18 (1979)

66. Kisel, D.V., Korablev, G.S., Navalyev, V.G., Petelin, M.I., Tsimring, S.Y.: Radio Eng. Electron. Phys. **19**(4) 781–788 (1974)

67. Kishek, R.A., Lau, Y.Y., Ang, L.K., Valfells, A., Gilgenbach, R.M.: Multipactor discharge on metals and dielectrics: Historical review and recent theories. Phys. Plasmas **5**(5), 2120–2126 (1998). http://dx.doi.org/10.1063/1.872883

68. Kowalski, E., Tax, D., Shapiro, M., Sirigiri, J., Temkin, R., Bigelow, T., Rasmussen, D.: Linearly polarized modes of a corrugated metallic waveguide. MIT report, PSFC/JA-10-61 (2010). https://dspace.mit.edu/bitstream/handle/1721.1/94403/10ja061_full.pdf?sequence=1

69. Krall, N., Trivelpiece, A.: Principles of Plasma Physics. Volumes 0-911351. International Series in Pure and Applied Physics. McGraw-Hill, New York (1973). https://books.google.com/books?id=b0BRAAAAMAAJ

70. Kuprianvov, S.E., Tunitski, N.N., Pyerov, A.A.: Studies of dissociation of D^+ ions from molecular collisions in the area of energies $3.5 \rightarrow 100\,\mathrm{keV}$. Zh. Tekh. Fiz. **33**, 1252 (1963)

71. Lamalle, P.U., Beaumont, B., Gassmann, T., Kazarian, F., Arambhadiya, B., et al.: Status of the ITER IC H&CD System. AIP Conf. Proc. **1187**, 265 (2009). http://aip.scitation.org/doi/abs/10.1063/1.3273744

72. Landau, L.D.: On the vibrations of the electronic plasma. J. Phys. (USSR) **10**, 25 (1946)

73. Landau, L.D., Lifshitz, E.M.: The classical theory of fields. Addison-Wesley, Boston (1951). Translated by M. Hamermesh

74. Leung, K., Bachman, D., Herz, P., McDonald, D.: RF driven multicusp ion source for pulsed or steady-state ion beam production. Nucl. Instrum. Methods Phys. Res. Sect. B Beam Interact. Mater. Atoms **74**(1), 291–294 (1993). http://dx.doi.org/10.1016/0168-583X(93)95063-B; http://www.sciencedirect.com/science/article/pii/0168583X9395063B

75. Lloyd, B.: Overview of ECRH experimental results. Plasma Phys. Controll. Fusion **40**(8A), A119 (1998). http://stacks.iop.org/0741-3335/40/i=8A/a=010

76. Lotz, W.: Electron-impact ionization cross-sections and ionization rate coefficients for atoms and ions from hydrogen to calcium. Z. Phys. **216**(3), 241–247 (1968). http://dx.doi.org/10.1007/BF01392963

77. Malmberg, J.H., Wharton, C.B.: Dispersion of electron plasma waves. Phys. Rev. Lett. **17**, 175–178 (1966). https://link.aps.org/doi/10.1103/PhysRevLett.17.175

78. Manheimer, W.M., Boris, J.P.: Self-consistent theory of a collisionless resistive shock. Phys. Rev. Lett. **28**, 659–662 (1972). https://link.aps.org/doi/10.1103/PhysRevLett.28.659

79. Marcuzzi, D., Palma, M.D., Pavei, M., Heinemann, B., Kraus, W., Riedl, R.: Detailed design of the RF source for the 1MV neutral beam test facility. Fusion Eng. Des. **84**(7), 1253–1258 (2009). Proceeding of the 25th Symposium on Fusion Technology, http://dx.doi.org/10.1016/j.fusengdes.2008.12.084; http://www.sciencedirect.com/science/article/pii/S0920379608005413.

80. McClure, G.W.: Charge exchange and dissociation of H^+, H_2^+, and H_3^+ ions incident on H_2 gas. Phys. Rev. **130**, 1852–1859 (1963). https://link.aps.org/doi/10.1103/PhysRev.130.1852

81. McClure, G.W.: Differential angular distribution of H and H^+ dissociation fragments of fast H_2^+ ions incident on H_2 gas. Phys. Rev. **140**, A769–A778 (1965). https://link.aps.org/doi/10.1103/PhysRev.140.A769

82. McClure, G.W.: Dissociation of $H_2{}^+$ ions in collision with H Atoms: 3 to 115 kev. Phys. Rev. **153**, 182–183 (1967). https://link.aps.org/doi/10.1103/PhysRev.153.182

83. McNeely, P., Falter, H.D., Fantz, U., Franzen, P., Fröschle, M., Heinemann, B., Kraus, W., Martens, C., Riedl, R., Speth, E.: Development of a rf negative-ion source for ITER neutral beam injection. Rev. Sci. Instrum. **77**(3), 03A519 (2006). http://dx.doi.org/10.1063/1.2166246

84. Mjølhus, E.: Coupling to z mode near critical angle. J. Plasma Phys. **31**(1), 7–28 (1984). https://doi.org/10.1017/S0022377800001392

85. Obiki, T., Mutoh, T., Adachi, S., Sasaki, A., Iiyoshi, A., Uo, K.: Alfvén-wave heating experiment in the Heliotron-D. Phys. Rev. Lett. **39**, 812–815 (1977). https://link.aps.org/doi/10.1103/PhysRevLett.39.812

86. Pamela, J.: A model for negative ion extraction and comparison of negative ion optics calculations to experimental results. Rev. Sci. Instrum. **62**(5), 1163–1172 (1991). http://dx.doi.org/10.1063/1.1141995

87. Park, J.K., Goldston, R., Crocker, N., Fredrickson, E., Bell, M., Maingi, R., Tritz, K., Jaworski, M., Kubota, S., Kelly, F., Gerhardt, S., Kaye, S., Menard, J., Ono, M.: Observation of EHO in NSTX and theoretical study of its active control using HHFW antenna. Nucl. Fusion **54**(4), 043013 (2014). http://stacks.iop.org/0029-5515/54/i=4/a=043013

88. Paul, J.W.M., Goldenbaum, G.C., Iiyoshi, A., Holmes, L.S., Hardcastle, R.A.: Measurement of electron temperatures produced by collisionless shock waves in a magnetized plasma. Nature **216**, 363 (1967). http://dx.doi.org/10.1038/216363a0

89. Pivovar, L.I., Tubaev, V.M., Novikov, M.T.: Dissociation of molecular hydrogen ions in collisions with gas molecules. J. Exp. Theor. Phys. **13**, 23 (1961)

90. Prater, R., Farina, D., Gribov, Y., Harvey, R., Ram, A., Lin-Liu, Y.R., Poli, E., Smirnov, A., Volpe, F., Westerhof, E., Zvonkov, A., the ITPA Steady State Operation Topical Group: Benchmarking of codes for electron cyclotron heating and electron cyclotron current drive under ITER conditions. Nucl. Fusion **48**(3), 035006 (2008). http://stacks.iop.org/0029-5515/48/i=3/a=035006

91. Riviere, A.: Penetration of fast hydrogen atoms into a fusion reactor plasma. Nucl. Fusion **11**(4), 363 (1971). http://stacks.iop.org/0029-5515/11/i=4/a=006

92. Ryutov, D.D., Derzon, M.S., Matzen, M.K.: The physics of fast z-pinches. Sandia National Laboratory Report, SAND98-1632 (1998). https://www.osti.gov/scitech/servlets/purl/291043

93. Scheuring, A., Probst, P., Stockhausen, A., Ilin, K., Siegel, M., Scherer, T.A., Meier, A., Strauss, D.: Dielectric rf properties of CVD diamond disks from sub-mm wave to THz frequencies. In: 35th International Conference on Infrared, Millimeter, and Terahertz Waves, Rome 2010, pp. 1–2. IEEE, Piscataway (2010). https://doi.org/10.1109/ICIMW.2010.5612543

94. Schmitt, J.P.M.: The magnetoplasma dispersion function: some mathematical properties. J. Plasma Phys. **12**(1), 51–59 (1974). https://doi.org/10.1017/S0022377800024922

95. Shevchenko, V., O'Brien, M., Taylor, D., Saveliev, A., team, M.: Electron Bernstein wave assisted plasma current start-up in mast. Nucl. Fusion **50**(2), 022004 (2010). http://stacks.iop.org/0029-5515/50/i=2/a=022004

96. Soumagne, G., Alberti, S., Hogge, J.P., Pedrozzi, M., Siegrist, M.R., Tran, M.Q., Tran, T.M.: Measurement of the parallel velocity distribution function of the electron beam in a quasi-optical gyrotron by electron cyclotron emission. Phys. Plasmas **3**(9), 3501–3506 (1996). http://dx.doi.org/10.1063/1.871500

97. Speth, E., Falter, H., Franzen, P., Fantz, U., Bandyopadhyay, M., Christ, S., Encheva, A., Fröschle, M., Holtum, D., Heinemann, B., Kraus, W., Lorenz, A., Martens, C., McNeely, P., Obermayer, S., Riedl, R., Süss, R., Tanga, A., Wilhelm, R., Wünderlich, D.: Overview of the RF source development programme at IPP Garching. Nucl. Fusion **46**(6), S220 (2006). http://stacks.iop.org/0029-5515/46/i=6/a=S03

98. Spitzer, L., Härm, R.: Transport phenomena in a completely ionized gas. Phys. Rev. **89**, 977–981 (1953). http://link.aps.org/doi/10.1103/PhysRev.89.977

99. Stix, T.: The Theory of Plasma Waves. McGraw-Hill Advanced Physics Monograph Series. McGraw-Hill, New York (1962). https://books.google.com/books?id=eZ48AAAAIAAJ

100. Stix, T.H.: Radiation and absorption via mode conversion in an inhomogeneous collision-free plasma. Phys. Rev. Lett. **15**, 878–882 (1965). https://link.aps.org/doi/10.1103/PhysRevLett.15.878

101. Stix, T.: Waves in Plasmas. American Institute of Physics, Melville (1992). https://books.google.com/books?id=OsOWJ8iHpmMC

102. Stix, T.H., Palladino, R.W.: Observation of ion cyclotron waves. Phys. Fluids **3**(4), 641–647 (1960). http://aip.scitation.org/doi/abs/10.1063/1.1706099

103. Stratton, J.: Electromagnetic Theory. International Series in Pure and Applied Physics. McGraw-Hill book company, New York (1941). https://books.google.com/books?id=LiZRAAAAMAAJ

104. Strauss, D., Aiello, G., Chavan, R., Cirant, S., deBaar, M., Farina, D., Gantenbein, G., Goodman, T., Henderson, M., Kasparek, W., Kleefeldt, K., Landis, J.D., Meier, A., Moro, A., Platania, P., Plaum, B., Poli, E., Ramponi, G., Ronden, D., Saibene, G., Sanchez, F., Sauter, O., Scherer, T., Schreck, S., Serikov, A., Sozzi, C., Spaeh, P., Vaccaro, A., Zohm, H.: Preliminary design of the ITER ECH upper launcher. Fusion Eng. Des. **88**(11), 2761–2766 (2013). https://doi.org/10.1016/j.fusengdes.2013.03.040; http://www.sciencedirect.com/science/article/pii/S0920379613003347

105. Strauss, D., Aiello, G., Bruschi, A., Chavan, R., Farina, D., Figini, L., Gagliardi, M., Garcia, V., Goodman, T., Grossetti, G., Heemskerk, C., Henderson, M., Kasparek, W., Krause, A., Landis, J.D., Meier, A., Moro, A., Platania, P., Plaum, B., Poli, E., Ronden, D., Saibene, G., Sanchez, F., Sauter, O., Scherer, T., Schreck, S., Serikov, A., Sozzi, C., Spaeh, P., Vaccaro, A., Weinhorst, B.: Progress of the ECRH upper launcher design for ITER. Fusion Eng. Des. **89**(7), 1669–1673 (2014). Proceedings of the 11th International Symposium on Fusion Nuclear Technology-11 (ISFNT-11) Barcelona, Spain, 15–20 September 2013, https://doi.org/10.1016/j.fusengdes.2014.02.045; http://www.sciencedirect.com/science/article/pii/S0920379614001422

106. Swain, D., Goulding, R., Rasmussen, D.: Status of ITER ICH matching system design. AIP Conf. Proc. **1187**(1), 293–296 (2009). http://aip.scitation.org/doi/abs/10.1063/1.3273751

107. Swanson, D.: Theory of Mode Conversion and Tunneling in Inhomogeneous Plasmas. A Wiley-Interscience publication. Wiley, New York (1998). https://books.google.com/books?id=ZakemvM28ssC

108. Sweetman, D.R.: The dissociation of fast H_2^+ ions by hydrogen. Proc. R. Soc. Lond. A Math. Phys. Eng. Sci. **256**(1286), 416–426 (1960). https://doi.org/10.1098/rspa.1960.0116; http://rspa.royalsocietypublishing.org/content/256/1286/416

109. Sweetman, D.: Ignition condition in tokamak experiments and role of neutral injection heating. Nucl.Fusion **13**(2), 157 (1973). http://stacks.iop.org/0029-5515/13/i=2/a=002

110. Taylor, G., Bonoli, P.T., Green, D.L., Harvey, R.W., Hosea, J.C., Jaeger, E.F., LeBlanc, B.P., Maingi, R., Phillips, C.K., Ryan, P.M., Valeo, E.J., Wilson, J.R., Wright, J.C.: HHFW heating and current drive studies of NSTX H-mode plasmas. AIP Conf. Proc. **1406**(1), 325–332 (2011). http://aip.scitation.org/doi/abs/10.1063/1.3664985

111. Trubnikov, B.A.: Particle interaction in a fully ionized plasma. In: Leontovich, M.A. (ed.) Reviews of Plasma Physics, vol. 1, pp. 105–204. Consultants Bureau, New York (1965)

112. Twiss, R.Q.: Radiation transfer and the possibility of negative absorption in radio astronomy. Aust. J. Phys. **11**, 564 (1958). https://doi.org/10.1071/PH580564

113. Uo, K., Iiyoshi, A., Akimune, H., Obiki, T., Morimoto, S., Wakatani, M., Sasaki, A., Kondo, K., Motojima, O., Sato, M., Mutoh, T., Ohtake, I., Nakasuga, M., Mizuuchi, T., Kinoshita, S., Hanatani, K., Amano, T., Hamada, S.: RF heating experiments on heliotron devices and analysis of equilibrium and stability of straight helical heliotron plasma. In: Plasma Physics and Controlled Nuclear Fusion Research 1978, vol. 2, pp. 323–334 (1979)

114. Verdeyen, J.T.: Laser Electronics, 2nd edn. Prentice Hall, Englewood Cliffs (1989)

115. Wasow, W.: A study of the solutions of the differential equation $y'''' + \lambda^2(xy'' + y) = 0$ for large values of λ. Ann. Math. **52**(2), 350–361 (1950). http://www.jstor.org/stable/1969474

116. Watson, G.: A Treatise on the Theory of Bessel Functions. Cambridge Mathematical Library. Cambridge University Press (1995). https://books.google.com/books?id=Mlk3FrNoEVoC

117. Weinberg, S.: Eikonal method in magnetohydrodynamics. Phys. Rev. **126**, 1899–1909 (1962). https://link.aps.org/doi/10.1103/PhysRev.126.1899

118. Weitzner, H., Batchelor, D.B.: Conversion between cold plasma modes in an inhomogeneous plasma. Phys. Fluids **22**(7), 1355–1358 (1979). http://aip.scitation.org/doi/abs/10.1063/1.862747

119. Williams, J.F., Dunbar, D.N.F.: Charge exchange and dissociation cross sections for H_1^+, H_2^+, and H_3^+ ions of 2- to 50-kev energy incident upon hydrogen and the inert gases. Phys. Rev. **149**, 62–69 (1966). https://link.aps.org/doi/10.1103/PhysRev.149.62

120. Witherspoon, F., Prager, S., Sprott, J.: Shear Alfvén resonances in Tokapole II. In: Gormezano, C., Leotta, G., Sindoni, E. (eds.) Heating in Toroidal Plasmas 1982, pp. 197–201. Pergamon, Oxford (1982). https://doi.org/10.1016/B978-1-4832-8428-6.50029-1; http://www.sciencedirect.com/science/article/pii/B9781483284286500291

121. Wong, K.L., Horton, R., Ono, M.: Current generation by unidirectional lower hybrid waves in the ACT-1 toroidal device. Phys. Rev. Lett. **45**, 117–120 (1980). https://link.aps.org/doi/10.1103/PhysRevLett.45.117

122. Wong, R.K., Morse, E.C.: Study of a quasi-optical electron cyclotron maser with output coupling mirrors. Int. J. Electron. **69**(2), 291–303 (1990). http://dx.doi.org/10.1080/00207219008920314

Chapter 10
Inertial Fusion

10.1 Introduction and History

A historical survey of advances in the inertial confinement fusion approach was
given in Chap. 1. Here we amplify one key point. Following a series of promising
early experiments using lasers to compress and heat small DT fusion targets to
near-thermonuclear conditions, a series of nuclear weapon tests were carried out
at the Nevada, U. S. nuclear test site. These tests, code-named Centurion and
Halite, (Centurion for Los Alamos shots and Halite for Livermore shots), used
intermediate-scale fusion targets to demonstrate the capability to ignite these targets
in the radiation field of a nuclear device. The data from these tests remains classified,
but the results of these tests were summed up in a review of the inertial confinement
fusion program by the National Academy of Sciences in 1990[35].

> "Further work in the Centurion/Halite program of underground experiments has shown
> qualitatively that the basic concept behind ICF is sound."

The general idea is that unlike the case with magnetic fusion, where the move
towards larger experiments is exploration into uncharted territory, i.e. motivated by
an upward extrapolation, next steps in inertial fusion are motivated by downward
extrapolation between drive energies in the $10^5 \rightarrow 10^6$ J range to the $10^{12} \rightarrow 10^{14}$ J
range available from nuclear explosives.

10.1.1 Differences Between Nuclear Explosive-Driven and
Laboratory Fusion Experiments

A critical analysis of the inertial confinement fusion (ICF) concept thus starts
with a discussion of the differences between a laboratory ICF experiment and a

© Springer Nature Switzerland AG 2018   345
E. Morse, *Nuclear Fusion*, Graduate Texts in Physics,
https://doi.org/10.1007/978-3-319-98171-0_10

"downhole" test using nuclear explosives. Some simple physics examples illuminate these differences.

1. The need for compression

Firstly, as outlined in Chap. 3, the key parameter in establishing thermonuclear gain using a fusion target independent of its size is the areal density ρR in the hot fuel. As shown earlier, $\rho R > 0.3 \, \text{g} \, \text{cm}^{-2}$ is a requirement for a gain $Q \sim 1$; reactor-grade performance requires $\rho R \sim 1 \rightarrow 10 \, \text{g} \, \text{cm}^{-3}$. The yield Y is given by the burnup fraction f_B and the compressed fuel mass $m_f^{compressed}$:

$$Y = Y_0 f_B m_f^{compressed} \tag{10.1}$$

with $Y_0 = 3.39 \times 10^{11} \, \text{J} \, \text{g}^{-1}$ for D-T. Since burnup can be written as a function of ρR (and temperature: we assume that the optimum temperature of around 10 keV for DT is used), we can write the fusion yield as

$$Y = Y_0 f_B(\rho R) \cdot (4/3)\pi \rho R^3 = Y_0 f_B(\rho R) \cdot (4/3)\pi \frac{(\rho R)^3}{\rho^2}. \tag{10.2}$$

Thus the yield scales as $1/\rho^2$ for constant ρR. Suppose, for example, we had a target with solid DT ice density of $0.25 \, \text{g} \, \text{cm}^{-3}$ which was taken to thermonuclear temperature with $\rho R = 1 \, \text{g} \, \text{cm}^{-2}$ without compression, i.e. $R = 4 \, \text{cm}$ before and after the heating without any mass loss. Then the mass of the target would be $(4/3)\pi \rho R^3 = 67 \, \text{g}$. If we assume that the burnup fraction $f_B = \rho R/(6 + \rho R) = 1/7$ as the most optimistic case (see Chap. 3), this gives a yield of $Y = 3.25 \times 10^{12}$ J, just under one kiloton (note that $1.0 \, \text{kT} = 4.18 \times 10^{12}$ J). Clearly this is too large a yield to contain in a reactor, and also this would require a heating source energy on the order of 10^{11} J to start up. Therefore, compression is required to give a modest gain with a workable yield and driver size. Typically compression ratios $\rho_f/\rho_i \sim 10^3 \rightarrow 10^4$ are considered desirable for ICF experiments, while reactors with capsule yields on the order of 10^8 J might require multi-megajoule drivers with compression ratios on the order of 10^2.

2. Drive symmetry, mix, and preheat

Next, DT capsules driven by nuclear explosives are likely to see different require-ments for the symmetry of the radiation field impinging on the capsule and the tolerable amount of "mix," i.e. the incorporation of non-fuel components such as the capsule shell (known as the "ablator" in ICF experiments) and the dependence on the time history of the heating pulse. Laboratory-scale ICF experiments, with their demanding requirements on high compression ratios, are susceptible to hydrodynamic instabilities which can cause a high degree of mixing of the fuel and ablator.

The current ICF experiments assume that a carefully controlled drive power vs. time curve can be obtained, in order to compress and heat the target in the most efficient way. For example, the strategy for high yields during the National Ignition Campaign (NIC), conducted until 2012, called for the laser energy waveform to be adjusted so that four shock waves could be launched so that they would converge in the center of the target at just the right time, which meant that the timing of the steps in the laser power must be controlled to tens of picoseconds. (Later experiments have used three or even two shocks, but the timing of these is still a critical issue.) It is especially important in laboratory ICF experiments to avoid "preheat," meaning the avoidance of heating the fuel before the bulk of the compression occurs, which requires extra PdV work to be done on the target to obtain the compression.

3. Lasers and heavy ion drivers

Finally, and perhaps the most important in the long run, is the difference in the way that X-rays are generated in indirect drive laser fusion targets vis-a-vis the production of X-rays in a nuclear explosive. The use of powerful lasers to heat the inside wall of a "hohlraum," as described in Chap. 1, opens the possibility of plasma instabilities in the strongly driven plasma at the hohlraum wall, generating losses due to parametric instabilities, and leading to asymmetry. These laser–plasma interactions (LPI) obviously do not arise in nuclear explosive-driven ICF tests, where the X-rays come directly from energy released in the driver explosive. Experiments where the lasers impinge directly on the capsule ("direct-drive") can also suffer from LPI issues, but at the capsule itself rather than at a hohlraum wall. (Heavy-ion driven experiments have not progressed to the point where possible ion beam–plasma interactions have been identified.)

Thus this chapter is organized along the lines of exploring these major differences between laboratory ICF experiments and the successful Centurion/Halite tests. The physics issues will be quantitatively assessed, and the physics and technology demands for successful high-yield ICF experiments will be outlined. Then we shall show the implications of these issues on the development of inertial fusion reactors in the future.

10.2   Direct vs. Indirect Drive

(Several alternatives to the basic ICF concepts have been proposed. These include magnetized target fusion [66], dense Z-pinches [23], and fast ignition [72]. These are not discussed here.)

As described in Chap. 1, two different "mainstream" design strategies have been employed for inertial fusion with lasers. These strategies are distinguished by the way in which the laser energy is coupled into the capsule containing the fusion fuel. Figures 1.13 and 10.1 show schematics illustrating the two different concepts [62]. The key difference between these concepts is the spectrum of the photons hitting the target. In the direct drive case, photons at the wavelength of the laser, typically

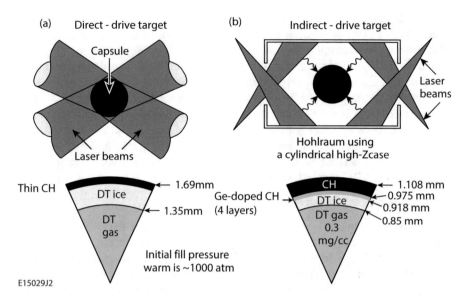

Fig. 10.1 Target designs for (**a**) direct-drive and (**b**) indirect-drive laser inertial fusion targets. The capsule design for (**b**) has been modified to reflect the NIC targets designed in 2010 (see [22] for details). From [62]. Used by permission, International Atomic Energy Agency

$0.2 \rightarrow 1.06\,\mu\text{m}$, impinge directly on the target. The ablation of the material on the outside of the target capsule causes compression of the capsule, and PdV work done by the compression also causes heating of the target fuel. In the indirect drive concept, laser energy impinges on the inside of the "hohlraum," where the main objective is heating of the hohlraum material, which is typically a high-Z material such as gold or uranium, to the point at which X-rays are emitted. The X-ray spectrum, which can be absorbed and re-emitted elsewhere in the hohlraum, bathes the target in high-energy photons that can subsequently cause ablation of the outer material in the target capsule, heating and compressing the inside of the capsule [42, 43]. The difference in the photon drive at the target between these two schemes has resulted in different optimal designs for the target, as shown in Fig. 10.1.

To compare and contrast these two schemes, one should first note that the difficulty in getting the laser light to penetrate at electron densities higher than the critical density n_{crit} (the point where $\omega = \omega_{pe}$ (see Chap. 3), or $n_{crit} = \omega_L^2 m_e \epsilon_0 / q_e^2$, where ω_L is the drive laser frequency) is not a problem in indirect drive, since the X-ray frequencies are such that the critical density at a typical X-ray photon energy, say 300 eV, are far higher than even the most compressed DT fuel would reach. Rather, the critical density issue is moved out to the hohlraum wall, where the plasma is formed with the high-Z materials there. The surface area of the hohlraum wall is larger than that of the capsule, and this may be used to advantage to mitigate the appearance of nonlinear effects such as stimulated Raman scattering (SRS) and stimulated Brillouin scattering (SBS), as will be explained later in the chapter. Also, using frequency-tripled light is advantageous for indirect drive for much the same reason as in direct drive, that is, higher values of n_{crit} can be obtained. However, energy is lost in this laser-to-X-ray conversion process. Some of the laser heating of the hohlraum wall goes into nonrecoverable heat loss in the hohlraum wall (caused by heat conduction and fast electrons), and some of the X-ray photons escape through the laser entrance holes (LEHs) at the ends of the hohlraum. Also, ablation and debris from the hohlraum wall is another source of energy loss. Other loss terms include backscattering due to SRS and SBS, which results in transport of light photons out of the hohlraum, along with phenomena associated with the closure of the laser entrance hole (LEH) with plasma, which include a "cross-beam energy transfer" (CBET) which can un-balance the energy delivered from each beam to the hohlraum wall. Therefore some inefficiency remains in converting the laser light to X-ray photons in the indirect-drive process.

In theory, the establishment of a "photon temperature" and a Planckian distribution of photon energies inside the hohlraum should happen quickly (picoseconds), and a uniform radiation field should arise. However, several processes can happen that limit the production of a uniform "photon gas" in the hohlraum, and the capsule may not experience a fully spherically symmetric radiation drive. The exposure of the spherical capsule's ablation shell to sections of a cylindrical wall presents an intrinsic geometrical problem. Also, the beams on the hohlraum wall do not overlap, but rather make a series of spots along azimuthal belt-lines on the hohlraum wall. As the hohlraum wall is heated, the wall material is driven off of the surface by expansion, and may block or refract some of the beams. There can be changes in opacity with varying wall temperatures due to hot spots and parametric instabilities. In short, asymmetries in the radiation drive on the capsule are unavoidable.

The deformation of the capsule's constant-density surfaces $r(n, \theta)$ for a given density n, as well as the radiation field $\Phi_{rad}(\theta)$ on the capsule, can be described as a summation of Legendre polynomials P_{ℓ}:

$$\{\Phi_{rad}(\theta), r(n, \theta)\} = \sum_{\ell} \left\{ \Phi_{rad}^{\ell}, r(n, \ell) \right\} P_{\ell}(\cos \theta), \qquad (10.3)$$

and low mode numbers ℓ such as $\ell = 2$ and $\ell = 4$ can point to asymmetries in the capsule performance that can limit the overall compression and heating that may occur. Figure 10.2 illustrates the perturbation of the constant density surfaces in the presence of P_2 and P_4 modes. Also, since the fusion power released is $\propto n^2 < \sigma v >_f$, with $< \sigma v >_f$ a strong function of temperature, contrary adjustments to the temperature and density profiles may have a drastic effect on the thermonuclear output.

Direct drive inertial confinement schemes may also suffer from asymmetries. But typically low mode number asymmetries are associated only with target positioning errors and perturbations caused by support structures positioning the target in place. Also, at least in theory, direct drive targets have radiation fields imposed

Fig. 10.2 Constant-density surfaces for spherical perturbations with mode numbers $\ell = 2$ and $\ell = 4$

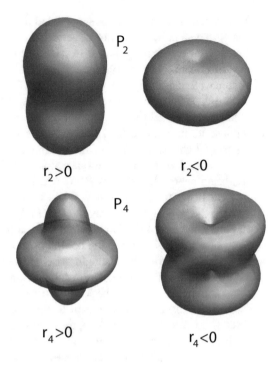

on them with more nonuniformity caused by the finite number of drive beams (at intermediate mode numbers), whereas in indirect drive some smoothing of the radiation flux impinging on the target is expected from the thermodynamics of the blackbody radiation within the hohlraum cavity.

One should note a technological development which has made it possible to smooth out nonuniformities in the laser beams in both indirect drive and direct drive applications. These are "phase plates," which are dielectric slabs in the beam path with a surface with a random thickness profile, causing the beam to resemble a set of random speckles, generating a flat beam profile (with some statistical noise) over most of the beam width, and tapering off in a half-gaussian sort of way at the edges, a "super-gaussian" profile. These were first investigated by Kato in 1984 [28] and adapted for NIF use in 2004 [77]. The NIF design uses 40×40 cm phase plates in the final optical assembly (FOA) which have rounded boundaries at places of phase maxima and minima to avoid breakdown, a design called a "continuous phase plate," or CPP.

10.3   Interaction of Laser Light with Matter

The overall physics picture for interaction of laser light with matter has many different aspects, including both linear and nonlinear interactions. Figure 10.3 shows an overall map of the zone of interaction of laser light with matter. The release

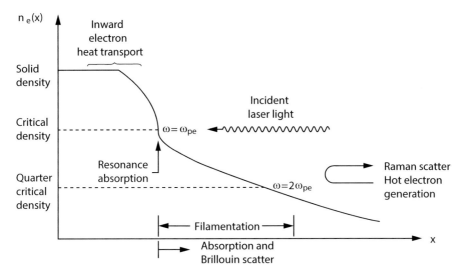

Fig. 10.3 Schematic of electron density vs. distance from target center for laser-induced plasma, showing zones for different types of laser–plasma interactions. (Note that the axes are not to scale.) From [46]. Lawrence Livermore National Laboratory

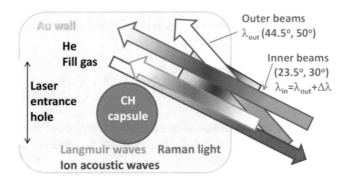

Fig. 10.4 Schematic of laser–plasma interaction processes in the NIF hohlraum. From[70]. Used with permission, American Physical Society

of material from the solid substrate, whether that is the capsule material (direct drive) or the hohlraum wall (indirect drive) causes a plume of plasma to emerge with electron densities from very low away from the solid surface up to the solid density of the material. The behavior of electromagnetic interactions in this zone involves a rich mixture of plasma phenomena. The laser–plasma interactions in the NIF hohlraum are shown in Fig. 10.4 .

10.3.1   Inverse Bremsstrahlung

The basic classical process for interaction of laser light with matter is described by
a process known as inverse bremsstrahlung. This process was elucidated in a book
by Ginsburg [19] in 1960 and is explored in detail in a report by Max [46].

Consider electrons experiencing the electric field of a plane wave $\mathbf{E} =
\mathbf{E}_0 \exp(i(\mathbf{k} \cdot \mathbf{x} - \omega t))$. We use a simple equation of motion for the electrons,
but include a drag term on the electrons caused by collisions with ions. (We
approximate the drag as a friction-like drag with no velocity dependence, called
the Langevin approximation.) With a Langevin-type momentum equation for the
electrons, we have:

$$m_e \frac{d\mathbf{v}_e}{dt} + \nu_{ei}\mathbf{v}_e = q_e\mathbf{E}$$

and then solving Maxwell's equation for the electric field $\mathbf{E}$, assuming that $Im\ \mathbf{k} \approx 0$
to first order:

$$-\mathbf{k} \times (\mathbf{k} \times \mathbf{E}) = -\mu_0 \left(\frac{\partial \mathbf{J}}{\partial t} + \epsilon_0 \frac{\partial^2 \mathbf{E}}{\partial t^2} \right).$$

Taking the time dependence to be $\sim \exp(-i\omega t)$, substituting $\mathbf{J} = n_e q_e \mathbf{v}_e$ and using
$\mathbf{k} \cdot \mathbf{E} = 0$ then gives a dispersion relation

$$\frac{k^2 c^2}{\omega^2} = 1 - \frac{\omega_{pe}^2}{\omega(\omega + i\nu_{ei})} \approx 1 - \frac{\omega_{pe}^2}{\omega_L^2} + i \left(\frac{\nu_{ei}}{\omega_L} \right) \left(\frac{\omega_{pe}^2}{\omega_L^2} \right). \tag{10.4}$$

In the last expression, we have replaced ω with ω_L, the laser drive frequency. If we
assume that the collision rate $\nu_{pe} \ll \omega_{pe}$, we can solve this quadratic equation in k
approximately with this assumption as

$$k \approx \pm \frac{\omega_L}{c} \left(1 - \frac{\omega_{pe}^2}{\omega_L^2} \right)^{1/2} \left\{ 1 + i \left(\frac{\nu_{ei}}{2\omega_L} \right) \left(\frac{\omega_{pe}^2}{\omega_L^2} \right) \frac{1}{1 - \omega_{pe}^2/\omega_L^2} \right\}. \tag{10.5}$$

The damping rate $\kappa_{ib} = 2\ Im k$ since the intensity $\propto |\mathbf{E}|^2$. Taking the imaginary
part of k from Eq. (10.5) gives

$$\kappa_{ib} = \frac{\nu_{ei}}{c} \frac{\omega_{pe}^2}{\omega_L^2} \frac{1}{\left(1 - \omega_{pe}^2/\omega_L^2 \right)^{1/2}}. \tag{10.6}$$

We now use a simple profile for the density such that the laser path encounters
a density $n_e = 0$ at $x = 0$ which climbs linearly to $n_e = n_{crit}$ at $x = L$, where it
is reflected back through the underdense plasma and exits into free space at $x = 0$

propagating in the $-\hat{x}$ direction. We note that the collision frequency is $\propto n_e \propto x$ in this case, so that $\nu_{ie}(x) = \nu_{ei}(n = n_{crit})(x/L)$. Using a variable substitution $u = x/L$ then gives the integrated logarithmic attenuation of the laser light $\ln(\alpha_T)$ over the path length as

$$- \ln(\alpha_T) = 2 \int_0^L \kappa_{ib} \, dx = 2 \frac{\nu_{ei}(n_{crit})}{c} L \int_0^1 \frac{u^2 du}{\sqrt{1-u}} = \frac{32}{15} \frac{\nu_{ei}(n_{crit})}{c} L.$$

(10.7)

Thus the total reflected energy fraction is $\alpha_{abs} = 1 - \alpha_T$ and thus the absorbed fraction is

$$\alpha_{abs} = 1 - \exp\left(-\frac{32}{15} \frac{\nu_{ei}(n_{crit})}{c} L\right).$$

(10.8)

Since $32/15 \approx 2$ there is a simple way to remember this: the total path length through the underdense plasma is $2L$ and thus the transit time without the plasma present would be $2L/c$; thus, the logarithmic energy transmission is roughly given by the light transit time divided by the mean time for a $90°$ scatter of the electrons on the ions $\tau_{ei} = 1/\nu_{ei}$. Since the collision frequency scales as $ZT_e^{-3/2}$, this shows that energy transfer to the underdense plasma drops as the electron temperature of the plasma increases, but increases as the atomic number Z is raised. This expression is also implicitly a function of the laser wavelength used: since $\nu_{ei} = n_e/(n_e\tau_{ei})$ and $n_e\tau_{ei}$ is a function of temperature only (ignoring its log Λ dependence), then shorter wavelengths imply higher n_{crit} and thus higher absorption through inverse bremsstrahlung.

The density profile in the underdense plasma is likely not a linear function of distance from the critical density location, but the approximate form given here is a good starting point. Other models take into consideration more realistic density profiles [46], and also the change in the collision rate as the plasma electron velocity distribution departs from maxwellian due to the strong electric fields [38, 67].

10.3.2 Resonance Absorption

Another classical linear absorption process is found by studying the behavior of light waves as they approach the $\omega = \omega_{pe}$ cutoff at oblique angles. If we consider a slab geometry where the density $n_e \propto \omega_{pe}^2$ is a function of x only, with propagation in the xz plane, then using $dn_e/dx = 0$ means that the wavenumber in the symmetry direction y is conserved, i.e. $k_y =$ constant. The dispersion relation is $\omega^2 = \omega_{pe}^2 + (k_y^2 + k_z^2)c^2$ for high frequency waves in unmagnetized plasma, and with an initial launch angle $\theta = \arctan(k_{y0}/k_{z0})$, we have a reflection point where $\omega_{pe}^2 = \omega^2 \cos^2 \theta$, i.e. a density such that $n = n_{crit} \cos \theta$. In the eikonal limit, the ray trajectory is a parabolic path when the density is a linear function of z, and there is no energy loss as the laser light encounters the density gradient. However, if we focus on the exact solution to the electromagnetic scattering problem without using ray optics, we find that there is an evanescent zone of electromagnetic fields inside the cutoff density at $n_e = n_{crit} \cos\theta$. The actual electric and magnetic fields in this zone are described by Airy functions, and a complete derivation is given in 1957 by Denisov [11] for the case where the electron collision frequency was taken as a constant. (Denisov was interested in ionospheric propagation and the collisions were between electrons and neutrals; taking the collisions to be between electrons and ions does not change the result if the collision frequency is a simple constant independent of density.) The coupling process is different for the case where **E** lies in the plane of incidence (*p*-polarization) from the case where **E** is normal to the plane of incidence. Figure 10.5 shows the electric fields for these two cases. Since the plasma wave equation is approximately

$$\nabla \cdot (\epsilon \mathbf{E}) = 0. \tag{10.9}$$

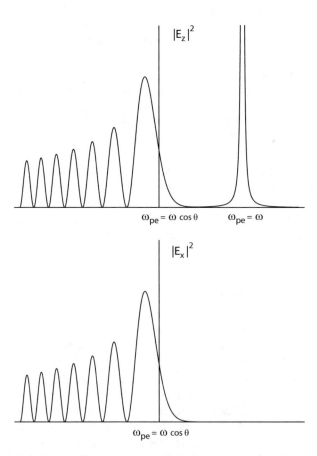

Fig. 10.5 Electric fields for oblique incidence of a plane wave near the critical density for two polarizations. Upper graph: *p*-polarization with **E** in the plane of incidence; lower graph: *s*-polarization with **E** normal to the plane of incidence

We can solve by distributing the $\mathbf{V}$ operator to obtain [36]:

$$\nabla \cdot (\epsilon \mathbf{E}) = \epsilon \nabla \cdot \mathbf{E} + \mathbf{E} \cdot \nabla \epsilon, \tag{10.10}$$

and thus

$$\nabla \cdot \mathbf{E} = -\frac{1}{\epsilon} \frac{\partial \epsilon}{\partial z} E_z. \tag{10.11}$$

The $1/\epsilon$ term on the right of Eq. (10.11) shows the resonant nature, since $\epsilon = 1 - \omega_{pe}^2/\omega^2$ goes to zero at $\omega = \omega_{pe}$, but only if there is an electric field in the $z-$ direction. For the $s-$ polarization, there is no z component to the electric field, and thus there is no electric field maximum at $\omega = \omega_{pe}$, as shown in Fig. 10.5. However, for the p-polarization, there is a z-component to the electric field, and the tunnelling of the electric field from the cutoff point where $\omega_{pe} = \omega \cos \theta$ becomes apparent if the gradient length is not too large, as shown in Fig. 10.5. Reference [11] gives an estimate of the absorbed fraction of the incident laser energy for p-polarization in terms of a function $\Phi(\tau)$ defined by

$$\Phi(\tau) = \frac{4 \left(\tau K_{\frac{1}{3}} \left(\frac{2\tau^3}{3} \right) \right)^{3/2}}{\sqrt{3\pi} \sqrt{K_{\frac{2}{3}} \left(\frac{2\tau^3}{3} \right)}}. \tag{10.12}$$

Here K is the modified Bessel function (these functions of order $1/3$ and $2/3$ are also known as Airy functions). The absorbed energy fraction is

$$f_{abs} = \Phi(\tau)^2/2, \tag{10.13}$$

with

$$\tau \equiv (L\omega_L/c)^{1/3} \sin \theta. \tag{10.14}$$

Here the scale length L is taken to mean that the density $n_e/n_{crit} = 1 - |z|/L$ in the underdense zone of the ablation plasma. A graph of the function $\Phi(\tau)$ is given as Fig. 10.6. Thus one can see that the absorbed energy fraction has a theoretical maximum of approximately 70%.

Resonance absorption can become the dominant mechanism for coupling of laser light into the plasma in some cases. It is important to remember that this process is a *linear* mechanism, and is thus not a function of field amplitude: as seen from the references, it was originally calculated to deal with absorption of radio waves in the ionosphere, which is definitely a problem in the low-amplitude regime. In addition to the energy loss by resonant absorption through collisions,

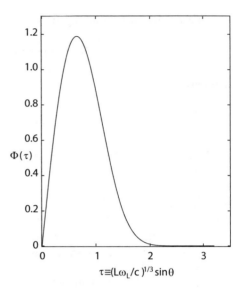

Fig. 10.6 Absorption function $\Phi(\tau)$ for p-polarization, from the theory of Denisov [11]

$\tau \equiv (L\omega_L/c)^{1/3} \sin\theta$

there is also the possibility of the excitation of Langmuir-type warm plasma waves at the $\omega = \omega_{pe}$ surface, but this is still a linear process. That the coupling is the strongest for oblique incidence on the interacting plasma is fortuitous for indirect drive fusion application, where all of the incident light on the hohlraum comes at oblique incidence.

10.3.3 Nonlinear Effects: Motion of Electrons in High-Amplitude Electromagnetic Fields

In order to understand the possible nonlinear interactions of laser light with plasma, we must first examine carefully the nature of the dynamics of electrons in large-amplitude electromagnetic fields. Consider the equation of motion of an electron in a plane electromagnetic wave field given by

$$\mathbf{E} = E_0 \hat{y} e^{i(kx - \omega t)}, \qquad \mathbf{B} = \frac{1}{c} \hat{x} \times \mathbf{E}.$$

We will start by taking the real part of this complex dependence, so that at $x = 0$, the electric field is $E = E_0 \hat{y} \cos \omega t$.

The non-relativistic equation of motion is

$$m_e \frac{d\mathbf{v}}{dt} = q_e \left(\mathbf{E} + \mathbf{v} \times \mathbf{B}\right)$$

$$\frac{d\mathbf{r}}{dt} = \mathbf{v}.$$

Define a dimensionless parameter a_0:

$$a_0 = \frac{|q_e| E_0}{m\omega c}. \tag{10.15}$$

If $a_0 \ll 1$, then $|q_e \mathbf{v} \times \mathbf{B}| \ll |q_e \mathbf{E}|$, and we shall ignore the magnetic field to first order. We can find the second-order effect of the $\mathbf{v} \times \mathbf{B}$ term by writing $\mathbf{v} = \mathbf{v}^{(1)} + \mathbf{v}^{(2)}$, where $\mathbf{v}^{(1)}$ is the solution to the equation of motion without the magnetic field term:

$$m_e \frac{d\mathbf{v}^{(1)}}{dt} = q_e \mathbf{E}$$

$$\frac{d\mathbf{r}^{(1)}}{dt} = \mathbf{v}. \tag{10.16}$$

The first-order solution near $x = 0$ is

$$\mathbf{v}^{(1)}(t) = \hat{y} \frac{q_e E_0}{m\omega} \sin \omega t = \hat{y} a_0 c \sin \omega t. \tag{10.17}$$

We then solve for $\mathbf{v}^{(2)}$ by adding in the $\mathbf{v} \times \mathbf{B}$ term, using the first-order expression for $\mathbf{v}$:

$$m_e \frac{d\mathbf{v}^{(2)}}{dt} = q_e \left(\mathbf{v}^{(1)} \times \mathbf{B}\right) = \hat{x} \frac{q_e^2 E_0^2}{m\omega c} \sin \omega t \cos \omega t = \hat{x} \frac{q_e^2 E_0^2}{2m\omega c} \sin 2\omega t, \tag{10.18}$$

which has a solution

$$\mathbf{v}^{(2)}(t) = -\hat{x} \frac{q_e E_0^2}{4m_e^2 \omega c} \cos 2\omega t = -\hat{x} a_0^2 \frac{c}{4} \cos 2\omega t. \tag{10.19}$$

The position of the electron in the y direction is given to first order by integrating $v_y^{(1)}(t)$ over time. This gives

$$y(t) = -\frac{q_e E_0}{m\omega^2} \cos \omega t = -a_0 \frac{c}{\omega} \cos \omega t, \tag{10.20}$$

and the lowest v_x term is from the correction $v_x^{(2)}(t)$, which gives upon integrating:

$$x(t) = -\frac{q_e^2 E_0^2}{8m_e^2 c}\sin 2\omega t = -\frac{a_0^2}{8}\frac{c}{\omega}\sin 2\omega t. \tag{10.21}$$

Notice that including the second-order terms in a_0 results in the inclusion of second-harmonic terms in $\cos 2\omega t$ and $\sin 2\omega t$, resulting from nonlinear terms in E^2. A drawing of the electron motion in the xy plane for different values of a_0 is shown as Fig. 10.7. Keeping higher-order terms, and incorporating a relativistic equation of motion, results in a more complex behavior from terms like E^3, E^4, and so forth, which can be shown to cause higher harmonics to be present from simple trigonometric identities.

10.3.4   Nonlinear Wave Interaction and Parametric Instability

We now consider a second and third electromagnetic wave being present in addition to the original electromagnetic wave. We will only use the electric fields for purposes of determining the nonlinear response. Let

$$\mathbf{E}_1 = \hat{y}E_1 e^{i(k_2 x - \omega_2 t)}$$

$$\mathbf{E}_2 = \hat{y}E_2 e^{i(k_3 x - \omega_3 t)}.$$

Calculate $\mathbf{v}^{(1)}$ as before:

$$\mathbf{V}^{(1)}(t) = \hat{y}\frac{q_e}{m\omega}\left(E_0 \sin \omega t + E_1 \sin \omega_2 t + E_2 \sin \omega_3 t\right),$$

and find the second-order velocities using the same $\mathbf{B}$ as before:

$$\frac{d\mathbf{v}^{(2)}}{dt} = q_e \left(\mathbf{v}^{(1)} \times \mathbf{B}\right)$$

$$= \hat{x}\frac{q_e^2 E_0}{m\omega c}\left(E_0 \sin \omega t \cos \omega t + E_1 \cos \omega t \sin \omega_2 t\right.$$

$$\left.+E_2 \cos \omega t \sin \omega_3 t\right) + \dots. \tag{10.22}$$

We now have terms in the force equation $\propto E_0^2$, E_1^2, E_2^2, $E_0 E_1$, $E_1 E_2$ and $E_0 E_2$ with their respective frequencies. Trigonometric identities can then be used to show that the new terms have frequencies $2\omega_0$, $2\omega_1$, $2\omega_2$, $\omega_0 \pm \omega_1$, $\omega_0 \pm \omega_2$, and $\omega_1 \pm \omega_2$. Now suppose that we have a condition where

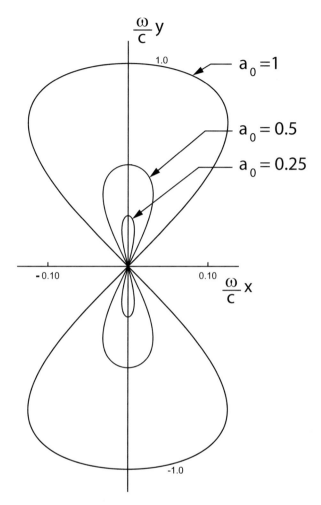

Fig. 10.7 Electron motion for an electron in a high-amplitude electromagnetic field, shown for three different values of the parameter a_0

$$\omega_0 = \pm\omega_1 \pm \omega_2. \qquad (10.23)$$

Then the perturbed velocity from two of the waves will have frequency components matching the third, since

$$\cos A \sin B = (1/2)\sin(A+B) + (1/2)\sin(A-B).$$

Energy present in the pump wave E_0 can be channeled into waves E_1 and E_2. For this process to be synchronized over space, we also need matching conditions

$$k_0 = \pm k_1 \pm k_2. \tag{10.24}$$

The matching conditions on ω and k, taken together, are called the Stokes relations. When the Stokes relations are met for some selection of frequencies for the waves as well as their respective wavenumbers, nonlinearity can result in wave energy being transferred from one mode to two others.

10.3.5   Model Equations for Three-Wave Parametric Instability

We now look for some general features of the three-wave interaction model. We can replace the wave equations involved with a model where each wave is represented by a simple harmonic oscillator with frequencies ω_0, ω_1, and ω_2 as appropriate [60]. We then model the nonlinear term pumping energy into or out of each oscillator (i.e., wave system) by a product of the two waves, emulating the nonlinear electron response shown above. For this simple model, we ignore the dissipation effects. Then a model system of equations results, given by:

$$\frac{d^2 A_1(t)}{dt^2} + \omega_1^2 A_1(t) = A_2(t) A_3(t)$$

$$\frac{d^2 A_2(t)}{dt^2} + \omega_2^2 A_2(t) = A_1(t) A_3(t) \tag{10.25}$$

$$\frac{d^2 A_3(t)}{dt^2} + \omega_3^2 A_3(t) = A_1(t) A_2(t).$$

This set of coupled second-order equations is simple to evaluate with *Mathematica* or some other mathematical simulation program such as *Matlab*, etc. Here is a *Mathematica* script which will generate solutions to this system with simple choices for the frequencies and initial conditions:

```
omega1=3;omega2=2;omega3=1;
sol=NDSolve[{A1"[t]+omega1∧2
A1[t]==A2[t]A3[t],
A2"[t]+omega2∧2 A2[t]==A1[t]A3[t],
A3"[t]+omega3∧2 A3[t]==A1[t]A2[t],
A1[0]==1,A2[0]==0.01,A3[0]==0.01,
A1'[0]==0,A2'[0]==0,A3'[0]==0},
{A1[t],A2[t],A3[t]},{t,0,100}]
```

Figure 10.8 shows the amplitudes A_1, A_2, and A_3 as the output for this simple example. Note that this system is conservative, and there is no dissipation, so that the growth and decay of the mode energies repeats itself perpetually.

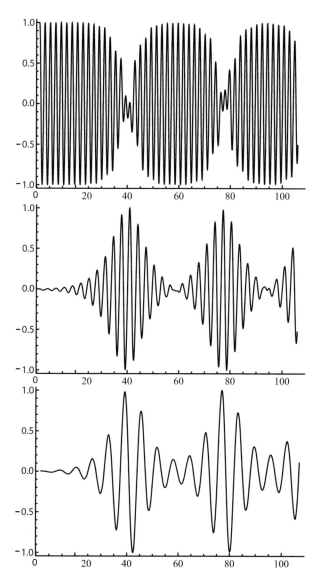

Fig. 10.8 Amplitudes A_1, A_2, and A_3 (from top to bottom) for model of parametric instability

10.3.6 The Stokes Diagram

The matching conditions for parametric instabilities, as shown above in Eqs. (10.23) and (10.24), $\omega_0 = \pm\omega_1 \pm \omega_2$ and $k_0 = \pm k_1 \pm k_2$, are reminiscent of the "force parallelogram"-type vector diagrams used in elementary physics. In one spatial dimension, the vectors in the two-dimensional space (ω, k) replace the two-

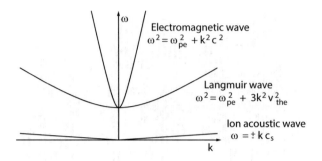

Fig. 10.9 Three candidate waves for parametric instabilities

dimensional forces (F_x, F_y) in elementary two-dimensional statics. In order to find matching conditions for the waves, we can look at the dispersion curves ω vs. k for the possible waves in the plasma, and find the graphical solutions to the Stokes relations representing possible three-wave interactions. For this purpose, we can look at three candidate waves, with their dispersion relations given (see Fig. 10.9):

1. **Electromagnetic waves ("photons"):**

$$\omega^2 = \omega_{pe}^2 + k^2 c^2, \tag{10.26}$$

2. **Electron Langmuir waves ("plasmons"):**

$$\omega^2 = \omega_{pe}^2 + 3k^2 v_{the}^2, \tag{10.27}$$

3. **Ion sound waves ("phonons"):**

$$\omega = \pm k c_s. \tag{10.28}$$

Here the electron thermal velocity $v_{the} = (T_e/m_e)^{1/2}$ and the ion-acoustic wave speed $c_s = ((ZT_e + \gamma T_i)/m_i)^{1/2}$ as used earlier.

Nonlinear coupling of these waves then creates various parametric instabilities, known as "scattering" processes in analogy to particle kinetics, as given below.

10.3.7   Stimulated Raman Scattering (SRS)

Raman scattering is the process of converting an incoming electromagnetic (EM) wave into an outgoing EM wave and a forward-traveling Langmuir wave, also known as a plasmon. The dispersion relations for the two types of wave are:

Fig. 10.10 Stokes diagram
for Raman scattering

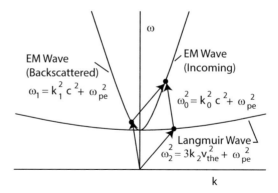

$$\omega_{0,1}^2 = k_{0,1}^2 c^2 + \omega_{pe}^2 \qquad \text{(EM)}$$
$$\omega_2^2 = 3k_2^2 v_{the}^2 + \omega_{pe}^2 \qquad \text{(Langmuir).} \qquad (10.29)$$

A Stokes diagram for the stimulated Raman scattering process is shown as Fig. 10.10. Notice that here we have $\omega_0 = \omega_1 + \omega_2$ and $k_1 = -(k_2 - k_0)$, with $k_2 < 0$. The wave (ω_2, k_2) has $k_2 < 0$ and is thus a backscattered wave, i.e. it represents energy being carried back along the path of the wave (ω_0, k_0), the incoming electromagnetic wave from the laser. The wave (ω_2, k_2) carries energy in the forward direction, but with a fraction of the incoming energy and with a slower group velocity (note that the group velocity is $d\omega/dk$ and is thus the slope of the dispersion curve in (ω, k) space).

One can notice a property of SRS from the Stokes diagram: since both daughter waves have $\omega_{1,2} > \omega_{pe}$, the Stokes relation shows that $\omega_0 > 2\omega_{pe}$. Since the density $\propto \omega_{pe}^2$, this shows that SRS can only happen in underdense plasma where $n_e < n_{crit}/4$, as shown in Fig. 10.3.

If we express the dispersion relations in dimensionless form and use the Stokes relations along with the dispersion relations, we can calculate the frequencies and wavenumbers for all three waves, if the local density and temperature for the interaction are known. Let $x = \omega_{pe}^2/\omega_0^2 = n_e/n_{crit}$ (< 0.25) and let $M = c/\sqrt{3v_{the}}$ be a light-to-electron thermal speed ratio. Taking $y = \omega_1/\omega_0$ (and $1 - y = \omega_2/\omega_0$ from the Stokes relation) then gives

$$x =$$
$$\frac{M^4 y^2 - 2M^4 y + M^4 - 3M^2 y^2 + 4M^2 y - 3M^2 + 2\sqrt{-2M^4 y^3 + 5M^4 y^2 - 2M^4 y + M^2 y^4 - 2M^2 y^2 + M^2}}{M^4 - 4M^2},$$
$$(10.30)$$

and this can be solved explicitly for y, either from a rather complex analytical relation (exact) or from an approximate form:

$$x = (y-1)^2 + \left(\frac{1}{M}\right)^2 \left(y^2 + 2\sqrt{-y\left(2y^2 - 5y + 2\right)} - 4y + 1\right) + O\left(\left(\frac{1}{M}\right)^4\right).$$

$$(10.31)$$

Since $M \approx 30$ for typical SRS with $T_e \sim 300\,\text{eV}$, the Raman process typically happens at a point where the density is slightly less than $n_{crit}/4$.

10.3.8   Stimulated Brillouin Scattering (SBS)

For Brillouin scattering, the Langmuir wave in the Raman scattering process is replaced by an ion acoustic wave. The dispersion relations for the three waves are:

$$\omega_{0,1}^2 = k_{0,1}^2 c^2 + \omega_{pe}^2 \qquad \text{(EM)}$$
$$\omega_2 = k_2 c_s + \omega_{pe}^2 \qquad \text{(Ion acoustic).}$$

$$(10.32)$$

Here the ion-acoustic wave speed is given by $c_s = ((ZT_e + \gamma_i m_i)/m_i)^{1/2}$. Here the electrons are taken as isothermal and the ions are an ideal gas with an adiabatic index γ_i, typically taken as 5/3. Note that Z can refer to the average charge state $< Z >$ when the plasma is a high-Z plasma such as typically results from ionization of a hohlraum wall.

A Stokes diagram for the Brillouin scattering process is shown as Fig. 10.11.

10.3.9   Two-Plasmon Decay

Two-plasmon decay is similar to SRS, but with the backscattered wave a Langmuir wave rather than an electromagnetic wave: The dispersion relations are then

Fig. 10.11 Stokes diagram for Brillouin scattering

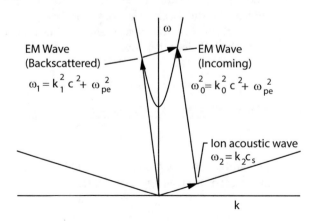

Fig. 10.12 Stokes diagram
for two-plasmon decay

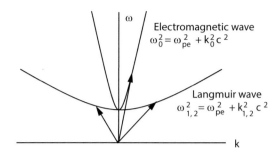

$$\omega_0^2 = k_0^2 c^2 + \omega_{pe}^2 \qquad \text{(EM)}$$

$$\omega_{1,2}^2 = 3k_{1,2}^2 v_{the}^2 + \omega_{pe}^2 \qquad \text{(Langmuir).}$$

(10.33)

Using the dimensionless variables $x = \omega_{pe}^2/\omega_0^2 = n_e/n_{crit}$, $M = c/\sqrt{3v_{the}}$ as in the Raman scattering case, an analytic solution is given by

$$y = \frac{1}{2} \pm \frac{\sqrt{(x-1)\left(M^4(4x-1) + M^2\left(4x^2 - 6x + 2\right) - (x-1)^2\right)}}{2M\left(M^2 + x - 1\right)}. \qquad (10.34)$$

Figure 10.12 shows the Stokes diagram for the two-plasmon decay. Note that as in the Raman case the frequencies ω_1 and ω_2 are greater than ω_{pe} and thus $\omega_1 > 2\omega_{pe}$, or $n \leq (1/4)n_c$ for this process to occur. As a typical example, take $x = 0.23$. This gives $y = 0.495884$ for the upper root and $y = 0.504116$ for the lower root. It is also interesting to note that the Landau damping parameter $k\lambda_{De}$ is given by

$$k\lambda_{De} = \frac{kv_{the}}{\omega_{pe}} = \frac{\sqrt{y^2 - x}}{\sqrt{3x}}, \qquad (10.35)$$

and this gives $k\lambda_{De} = 0.15$ and 0.19 for these conditions, meaning that the Landau damping is not particularly strong here.

10.3.10   Intra-Beam Transfer

Intra-beam transfer (IBT, also called crossed-beam energy transfer (CBET)) is a variation on SBS in that two electromagnetic waves and one ion acoustic wave are involved in a parametric scattering process. In this case, however, the two EM waves are crossing over each other near the laser entrance hole in the hohlraum, and each can be considered a pump wave since they have roughly equal intensities. Each of the two beams contains up to four frequencies staggered by a small amount, e.g. $\delta\lambda \approx 1 \rightarrow 5\text{Å}$ for $\lambda = 0.35\mu$ [32, 37, 70]. The purpose of this intentionally broadened frequency spectrum is to avoid a "filamentation" instability where self-focussing can occur when the beam crosses through plasma. However, this strategy can lead to a parametric instability where the frequency separations $\delta\omega$ couple to ion acoustic waves. The IBT process can then cause a shift of energy from one beam angle to another, resulting in asymmetry to the radiation drive in the hohlraum wall. The IBT process can also result in reflection of some of the EM wave energy as well, thus reducing the available drive power.

A simple Stokes diagram similar to the Stokes diagram for SBS (Fig. 10.11) cannot be drawn for IBT. The analysis is complicated by complexities in the dispersion characteristics caused by wave damping (which adds an imaginary part to the wavenumber $\mathbf{k}$), and by plasma flow in the interaction region, which causes a Doppler shift term in the dispersion relations.

Symmetry can be improved by adjusting the wavelength shift $\delta\lambda$ between the beams in the inner and outer cones [20, 48]. It may be possible to "tune out" the drive asymmetries by judicious use of the wavelength shift $\Delta\lambda$ between the interacting beam lines.

10.4   Implosion Hydrodynamics

10.4.1   Hydrodynamic Efficiency: The Rocket Equation

An imploding shell in three dimensions can be compared to the process of rocket acceleration, where forward momentum of the rocket is balanced by the reverse momentum of the rocket's thrust. The classic rocket momentum balance is shown as Fig. 10.13.

We see from Fig. 10.13 that the rate of change of forward momentum is determined by the momentum rate leaving the rocket through its exhaust (in the frame of the rocket). Assuming a constant exhaust velocity u_{ex}, this gives

$$m\frac{dv}{dt} = u_{ex}\dot{m} = -u_{ex}\frac{dm}{dt}, \tag{10.36}$$

and thus $-u_{ex}dm/m = dv$, so that

$$v(t) = u_{ex}\ln\left(\frac{m(0)}{m(t)}\right). \tag{10.37}$$

Fig. 10.13 Rocket momentum balance

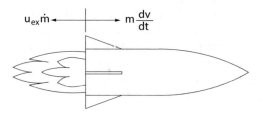

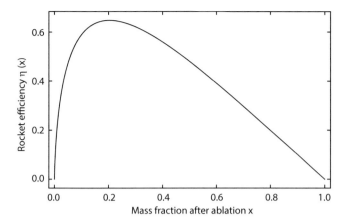

Fig. 10.14 Rocket efficiency $\eta(x)$ as a function of post-ablation mass fraction x

The rocket efficiency η_{rocket} is defined as the ratio of the kinetic energy attained by the rocket after the burn phase is over, when the mass of the rocket is depleted to a fraction x of its initial mass, thus $K.E. = (1/2)xm(0)v_f^2$, to the energy released W_{ex} by the rocket exhaust, which is given by $(1/2)\dot{m}u_{ex}^2$ per unit time, integrated over the burn, i.e. $(1/2)\Delta m u_{ex}^2 = (1/2)m(0)(1-x)u_{ex}^2$. Thus the rocket efficiency is

$$\eta_{rocket} = \frac{x(\ln x)^2}{1 - x}. \tag{10.38}$$

Figure 10.14 shows the rocket efficiency as a function of x. Note that maximum efficiency is obtained when $x \approx 0.2$, where $\eta \approx 0.65$.

Applying this concept to the problem of compressing ICF targets, we can see that the rocket efficiency sets an upper bound on the overall hydrodynamic efficiency, which is lower due to other energy loss mechanisms present in the ICF scenarios. In the case of ablation due to either X-ray heating (indirect drive) or laser (direct drive) heating, the energy invested in the heating and ionization of the ablated material needs to be taken into account, and there is an energy sink in fast electrons generated at the ablation front as well as re-radiation from the ablation front due to the emissivity (albedo) of the heated target.

10.5 Hydrodynamic Instabilities

Hydrodynamic instabilities form a limit on the ability for an imploding ICF target to retain its spherical symmetry, and thus limit the maximum achievable compression of the target and lead to mixing of the fuel and ablator material. As is the case with any instability, one can relate the instability to a source of free energy caused by

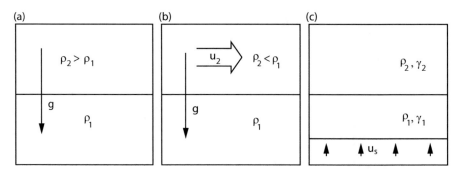

Fig. 10.15 Hydrodynamic instabilities of interest in ICF. (**a**) Rayleigh-Taylor instability for heavier fluid above a lighter fluid, (**b**) Kelvin-Helmholtz instability with relative velocity between the fluids, (**c**) Richtmeyer-Meshkov instability with shock passing through dissimilar fluids

a gradient of some kind. In the case of some simple hydrodynamic instabilities, the gradients are in density, fluid flow, and adiabatic exponent. These gradients work in concert with acceleration, both smooth and shock-induced, to produce growing modes. Three major fluid instabilities of interest in ICF are the Rayleigh-Taylor, Kelvin-Helmholtz, and Richtmeyer-Meshkov instabilities. The conditions for one-dimensional models of fluid equilibria with the associated discontinuities and driving forces are shown in Fig. 10.15. The growth rates for these instabilities form a boundary condition for target design, with the overall goal of obtaining adequate mix-free target compression on a fast enough timescale so that these instabilities do not adversely affect the performance of the implosion, ignition, and burn.

10.5.1   Rayleigh-Taylor Instability

Rayleigh-Taylor (RT) instability is defined as a fluid instability arising when a heavier fluid is atop a lighter fluid with gravity (or similar body force) pointed downward [56, 73]. The classic derivation is found in Chandrasekhar [9]. For the case of no viscosity and a one-dimensional planar geometry, with perturbation modes given spatial and time dependence as $\exp(ikx)$ and $\exp(int)$, we find a solution to the growth of a perturbation at the fluid interface given by

$$n^2 = gk\frac{\rho_1 - \rho_2}{\rho_1 + \rho_2}. \tag{10.39}$$

Here the dimensionless ratio $\rho_1 - \rho_2/\rho_1 + \rho_2$ is called the Atwood number, and the system is unstable if $\rho_2 > \rho_1$, i.e. the heavier fluid is on top of the lighter fluid. In the context of inertial fusion, one can think of the "gravity" g representing the acceleration of the imploding shell inward, which in the (non-inertial) reference

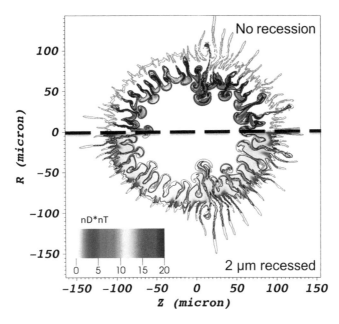

Fig. 10.16 Hydrodynamic simulation (using the two-dimensional ARES code) of Rayleigh-Taylor growth for imploding ICF capsules in a series of experiments where a plastic (CH) ablator was filled with pure tritium gas and a layer of deuterated plastic (CD) was placed either at the fuel-ablator interface (top) or recessed into the plastic ablator by 2.27 microns (bottom). The plots are of the quantity $n_D n_T$, thus creating a neutron signal only from mixing. From [69]. Used by permission, American Physical Society

frame of the shell causes the gravity vector to appear in the $+\hat{r}$ direction, and the density of the ablating material $\rho_1 \ll \rho_2$. Then the growth rate $\gamma \approx \sqrt{gk}$, showing that the higher k wavenumbers (shorter wavelengths) have the fastest growth times. The shortest wavelengths to be considered are on the order of the shell thickness ΔR. Figure 10.16 shows the result of a fluid simulation of the RT instability for the case of an imploding ICF target containing only tritium in the interior and a deuterated plastic layer in the ablator, thus producing neutrons only from the fuel-ablator mixing caused by RT instabilities. Note that nonlinearity appears quickly as the perturbations become of order ΔR in amplitude, and can result in mixing of the DT fuel layer with the target material, thus lowering the yield.

10.5.2   Kelvin-Helmholtz Instability [9, 29, 74, 76]

Another type of instability involves the case where there is a change in density across a fluid interface, similar to the Rayleigh-Taylor case, but now with $\rho_2 < \rho_1$ (typically) so that the Rayleigh-Taylor instability is not present, but instead there is a difference in velocity across the fluid interface, with gravity also present as in the Rayleigh-Taylor case. (As in the Rayleigh-Taylor case, "gravity" can stand for generalized fluid forces acting normal to the interface.) The velocity component in one direction of the fluid interface, i.e. perpendicular to the normal of the fluid interface plane, is the destabilizing component. Again, Ref. [9] provides the complete mathematical description of the instability, including the effects of surface tension and viscosity, as well as cases where a gradient in density and velocity are present without a sharp interface. For the simplest case with two uniform fluids with no surface tension or viscosity, the dispersion relation for small perturbations at the fluid interface is given, with time and spatial dependence given as $\exp(int)$ and $\exp(ikx)$ as before, and with x in one direction along the fluid interface, as:

$$n = -k(\alpha_1 u_1 + \alpha_2 u_2) \pm \left[gk\,(\alpha_1 - \alpha_2) - k^2 \alpha_1 \alpha_2\,(u_1 - u_2)^2 \right]^{1/2}. \qquad (10.40)$$

Here $\alpha_1 = \rho_1/(\rho_1 + \rho_2)$ and $\alpha_2 = \rho_2/(\rho_1 + \rho_2)$. Without loss of generality, we can set $u_1 = 0$ to find the growth rate as only $\Delta u = u_1 - u_2$ enters in to the potentially complex part of Eq. (10.40), i.e. the square root term. The threshold for instability is found by setting the square root term of Eq. (10.40) to zero, which generates an expression for the minimum value of k for instability:

$$k_{\min} = \frac{g\,(\alpha_1 - \alpha_2)}{\alpha_1 \alpha_2 \Delta u^2}. \qquad (10.41)$$

An interesting feature of the Kelvin-Helmholtz instability is that an instability exists for any value of the relative velocity Δu between the fluids, although the minimum wavenumber may be unrealistically high in some geometries, requiring a more refined physical picture to treat the instability. Replacing the sharp boundary with a finite gradient in density and velocity results in the appearance of stable solutions for all wavenumbers in some situations. For the case of a single fluid with finite density gradients and velocity shear, the instability threshold is found from the Richardson number

$$\mathrm{Ri} = -\frac{g}{\rho} \frac{dp/dz}{(du/dz)^2}. \qquad (10.42)$$

A sufficient condition for stability is that $\mathrm{Ri} > 1/4$. Specialized density and velocity distributions may widen the parameter space for stability, however. Reference [9] gives some examples of some special cases.

Kelvin-Helmholtz instabilities may arise in ICF applications due to sideways motion of the fluid caused by asymmetries in the drive, or in the late stages of a fully developed Rayleigh-Taylor instability due to three-dimensional motion. Kelvin-Helmholtz instabilities can be present in "fast ignition" ICF schemes, where the goal is to provide tightly focussed beams of very high intensity to penetrate deeply into the fuel and start a burn wave from the inside of the fuel before heat conduction

Fig. 10.17 Experimental configuration used to generate Kelvin-Helmholtz instability in a sheared flow caused by laser ablation at the Omega laser facility. From [24]. Used by permission, American Physical Society

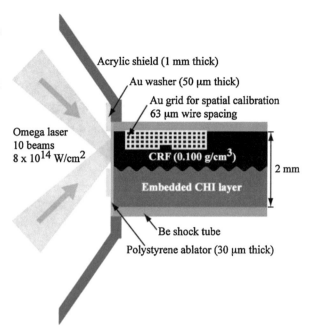

Acrylic shield (1 mm thick)

Au washer (50 μm thick)

Au grid for spatial calibration 63 μm wire spacing

Omega laser 10 beams 8 x 10^14 W/cm²

CRF (0.100 g/cm³)

2 mm

Embedded CHI layer

Be shock tube

Polystyrene ablator (30 μm thick)

and radiative losses become factors. Figures 10.17 and 10.18 show an experiment performed at the Omega laser facility where a Kelvin-Helmholtz instability was intentionally excited at a fixed wavenumber k in order to study its growth and nonlinear saturation characteristics.

10.5.3 Richtmeyer-Meshkov Instability [47, 59]

The Richtmeyer-Meshkov instability is an analog of the Rayleigh-Taylor instability applicable to the case where an interface between a lighter fluid and a heavier fluid is subject to shock acceleration. Figure 10.19 illustrates some measurements by Meshkov where two dissimilar fluids have been subjected to a shock wave which travels through the interface.

The theory of the Richtmeyer-Meshkov instability follows as an extension of the Rayleigh-Taylor instability. For an interface between two fluids with no viscosity or surface tension, the amplitude η for perturbations at the fluid interface is given by [8]:

$$\frac{d^2\eta(t)}{dt^2} = kgA\eta(t), \qquad (10.43)$$

Fig. 10.18 X-ray microphotographs of the ablation-driven Kelvin-Helmholtz experiment shown in Fig. 10.17. Shown for times (**a**) $t = 25$ ns, (**b**) $t = 45$ ns, and (**c**) $t = 75$ ns. From [24]. Used by permission, American Physical Society

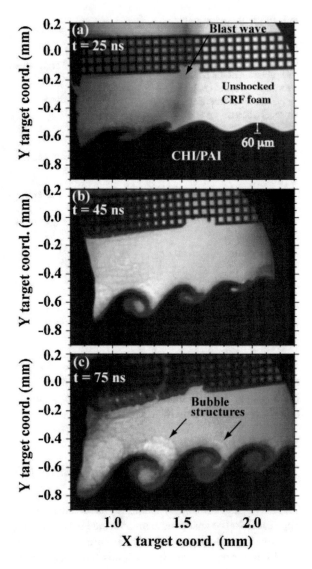

with the wavenumber k taken along the interface and $A = (\rho_2 - \rho_1)/(\rho_2 + \rho_1)$ is the Atwood number. Here g is the acceleration as before, but we are interested in the case where g approaches a delta function in time such that

$$g(t) = [u]\delta(t). \tag{10.44}$$

Then integrating Eq. (10.43) in time gives

$$\dot{\eta}(t) = k[u]A\eta(0). \tag{10.45}$$

Fig. 10.19 Interface instability at a shock front between two gases at atmospheric pressure. Gases (1,2) are: (**a**) Air/$CO_2(\rho_2/\rho_1 = 1.54)$, (**b**) He/Freon-22 $(\rho_2/\rho_1 = 21.70)$, (**c**) Freon-22/He $(\rho_2/\rho_1 = 1/21.70)$. Time intervals are 64 μs. From [47]. Used by permission, Springer

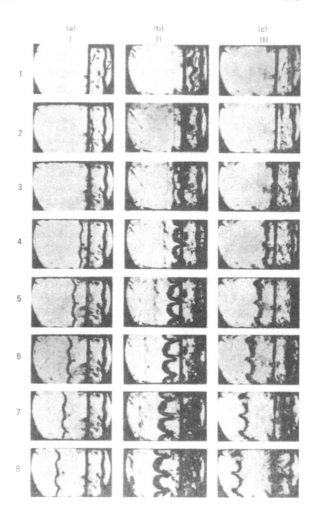

For the case of a pure impulse, we see that the interface perturbation amplitude η grows linearly in time from its initial value, with higher wavenumbers growing faster. It is also interesting to note that both signs of the Atwood number, i.e. either $\rho_2 > \rho_1$ or $\rho_1 > \rho_2$, grow from an initial amplitude, although the case of a shock through a lighter fluid into a heavier fluid ($A < 0$) causes the perturbation to reverse itself and then grow in the opposite direction. It should also be noted that it is only the jump in velocity $[u]$ that appears in the growth expression.

10.5.4   *Hydrodynamic Mix [44, 57]*

The fluid instabilities outlined above have two primary detrimental effects on
ICF target performance. One is that with these instabilities growing to significant
amplitudes during compression, the overall compression ratio achievable is limited.
A simple estimate of the achievable compression is given by

$$\frac{\rho_2}{\rho_1} \approx \left(\frac{R}{\Delta R}\right)^3, \tag{10.46}$$

where ΔR is the radial size of the perturbation amplitude near maximum compression. Clearly this points to the need for compression that is fast enough to avoid
substantial growth of the hydrodynamic instabilities outlined above. In addition to
hydrodynamic instabilities limiting the overall compression achievable, they also
cause mixing of the ablator material with the fuel, which can increase the radiation
loss from the hot fuel and drain thermal energy otherwise available for the fuel
ions. Studies of "mix" have been thoroughly investigated using three-dimensional
fluid codes. An example of the effect of mix on capsule performance is shown as
Fig. 10.20. Here the neutron output from the compressed target was compared to the
mass of ablator mixed in with the fuel, which was inferred from the X-ray output.

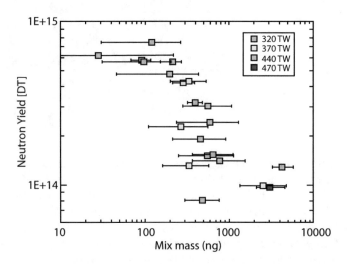

Fig. 10.20 Plot of experimentally observed neutron yield for shots on the National Ignition
Facility vs. amount of ablator material mixing in with the fuel. From [44]. Used with permission,
American Physical Society

Fig. 10.21 Schematic of
planar shock wave. Shock is
progressing left to right,
region 2 is behind the shock
("upstream," and region 1 is
in front of the shock
("downstream"). After [2]

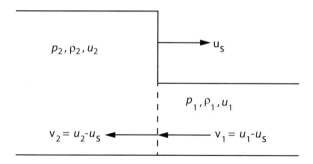

10.5.5 Shocks

Shocks are discontinuous solutions to the hydrodynamic equations which are
important for implosion physics in ICF. The distinguishing feature of a shock
wave, as opposed to a less violent ablation process, is that material enters an
interface zone between two distinctly different states of a fluid with supersonic flow
conditions. The fluid state on the two sides of the shock are matched by energy
and momentum conservation, and in general shocks are non-adiabatic. Whereas
adiabatic compression has no limits imposed on the density ratios between the
material before and after the compression, with shocks the ratio of the density of
the material that the shock has already passed through ("downstream") with the
material in front of the shock ("upstream") has a finite, limiting value at very high
shock velocities. Figure 10.21 gives the basic parameters for consideration of the
case of a one-dimensional (planar) shock.

Using the nomenclature of Fig. 10.21, we can write down equations representing
conservation of mass, momentum, and energy at the shock interface. These are
known as the Rankine-Hugoniot equations [2, 82]. We use a set of velocity variables
which are given in the reference frame of the shock front, which is traveling at a
speed u_s, thus:

$$v_{1,2} = u_{1,2} - u_s.$$

The mass flow $\mathbf{j} = \rho \mathbf{v}$ is a constant in any rest frame, and here we use the reference
frame of the shock, so that in one dimension

$$j = \rho_1 v_1 = \rho_2 v_2 \qquad (10.47)$$

is the equation for conservation of mass. Conservation of momentum is given by
summing the kinetic pressure p and Bernoulli pressure ρv^2 on the two sides of the
shock front and setting them equal:

$$\rho_1 v_1^2 + p_1 = \rho_2 v_2^2 + p_2. \qquad (10.48)$$

Energy conservation comes about from setting the sum of the enthalpy h and the kinetic energy $\rho v^2/2$ across the shock boundary:

$$\left(v_1^2/2 + h_1 \right) \rho_1 v_1 = \left(v_2^2/2 + h_2 \right) \rho_2 v_2, \tag{10.49}$$

or, using Eq. (10.47):

$$v_1^2/2 + h_1 = v_2^2/2 + h_2. \tag{10.50}$$

Equations (10.47), (10.48), and (10.50) together form the Rankine-Hugoniot equations. Note especially the role of the enthalpy h here: this is the specific enthalpy (per unit mass) $h = e + pV$, where e is the internal energy per unit mass and $V = 1/\rho$ is the volume per unit mass, so that $h = e + p/\rho$. These equations can be manipulated to give some additional equations describing the shock. Using Eqs. (10.47) and (10.48) then yields

$$j = \sqrt{(p_2 - p_1)/(V_1 - V_2)} \tag{10.51}$$

and using Eq. (10.50) then gives

$$h_2 - h_1 = \frac{1}{2}(V_1 + V_2)(p_2 - p_1), \tag{10.52}$$

and using $h = e + pV$ gives an alternative form

$$e_2 - e_1 = \frac{1}{2}(V_1 - V_2)(p_2 + p_1). \tag{10.53}$$

The jump condition on the velocity can be found from these conditions as

$$u_2 - u_1 = v_2 - v_1 = \sqrt{(p_2 - p_1)(V_1 - V_2)} \tag{10.54}$$

and the shock velocity u_s can be determined from the mass flow as

$$u_s = u_1 + |j|V_1 = u_2 + |j|V_2. \tag{10.55}$$

The overpressure $p_2 - p_1$ is also easily solved for:

$$p_2 - p_1 = \rho_1 (u_2 - u_1)(u_s - u_1). \tag{10.56}$$

The solution of the Rankine-Hugoniot equations depends on the equation of state for the fluid. In general, equations of state can be relatively complex because of phase changes in the fluid as the temperature and density increase. For example, deuterium at low temperatures can start out as a simple liquid, but under shock conditions becomes a gas of dimers D_2, then a gas of monomers D, and then a plasma with some neutral fraction, and finally a fully ionized $D^+ + e^-$ plasma. However, the special case of an ideal gas with a ratio of specific heats γ=constant has a simple equation of state and thus simple Hugoniot relations. If we take

$$e = \frac{pV}{\gamma - 1},$$

the Hugoniot relations become

$$\frac{\rho_2}{\rho_1} = \frac{V_1}{V_2} = \frac{(\gamma + 1)p_2 + (\gamma - 1)p_1}{(\gamma + 1)p_1 + (\gamma - 1)p_2} \qquad (10.57)$$

or equivalently

$$\frac{p_2}{p_1} = \frac{V_1}{V_2} = \frac{(\gamma + 1)V_1 - (\gamma - 1)V_2}{(\gamma + 1)V_2 - (\gamma - 1)V_1}. \qquad (10.58)$$

Equation (10.57) is especially illuminating in showing the achievable compression ratio ρ_2/ρ_1 for a shock. If we take the ratio of pressures $p_2/p_1 \to \infty$, then the maximum density ratio achievable is

$$\left.\frac{\rho_2}{\rho_1}\right|_{max} = \frac{\gamma + 1}{\gamma - 1} \qquad (10.59)$$

which for $\gamma = 5/3$ for fully ionized plasma gives a maximum compression ratio of 4. Even four consecutive shocks, which are the number of shocks designed for in the NIF experiment, are only sufficient to cause a $4^4 = 256$ compression ratio, which shows that shocks must be combined with some other heating and compression processes to get to the desired $> 1000\times$ compression. However, this result is only valid for planar shocks, and for the more realistic spherical case this analysis requires some adjustments when the shell aspect ratio $R/\Delta R$ becomes of order unity. An early attempt at modeling the effects of a spherically imploding shock in spherical geometry was given by Guderley in 1942 [21]. Guderley's analysis used a self-similar solution to the shock equations in spherical geometry for a constant-density sphere, i.e. not for a shell. While this solution shows high compression ratios with just a single shock, the pressure and density profiles for this solution are not consistent with the profiles generated from realistic ablation scenarios. Subsequent analysis has allowed treatment of cylindrically and spherically convergent shocks with a range of adiabatic indices γ and including spherical shells [14, 30]. See Figs. 10.22 and 10.23. In these studies, self-similar solutions to the shock equations were still used, and the time integration was continued after the shock wave reached the center of the target, and re-expansion followed as the shock wave progressed outward again.

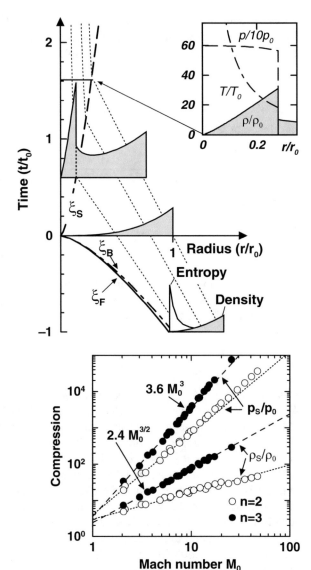

Fig. 10.22 Diagram of spherically imploding shell using the self-similar model. here $t = 0$ corresponds to the time when the inside of the shell reaches $r = 0$; negative times are during the start of the implosion. Characteristic curves following fluid elements starting at different initial radii are shown. The secondary shock wave after $t = 0$ shows the re-expansion of the material. From [30]. Used by permission, American Physical Society

Fig. 10.23 Peak pressure ratio P_s/P_0 and density compression ρ_s/ρ_0 as a function of the initial Mach number $M_0 = u/c_s(0)$ for shock compression of annular shells in cylindrical ($n = 2$) and spherical ($n = 3$) geometry using a self-similar model. From [30]. Used by permission, American Physical Society

For the case of N multiple shocks with constant pressure ratios $p_{i+1}/p_i \equiv \xi$, the overall final-to-initial pressure ratio is given by

$$\frac{P_f}{P_i} = \xi^N = \left(\frac{2\gamma M^2 - (\gamma - 1)}{\gamma + 1}\right)^N = \left(\frac{5}{4}M^2 - \frac{1}{4}\right)^N \quad \text{for } \gamma = \frac{5}{3}. \quad (10.60)$$

Here M is the upstream Mach number ($= u_2/c_s$ in the notation used here, with $c_s = (\gamma p_2/\rho_2)^{1/2}$.) Similarly, Eq. (10.56) for the density ratio $\eta \equiv \rho_2/\rho_1$ can be written as

$$\eta = \frac{\xi \eta_\infty - 1}{\xi + \eta_\infty},\tag{10.61}$$

where η_∞ is the density ratio for a single shock with $p_2/p_1 = \infty$, which equals 4 for $\gamma = 5/3$, Then the overall density ratio is just

$$\frac{\rho_f}{\rho_i} = \eta^N.\tag{10.62}$$

The single-shock temperature ratio can also be derived from the Rankine-Hugoniot equations for an ideal gas with adiabatic coefficient γ as

$$\frac{T_1}{T_2} = \frac{(\gamma + 1)^2 M^2}{2(\gamma - 1)\left(1 + \frac{\gamma - 1}{2}M^2\right)\left(\frac{2\gamma}{\gamma - 1}M^2 - 1\right)} = \frac{16M^2}{(3 + M^2)(5M^2 - 1)} \quad \text{for } \gamma = \frac{5}{3},\tag{10.63}$$

and thus the overall temperature ratio $T_f/T_i = (T_2/T_1)^N$. The entropy change per shock is given by

$$\Delta s = \frac{5}{2}R \ln \frac{T_2/T_1}{(p_2/p_1)^{(\gamma - 1)/\gamma}}$$

$$= 0.96 \times 10^8 \left[\ln (T_2/T_1) - 0.4 \ln (p_2/p_1)\right] \text{J keV}^{-1} \text{g}^{-1} \quad \text{for } \gamma = \frac{5}{3},\tag{10.64}$$

and the overall entropy change for N shocks is $\Delta s_{i \to f} = N \Delta s_{1 \to 2}$. Figure 10.24 shows the entropy change Δs resulting from a series of shocks, each with the same pressure ratio ξ, such that the total final pressure ratio is held fixed and the number of shocks N is varied to achieve this with a series of shocks with a smaller pressure ratio per shock. The entropy production for a single shock with a varying pressure ratio is also shown. Note that as the pressure ratio ξ for each individual shock becomes smaller (and N increases), the total entropy production approaches zero, i.e. the process becomes adiabatic.

10.6   Equation of State and Shock Timing

10.6.1   Fermi Degeneracy

ICF conditions require that issues regarding quantum-mechanical effects on the equation of state must be examined. The discussion begins with a simple model

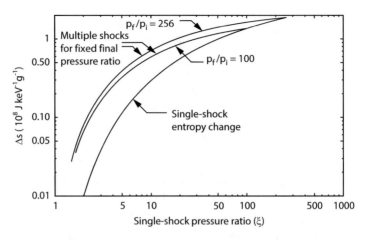

Fig. 10.24 Entropy change Δs for a series of shocks with identical pressure ratios ξ for a fixed overall pressure ratio ($p_2/p_1 = 100$ or 256). The entropy change for a single shock is also shown. After [43]

for the equation of state for a very dense system of electrons at low energy. Here we ignore the effects of molecular and ionic bonding, and essentially ignore the ions except in their role of giving the system charge neutrality and adding mass. Electrons are particles with a spin of $\pm\hbar/2$, and as such are considered fermions. Fermions obey the Pauli exclusion principle, which is to say that no two electrons can be in exactly the same state. The distribution of electron energies at low temperatures is dictated by the finite number of discrete quantum mechanical states available for the electrons to populate.

We start with a simple theoretical model, where all of the electrons are confined to a rectangular box of dimensions $L_x \times L_y \times L_z$. (Neither the dimensions of the box nor choosing it to be rectangular turns out to be relevant.) Schrodinger's equation for each electron is

$$-\frac{\hbar^2}{2m_e}\nabla^2\Psi = E\Psi,\tag{10.65}$$

where E is the energy of the free particle and we take the potential energy $\mathscr{V} = 0$ inside the box, with $\mathscr{V} \to \infty$ at the boundaries. The solution to this equation on the domain $\{0, L_x\} \otimes \{0, L_y\} \otimes \{0, L_z\}$ is simply

$$\Psi_{n_x n_y n_z} = \frac{1}{\sqrt{2V}}\sin\left(\frac{n_x\pi x}{L_x}\right)\sin\left(\frac{n_y\pi y}{L_y}\right)\sin\left(\frac{n_z\pi z}{L_z}\right)\tag{10.66}$$

and the energy of the particle is $E = \hbar^2 k^2/(2m_e)$, where

$$k^2 = k_x^2 + k_y^2 + k_z^2 = \frac{n_x^2\pi^2}{L_x^2} + \frac{n_y^2\pi^2}{L_y^2} + \frac{n_z^2\pi^2}{L_z^2}.\tag{10.67}$$

Fig. 10.25 Eigenvalues in **k** for free electrons in a box. Energy surfaces are spheres in **k**-space

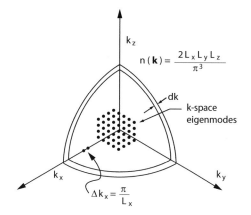

$$n(\mathbf{k}) = \frac{2L_xL_yL_z}{\pi^3}$$

dk

k-space eigenmodes

k_z k_x k_y

$$\Delta k_x = \frac{\pi}{L_x}$$

We now fill the box with electrons in order of ascending energy. Figure 10.25 shows the eigenstates in $\{k_x, k_y, k_z\}$ space, and how the counting of filled states works to some energy E. Since $E \propto k_x^2 + k_y^2 + k_z^2$, surfaces of constant E are spherical surfaces in k-space. The density of states in the three-dimensional k-space is the inverse of the unit cell size $1/(\Delta k_x \Delta k_y \Delta k_z)$, multiplied by two since the half-integer spin electrons have $2s + 1 = 2$ possible independent states for each wavenumber triplet, thus:

$$n(\mathbf{k}) = \frac{2L_xL_yL_z}{\pi^3}. \tag{10.68}$$

For the case of electrons at low temperature, all states are filled to some critical energy (the "Fermi energy") $E_f \propto k^2$. The total number of states up to some critical value of k is then given by the volume in k-space that this represents multiplied by the density of states given above. Note that only the volume of the spherical octant with $k_x, k_y, k_x > 0$ is used, because negative integer values of the quantum numbers n_x, n_y, n_z have the same wavefunctions as their positive-integer counterparts except for a change of sign, so that these are not independent of the positive integer states and must be excluded. Then the spherical octant volume is $(4\pi/(8\cdot3))k_{max}^3$, and thus

$$N = \frac{L_xL_yL_z}{3\pi^2}k_{max}^3. \tag{10.69}$$

Since $E_f = E_{max} = \hbar^2 k_{max}^2/(2m_e)$, this expression can be re-written as

$$N = \frac{V}{3\pi^2}\left(\frac{2m_eE_f}{\hbar^2}\right)^{3/2} \tag{10.70}$$

Here we have used the volume $V = L_xL_yL_z$, and note that the total number of states N is independent of the box dimensions, aside from the total volume. The number of states required for the electrons $= n_e V$, and then E_f can be found by inverting Eq. (10.69)

$$E_f = \frac{\hbar^2 \left(3\pi^2 n_e\right)^{2/3}}{2m_e}. \tag{10.71}$$

It is then a simple matter to find the average energy per electron from the Fermi statistical model. Since the density of states in $k-$ space is a constant, and as Fig. 10.25 shows, the number of states in an increment dk of a spherical shell is $\propto k^2 dk$, and then

$$< k^2 > = \frac{\int_0^{k_{max}} k^4 dk}{\int_0^{k_{max}} k^2 dk} = \frac{3}{5} k_{max}^2 \tag{10.72}$$

and then

$$< E > = \frac{3}{5} E_f. \tag{10.73}$$

The pressure is given as in the non-degenerate case by $p = (2/3)n_e < E >$, thus:

$$p_f = \frac{\hbar^2 \left(3\pi^2\right)^{2/3} n_e^{5/3}}{5m_e}. \tag{10.74}$$

We can evaluate this expression for equimolar DT by taking the electron density as $n_e = n_i = 10^6 \rho N_A / M \text{ m}^{-3}$, with Avogadro's number $N_A = 6.02 \times 10^{23}$ per g-mole and ρ in g cm^{-3}, $M = 2.5$, and MKS units otherwise, gives

$$p_f = 2.16 \times 10^{11} \rho^{5/3} \text{ Pa} = 2.16 \, \rho^{5/3} \text{ MBar}. \tag{10.75}$$

The Fermi pressure thus gives a limit on the compressibility of the free electrons in a material at low temperatures. Unlike the ideal Maxwellian gas, the Fermi pressure does not go to zero at low temperatures, but rather represents a real effect from the quantum-mechanical effects of Pauli exclusion.

For finite temperature, the occupancy factor for states above and below the Fermi energy are neither zero nor one but follow a Boltzmann-like law which follows from the quantum statistics for the density of states. The occupancy factor $f(E)$ is given by

$$f(E) = \frac{1}{1 + \exp\left[(E - \mu)/T\right].} \tag{10.76}$$

Here μ is a chemical potential (yet to be determined). A plot of $f(E)$ vs. μ is shown as Fig. 10.26.

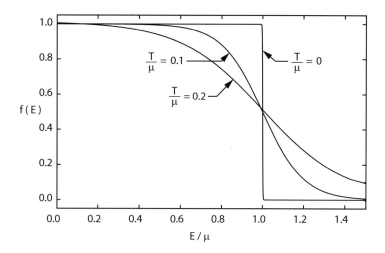

Fig. 10.26 The Fermi occupancy factor $f(E)$ as a function of the normalized energy E/μ for three values of T/μ

Now the chemical potential is found by the normalization of the distribution function, which includes the density of states derived above and the Fermi occupancy factor. This can be written as

$$N_e = n_e V = \int \frac{2}{h^3} d^3 x d^3 p f(E) = \frac{2V}{\lambda_{th}^3} I_{1/2}\left(\frac{\mu}{T}\right) \tag{10.77}$$

where λ_{th} is the thermal deBroglie wavelength given by

$$\lambda_{th}^2 = 2\pi\hbar^2/(m_e T) \tag{10.78}$$

(This follows directly from the deBroglie equation $\lambda = h/p$, where p is the momentum of a particle and h is Planck's constant, one of the famous relations that started quantum theory.) The Fermi integral $I_n(x)$ is given by

$$I_n(x) = \frac{1}{n!} \int_0^\infty dy \frac{y^n}{1 + \exp(y - x)} = -\text{PolyLog}(n + 1, -e^x). \tag{10.79}$$

(The PolyLog function used in the latter expression is perhaps more convenient because it is available in Mathematica and like programs.) For cold, dense plasma the argument $\mu/T \gg 1$, and the asymptotic expansion of the Fermi integral $I_{1/2}$ is given by

$$I_{1/2}(x) = \frac{4}{3\sqrt{\pi}} x^{3/2}\left[1 + \frac{\pi^2}{8x^2} + \dots\right]. \tag{10.80}$$

The chemical potential μ is then implicitly given by the normalized temperature $\Theta \equiv T/E_f$:

$$- \text{PolyLog}\left(\frac{3}{2}, -\exp\left(\frac{\mu}{T}\right)\right) = \frac{4}{3\sqrt{\pi}\Theta^{3/2}}. \qquad (10.81)$$

This asymptotic expression for $I_{1/2}$ then yields a relationship between the potential μ and the Fermi energy E_f:

$$\mu = E_f\left(1 - \frac{\pi^2}{12}\Theta^2 + \dots\right). \qquad (10.82)$$

The pressure is given by

$$p = nT\frac{\text{PolyLog}(5/2, -z)}{\text{PolyLog}(3/2, -z)} \approx \frac{2}{5}nE_f\left(1 + \frac{5\pi^2}{12}\Theta^2 + \dots\right). \qquad (10.83)$$

Here we have used $z = \exp(\mu/T)$, sometimes referred to as the "fugacity." From Eq. (10.83), it is apparent that in the low-temperature limit, higher pressures lead to higher degeneracy factors because of the linear relationship between p and E_f. The equations given in (10.84) give the exact expressions for the total energy E, the free energy F, and the entropy $S = (E - F)/T$. Also shown are these functions in the limit of $\theta \ll 1$:

$$E = 3/2pV \qquad\qquad = \frac{3}{5}NE_f\left(1 + \frac{5\pi^2}{12}\Theta^2 + \dots\right)$$

$$F = NT\left(\ln z - \frac{\text{PolyLog}(5/2, -z)}{\text{PolyLog}(3/2, -z)}\right) \quad = \frac{3}{5}NE_f\left(1 - \frac{5\pi^2}{12}\Theta^2 + \dots\right)$$

$$S = N\left(-\ln z + \frac{5}{2}\frac{\text{PolyLog}(5/2, -z)}{\text{PolyLog}(3/2, -z)}\right) = N\left(\frac{\pi^2}{2}\Theta + \dots\right).$$

$$(10.84)$$

Figure 10.27 shows the curves for chemical potential μ, the pressure p, the free energy F, and the entropy S as functions of the normalized temperature $\Theta \equiv T/E_f$. At high temperatures, the pressure reverts to the ideal gas law $p = nT$. The overall pressure equation of state, as Θ ranges from 0 to ∞ can also be written in approximate form [27]:

$$\beta \equiv \frac{p}{p_f} \approx \frac{X\Theta^{-y} + Y\Theta^{(y-1)/2}}{X\Theta^{-y} + 1} + \frac{5}{2}\Theta \qquad (10.85)$$

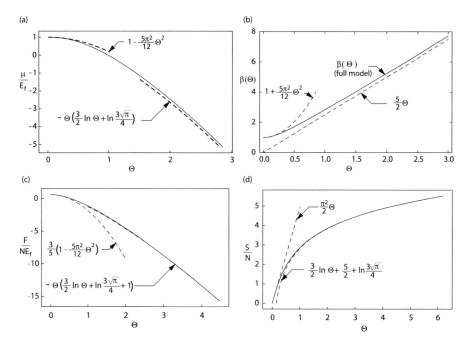

Fig. 10.27 Thermodynamic quantities from Fermi model, with asymptotic values for $\Theta \ll 1$ and $\Theta \gg 1$ shown. (**a**) Chemical potential μ vs. normalized temperature $\Theta \equiv T/E_f$, (**b**) Normalized pressure $\beta = p/p_f$ vs. normalized temperature Θ, (**c**) normalized free energy $F/(NE_f)$ vs. Θ, (**d**) normalized entropy S/N vs. Θ

with coefficients $X = 0.272332$, $Y = 0.145$, and $y = 1.044$. A graph of the pressure ratio β as a function of normalized temperature , with the asymptotic values shown, is shown as Fig. 10.27. (The approximate form given in 10.85 is accurate to within 0.1% and cannot be distinguished from the exact result in this figure.)

The degeneracy factor β is central to the design of an ignition-capable capsule. Lindl [43] gives a set of scaling laws for capsule radius, implosion time, and drive power as functions of β, hohlraum radiation temperature, and total absorbed energy in the capsule. The capsule's initial value of β sets the trajectory in pressure vs. density space that the capsule follows during compression, as will be shown later.

The Fermi model does not take into account the effects of chemical bonding in a solid or liquid material. One has only to consider the predicted Fermi pressure for ordinary materials, such as copper ($A = 63.546$), at room temperature and density ($\rho = 8.97\,\mathrm{g\,cm}^{-3}$). If we assume one free electron per atom in Cu ($n_e = 8.498 \times 10^{22}\,\mathrm{cm}^{-3}$), we obtain a Fermi pressure of $p_{Fermi} = 0.38\,\mathrm{MBar}$, or around 400,000 atmospheres. So why don't pennies explode? The answer is that the proper quantum mechanical treatment of the ionic lattice and the free energy associated with the free electrons traveling in that lattice are not included in the simple Fermi model. The overall equation of state (EOS) must take these factors into account. More sophisticated EOS models that work down to more everyday temperatures and

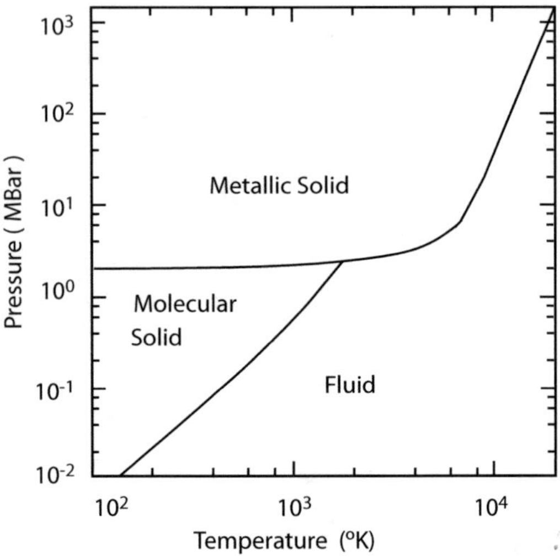

Fig. 10.28 Phase diagram
for deuterium at low
temperatures and pressures.
From[31]. Used with
permission, Elsevier

Fig. 10.29 Pressure vs.
normalized density
($\rho_0 = 0.17\,\text{g cm}^{-3}$) for
deuterium at low
temperatures and pressures.
From[31]. Used with
permission, Elsevier

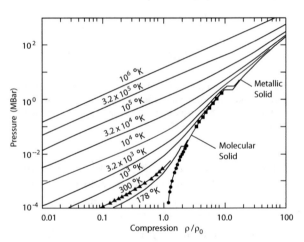

pressures ("quotidian EOS")(QEOS) must include the physics necessary to match
the known phenomena such as melting and vaporization. Several attempts have been
made to concoct EOS models which correctly predict the properties over a broad
range of parameter space covering ICF conditions. The most well-known model
is available from a code called SESAME [31], maintained by Los Alamos. This
code is unclassified and distributed openly. An example of a SESAME EOS model
for deuterium at low temperatures is shown as Fig. 10.28, which shows the phase
diagram, and Figure 10.29, which shows the computed isotherms.

It is important to note that for simulation of shock heating and other purposes
not only are values for one master thermodynamic function, such as the free
energy $F(\rho, T) = U - TS$ required, but also derivatives of the function required.
For example, three relations give thermodynamic quantities p, S, and U derived

from F:

$$p = \rho^2 \frac{\partial F}{\partial \rho}$$

$$S = -\frac{\partial F}{\partial T} \qquad (10.86)$$

$$U = F + TS.$$

Thus the data for $F(\rho, T)$ must be smooth enough to allow for differentiation. Two derivatives on $F(\rho, T)$ are required to obtain quantities such as the specific heats c_p and c_V and thus the adiabatic index γ. In fact, all of the Maxwell relations involve mixed second partial derivatives, so the data must not have any step-type discontinuities, or these relations will not hold numerically. The EOS must fit the Fermi model for the electrons at intermediate temperatures and pressures where degeneracy effects are important. The Fermi model should be modified, however, to take account of the electrostatic potentials for the ions in the material, and this results in an additional energy term $eV(r)$ in the Boltzmann energy distribution along with the chemical potential μ discussed earlier. Equation (10.78) is then written as

$$n_e(r) = \frac{2}{\lambda_{th}^3} I_{1/2} \left(\frac{\mu + eV(r)}{T} \right). \qquad (10.87)$$

The strength of this term depends on the equivalent spherical radius of the volume per atom, such that $n_{ion} = 1/((4/3)\pi r_s^3)$, where r_s is called the Wigner-Seitz radius. Then a measure of the strength of the perturbation $eV(r)$ is given by the parameter [26]:

$$\Gamma = \frac{e^2}{4\pi \epsilon_0 r_s T}. \qquad (10.88)$$

Physically, Γ is the ratio of electrostatic Coulomb energy to the kinetic energy of the particles. Using the Fermi statistical model with this additional term results in an EOS known as the Thomas-Fermi EOS. This extension of the simple Fermi model is important where $\Gamma > 1$, i.e. at high densities and low temperatures.

10.6.2   Ion Equations of State: Solid Phase

The ion equation of state in a solid at low temperatures comes about from an understanding of the lattice vibrations. Lattice vibrational quanta are phonons and have integer spin, and thus follow Bose-Einstein statistics. (See a standard textbook on statistical physics such as Reif [58] for a full derivation.) The atomic lattice vibration spectra are characterized by a Debye temperature Θ_D, which is typically

$100 \rightarrow 500\,^\circ$K for most metals. The free energy per unit mass for the ionic lattice material is then given by [50]:

$$F_i = \frac{T_i}{Am_p} f(\Theta_D/T) \tag{10.89}$$

with A being the atomic mass, m_p the mass of a proton, and

$$f(x) = \frac{9}{x^3} \int_0^x u^2 \left[u/2 + \ln\left(1 - e^{-u}\right) \right] du$$

$$= \frac{9}{x^3} \left[-x^2 \mathrm{Li}_2\left(e^x\right) + 2x\mathrm{Li}_3\left(e^x\right) - 2\mathrm{Li}_4\left(e^x\right) + \frac{x^4}{8} \right.$$

$$\left. + \frac{1}{180}\left(15x^4 + 60x^3\left(\ln(\sinh(x) - \cosh(x) + 1\right) - \ln\left(1 - e^x\right)\right) + 4\pi^4\right) \right].$$

$$\tag{10.90}$$

Here the logarithmic integral function $\mathrm{Li}(x)$ is given by

$$\mathrm{Li}(x) = \int_0^x \frac{dt}{\ln t}.$$

A graph of the function $f(x)$ is shown as Fig. 10.30. Note the logarithmic singularity at $x = 0$. A series expansion, valid for small x, is

$$f(x) \approx -1 + 3\ln x + 3x^2/40 - x^4/2240, \tag{10.91}$$

and for large x:

$$f(x) \approx \frac{9x}{8} + 3\ln\left(1 - e^{-x}\right) - \frac{\pi^4}{5x^3} + e^{-x}\left(3 + \frac{9}{x} + \frac{18}{x^2} + \frac{18}{x^3}\right). \tag{10.92}$$

Here the physical meaning of the integration variable u is as the normalized frequency $u = \hbar\omega/T$ for the vibrations. The expressions for the pressure and the entropy and internal energy per unit mass are given by

Fig. 10.30 Free energy function $f(x)$ associated with phonon distribution

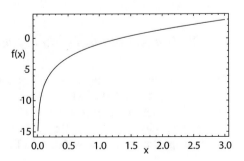

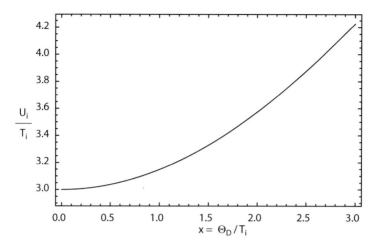

Fig. 10.31 Internal energy U_i per ion as a function of inverse normalized temperature Θ/T_i

$$p_i = \frac{1}{Am_p} \rho^2 \frac{d\Theta_D}{d\rho} f'(x)$$

$$S_i = \frac{1}{Am_p} \left[x f'(x) - f(x) \right] \qquad (10.93)$$

$$U_i = \frac{\Theta_D}{Am_p} f'(x).$$

The internal energy per ion in the limit of high temperatures ($T_i \gg \Theta_D$), using Eqs. (10.91) and (10.92), becomes

$$U_i = 3T_i \left(1 + \frac{3}{30} (\Theta_D/T_i)^2 - \frac{1}{560} (\Theta_D/T_i)^4 + \ldots \right), \qquad (10.94)$$

and this is consistent with the Dulong-Petit law $U_i = 3T_i$ at very high temperatures. A graph of the internal energy function is shown as Fig. 10.31. At low temperatures, Eq. (10.92) becomes $f(x) \approx 9x/8 - \pi^2/(5x^3)$, and the specific heat approaches

$$\frac{dU_i}{dT} = \frac{12\pi^4}{5} \left(\frac{T}{\Theta_D} \right)^3, \qquad (10.95)$$

and then the specific heat and entropy both approach zero as $T \to 0$, as required by Nernst's theorem. The two asymptotic forms for $f'(x)$ match at $x \equiv \Theta_D/T \approx 3$, so this can be taken as the dividing point between the high-temperature form, or Gruneisen EOS, and the low-temperature, Nernst-theorem conforming form, as shown in Fig. 10.33.

The relation between pressure and energy, using Eq. (10.91), can be written as

$$p_i = \gamma_s \rho U_i \tag{10.96}$$

with

$$\gamma_s = \frac{d \ln \Theta_D}{d \ln \rho}. \tag{10.97}$$

A model for the Debye temperature Θ_D is needed in order to obtain the material equations of state. An empirical model, developed by Cowan and published by More [49], is:

$$\Theta_D = \frac{1.68}{Z + 22} \frac{\xi^{b+2}}{(1 + \xi^2)} \text{ eV}, \tag{10.98}$$

with $\xi = \rho/\rho_r$ and $\rho_r = A/(9Z^{0.3}) \text{ g cm}^{-3}$ and $b = 0.6Z^{1/9}$. The Gruneisen coefficient γ_s is then given by

$$\gamma_s = b + \frac{2}{1 + \xi}. \tag{10.99}$$

The inclusion of the extra free energy terms representing molecular/lattice bonding then provides a more complete picture of the equation of state, good down to "quotidian" conditions. Figure 10.32 shows a calculation of the cold compression of aluminum from standard pressure and temperature up to megabar pressures, using five different models: the Thomas-Fermi (TF) result as outlined above, the Dirac modification of the Thomas-Fermi theory [12, 16, 40, 61, 68], then the same two models with bonding energy terms added consistent with the Gruneisen model, and finally, some point calculations using a quantum mechanical simulation incorporating an augmented plane wave (APW) potential function [1, 10]. Without the bonding energy terms, pressure at standard density for Al at room temperature (2.7 g cm^{-3}) is unrealistically high.

10.6.3   Ion Equation Of State: Fluid Phase

Melting occurs at higher temperatures than the solid phase in all materials. A simple model for the melting temperature T_m is given by Lindemann's law [41]

$$T_m(\rho) = \alpha_L \Theta_D(\rho)^2 \rho^{-2/3}, \tag{10.100}$$

where α_L depends on the material but not on its pressure or density. A theoretical justification for Lindemann's melting law is given in [39]. Melting usually occurs in

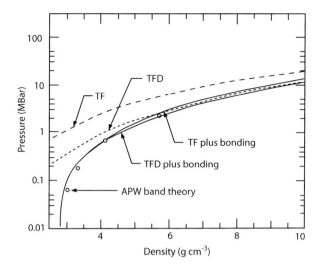

Fig. 10.32 Five models for the EOS of aluminum at low temperature. Thomas-Fermi (TF) and Thomas-Fermi-Dirac (TFD) model, and same models with bonding terms added. Also quantum calculation points shown using augmented plane wave (APW) potentials. From [49]

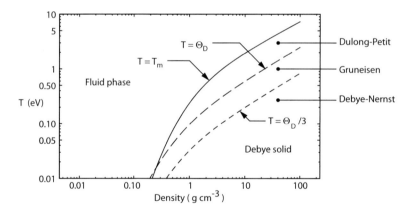

Fig. 10.33 Plots of the melting temperature, Debye temperature, and one-third of the Debye temperature for DT from Cowan's EOS model

a solid crystalline material when the rms amplitude of the lattice vibrations reaches approximately 8% of the interatomic spacing d.

An empirical law for the Lindemann coefficient α_L is given by More [49] as

$$\alpha_L = 0.0262A^{2/3}(Z+22)^2/Z^{0.2} \left(g\,cm^{-3}\right)^{2/3} eV^{-1}. \qquad (10.101)$$

A plot of the predicted melting temperature for DT is given as Fig. 10.33. Using the expression for the Debye temperature given in Eq. (10.98), Eq. (10.101) can also be written as:

$$T_m = 0.32 \frac{\xi^{2b+10/3}}{(1+\xi)^4} \text{ eV.}$$ (10.102)

The free energy in the fluid phase can be modeled as [49]:

$$F_i = T_i f(x, y)$$ (10.103)

with

$$f(x, y) = -\frac{11}{2} + \frac{9}{2} y^{1/3} + \frac{3}{2} \ln\left(x^2/y\right),$$ (10.104)

with $x = \Theta_D/T$ as before and $y = T_m/T_i$. Since this model EOS is only used for the fluid state, $0 < y < 1$ is the range of validity, with x taking on all positive values. The pressure, entropy (per ion), and internal energy are:

$$p_i = n_i T_i \left(1 + \gamma_f y^{1/3}\right)$$

$$S_i = 7 - 3y^{1/3} - \frac{3}{2} \ln \frac{x^2}{y}$$ (10.105)

$$U_i = n_i \frac{3}{2} T_i \left(1 + y^{1/3}\right).$$

Here the dimensionless coefficient γ_f is defined :

$$\gamma_f = \frac{3}{2} \frac{d \ln T_m}{d \ln \rho},$$ (10.106)

which can also be written in terms of the Gruneisen coefficient γ_s given earlier by:

$$\gamma_f = 3\gamma_s - 1.$$ (10.107)

In this model, no distinction is made between liquid and gaseous phases. From Eqs. (10.105), for temperatures well above the melting temperature where $y \to 0$, we find that $p_i \to n_i T_i$ and $U \to (3/2)n_i T_i$, the equations of state for an ideal gas. The entropy term, less obviously, also returns to the classical form when Eq. (10.100) is used to eliminate the Debye temperature Θ_D from the entropy equation in Eqs. (10.105); then the entropy terms become

$$S = \frac{1}{A m_p} \left[S_0 + \frac{3}{2} \ln\left(\frac{T}{\rho^{2/3}}\right) \right],$$ (10.108)

with S_0 given by

$$S_0 = \frac{5}{2} + \frac{3}{2}\ln\left(g\,Am_p\right)\frac{1}{Am_p} - \frac{3}{2}\ln\left(\frac{h^2}{2\pi\,Am_p}\right). \tag{10.109}$$

Here $g = 2s + 1$ is the nuclear spin statistical factor. The equation for S_0 is the celebrated Sackur-Tetrode law [15]. A simplified form for Eq. (10.109) is given in [49]:

$$S_0 = 8.4662 + \ln\left(g\,A^{5/2}\right), \tag{10.110}$$

for T in eV and cgs units for the other quantities. Notice the dependence of the entropy on the nuclear spin. This presents a complexity for DT, which is a mixture of $s = 1$ nuclei (deuterons) and $s = 1/2$ particles (tritons). The entropies can be computed separately for these two species, and then the mixing rules for entropy, pressure, and internal energy can be applied in a straightforward manner.

10.6.4 Metallic Hydrogen

The existence of a metallic phase for hydrogen was first proposed by Wigner in 1935 [79]. Estimates of the pressure required for stable existence of this phase have been in the megabar range. Several experimental attempts at producing metallic hydrogen in anvil cells and with gas guns have been undertaken, such as one at Livermore in 1996 [78], but in general the results have not been corroborated between laboratories or between different methodologies. It is probable that the band gap E_b of the liquid phase has gotten close enough to zero so that $T \sim E_b$ in the experiments, providing a free electron population large enough to show some metal-like conduction. At the time of this writing, no definitive experimental results or robust theoretical models exist that conclusively confirm or deny the existence of metallic hydrogen. Perhaps we will have to wait for a trip to Jupiter!

10.6.5 Dissociation of Hydrogen Dimers

Of more practical interest for inertial fusion is the equation of state in the region where dissociation of hydrogen isotopes is happening, typically around $1\,\text{eV}$ temperatures. A recent article (2015) describes progress in resolving the EOS in this region [7]. Here quantum mechanical calculations are carried out using a modified Buckingham potential:

Fig. 10.34 "Exp-6" potential functions for D + D and D_2+D_2 interactions. From [7]

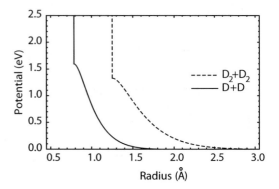

$$
\Phi_{ij} =
\begin{cases}
\dfrac{\epsilon_{ij}}{\alpha_{ij} - 6} \left(6 \exp\left[\alpha_{ij} \left(1 - \dfrac{r}{r_{m,ij}} \right) \right] - \alpha_{ij} \left(\dfrac{r_{m,ij}}{r} \right)^6 \right) & r > c_{ij}, \\[20pt]
\infty & r < c_{ij}.
\end{cases}
$$

$$\tag{10.111}$$

Here c_{ij} is the value of r at the maximum of the function, so that the potential is always positive and goes to infinity at the maximum value. This potential is called the "Exp-6" potential. For this model, [7] used the values $\alpha = 11.2$, $\epsilon = 30.4/11608\,\mathrm{eV}$, and $r_m = 3.49\,\text{Å}$ for D_2 molecules interacting with themselves, and similarly $\alpha = 11.0$, $\epsilon = 50.0/11608\,\mathrm{eV}$, and $r_m = 2.1\,\text{Å}$ for deuterium atoms interacting with themselves. A plot of this function is shown as Fig. 10.34.

Molecular dynamic (MD) simulations using these potential functions, for the gas mixtures present, can then yield a model equation of state which in turn can generate Hugoniot curves for shocks.

10.6.6   Ionization and Plasma EOS

High-temperature EOS relations are determined primarily by the Fermi model for the electrons and ideal-gas limits for the ions. The remaining question is simply to determine how many ions and electrons are present. Various models also exist for determining the degree of ionization of the material. At intermediate densities, this can be determined from the Saha equation given in Chap. 5. At progressively higher densities, however, where the interatomic spacing (the Seitz radius r_S mentioned earlier) is on the order of the Bohr atomic radius, the ionization states are more difficult to calculate. Figure 10.35 shows a recent calculation of the degree of ionization arising in a deuterium fluid with a mass density $\rho = n_0 m_{D_2}$ at a variety of temperatures. Note the intermediate region in mass density where ionization

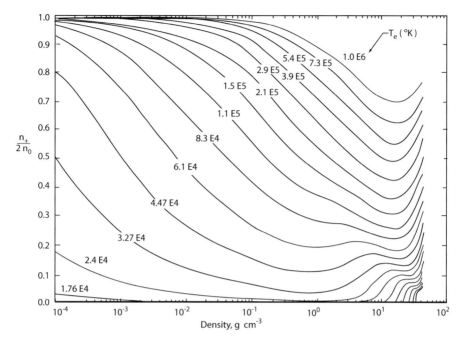

Fig. 10.35 Ionization fraction $n_+/(2n_0)$ for deuterium as a function of mass density $n_0 m_{D_2}$ at different temperatures. Temperatures are in $^\circ$K (note that $1\,\text{eV}=11{,}608\,^\circ$K). Data from [80]

is minimum. This is because of the interplay between the Saha equation, which favors ionization as the density increases, and the lowering of the ionic potential as $a_0/r_S \to 1$, related to the number of particles in a sphere with radius of the Bohr radius. Similar complexity exists in the atom-to-dimer ratio n_{D_0}/n_{D_2}.

10.6.7   Hugoniots from Equation of State [13, 25]

The equations of state generated from the models given above generate curves for the trajectory of a shock wave through pressure vs. mass density space, often simply called the Hugoniot for the system. As shown earlier in the description of shocks, the ultimate compression ratio for a shock in an ideal gas with adiabatic index $\gamma = 5/3$ is $\rho_\infty/\rho_0 = (\gamma + 1)/(\gamma - 1) = 4.0$. In the intermediate range of pressures corresponding to temperatures between 0.1 and 10.0 eV, the Hugoniot for deuterium (or DT) swings out past the ideal-gas limit, and returns to that limit at very high pressures (and thus temperatures). This "softening" of the material is due to departure from the ideal-gas adiabatic index due to dissociation and ionization. The difficulty in measuring or calculating the Hugoniot in this region is fairly obvious from Fig. 10.36, showing calculated and measured Hugoniot curves for deuterium from various experiments and theoretical efforts.

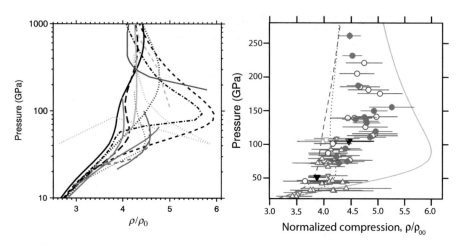

Fig. 10.36 Hugoniot curves for fluid deuterium at pressures between 45 and 220 GPa, from experiments and theoretical models. From [25]. Used by permission, American Physical Society

A major impetus in developing the " "quotidian" equation of state (QEOS)," or QEOS [49], has been to find a model that is fairly consistent with the behavior of the Hugoniot around the "softening" between 100 and 150 GPa. It is fair to say, however, that developing a good EOS for deuterium and DT is still an active area of research.

10.6.8   Shock Timing

The velocities of the shock waves through an ICF capsule are also contingent upon the EOS. The convergence of four shock waves at the compressed core of the capsule, at the so-called "bang time," critically depends on the progression of the shocks in time. The margin of error for the convergence of these shocks is on the order of tens of picoseconds. The conditions for the shock wave velocity and arrival time are also dependent on "preheat," i,e. heating of the upstream fuel caused by electrons, and on "mix," i.e. combination of the DT fuel with ablator and other capsule materials. In turn these are dependent on the growth of fluid instabilities as described earlier, and on uniformity of the radiation flux arriving at the ablation front, partially due to laser–plasma interaction issues at the hohlraum wall. A sort of "error budget" can be developed that can be used to discuss performance tradeoffs in the ultimate target and hohlraum design. The design of a superior-performance ICF target thus becomes a sort of black art as all the pieces are fit together.

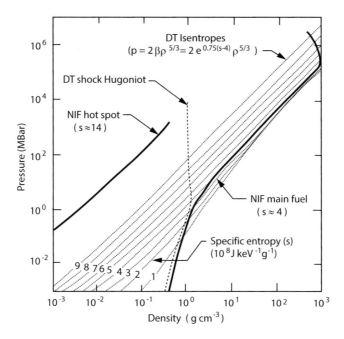

Fig. 10.37 Pressure vs. density for NIF reference design. Data from [43]

10.6.9   Ignition Strategy for NIF

A recent review of the status and history of the ICF program, leading up to the current NIF experiment, is given in [6].

The NIF ignition campaign baseline design called for four timed shocks, each with a pressure ratio of around 4.0. The first shock takes the imploding fuel up through the isentropes rather abruptly, as shown in the pressure vs. density diagram for DT in Fig. 10.37. It is important to note in this figure that the target composition, as shown in Fig. 10.1, has two distinct fuel phases. The outer shell of fuel is kept at cryogenic temperatures and is a solid form at a typical pressure of tens of bars before the shot. The interior of the target is deuterium-tritium gas in a pressure equilibrium with the solid fuel. (It is important to note that the tritium present in the fuel presents a constant source of heating due to the beta decay, and this heating, combined with external cooling of the fuel, creates an active thermodynamic balance between the gas and solid phases.) The gaseous phase inside of the solid DT shell is on a very different trajectory in pressure-density space than the solid phase. This trajectory is also shown in Fig. 10.37.

Subsequent laser power drive levels into the hohlraum create higher radiation temperatures at the hohlraum wall and thus drive stronger X-ray radiation at the ablation front of the material outside the fuel shell. The design idea in this part of the compression cycle is to stay close to the entropy imparted to the fuel after the first shock has passed through, and "ride the adiabat" to the point of stagnation, where momentum balance dictates that the fuel layer is no longer moving inward. The design pressure ratio for the second, third, and fourth shocks (around 4:1) is a compromise value which gives close to the ultimate achievable density ratio for each shock, which would be four (by coincidence) for an adiabatic index of $5/3$. The equation of state (EOS) drives the actual pressure vs. density characteristics for the shock Hugoniot, and Fig. 10.37 shows that a multiplier β is used from the EOS model to show the ratio of the pressure to the theoretical Fermi pressure of the electrons at each point along the curve; typically $\beta \leq 2$ for the pressure-density trajectory desired.

At some point one has to ask what the departures from the planar shock model are from the realistic spherical case. The shock waves, as the dense, cold fuel in the shell reaches stagnation, suffer some rarefaction at the interface between the high-density and low density regions formerly associated with the solid shell and the interior gas. The shock waves from all four pulses are chosen so as to have a convergence at the center of the spherical target. The pdV work from these shock waves creates very high temperatures at the central point of the compressed core. The temperatures involved are in the thermonuclear regime, i.e. ion temperatures are above 4.0 keV. A schematic drawing showing the density and temperature of the central core plasma and the dense colder material surrounding it is shown in Fig. 10.38. As the temperature in the core rises into the thermonuclear regime, a thermonuclear ignition burn wave is initiated that travels radially outward. The central, hot-ion core is thus the "spark plug" for ignition of the bulk of the high-density, but relatively cold, fuel outside of the burning core, which is then heated primarily by fusion alpha particles generated in the core.

It is interesting to note that three phenomena are simultaneously changed by achieving a target parameter $\rho R > 1$ (cgs): (1) The Lawson parameter, as discussed in Chap. 3, is satisfied to give a fusion gain $Q > 1$, (2) the alpha particle range ρR_α also is of order unity or greater, ensuring that the fusion energy generated in the central hot burning core is absorbed in the imploded fuel shell nearby, and (3) the density-weighted electron–ion equipartition time $n\tau_{ei}$ scales with the density-

Fig. 10.38 Schematic of density and temperature conditions just as the thermonuclear burn wave is about to start in the NIF ignition scenario

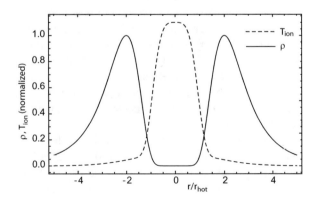

weighted assembly time $\rho R/c_s$, and thus the shock energy and alpha heating energy is shared between the electrons and ions, allowing the ion fusion reactivity to increase to significant levels in the high-density fuel. Thus the "hot-spot ρR" becomes the single most important parameter in describing the success or failure of capsule ignition.

Designing the implosion trajectory is done both with the capsule design and the selection of a power pulse waveform for the laser to deliver energy to the hohlraum. The potentially deleterious effects include (1) asymmetry of the X-ray photon flux on target, (2) effects of nonsymmetric perturbations in the capsule due to its design, including a fill tube connecting the spherical capsule to the cryogenic DT source (think of a golf ball on a tee) and/or the structure used to center the capsule in the hohlraum, (3) limits to compression caused by hydrodynamic instabilities as outlined earlier, and (4) mixing of the fuel and the capsule material (the ablator), also caused by hydrodynamic instabilities. Capsule design affects the density profile during compression and thus the Atwood number driving the hydrodynamic instabilities. The laser power pulse programming affects the time available for the hydrodynamic instabilities to grow, as well as the time for SRS, SBS, and IBT instabilities to grow from laser–plasma interactions. Thus the strategy required for a successfully ignited ICF capsule at NIF is not a trivial endeavor.

10.7 Laser Technology

This section will describe some aspects of laser engineering unique to inertial fusion experiments. It is not intended as a general background in lasers. For that, the reader is referred to general textbooks on the subject such as Verdeyen [75], Koechner [34], or Svelto[71].

10.7.1 Frequency Doubling and Tripling

As shown earlier in this chapter, laser–plasma interaction issues are characterized by the value of $I\lambda^2$, where I is the laser intensity (W m^{-2}) and λ is the wavelength of the incident light. For this reason it is desirable to convert the light output from relatively long-wavelength lasers such as those using Nd-doped glass to higher harmonics. One way to accomplish this is by use of a nonlinear optical element. In fact second harmonic generation was first observed by Franken *et al.* in 1961 [17] using a ruby laser and quartz as the nonlinear element. However, potassium dihydrogen phosphate (KDP) and deuterated KDP (DKDP) have become the materials of choice as predicted from their crystalline structure [33]. KDP has been used since the 1960s as an electro-optical element, especially for Pockels cells, enabling the development of Q-switched lasers. (See below for a description of Pockels cells.). However, use of this material as an efficient frequency doubler at

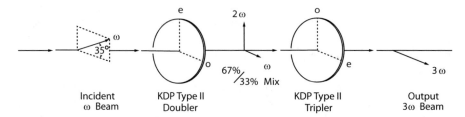

Fig. 10.39 Schematic of 3ω frequency tripling using KDP crystals. From [65]. Used with permission, Elsevier

high power levels did not become feasible until the late 1970s [65]. The general idea behind frequency conversion is the appearance of a nonlinear dielectric polarization tensor such that

$$P'_k = \sum_{i,j} d'_{ijk} E_i E_j, \qquad (10.112)$$

where $\mathbf{P} = \mathbf{P}_L + \mathbf{P}'$ is the overall dielectric polarization, with $\mathbf{P}_L$ representing the linear response. The bilinear form represented by Eq. (10.112) means that sum and difference frequencies are generated by a mixture of light at two different frequencies, and second harmonic generation comes about from the trigonometric expansion of $\sin^2 \omega t = 1/2 - 1/2 \cos 2\omega t$. Third harmonic generation is effected by putting a combination of light at frequencies ω and 2ω through a second nonlinear element, as shown in Fig. 10.39. High efficiency requires careful attention to the orientation of the electric field with respect to the ordinary (o) and extraordinary (e) axes within the crystal. Total conversion efficiencies in excess of 80% are possible with optimized designs.

In order to provide optimum conversion conditions and to avoid optical damage from excessive photon flux, the sizes required for the frequency-conversion crystals at NIF and LMJ are very large, up to $50 \times 50 \times 50$ cm. Development of techniques to grow such large crystals was started at Moscow State University by Rashkovich [55] and brought to Livermore by Zaitseva [81]. Crystal growth rates up to 20 mm per day are now possible, with adequate quality for use in the NIF final optics assembly (FOA). Figure 10.40 shows an example of one of the production NIF frequency conversion crystals.

10.7.2   Deformable Mirrors

Maintaining sufficient amplitude and phase control over the laser beams during amplification, frequency conversion, and subsequent focusing in the large-cross section beams requires the use of adaptive optics. The ultimate spot size of the NIF

Fig. 10.40 Large format
KDP crystal for NIF.
Lawrence Livermore National
Laboratory

Fig. 10.41 Deformable
mirror used on the NIF.
Lawrence Livermore National
laboratory

beams after final focus must be under $100\,\mu m$ in order to assure optimum X-ray
production in the hohlraum. Consequently the $40 \times 40\,cm$ footprint of the unfocused
beams have very tight tolerances on beam quality. At NIF, this is achieved by using
deformable mirrors in a computer-controlled system. Each of the 192 beamlines
has one adjustable mirror with thirty-nine actuators attached to each mirror. These
adjustable mirrors can be programmed to respond to optical distortion elsewhere in
the system caused by thermal effects, optical flatness irregularities, defects in optical
materials, and changes in flashlamp performance due to heat and aging. Figure 10.41
shows one of the NIF deformable mirrors.

10.7.3 Plasma Electrode Pockels Cells

Pockels cells are electro-optical components that can rotate the electric field of the
incident photon beam in the plane of polarization through the application of an
electric field across the crystal, in the direction of beam propagation. The rotation

of the beam electric field is used to prevent optical gain in the active laser medium until the laser amplification is required and the active medium is fully pumped by flashlamps. The basic physical phenomenon was first described by Pockels in 1906 [53] and practical features of KDP as a Pockels cell material were described by Mason in 1946 [45]. Adapting the Pockels cell concept to the large format required for NIF required some ingenuity, since the usual metallic ring or thin-film electrodes are impractical with the flat-slab geometry and the high optical power. A design innovation for NIF was to provide a cover plasma on both sides of the KDP crystal before the voltage pulse is applied, resulting in a high-conductivity conducting sheet on both sides of the crystal which is transparent to the laser light. The resulting plasma electrode Pockels cells (PEPCs) are packed four to a module to accommodate adjacent amplifiers. Figure 10.42 shows a PEPC module used at the National Ignition Facility.

Fig. 10.42 Plasma-electrode Pockels cell module for NIF. Lawrence Livermore National Laboratory

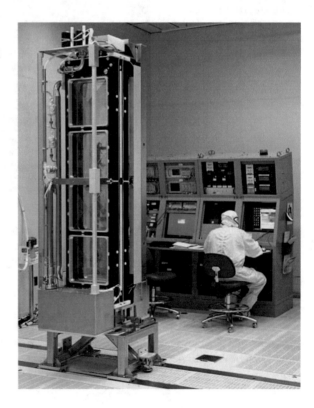

10.8   Advanced Driver Concepts

10.8.1   KrF Lasers

Glass laser technology for inertial fusion requires frequency conversion to enable operation at lower values of $I\lambda^2$ for the required intensity levels. An alternative technology exists utilizing krypton fluoride (KrF) as a lasing medium. This gaseous laser technology has been pursued at the U. S. Naval Research Laboratory for the past 20 years [52]. The concept is an outgrowth of the original Xe_2 excimer laser by Basov et al. [4] and may be called an "exciplex" laser because of the two different atoms (Xe and F) used in the excited complex. Excitation is provided by an electron beam, and the device operates at a wavelength of 248 nm, well below even the tripled-frequency Nd glass laser. A diagram of the current generation of KrF laser is shown in Fig. 10.43. Advantages of the system are simplified phase control for the beams produced, and the inherent ability to operate at reactor-relevant pulse rates ($1 \rightarrow 10$ Hz). Disadvantages are degradation of optical materials due to chemical compatibility issues with the lasing medium (especially if water is present, which can result in etching of the SiO_2 surfaces by HF), and mismatching in the time history of a KrF discharge with the $\sim$ few ns pulse length requirement for the ICF capsule. Since the record energy output is on the order of 700 J, KrF technology has not experienced the scale-up to megajoule-class laser systems as might be needed for an ignition experiment.

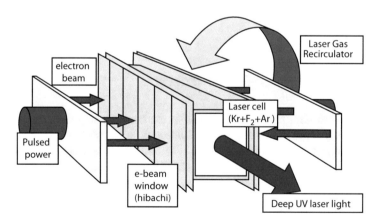

Fig. 10.43   KrF laser design from Naval Research Laboratory. From [52]

10.8.2   Diode-Pumped Solid State Lasers

The current NIF laser configuration is based upon Nd-doped glass slabs which are very large (sixteen $3.4 \times 46 \times 81$ cm slabs for each of the 192 amplifiers). These are pumped with xenon-filled flashlamps. Because of the size and thickness of these slabs, repetition rates are very low, limited by the cooling time of these slabs. Also, flashlamps have a limited lifetime due to thermal and other aging effects, and have a light output spectrum which is too broad to achieve high efficiency operation as a solid state laser pump. Extrapolating laser systems like these to a reactor, where pulse rates of many Hz are required, is impractical at best. Efforts to develop laser technology that could be used at higher repetition rates is a necessary step to get to a laser fusion reactor someday. To this end, several studies have been conducted into alternative solid-state laser technologies. One obvious choice would be to replace the laser flashlamps with solid-state light diodes. (As anyone who has changed a light bulb in the past decade would certainly think about!) Diodes can be designed with fairly narrow wavelength spectra which are optimized for the laser media being used. Laser media other than Nd glass may also be chosen for compatibility with the pump diodes and for higher cooling rates and shorter stress cycles. The Mercury study, conducted at LLNL, is an example of a design and development effort aimed at improving on the flashlamp/Nd glass systems [5]. In this study, prototype lasers were built utilizing solid-state light bars which illuminated ytterbium doped strontium fluorapatite (Yb:S-FAP, or $Yb^{3+}:Sr_5(PO_4)_3F$) crystals, and demonstrated reliable operation at 10 Hz rep-rate. While the energy per pulse was only about 60 J, technologies like this could pave the way towards higher-energy systems in the future.

10.8.3   Heavy Ion Accelerator Drivers

An alternative to laser technology for ICF is the use of heavy ions, accelerated to $3 \rightarrow 10$ GeV energy and focused on a hohlraum target similar to the ones used in indirect drive laser fusion experiments [3]. This approach has been pursued at Lawrence Berkeley National Laboratory (LBNL) as well as at GSI Darmstadt and other places. Figure 10.44 shows a schematic of a potential reactor driver design utilizing heavy ions.

There are several differences in the heavy ion approach compared to laser ICF. For one thing, the hohlraum can be enclosed and completely symmetrical, since the heavy ion beams can pass through the metallic hohlraum wall, depositing their energy near the inside surface due to the Bragg effect. Secondly, since the ions are not moving at the speed of light, the accelerating potentials can be programmed in time to allow for compression of the beams in time, so that the beam currents remain at acceptable levels in the accelerating structures and then peak only as the beams enter the capsule. Pulse power technology for particle beams is highly developed,

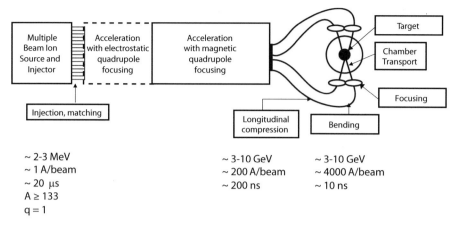

Fig. 10.44 Schematic of a heavy ion fusion accelerator design, giving some key parameters. From [63]. American Physical Society

and systems operating at repetition rates of $1 \rightarrow 10\,$Hz needed for a reactor are not expected to be difficult to design and build. While the final optics assembly (FOA) of laser-based ICF systems is vulnerable to damage by explosive debris, heavy ion-based systems employ an electromagnet for ion beam steering and focusing and are less prone to damage. Also, highly efficient ($> 30\%$ in induction linacs and a few percent in RF-based accelerators) accelerator technology already exists, and the cost of the accelerator scales with the energy as $E^{0.5}$ through the use of multiple beams. The strengths and weaknesses in the heavy ion approach have been described in a National Research Council report in 2013 [51].

While the investment in GeV ion driver technology has not yet been made, demonstration of the "low voltage" (2 MV) injector has been done at LBNL using test stands HCX (for "High Current Experiment") [54] and NDCX (for "Neutralized Drift Compression Experiment"). The upgraded test stand NDCX-II, designed in 2009 [18] is now operating at LBNL [64]. Primary experimental milestones are the control of emittance growth due to electrostatic self-fields and neutral gas interactions and achievement of high time-compression ratios. Future technical issues will include the demonstration of stable operation with multiple beams, control of "halo" in a high repetition rate environment, and control of emittance growth during beam bending at high energies.

Problems

10.1 Suppose that a DT ICF plasma at a density of $1000\times$ liquid density ($\rho_{Liq} = 0.25\,$g cm^{-3}) is exposed to a blackbody radiation spectrum with a photon temperature of 350 eV. Find the fraction of the blackbody radiation spectrum's energy that can pass through the critical density of the target and thus couple to the ablation front occurring at the target's outer surface.

10.2 Consider a planar shock in a medium with a ratio of specific heats $\gamma = 5/3$.

a. Find the compression ratio ρ_2/ρ_1 and pressure ratio p_2/p_1 for a shock with a Mach number $M = 4$ for this material.
b. Find the minimum Mach number M for each of four shocks to pass through the material, leaving the final density compressed to $100\times$ the initial density.

10.3 Backscattered light is observed emanating from a laser-driven depleted uranium ($M = 238$) hohlraum. The drive is from a frequency-tripled neodymium glass laser at $\lambda = 0.35 \ \mu$. It is suspected that the light is produced by laser–plasma interaction (LPI) at a place where the density is $0.22\times$ the critical density for this drive frequency. It is suspected that stimulated Raman scattering (SRS) is involved here.

a. Calculate the electron density for the conditions stated above.
b. Find the free-space wavenumber k_0 for the drive laser light in vacuum, and the wavenumber k_1 for the drive laser light in the plasma.
c. Draw a Stokes diagram for this process, labeling the incoming wave as (ω_1, k_1) and the outgoing EM wave as (ω_2, k_2) and the forward-going daughter wave as (ω_3, k_3). Write down the dispersion relations next to each curve.
d. For an electron temperature of $350\,\text{eV}$, find ω_2, ω_3, k_2, and k_3.
e. Assuming the average charge state $< Z >= 30$ for uranium, find the phase velocity of an ion acoustic wave in these conditions.

10.4 A baseline design for a NIF capsule/hohlraum combination is given below. Note that the "cocktail" hohlraum is a combination of gold ($Z = 79$) and uranium ($Z = 92$). The X-ray production energy is split in this table into the heating of the hohlraum wall ("Wall Loss"), loss through the laser entrance holes ("Hole Loss"), and that remaining which couples into the target capsule. Energies are in MJ.

Laser light	1.0
Absorbed in wall	0.9
X-rays	0.765
Wall loss	0.405
Hole loss	0.195
Capsule	0.165
Efficiency	16.5%

a. Using the published value for hole loss, find the blackbody photon temperature, assuming that the laser pulse is a flat-top pulse of width 3.5 ns. Note that there are two laser entrance holes each of 2.5 mm diameter.

b. Explain why the transfer of energy from the laser to the capsule might be more efficient for the "cocktail" of Au and U rather than for pure Au or pure U hohlraums.

c. This design has a quoted yield estimate of 13 MJ. The total DT fuel mass is 0.161 mg. Find the burnup fraction f_B.

d. Estimate the value of ρR required to achieve this yield, in $\mathrm{g\,cm^{-2}}$.

e. What physics processes might have caused the experimentally observed yields to be much less than the predicted performance for this design?

10.5 Suppose that a hohlraum wall is irradiated with a laser spot with a spot radius r_0 of 20 μm and an intensity consistent with one-twenty-fourth of a 1.8 MJ laser at wavelength 0.35 μm in a 3.5 ns flat-top pulse.

a. Find the peak vacuum electric field E_0, assuming that the spatial footprint of the beam is gaussian with the $1/e$ power profile given by the spot size r_0.

b. Find the quiver velocity of an electron in this field and the dimensionless parameter a_0 comparing this to the speed of light. Would a full relativistic treatment for the electron motion be required, or could one use the perturbation approach outlined in the text?

c. Show how the parameter a_0 scales with the wavelength-intensity product $I\lambda^2$.

References

1. Abrahams, A.M., Shapiro, S.L.: Cold equation of state from Thomas-Fermi-Dirac-Weizsacker theory. Phys. Rev. A **42**, 2530–2538 (1990). https://link.aps.org/doi/10.1103/PhysRevA.42.2530

2. Atzeni, S., Meyer-ter Vehn, J.: The Physics of Inertial Fusion: Beam-Plasma Interaction, Hydrodynamics, Hot Dense Matter. International Series of Monographs on Physics. OUP, Oxford (2004). https://books.google.com/books?id=BJcy_p5pUBsC

3. Bangerter, R.O., Faltens, A., Seidl, P.A.: Accelerators for inertial fusion energy production. Rev. Accel. Sci. Technol. **06**, 85 (2013). https://doi.org/10.1142/S1793626813300053

4. Basov, H., Danilychev, V., Popov, Y., Khodkevich, D.: Laser operating in the vacuum region of the spectrum by excitation of liquid xenon with an electron beam. JETP Lett. **12**, 329 (1970)

5. Bayramian, A., Armstrong, P., Ault, E., Beach, R., Bibeau, C., Caird, J., Campbell, R., Chai, B., Dawson, J., Ebbers, C., Erlandson, A., Fei, Y., Freitas, B., Kent, R., Liao, Z., Ladran, T., Menapace, J., Molander, B., Payne, S., Peterson, N., Randles, M., Schaffers, K., Sutton, S., Tassano, J., Telford, S., Utterback, E.: The mercury project: a high average power, gas-cooled laser for inertial fusion energy development. Fusion Sci. Technol. **52**(3), 383–387 (2007). https://doi.org/10.13182/FST07-A1517

6. Betti, R., Hurricane, O.A.: Inertial-confinement fusion with lasers. Nature Phys. **12**, 435 (2016). http://dx.doi.org/10.1038/nphys3736

7. Bogdanova, Y., Gubin, S., Anikeev, A., Victorov, S.: Dissociation of shock-compressed liquid hydrogen and deuterium. Phys. Procedia **72**(Supplement C), 329–332 (2015). Conference of Physics of Nonequilibrium Atomic Systems and Composites, PNASC 2015, 18–20 February 2015 and Conference of Heterostructures for Microwave, Power and Optoelectronics: Physics, Technology and Devices, 19 February 2015, https://doi.org/10.1016/j.phpro.2015.09.104; http://www.sciencedirect.com/science/article/pii/S1875389215012742

8. Brouillette, M.: The Richtmyer-Meshkov instability. Annu. Rev. Fluid Mech. **34**, 445–468 (2002)
9. Chandrasekhar, S.: Hydrodynamic and Hydromagnetic Stability. Clarendon Press, Oxford (1961)
10. Cox, P.A.: The Electronic Structure and Chemistry of Solids. Oxford Science Publications. Clarendon Press, Oxford (1987)
11. Denisov, N.: On a singularity of the field on an electromagnetic wave propagated in an inhomogeneous plasma. JETP **4**(4), 544 (1957)
12. Dirac, P.A.M.: Note on exchange phenomena in the Thomas atom. Math. Proc. Camb. Philos. Soc. **26**(3), 376–385 (1930). https://doi.org/10.1017/S0305004100016108
13. Drake, R.P.: High-Energy-Density Physics: Fundamentals, Inertial Fusion, and Experimental Astrophysics. Shock Wave and High Pressure Phenomena. Springer, Berlin (2006). https://link.springer.com/book/10.1007%2F3-540-29315-9
14. Dyke, M.V., Guttmann, A.J.: The converging shock wave from a spherical or cylindrical piston. J. Fluid Mech. **120**, 451–462 (1982)
15. Fermi, E.: Thermodynamics. Dover books in Physics and Mathematical Physics. Dover Publications, New York (1956). https://books.google.com/books?id=VEZ1ljsT3IwC
16. Feynman, R.P., Metropolis, N., Teller, E.: Equations of state of elements based on the generalized Fermi-Thomas theory. Phys. Rev. **75**, 1561–1573 (1949). https://link.aps.org/doi/10.1103/PhysRev.75.1561
17. Franken, P.A., Hill, A.E., Peters, C.W., Weinreich, G.: Generation of optical harmonics. Phys. Rev. Lett. **7**, 118–119 (1961). https://link.aps.org/doi/10.1103/PhysRevLett.7.118
18. Friedman, A., Barnard, J., Briggs, R., Davidson, R., Dorf, M., Grote, D., Henestroza, E., Lee, E., Leitner, M., Logan, B., Sefkow, A., Sharp, W., Waldron, W., Welch, D., Yu, S.: Toward a physics design for NDCX-II, an ion accelerator for warm dense matter and hif target physics studies. Nucl. Instrum. Methods Phys. Res. Sect. A Accel. Spectrom. Detect. Assoc. Equip. **606**(1), 6–10 (2009). Heavy Ion Inertial Fusion, https://doi.org/10.1016/j.nima.2009.03.189; http://www.sciencedirect.com/science/article/pii/S0168900209005403
19. Ginzburg, V.: Propagation of Electromagnetic Waves in Plasma. Russian Monographs and Texts on Advanced Mathematics and Physics. Gordon and Breach, Philadelphia (1962). https://books.google.com/books?id=lwYvAAAAIAAJ
20. Glenzer, S.H., MacGowan, B.J., Michel, P., Meezan, N.B., Suter, L.J., Dixit, S.N., Kline, J.L., Kyrala, G.A., Bradley, D.K., Callahan, D.A., Dewald, E.L., Divol, L., Dzenitis, E., Edwards, M.J., Hamza, A.V., Haynam, C.A., Hinkel, D.E., Kalantar, D.H., Kilkenny, J.D., Landen, O.L., Lindl, J.D., LePape, S., Moody, J.D., Nikroo, A., Parham, T., Schneider, M.B., Town, R.P.J., Wegner, P., Widmann, K., Whitman, P., Young, B.K.F., Van Wonterghem, B., Atherton, L.J., Moses, E.I.: Symmetric inertial confinement fusion implosions at ultra-high laser energies. Science **327**(5970), 1228–1231 (2010). https://doi.org/10.1126/science.1185634; http://science.sciencemag.org/content/327/5970/1228
21. Guderley, G.: Starke kugelige und zylindrische verdichtungsstosse in der nahe des kugelmittelpunktes bzw. der zylinderachse. Luftfahrtforschung **19**, 302–312 (1942)
22. Haan, S.W., Lindl, J.D., Callahan, D.A., Clark, D.S., Salmonson, J.D., Hammel, B.A., Atherton, L.J., Cook, R.C., Edwards, M.J., Glenzer, S., Hamza, A.V., Hatchett, S.P., Herrmann, M.C., Hinkel, D.E., Ho, D.D., Huang, H., Jones, O.S., Kline, J., Kyrala, G., Landen, O.L., MacGowan, B.J., Marinak, M.M., Meyerhofer, D.D., Milovich, J.L., Moreno, K.A., Moses, E.I., Munro, D.H., Nikroo, A., Olson, R.E., Peterson, K., Pollaine, S.M., Ralph, J.E., Robey, H.F., Spears, B.K., Springer, P.T., Suter, L.J., Thomas, C.A., Town, R.P., Vesey, R., Weber, S.V., Wilkens, H.L., Wilson, D.C.: Point design targets, specifications, and requirements for the 2010 ignition campaign on the National Ignition Facility. Phys. Plasmas **18**(5), 051001 (2011). https://doi.org/10.1063/1.3592169
23. Haines, M.G.: A review of the dense z-pinch. Plasma Phys. Controll. Fusion **53**, 093001 (2011). https://doi.org/10.1088/0741-3335/53/9/093001
24. Harding, E.C., Hansen, J.F., Hurricane, O.A., Drake, R.P., Robey, H.F., Kuranz, C.C., Remington, B.A., Bono, M.J., Grosskopf, M.J., Gillespie, R.S.: Observation of a Kelvin-

Helmholtz instability in a high-energy-density plasma on the Omega laser. Phys. Rev. Lett. **103**, 045005 (2009). https://link.aps.org/doi/10.1103/PhysRevLett.103.045005

25. Hicks, D.G., Boehly, T.R., Celliers, P.M., Eggert, J.H., Moon, S.J., Meyerhofer, D.D., Collins, G.W.: Laser-driven single shock compression of fluid deuterium from 45 to 220 GPa. Phys. Rev. B **79**, 014112 (2009). https://link.aps.org/doi/10.1103/PhysRevB.79.014112

26. Hu, S.X., Militzer, B., Goncharov, V.N., Skupsky, S.: First-principles equation-of-state table of deuterium for inertial confinement fusion applications. Phys. Rev. B **84**, 224109 (2011). https://link.aps.org/doi/10.1103/PhysRevB.84.224109

27. Ichimaru, S.: Statistical Plasma Physics: Condensed plasmas. Frontiers in Physics. Westview Press, Boulder (2004). https://books.google.com/books?id=T2ssAAAAYAAJ

28. Kato, Y., Mima, K., Miyanaga, N., Arinaga, S., Kitagawa, Y., Nakatsuka, M., Yamanaka, C.: Random phasing of high-power lasers for uniform target acceleration and plasma-instability suppression. Phys. Rev. Lett. **53**, 1057–1060 (1984). https://link.aps.org/doi/10.1103/PhysRevLett.53.1057

29. Kelvin, L.: Hydrokinetic solutions and observations and on the motion of free solids through a liquid. Philos. Mag. **4**, 362 (1871)

30. Kemp, A., Meyer-ter Vehn, J., Atzeni, S.: Stagnation pressure of imploding shells and ignition energy scaling of inertial confinement fusion targets. Phys. Rev. Lett. **86**, 3336–3339 (2001). https://link.aps.org/doi/10.1103/PhysRevLett.86.3336

31. Kerley, G.I.: Equation of state and phase diagram of dense hydrogen. Phys. Earth Planet. Inter. **6**(1), 78–82 (1972). https://doi.org/10.1016/0031-9201(72)90036-2; http://www.sciencedirect.com/science/article/pii/0031920172900362

32. Kirkwood, R.K., Afeyan, B.B., Kruer, W.L., MacGowan, B.J., Moody, J.D., Montgomery, D.S., Pennington, D.M., Weiland, T.L., Wilks, S.C.: Observation of energy transfer between frequency-mismatched laser beams in a large-scale plasma. Phys. Rev. Lett. **76**, 2065–2068 (1996). https://link.aps.org/doi/10.1103/PhysRevLett.76.2065

33. Kleinman, D.A.: Nonlinear dielectric polarization in optical media. Phys. Rev. **126**, 1977–1979 (1962). https://link.aps.org/doi/10.1103/PhysRev.126.1977

34. Koechner, W.: Solid-State Laser Engineering, 5th edn. Springer Series in Optical Sciences. Springer, Berlin (1999). http://www.springer.com/us/book/9780387290942

35. Koonin, S.E., et al.: Final Report: National Academy of Sciences review of the Department of Energy's inertial confinement fusion program. National Academy of Sciences report, pp. 1–44 (1990)

36. Kruer, W.: The Physics of Laser Plasma Interactions. Frontiers in Physics. Addison-Wesley, Boston (1988). https://books.google.com/books?id=csDvAAAAMAAJ

37. Kruer, W.L., Wilks, S.C., Afeyan, B.B., Kirkwood, R.K.: Energy transfer between crossing laser beams. Phys. Plasmas **3**(1), 382–385 (1996). https://doi.org/10.1063/1.871863

38. Langdon, A.B.: Nonlinear inverse bremsstrahlung and heated-electron distributions. Phys. Rev. Lett. **44**, 575–579 (1980). https://link.aps.org/doi/10.1103/PhysRevLett.44.575

39. Lawson, A.: Physics of the Lindemann melting rule. Philos. Mag. **89**(22–24), 1757–1770 (2009). https://doi.org/10.1080/14786430802577916

40. Lewis, H.W.: Fermi-Thomas model with correlations. Phys. Rev. **111**, 1554–1557 (1958). https://link.aps.org/doi/10.1103/PhysRev.111.1554

41. Lindemann, F.A.: The calculation of molecular vibration frequencies. Z. Phys. **11**, 609 (1910)

42. Lindl, J.: Development of the indirect-drive approach to inertial confinement fusion and the target physics basis for ignition and gain. Phys. Plasmas **2**(11), 3933–4024 (1995). http://dx.doi.org/10.1063/1.871025

43. Lindl, J.D., Amendt, P., Berger, R.L., Glendinning, S.G., Glenzer, S.H., Haan, S.W., Kauffman, R.L., Landen, O.L., Suter, L.J.: The physics basis for ignition using indirect-drive targets on the National Ignition Facility. Phys. Plasmas **11**(2), 339–491 (2004). http://dx.doi.org/10.1063/1.1578638

44. Ma, T., Patel, P.K., Izumi, N., Springer, P.T., Key, M.H., Atherton, L.J., Benedetti, L.R., Bradley, D.K., Callahan, D.A., Celliers, P.M., Cerjan, C.J., Clark, D.S., Dewald, E.L., Dixit, S.N., Döppner, T., Edgell, D.H., Epstein, R., Glenn, S., Grim, G., Haan, S.W., Hammel, B.A.,

Hicks, D., Hsing, W.W., Jones, O.S., Khan, S.F., Kilkenny, J.D., Kline, J.L., Kyrala, G.A.,
Landen, O.L., Le Pape, S., MacGowan, B.J., Mackinnon, A.J., MacPhee, A.G., Meezan, N.B.,
Moody, J.D., Pak, A., Parham, T., Park, H.S., Ralph, J.E., Regan, S.P., Remington, B.A., Robey,
H.F., Ross, J.S., Spears, B.K., Smalyuk, V., Suter, L.J., Tommasini, R., Town, R.P., Weber,
S.V., Lindl, J.D., Edwards, M.J., Glenzer, S.H., Moses, E.I.: Onset of hydrodynamic mix in
high-velocity, highly compressed inertial confinement fusion implosions. Phys. Rev. Lett. **111**,
085004 (2013). https://link.aps.org/doi/10.1103/PhysRevLett.111.085004
45. Mason, W.P.: The elastic, piezoelectric, and dielectric constants of potassium dihydrogen
phosphate and ammonium dihydrogen phosphate. Phys. Rev. **69**, 173–194 (1946). https://link.
aps.org/doi/10.1103/PhysRev.69.173
46. Max, C.: Physics of Laser Fusion. Volume I. Theory of the Coronal Plasma in Laser-Fusion
Targets. Lawrence Livermore National Laboratory, Livermore (1981). https://doi.org/10.2172/
5630914; http://www.osti.gov/scitech/servlets/purl/5630914
47. Meshkov, E.E.: Instability of the interface of two gases accelerated by a shock wave. Fluid
Dyn. **4**(5), 101–104 (1969). https://doi.org/10.1007/BF01015969
48. Michel, P., Glenzer, S.H., Divol, L., Bradley, D.K., Callahan, D., Dixit, S., Glenn, S., Hinkel,
D., Kirkwood, R.K., Kline, J.L., Kruer, W.L., Kyrala, G.A., Pape, S.L., Meezan, N.B., Town,
R., Widmann, K., Williams, E.A., MacGowan, B.J., Lindl, J., Suter, L.J.: Symmetry tuning via
controlled crossed-beam energy transfer on the National Ignition Facility. Phys. Plasmas **17**(5),
056305 (2010). http://dx.doi.org/10.1063/1.3325733
49. More, R.M., Warren, K.H., Young, D.A., Zimmerman, G.B.: A new quotidian equation of state
(QEOS) for hot dense matter. Phys. Fluids **31**(10), 3059–3078 (1988). http://aip.scitation.org/
doi/abs/10.1063/1.866963
50. Mott, N., Jones, H.: The Theory of the Properties of Metals and Alloys. Dover Books
on Physics and Mathematical Physics. Dover Publications, New York (1958). https://books.
google.com/books?id=LIPsUaTqUXUC
51. National Research Council: An assessment of the prospects for inertial fusion energy. Technical
report, National Research Council of the National Academies (2013)
52. Obenschain, S., Lehmberg, R., Kehne, D., Hegeler, F., Wolford, M., Sethian, J., Weaver, J.,
Karasik, M.: High-energy krypton fluoride lasers for inertial fusion. Appl. Opt. **54**(31), F103–
F122 (2015). https://doi.org/10.1364/AO.54.00F103; http://ao.osa.org/abstract.cfm?URI=ao-
54-31-F103
53. Pockels, F.: Lehrbuch der Kristalloptik,. B.G. Teubners Sammlung von Lehrbüchern auf
dem Gebiete der mathematischen Wissenschaften mit Einschluss ihrer Anwendungen. B. G.
Teubner, Stuttgart (1906). https://books.google.com/books?id=AkcNAAAAYAAJ
54. Prost, L.R., Seidl, P.A., Bieniosek, F.M., Celata, C.M., Faltens, A., Baca, D., Henestroza, E.,
Kwan, J.W., Leitner, M., Waldron, W.L., Cohen, R., Friedman, A., Grote, D., Lund, S.M.,
Molvik, A.W., Morse, E.: High current transport experiment for heavy ion inertial fusion. Phys.
Rev. ST Accel. Beams **8**, 020101 (2005). https://link.aps.org/doi/10.1103/PhysRevSTAB.8.
020101
55. Rashkovich, L.N.: Rapid growth from solution of large crystals for nonlinear optics. Vestn.
Akad Nauk SSSR **9**, 15–19 (1984)
56. Rayleigh, L.: Investigation of the character of the equilibrium of an incompressible heavy fluid
of variable density. Proc. Lond. Math. Soc. **14**, 170 (1883)
57. Regan, S.P., Epstein, R., Hammel, B.A., Suter, L.J., Scott, H.A., Barrios, M.A., Bradley, D.K.,
Callahan, D.A., Cerjan, C., Collins, G.W., Dixit, S.N., Döppner, T., Edwards, M.J., Farley,
D.R., Fournier, K.B., Glenn, S., Glenzer, S.H., Golovkin, I.E., Haan, S.W., Hamza, A., Hicks,
D.G., Izumi, N., Jones, O.S., Kilkenny, J.D., Kline, J.L., Kyrala, G.A., Landen, O.L., Ma, T.,
MacFarlane, J.J., MacKinnon, A.J., Mancini, R.C., McCrory, R.L., Meezan, N.B., Meyerhofer,
D.D., Nikroo, A., Park, H.S., Ralph, J., Remington, B.A., Sangster, T.C., Smalyuk, V.A.,
Springer, P.T., Town, R.P.J.: Hot-spot mix in ignition-scale inertial confinement fusion targets.
Phys. Rev. Lett. **111**, 045001 (2013). https://link.aps.org/doi/10.1103/PhysRevLett.111.045001
58. Reif, F.: Fundamentals of Statistical and Thermal Physics. McGraw-Hill Series in Funda-
mentals of Physics. McGraw-Hill, New York (1965). https://books.google.com/books?id=
w5dRAAAAMAAJ

59. Richtmyer, R.D.: Taylor instability in shock acceleration of compressible fluids. Commun. Pure Appl. Math. **13**(2), 297–319 (1960). http://dx.doi.org/10.1002/cpa.3160130207

60. Sagdeev, R., Galeev, A.: Nonlinear Plasma Theory [by] R.Z. Sagdeev and A.A. Galeev: Rev. and Edited by T.M. O'Neil [and] D.L. Book. Frontiers in Physics. W.A. Benjamin, Los Angeles (1969). https://books.google.com/books?id=5gRlNwAACAAJ

61. Salpeter, E.E., Zapolsky, H.S.: Theoretical high-pressure equations of state including correlation energy. Phys. Rev. **158**, 876–886 (1967). https://link.aps.org/doi/10.1103/PhysRev.158.876

62. Sangster, T., McCrory, R., Goncharov, V., Harding, D., Loucks, S., McKenty, P., Meyerhofer, D., Skupsky, S., Yaakobi, B., MacGowan, B., Atherton, L., Hammel, B., Lindl, J., Moses, E., Porter, J., Cuneo, M., Matzen, M., Barnes, C., Fernandez, J., Wilson, D., Kilkenny, J., Bernat, T., Nikroo, A., Logan, B., Yu, S., Petrasso, R., Sethian, J., Obenschain, S.: Overview of inertial fusion research in the United States. Nucl. Fusion **47**(10), S686 (2007). http://stacks.iop.org/0029-5515/47/i=10/a=S17

63. Seidl, P.A., Barnard, J.J., Faltens, A., Friedman, A.: Research and development toward heavy ion driven inertial fusion energy. Phys. Rev. ST Accel. Beams **16**, 024701 (2013). https://link.aps.org/doi/10.1103/PhysRevSTAB.16.024701

64. Seidl, P., Barnard, J., Feinberg, E., Friedman, A., Gilson, E., Grote, D., Ji, Q., Kaganovich, I., Ludewigt, B., Persaud, A., et al.: Irradiation of materials with short, intense ion pulses at NDCX-II. Laser Part. Beams **35**(2), 373–378 (2017). https://doi.org/10.1017/S0263034617000295

65. Seka, W., Jacobs, S., Rizzo, J., Boni, R., Craxton, R.: Demonstration of high efficiency third harmonic conversion of high power Nd-glass laser radiation. Opt. Commun. **34**(3), 469–473 (1980). https://doi.org/10.1016/0030-4018(80)90419-8; http://www.sciencedirect.com/science/article/pii/0030401880904198

66. Siemon, R., Lindemuth, I., Schoenberg, K.: Why MTF is a low cost path to fusion. Comment. Plasma Phys. Controll. Fusion **18**(6), 363–386 (1999)

67. Silin, V.P.: Nonlinear high-frequency plasma conductivity. Sov. Phys.-JETP **20**, 1510 (1965)

68. Slater, J.C., Krutter, H.M.: The Thomas-Fermi method for metals. Phys. Rev. **47**, 559–568 (1935). https://link.aps.org/doi/10.1103/PhysRev.47.559

69. Smalyuk, V.A., Tipton, R.E., Pino, J.E., Casey, D.T., Grim, G.P., Remington, B.A., Rowley, D.P., Weber, S.V., Barrios, M., Benedetti, L.R., Bleuel, D.L., Bradley, D.K., Caggiano, J.A., Callahan, D.A., Cerjan, C.J., Clark, D.S., Edgell, D.H., Edwards, M.J., Frenje, J.A., Gatu-Johnson, M., Glebov, V.Y., Glenn, S., Haan, S.W., Hamza, A., Hatarik, R., Hsing, W.W., Izumi, N., Khan, S., Kilkenny, J.D., Kline, J., Knauer, J., Landen, O.L., Ma, T., McNaney, J.M., Mintz, M., Moore, A., Nikroo, A., Pak, A., Parham, T., Petrasso, R., Sayre, D.B., Schneider, M.B., Tommasini, R., Town, R.P., Widmann, K., Wilson, D.C., Yeamans, C.B.: Measurements of an ablator-gas atomic mix in indirectly driven implosions at the National Ignition Facility. Phys. Rev. Lett. **112**, 025002 (2014). https://link.aps.org/doi/10.1103/PhysRevLett.112.025002

70. Strozzi, D.J., Bailey, D.S., Michel, P., Divol, L., Sepke, S.M., Kerbel, G.D., Thomas, C.A., Ralph, J.E., Moody, J.D., Schneider, M.B.: Interplay of laser-plasma interactions and inertial fusion hydrodynamics. Phys. Rev. Lett. **118**, 025002 (2017). https://link.aps.org/doi/10.1103/PhysRevLett.118.025002

71. Svelto, O., Hanna, D.C.: Principles of Lasers. Springer, Berlin (2010). https://doi.org/10.1007/978-1-4419-1302-9; http://www.springer.com/us/book/9781441913012

72. Tabak, M., Hammer, J., Glinsky, M.E., Kruer, W.L., Wilks, S.C., Woodworth, J., Campbell, E.M., Perry, M.D., Mason, R.J.: Ignition and high gain with ultrapowerful lasers. Phys. Plasmas **1**(5), 1626–1634 (1994). https://doi.org/10.1063/1.870664

73. Taylor, G.I.: The instability of liquid surfaces when accelerated in a direction perpendicular to their planes. I. Proc. R. Soc. Lond. A Math. Phys. Eng. Sci. **201**(1065), 192–196 (1950). https://doi.org/10.1098/rspa.1950.0052; http://rspa.royalsocietypublishing.org/content/201/1065/192

74. Vandenboomgaerde, M., Bonnefille, M., Gauthier, P.: The Kelvin-Helmholtz instability in National Ignition Facility hohlraums as a source of gold-gas mixing. Phys. Plasmas **23**(5), 052704 (2016). https://doi.org/10.1063/1.4948468

75. Verdeyen, J.T.: Laser Electronics, 2nd edn. Prentice Hall, Englewood Cliffs (1989)
76. von Helmholtz, H.: Ueber discontinuirliche flüssigkeitsbewegungen. Mon. Rep. R. Prussian Acad. Philos. Berl. **233**, 215 (1868)
77. Wegner, P.J., Auerbach, J.M., Biesiada Jr., T.A., Dixit, S.N., Lawson, J.K., Menapace, J.A., Parham, T.G., Swift, D.W., Whitman, P.K., Williams, W.H.: NIF final optics system: frequency conversion and beam conditioning. In: Lane, M.A., Wuest, C.R. (eds.) Optical Engineering at the Lawrence Livermore National Laboratory II: The National Ignition Facility. Proceedings of the SPIE, vol. 5341, pp. 180–189 (2004). https://doi.org/10.1117/12.538481
78. Weir, S.T., Mitchell, A.C., Nellis, W.J.: Metallization of fluid molecular hydrogen at 140 GPa (1.4 Mbar). Phys. Rev. Lett. **76**, 1860–1863 (1996). https://link.aps.org/doi/10.1103/PhysRevLett.76.1860
79. Wigner, E., Huntington, H.B.: On the possibility of a metallic modification of hydrogen. J. Chem. Phys. **3**, 764–770 (1935). https://doi.org/10.1063/1.1749590
80. Zaghloul, M.R.: Dissociation and ionization equilibria of deuterium fluid over a wide range of temperatures and densities. Phys. Plasmas **22**(6), 062701 (2015). https://doi.org/10.1063/1.4921738
81. Zaitseva, N., Carman, L.: Rapid growth of KDP-type crystals. Prog. Cryst. Growth Charact. Mater. **43**(1), 1–118 (2001). https://doi.org/10.1016/S0960-8974(01)00004-3; http://www.sciencedirect.com/science/article/pii/S0960897401000043
82. Zel'dovich, Y.B., Raizer, Y.P.: Physics of Shock Waves and High-Temperature Hydrodynamic Phenomena. Academic Press, New York (1967)

Chapter 11
Fusion Technology

11.1 Materials Issues in Magnetic Fusion Reactors

11.1.1 Sources of Stress

Stress, and especially cyclic stress, is an important issue in fusion reactor wall design [22]. One may wonder why stress is such an important issue, since there are typically few moving parts in the reactor. The stresses on the first wall come about from three major sources: (1) thermal stress, (2) stress induced by the pressurized coolant, and (3) stress due to transient magnetic forces on the blanket structure. These will be addressed here.

Thermal Stress

The simple process of sending heat through a material induces stress. The stress arises from the propensity of a material to expand when heated. We can assign a volumetric coefficient α so that

$$\frac{\Delta V}{V} = \alpha \Delta T. \tag{11.1}$$

We note that the temperature rise ΔT is not uniform across a material passing heat but can be related to the thermal conductivity k such that, for a flat plate of thickness t passing a heat flux q'' and having an internal heating rate q'''

$$\Delta T = \frac{q'' t + q''' t^2 / 2}{k}. \tag{11.2}$$

© Springer Nature Switzerland AG 2018
E. Morse, *Nuclear Fusion*, Graduate Texts in Physics,
https://doi.org/10.1007/978-3-319-98171-0_11

The material must respond to the volumetric change ΔV given by Eq. (11.1) by generating a stress to cancel the expansion of the material, if it is constrained in one or more directions. For a flat plate constrained in both dimensions perpendicular to the thickness direction, the compensating stress is given by

$$\sigma_{th} = \pm \frac{1}{2} \frac{\alpha E}{1 - \nu} \frac{\Delta V}{V},$$ (11.3)

where ν is Poisson's ratio (typically around 0.3 for metals) and E is the modulus of elasticity for the material. Putting these equations together, one has an equation for the thermal stress given by [29]

$$\sigma_{th} = \pm \frac{1}{2} \frac{\alpha E}{k(1 - \nu)} \left(q'' t + q''' \frac{t^2}{2} \right).$$ (11.4)

One can see that an overall figure of merit for a given material from the standpoint of thermal stress is given by the ratio of the yield stress σ / σ_y to the thermal stress σ_{th} for some fixed value of heat flux, and ignoring the internal heating rate, which is a small contribution when the thickness t is small. This figure of merit M can be written as

$$M = \frac{2k(1 - \nu)\sigma_y}{\alpha E}.$$ (11.5)

Coolant Pressure and Stress

The coolant moving through a fusion reactor blanket must be pushed through the reactor blanket with a certain amount of pressure behind it. The pressure head on the coolant must be large enough to overcome the drag on the fluid due to viscosity and/or turbulence, as well as magnetic forces on the coolant if it is electrically conductive. In both cases, the forces on the fluid are proportional to its velocity (or to the velocity squared in the case of turbulent flow). Using helium gas as a coolant, for example, ducks the electromagnetically induced pressure drop but requires large velocities because of the low heat capacity per unit volume. Once the characteristic length L for the coolant path through the reactor is known, the coolant average velocity can be calculated from the desired temperature increase ΔT for the coolant in one pass through the blanket. The temperature rise ΔT is in turn determined by the thermodynamics of the energy conversion cycle and the maximum allowable temperature in the blanket materials, piping, and turbine blades. For example, nuclear steam cycles in fission reactors typically have an outlet temperature from the heat exchanger of around $330\,^{\circ}\text{C}$ and an inlet temperature of around $300\,^{\circ}\text{C}$, for a temperature rise of about $30\,^{\circ}\text{C}$. Keeping this same ΔT for a fusion blanket allows one to solve for the total mass flow:

$$\dot{M} = \frac{P_{th}}{c_p \Delta T}.$$ (11.6)

The mass flow $\dot{M}$ can be written in terms of the blanket dimensions L and $A = V/L$, the characteristic length and the summed area of the coolant channels (with blanket coolant volume V and the velocity v) as

$$\dot{M} = \rho v A = \rho v V/L \qquad (11.7)$$

and thus

$$v = \frac{P_{th} L}{\rho c_p \Delta T V}. \qquad (11.8)$$

An example calculation of the required coolant velocity is given in the following box.

Suppose that a fusion reactor blanket for a tokamak producing 3000 MWth is a shell between two tori with a common major radius $R = 8.0$ m, with the inner boundary at $a_1 = 8/3$ m and $a_2 = 8/3 + 1$ m with elongation $\kappa = 1.65$. Suppose that the coolant path length is 30.0 m and the coolant is lithium at 400 °C. Take the density as $\rho = 450$ kg m^{-3} and the specific heat as $c_p = 4.2 \times 10^3$ J kg^{-1} °C. What is the coolant velocity?

First find the volume.

$$V = 2\pi^2 \kappa R(a_2^2 - a_1^2) = 1650 \, \text{m}^3$$

Then the coolant velocity is

$$v = \frac{P_{th} L}{\rho c_p \Delta T V} = \boxed{0.96 \, \text{m s}^{-1}}$$

Now we look at the MHD pressure acting on this coolant. To do this, we must consider whether MHD forces dominate over viscous forces and this is determined by a calculation of the Hartmann number

$$\text{Ha} = B_\perp d(\sigma/\mu)^{1/2}, \qquad (11.9)$$

with viscosity μ, electrical conductivity σ, and effective diameter d of the coolant channel. For large values of Ha, MHD forces dominate. With the example given above, with $\mu = 2.5 \times 10^{-4}$ Pa s and $\sigma = 2 \times 10^6 \, \Omega^{-1} \, \text{m}^{-1}$, and taking $d = 0.5$ m, this gives Ha=45,000, well within the MHD regime called Hartmann flow. Hartmann flow features a flat-top velocity profile across the channel except for a thin boundary layer near the edge of the channel, justifying the average-velocity model given earlier. The pressure drop per unit length is determined by the magnetic field perpendicular to the channel, the conductivity of the coolant, its velocity, and the ratio of the resistivity across the coolant channel with the resistivity in the wall, which forms a return current for the overall electric circuit. This conductivity ratio is given by

$$C = \frac{2\sigma_w t_w}{\sigma_{Li} d}. \tag{11.10}$$

For a 5 mm thick Mo wall ($\sigma = 14.3 \times 10^6\,\Omega^{-1}\,m^{-1}$), this gives $C = 0.143$. The MHD pressure drop is given by

$$\frac{dP}{dz} = \sigma B_\perp^2 v \frac{C}{1+C}. \tag{11.11}$$

Calculate the MHD pressure drop and the required pumping power for the Li-cooled blanket structure with Mo walls with $t = 5$ mm.

Note that $\Delta p = (dp/dz)L$. Then using Eq. (11.11) gives a pressure drop of 240 kPa m^{-1} for this coolant/wall combination. A thirty-meter path through the reactor would thus have a pressure drop of $\boxed{7.2\,\text{MPa}}$ and a total pumping power of

$$P_{pump} = \dot{V}\Delta p = \frac{\dot{M}}{\rho}\Delta p,$$

which gives a total pumping power of $\boxed{382\,\text{MW}}$, more than a third of the reactor's electrical output! More judicious choices for coolant, wall materials, piping organization (to avoid perpendicular magnetic fields across the channel), and wall thickness between the channels can reduce this number by orders of magnitude.

Stress Due to Plasma Displacements

Plasma displacements are bulk plasma instabilities which can result in large mechanical stresses and heat loads on the surrounding plasma-facing components. Two types of displacement events have been seen in large tokamak experiments. The first of these is a major disruption (MD). Major disruptions are often viewed as extreme amplitude tilting-mode type MHD instabilities in which the conducting wall surrounding the plasma becomes part of the current path. In the JET experiment at Culham, these events have been observed to generate forces in excess of 1.0 MN and to distort the machine by some millimeters [12]. The resultant torquing of the machine can excite the resonant mechanical modes of the structure, which in the case of JET has been seen as a damped vibration with a frequency of around 14 Hz.

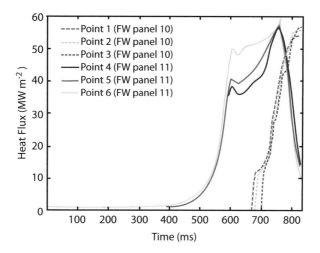

Fig. 11.1 Heat flux calculation for vertical displacement events (VDEs) in ITER. From [85]. Used with permission, Elsevier

While the stress analyses performed to simulate these events do not show that stress limits on the materials involved may be exceeded, there has been concern that a large number of these events (more than 1000) may cause metal fatigue, especially on welds around ports and the like.

The second type of event does not show the tilting behavior, but is more a sort of lack of vertical equilibrium control. These are called vertical displacement events (VDEs). While the VDEs do not in general generate large electromagnetic forces, they can generate large heat fluxes as plasma thermal energy can be transported to the walls and divertor structures. A thermal analysis of a simulated VDE on ITER is shown as Fig. 11.1. Note that the short timescale associated with a VDE may cause transient stress above that associated with a steady-state heat flux of the same magnitude, and again may lead to fatigue failure under repetition, like the MD events.

11.1.2 Fatigue

Important for fusion reactor materials is the ability of the material to avoid failure by fatigue. Fatigue in materials is determined by the number of cycles of stress/strain that the material can endure before failure, as a function of the applied stress. The applied stress may be of three types: (1) axial stress, i.e. simple compression and tension, (2) flexural stress, i.e. bending, and (3) torsional stress, i.e. twisting. Fatigue failure is typically related to the applied stress, which is a fraction of the one-cycle stress to failure. In many instances the mean number of cycles to failure N_f is related to the normalized stress through a simple power law suggested by Weibull [96]:

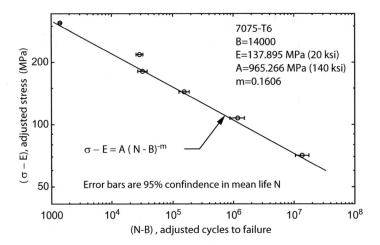

Fig. 11.2 Stress vs. number of cycles to failure from experimental data, with both axes shifted to fit an empirical model by Weibull[96]. Data from [84]

$$\frac{\sigma - E}{A} = \left(\frac{1}{N_f - B}\right)^m, \tag{11.12}$$

with exponent m typically small, $0.1 < m < 0.2$. A classic study by Sinclair and Dolan [84] on 7075-T6 aluminum is shown here in Fig. 11.2. In this and other studies, the statistical distribution of cycles to failure N vs. stress becomes rather broad as the stress is reduced and N increases. Reference [84] found that the distribution of the number of cycles to failure was modeled fairly well by a log-normal distribution, that is, the failure probability P had a gaussian distribution in $\log N'$ with a standard deviation in $\log N$, σ_{logN} inversely proportional to a fairly large exponent of the applied stress σ:

$$\sigma_{logN} = \frac{C}{\sigma^r}, \tag{11.13}$$

and the exponent r was found to be 3.19 in the 7075-T6 study in Ref. [84]. The N vs. σ curves thus have a broad probability distribution at low stress, as shown in Fig. 11.3.

11.1.3   Fracture Toughness

The concept of fatigue is closely allied with the ideas behind cracks and crack growth. The general idea is that some surface flaw, microscopic or otherwise, will grow with the application of stress. A conceptual drawing of a crack of type I is

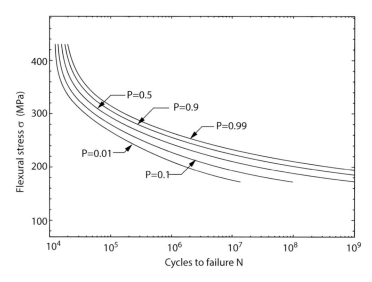

Fig. 11.3 Fatigue lifetime vs. applied stress, for 7075-T6 aluminum, showing curves of probability using a Weibull distribution for the number of cycles to failure. Data from [84]

shown in Fig. 11.4. Here the stress σ is normal to the plane of the crack, which is modeled as an ellipse with major axis a, considered to be large compared to the crack opening. An analysis of the elastic energy released by the crack, from linear elasticity theory, is given by Griffith [37]:

$$\Delta U_{vol} = -\frac{\pi \sigma^2 a^2 \delta}{2E}, \tag{11.14}$$

where E is the elastic modulus for the material. The surface energy added by the crack is given by the surface area times the surface energy γ_{surf}, hence $\Delta U_{surf} = 2a\delta\gamma_{surf}$. The crack is unstable, i.e. will grow, when

$$\frac{\partial}{\partial a}\left(\Delta U_{vol} + \Delta U_{surf}\right) = -\frac{\pi \sigma^2 a}{E} + 2\gamma_{surf} < 0 \qquad \rightarrow \qquad \sigma \geq \sqrt{\frac{2E\gamma_{surf}}{\pi a}}. \tag{11.15}$$

Further analysis of crack growth in materials suggested that the estimates of the critical stress were too low for the cases where the surface tension was known by other means, and this was considered to be due to the use of a linear elasticity model for the process. An additional term was added by Irwin [47] reflecting the additional energy from plastic flow:

$$\sigma \sqrt{\pi a} \geq \sqrt{2E(\gamma_{surf} + \gamma_{ps})} \equiv K_{Ic}. \tag{11.16}$$

Fig. 11.4 Model of type-I
crack for Griffith crack
propagation theory

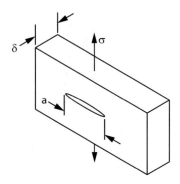

(More general models can be applied from elasticity theory to arrive at the
parameter K_{Ic} for a variety of crack geometry models. A general expression is
given by

$$K_{Ic} = \sigma_c \sqrt{\pi a} F(a/b) \tag{11.17}$$

where σ_c is the fracture stress, a is the crack length (notch depth), and $F(a/b)$ is a
dimensionless shape factor dependent on sample geometry [21].)

The quantity K_{Ic} is thus the overall figure of merit for the resistance of a material
to cracking, and is referred to as the fracture toughness parameter. A related quantity,
the J-integral, describes the rate of energy release from a crack per unit fracture
area. The relation between the J-integral and the fracture toughness parameter K_{Ic}
depends on the type of crack. For type I cracks,

$$J_{Ic} = K_{Ic}^2 \left(\frac{1 - \nu^2}{E} \right) \tag{11.18}$$

where E is the modulus of elasticity and ν is Poisson's ratio.

While there has been limited experience with crack growth and fatigue in 14 MeV
fusion neutron environments, there is a fair amount of experience with fracture
behavior in stainless steels with fast fission neutron spectra. Figure 11.5 shows
the range of values obtained from many observations in 304 and 316 stainless
steels, showing the reduction of fracture toughness with neutron fluence. It is also
interesting to note that "advanced" materials, i.e. newer alloys with high mechanical
strength and better high-temperature properties, often have lower values of K_{Ic} and
are thus more susceptible to failure by brittle fatigue, as shown in Fig. 11.6.

Fig. 11.5 Fracture toughness J-integral for stainless steel in fast fission reactor environments. Dashed lines show upper and lower recommended limits. From [16], U. S. Nuclear Regulatory Commission

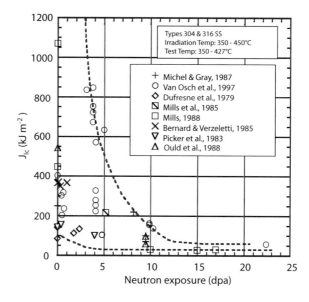

Fig. 11.6 Comparison of yield strength vs. K_{Ic} for common alloys and "advanced" alloys under consideration for fusion reactor structural materials, From [102]. Used by permission, Elsevier

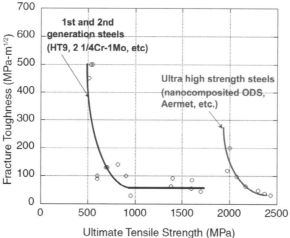

11.1.4 Radiation Damage: History and Theory

Probably the most significant threat to the feasibility of a fusion-based energy economy, assuming that the physics issues do not prevent a fusion reactor of economical size with adequate energy gain, is the lifetime of the materials with which the reactor is constructed. Of special importance in the magnetic fusion

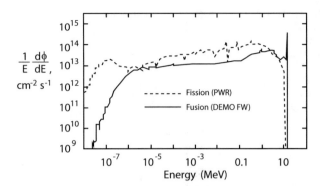

Fig. 11.7 Neutron Flux per unit lethargy, $1/E d\phi/dE$, for a typical fission reactor (pressurized water reactor type) and for the first wall of the DEMO reactor. Data from [36]

approach is the lifetime of the first wall to radiation damage. For the case of D-T fusion fuel, which is the obvious first choice for fueling because of its high reactivity, the bulk of the energy release comes from a fast neutron with 14.08 MeV energy. The fuel in a fusion reactor, in both inertial and magnetic fusion schemes, is fairly transparent to these neutrons and thus they emerge from the burning plasma with essentially full energy. Figure 11.7 shows a representative neutron spectrum in the blanket of a fusion reactor. One notices the presence of much higher energy neutrons than in a fission reactor, so that experience with radiation damage in fission reactors, especially light water-moderated reactors, is not directly applicable to fusion reactor radiation damage assessment.

An advantage of the inertial fusion approach is that ultra-high vacuum conditions are not required inside the chamber containing the fusion fuel, and the first surface can be (mostly) covered with a liquid metallic coolant such as lithium or a lithium-lead eutectic, or even with a molten salt such as FLiBe; however, the vapor pressure of these liquids is too high to consider in a magnetically confined fusion reactor's plasma chamber. Therefore magnetic fusion reactor conceptual design studies have usually featured a metallic or ceramic first wall. The choice of a first wall material then begs the question of how well it can survive the high flux of the neutrons coming out of the burning plasma.

With the caveat given above, a great deal of study of the nature of radiation damage has come out of the seven decades of experience with fission reactors, and the concept of radiation damage in general appears as a literary device in a 1909 novel by Wells [97]. The general idea behind models of neutron-induced radiation damage is that the interaction of fast neutrons with an atomic nucleus in a crystalline material can transfer enough energy to the atom that it can leave its lattice location. A typical displacement energy E_d for this to occur is on the order of $30 \rightarrow 50$ eV. The energy transferred to a nucleus with a mass of Am_p, where A is the atomic mass number and m_p is the proton mass can be derived from a simple calculation, assuming that an elastic collision takes place with a neutron in a head-on collision:

$$\left.\frac{E_{recoil}}{E_n}\right|_{max} = \frac{4A}{(A+1)^2} \equiv \Lambda. \qquad (11.19)$$

For example, for molybdenum with $A \approx 100$, Eq. (11.19) simplifies to $E_{recoil}(max) \approx (4/A)E_n \approx 560\,\text{keV}$ for $E_n = 14\,\text{MeV}$ and this is far above the minimum energy required for the atom to leave is location in the metal lattice. For arbitrary collision angles, we first consider collisions in the center-of-mass (CM) frame, with the angle θ being defined as the deflection angle of the neutron in the CM frame. Then Eq. (11.19) is modified to read

$$\frac{E_{recoil}}{E_n} = \frac{\Lambda}{2}(1 - \cos\theta), \tag{11.20}$$

with maximum energy transfer occurring when $\cos\theta = -1$. A relation between the CM angle θ and the angle of the recoiling atom ϕ_{recoil} in the lab frame is given by

$$\tan\phi_{recoil} = \frac{\sin\theta}{1 - \cos\theta}, \tag{11.21}$$

with $\phi_{recoil} = \pi/2$ for a grazing collision and $\phi_{recoil} = 0$ for a head-on collision, i.e. the recoil always travels forward in the lab frame.

Since the majority of the collisions will leave the recoiling atom, now referred to as the "primary knock-on atom," or PKA, with energy well above the displacement energy E_d, the PKA can also dislodge other lattice atoms as it collides with them as it travels through the lattice. This is important because although energy transfer for the neutron–PKA interaction is limited at a ratio $\approx 4/A$, the PKA is in the same "weight class" as the other lattice atoms and thus the PKA can impart zero to 100% of its energy to another lattice atom. Not all of the PKA's energy loss will go directly to elastic collisions with other lattice atoms, however, as there is electronic stopping to be considered as well. A "displacement cascade" is formed by the transfer of energy from the PKA to subsequent recoiling atoms, with two additional factors reducing the efficiency of this process from the simple billiard ball mechanics. The first factor κ is a factor determined by the efficiency by which the PKA travels through the lattice, with some losses going to lattice vibrations. A proposed model by Norgett, Robinson, and Torrens [72] (now called the "NRT model") uses a constant value of $\kappa = 0.8$. The second factor is known as the Lindhard factor $\xi_L(E)$ [58, 89] and it accounts for the energy loss to electronic stopping, which does not contribute to the radiation damage caused by atomic displacements. Ultimately the recoil energy of the atoms at the last stages of the cascade will be less than the displacement energy E_d, and further energy loss will be dissipated by phonons traveling through the material, i.e. as heat. Taking into account this energy threshold at the end of the cascade, we can write down an expression for the overall number of displacements $\nu(E)$ as a function of the initial PKA energy E as

$$\nu(E) = \kappa\frac{E}{2E_d}\xi_L(E), \tag{11.22}$$

assuming that the energy of the PKA $E > E_d$ to begin with. An approximate form for the Lindhard factor, for collisions between projectiles and lattice atoms both with atomic number Z and mass A, is given by [58, 59]:

$$\xi(E) = \frac{1}{1 + 0.1337 Z^{1/6} (Z/A)^{1/2} \left(3.4008\epsilon^{1/6} + 0.40244\epsilon^{3/4} + \epsilon\right)}. \tag{11.23}$$

Here the Thomas-Fermi reduced energy ϵ is given by

$$\epsilon = \frac{E}{2Z^2 e^2/a}, \tag{11.24}$$

with $e^2 = 14.4 \, \text{eV-Å}$ and a given in terms of the Bohr radius $a_B = 0.529177 \, \text{Å}$ and

$$a = \frac{(3\pi)^{2/3}}{4\sqrt[3]{2}} \frac{a_B}{Z^{1/3}} \approx 0.885 \frac{a_B}{Z^{1/3}} \tag{11.25}$$

and a_B is the Bohr radius ($0.529 \, \text{Å}$). A graph of the Lindhard damage efficiency function $\xi_L(E)$ as a function of the PKA energy E is shown as Fig. 11.8. The Lindhard model does not account directly for inelastic excitation of electrons in the material, however, and competition with electronic stopping is expected to increase when the PKA energy is over a certain threshold value E_c such that valence electrons can be knocked out of lattice sites by direct kinematic collisions with the PKAs. Two estimates have been given in the literature for this excitation energy. One estimate, given in a 1955 paper by Snyder and Neufeld [86], takes the limit as being when the PKA is moving at a velocity higher than the classical electron Bohr velocity αc, with $\alpha = e^2/(\hbar c) = 1/137.04$. This happens when

$$E > E_c = \frac{M}{m_e} \times 13.6 \, \text{eV}. \tag{11.26}$$

A lower estimate is taken where the energy imparted to an electron by a PKA is greater than the average energy I required to promote an electron from the valence band to the conduction band in a solid, which is around $2.4 \, \text{eV}$. Since the energy transfer to an electron from a head-on collision with a PKA atom of mass M is $M/(4m_e)E$, this limit becomes

$$E > E_c = \frac{M}{4m_e} I \, \text{eV}. \tag{11.27}$$

This energy estimate is about a factor of twenty lower than the Snyder-Neufeld estimate and will result in lower estimates of ξ_L if it is employed. However, making the assumption that all energy transfer is to electrons for $E > E_c$ is also flawed, so the actual situation may not be as drastically affected by these two choices. In any case, the PKA energy for a 14 MeV fusion neutron for one of the higher atomic

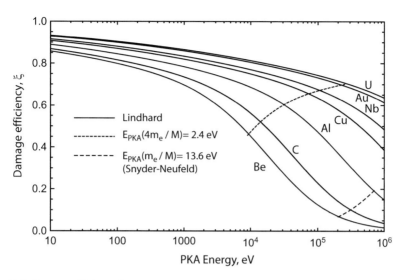

Fig. 11.8 Damage efficiency factor ξ vs. PKA energy for various crystalline materials. Limits are given for two upper bounds on the assumption of elastic scattering for the electronic stopping. After [81]. See also [95]

number alloys such as Mo ($A = 100$) is around 280 keV and not much above the more stringent estimate given by Eq. (11.27), so that use of the naive Lindhard estimate will result in only a slight overestimate of the damage function.

The number of displaced atoms, divided by the total number of atoms (the displacements per atom (dpa)) per primary neutron can then be turned into an equivalent damage cross section, the "dpa cross section," by convolving the energy spectrum of the PKAs produced with the Lindhard factor using the energy–angle relationship given above, and applying the Kinchin-Pease statistical factor $E/(2E_d)$ for the resulting PKA energy E. The overall damage cross section is defined so that the displacements per atom (dpa) are given by

$$\text{dpa} = \sigma_{dpa}(E_n)\phi(E_n)t, \tag{11.28}$$

where $\phi(E_n)t$ is the neutron fluence for a flux of monoenergetic neutrons with energy E_n. For elastic collisions as described above, the dpa cross section, including the angular dependence of the cross section, then becomes

$$\sigma_{dpa} = 2\pi \int_{E_d}^{\Lambda E_n} dE \left(\frac{d\sigma_{el}}{d\Omega}\right) \frac{\Lambda}{2}(1 - \cos\theta)\frac{E_n}{2E_d}\kappa\xi_L(E), \tag{11.29}$$

where the dpa energy $E = (\lambda/2)(1 - \cos\theta)$ from kinematics.

This simplified approach to the displacement calculation is called the Kinchin-Pease theory [53], and the resulting modifications by Lindhard et al. and later by Norgett et al. resulting in a sort of "unified" NRT theory. Several refinements need

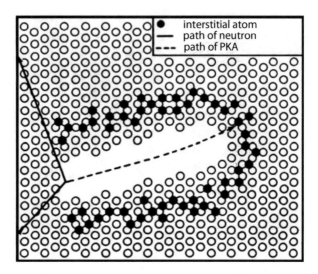

Fig. 11.9 Brinkman's [9] model for radiation damage displacement spike. From [95]. Used with permission, Springer

to be considered. Firstly, the displacement energy E_d is sensitive to the direction of the recoiling atoms as they move through the lattice, and appropriate averages need to be taken in order to obtain good estimates of the total number of displacements obtained. There can be a propensity for the displaced atoms to travel in symmetry directions between the lattice atoms in the material (called channeling) or to cause a line of collisions by the atoms, similar to a popular toy with several ball bearings suspended in a straight line (a process called focusing). These factors can depend on the particular crystalline structure involved. Secondly, the production of the displaced atoms can be visualized as a vacancy-interstitial pair, called a Frenkel pair. Some of these Frenkel pairs may go away by thermal diffusion on a short timescale. This diffusion is a strong function of the temperature at which the irradiation is taking place. The notion of the distribution of the vacancies and interstitials has evolved over time. Figure 11.9 shows an early depiction of the damage cascade. Figure 11.10 shows a modification of the model where effects of the crystalline structure have been more carefully taken into account. More recently, molecular dynamics (MD) simulations have been used to simulate the damage cascade in a manner more carefully representing the quantum-mechanical nature of the damage process. The important thing to realize as models change somewhat with the advent of more sophisticated modeling approaches is that the damage level, estimated by some calculated dpa cross section for a known neutron fluence, provides a better relative indicator than an absolute one, and is useful for comparing radiation effects in different materials if and only if the same dpa estimation methodology is used.

Two important aspects of the nuclear physics involved in high-energy neutron interactions must also be considered for radiation damage estimates based on the modified Kinchin-Pease theory. The first is to note that the energy transferred to

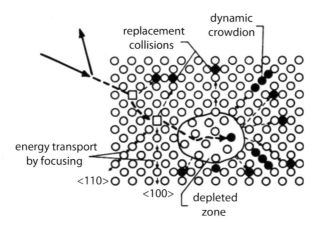

Fig. 11.10 Seegers's [82] modification of Brinkman's [9] model for radiation damage displace-ment spike, accounting for crystal structure effects on knock-on atom transport. From [95]. Used with permission, Springer

the PKA, as indicated in Eq. (11.20), is proportional to $1 - \cos\theta$, where θ is the CM angle. However, elastic scattering cross sections are largely forward-peaked at 14 MeV. A typical 14 MeV neutron elastic scattering cross section is shown as Fig. 11.11a. One notices the strong forward peak in the differential cross section. This is entirely a wavelike property, and is common to all medium-to-heavy nuclides at these neutron energies. If one calculates the deBroglie wavelength of a 14 MeV neutron $\lambda = h/p$, this gives $\lambda = 7.63$ fm and then the normalized nuclear size $ka = 2\pi a/\lambda$ gives $ka \approx 6.5$ for Mo, assuming that $a = 1.4(1 + A^{1/3})$ fm. The analogous problem of the scattering of sound waves with a spherical scatterer can be solved analytically and is given in the classic book by Morse and Feshbach [67]. The results for acoustical scattering from a hard sphere of size $ka = 6.5$ are shown as Fig. 11.11b, for comparison. For damage cross section calculations, the large peak at $\theta = 0$ is nulled out by the $1 - \cos\theta$ factor in the PKA energy calculation. Thus as the energy of the neutron increases, the contribution to the damage cross section from elastic scattering decreases.

A second issue is important at higher neutron energies and that is the emergence of inelastic reactions as the neutron energy is raised. These reactions include (n, γ), $(n, 2n)$, (n, p), and (n, α) reactions. The momentum balance is shifted in these reactions, and the nulling of the forward peak for the elastic reactions is alleviated to some extent. Thus substantially higher energies are available for the PKA involved in these reactions. Figure 11.12 illustrates the effect of these inelastic reactions on the overall damage cross section for Ni. Similarly, Fig. 11.13 shows dpa cross section data for 18Cr-10Ni stainless steel.

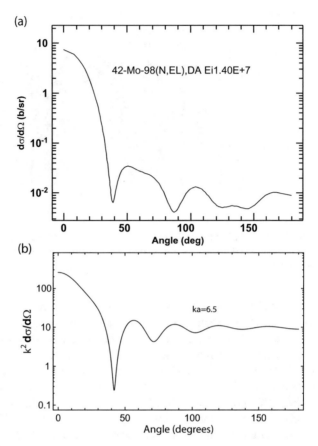

Fig. 11.11 (**a**) Elastic scattering cross section $d\sigma/d\Omega$ for ^{98}Mo at 14 MeV. from EXFOR IAEA data base, with data from [83]. (**b**) Acoustic scattering cross section for conditions similar to 14 MeV neutron scattering on ^{98}Mo ($ka = 6.5$). See [67, p. 1483]

11.1.5   Helium Production

In addition to displacement damage, gas production inside the matrix of a fusion structural material is also important. The two major production sources caused by neutron irradiation are (n, α) and (n, p) reactions. These two reactions will produce helium and hydrogen, which then can migrate in the material and coalesce in the form of gas bubbles. These reactions are threshold reactions, typically with neutron energy thresholds of $5 \rightarrow 10$ MeV. Because of these high energy thresholds, there is limited applicability of materials damage experience from fission reactor materials, since the fission neutron spectrum has very few neutrons with energies above 5 MeV. An example of a typical (n, α) cross section is shown for ^{58}Ni for various reaction channels giving He production is shown as Fig. 11.14.

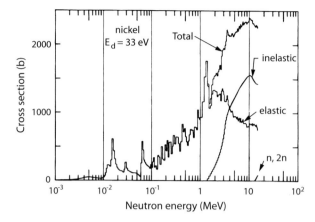

Fig. 11.12 Damage cross section for Ni as a function of neutron energy. From [95], with data from [23]

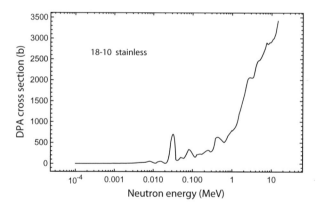

Fig. 11.13 Damage cross section for 18Cr-10Ni stainless steel as a function of neutron energy. Data from [23]

11.1.6 Material Degradation Due to Radiation Damage

It is now important to ask what the effects of irradiation of candidate fusion reactor materials are. The short answer is that the type of damage in any material is heavily influenced by the temperature that the irradiation occurs at and any annealing temperatures and times that might also be involved. Also of interest is the energy spectrum of the radiation and especially the calculated ratio of displacement damage to helium production expected.

Fig. 11.14 Cross sections for (n, α) and other reactions at high neutron energies in ^{58}Ni. From [56]. Used by permission, American Physical Society

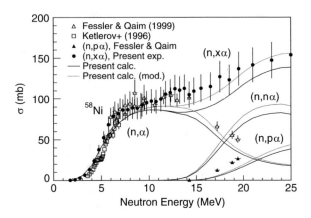

11.1.7 Embrittlement

At low irradiation temperatures (here we normalize temperature T to the melting temperature T_m measured from absolute zero), typically $0 < T/T_m < 0.3$, the dominant radiation damage mechanism is embrittlement. Structural materials are usually required to be ductile at their working temperature down to ambient temperature at shutdown. At some lower temperature, the material typically becomes brittle, meaning that very little energy is required to fracture it. The fracture energy can be determined using a simple test such as the Charpy V-notch test, as shown in Fig. 11.15. In this test, an axe-like pendulum is swung through a specimen of controlled dimensions, which includes a V-notch on its backside. The blunt blow delivered by the impact tool results in energy being absorbed during the fracture, which is determined by the maximum height of the forward swing through the material, such that the energy absorbed is $U_{fracture} = Mg(h - h')$. By testing samples at various temperatures, a transition is found between ductile failure at higher temperatures and brittle failure at lower temperatures as shown in Fig. 11.16. As the material is exposed to neutrons (or other damage-causing particles) the ductile-to-brittle transition temperature (DBTT) tends to shift upwards. When the material has sufficient fluence so that the DBTT is above the minimum temperature expected during thermal cycling of the material, such as during startup and shutdown conditions, thermal stresses induced at that time may cause the material to fail by brittle fracture.

Examples of materials experiencing DBTT shift threatening their use in a fusion reactor are the molybdenum alloys. The melting temperature of elemental Mo is $T_m \approx 2623\,^\circ\mathrm{C} = 2896\,^\circ\mathrm{K}$. While Mo alloys have otherwise superb characteristics for a fusion reactor first wall, they are notoriously susceptible to DBTT shift from radiation. Figure 11.17 shows the performance of two Mo alloys, both of which are Mo-0.2% TiC but with differing amounts of cold rolling during manufacture [54]. The samples were irradiated in two different conditions: one campaign irradiated the materials to 0.1 dpa with one cycle of temperature to 773 $^\circ$K($= 500\,^\circ$C) and

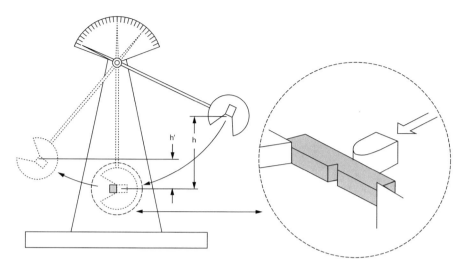

Fig. 11.15 Charpy V-notch test for determination of DBTT. After [100]

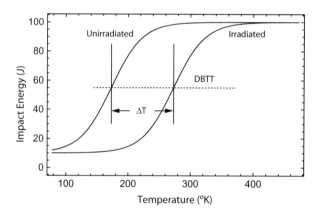

Fig. 11.16 Typical absorbed energy vs. temperature as determined by Charpy test for unirradiated and irradiated materials, showing DBTT shift

another irradiated the samples to 0.15 dpa with four cycles of temperature to $673\,°K(= 400\,°C)$. One can see the differing amounts of DBTT shift depending on the three variables of cold working, radiation temperature, and fluence. The authors of [54] report that a principal feature of the radiation exposure is that precipitates ("dispersoids") of Mo_2C, $(Mo,Ti)_2C$, $(Mo,Ti)C$, $(Ti,Mo)C$ and TiC of relatively large size form and may strengthen the grain boundaries, while making the material more brittle at the same time. Figure 11.18 shows electron microphotographs substantiating this observation.

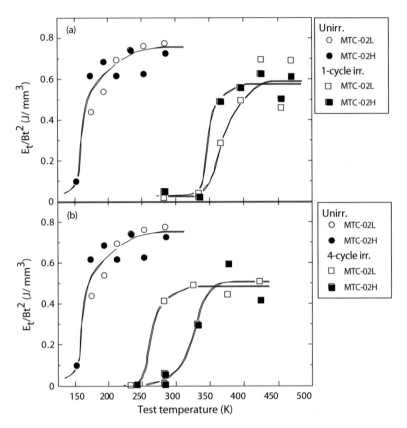

Fig. 11.17 Impact energy vs. temperature for two Mo-0.2 wt% TiC alloys, with two different temperature cycles. From [54]. Used by permission, Elsevier

11.1.8   Void Swelling

At higher temperatures ($0.3\,T_m < T < 0.55\,T_m$), embrittlement is supplanted by void swelling as the dominant failure mode caused by radiation damage. Void swelling is characterized by an overall increase ΔV in the volume V of a component. In almost every case the volume of the component in question is constrained in its dimension in one or more directions, and thus the result of void swelling is buckling or bursting of the component. A rupture-type of failure in a fusion reactor first wall may be especially catastrophic, as it can lead to the release of radioactive coolant in a manner that may or may not be controllable.

The root cause of void swelling is the generation of vacancy-interstitial pairs as described at the beginning of this section. A vacancy may be treated as a pseudoparticle and can be assigned a diffusion coefficient as well as a recombination rate with interstitials. Vacancies can thus coalesce, creating voids in the material on a scale many times the lattice constant. Interstitials can migrate to grain boundaries

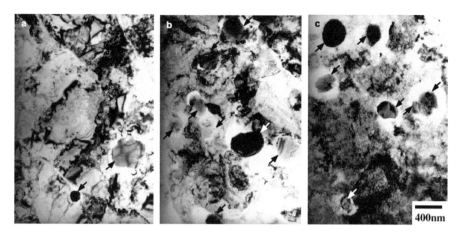

Fig. 11.18 Electron microphotographs of unirradiated and irradiated samples of the two Mo-0.2 wt% TiC alloys described in [54]. (**a**) as-rolled low cold-work alloy, (**b**) 4-cycle irradiated low cold-work alloy, (**c**) 4-cycle irradiated high cold-work alloy. Arrows indicate precipitates. Used by permission, Elsevier

where they can create extra rows and columns in the crystalline lattice. More likely, however, if the grain size is $>10\,\mu m$, is that the interstitials will nucleate into dislocations. (Dislocations are irregularities in the crystal structure, such as the appearance of an extra half-plane of atoms (an edge dislocation), or the side-shift of all the atoms in one half-plane and above by one lattice location (a screw dislocation). See [73] for a good introduction.) Dislocation structures typically form loops. Interstitials can be attracted to dislocation loops and nucleate into the loop. This is less likely for vacancies, however, and void cavities are formed by the coalescence of vacancies. The difference in energy for an interstitial to join a dislocation loop vs. that for a vacancy (e.g. $\epsilon_i \approx 4\,eV$ vs. $\epsilon_v \approx 1\,eV$ in stainless steels), along with the differences in diffusion rates, which also have a temperature-dependent form, contributes to the observed temperature dependence of void formation. With irradiation comes an increase in the production of vacancies and interstitials in quantities that can saturate the accommodation mechanisms available in equilibrium conditions, and thus void swelling typically shows a nonlinear response to radiation fluence, i.e. $\Delta V/V \propto (\Phi t)^n$, with exponent $n \approx 2$. Models of void growth as a function of temperature and fluence have been developed which show a fair amount of consistency with the observed data.

Figure 11.19 shows an example of void swelling in copper as a function of temperature. Note that the zero swelling points (182 and 500 °C) correspond to temperatures equal to $0.335 T_m$ and $0.569 T_m$, respectively. Similar behavior is found in materials with higher melting temperatures, and the window for void swelling tracks fairly closely to this band of normalized temperatures.

Fig. 11.19 $\Delta V/V$ vs.
temperature for neutron
irradiation of copper to
$1 \rightarrow 1.3$ dpa at a rate of
2×10^{-7} dpa s^{-1}. From
[103]. Used by permission,
Elsevier

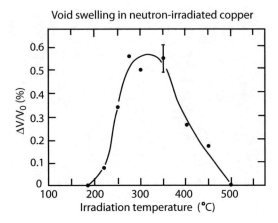

11.1.9 Irradiation Creep

Creep is a phenomenon that becomes important at threshold temperatures similar
to void swelling, i.e. $T > 0.3T_m$. Unlike void swelling, however, creep does
not subside at higher temperatures, but at $T > 0.55T_m$, thermally induced
creep, not due to radiation but also present in unirradiated materials, becomes
the dominant mechanism for material failure. Creep is characterized by a *rate*
of strain $\dot{\epsilon}$ as opposed to a simple strain ϵ and represents a plastic (irreversible)
deformation of the material. At creep-relevant temperatures, there is typically a
thermal creep happening which is independent of radiation dose as well as a non-
thermal component due to radiation damage. The dominant effect in thermal creep is
the organization of thermally generated vacancies and interstitials into dislocation
loops which move through the lattice in response to stress. The mobility of these
vacancies and interstitials is impeded by the presence of existing dislocation loops
in the material and by the dynamics of the dislocations at grain boundaries. For
intermediate values of stress (less than $5 \times 10^{-3}\mu$, where μ is the shear modulus)
and temperatures ($0.3T_m < T < 0.6T_m$), the thermal creep strain rate is given by a
power law in the form applied stress and an exponential law in temperature [94]:

$$\dot{\epsilon} = \frac{AD_0\mu b}{kT} \left(\frac{\sigma}{\mu}\right)^n \left(\frac{b}{d}\right)^p \exp(-Q/kT). \qquad (11.30)$$

Here D_0 is the temperature-invariant part of the diffusion coefficient for vacancies,
Q is the activation energy for diffusion of vacancies (in eV), d is the grain size, b is
the Burgers vector (characteristic displacement length), k is Boltzmann's constant
(such that $k \cdot 11608\,^\circ$ K $= 1$ eV), T is the temperature in $^\circ$ K, σ is the applied stress,
n is the stress exponent (typically between 3 and 10), p is the inverse grain size
exponent, and A is a dimensionless parameter.

Models for radiation-induced creep can have several components. Three of these are [25, 64, 94]:

1. Stress-Induced Preferential Nucleation of Loops (SIPN),
2. Stress-Induced Preferential Absorption (SIPA), and
3. Preferential Absorption Glide (PAG).

Stress-induced preferential nucleation of loops (SIPN) occurs due to the propensity of interstitial loops to nucleate on planes perpendicular to the applied stress, and for vacancy loops to nucleate on planes parallel to the applied stress. Both of these will cause the material to show movement in the direction of the applied stress, i.e. strain. The probability that a loop involving n interstitial atoms is proportional to a Boltzmann factor $\exp(\sigma_i n \Omega/(kT))$, where Ω is the unit cell volume occupied by each interstitial atom and σ_i is the component of the stress in the ith possible loop orientation. The rate processes involved lead to a creep displacement which is directly proportional to the void swelling, i.e. $\dot{\epsilon} = K\dot{S}$, where S is the amount of void swelling. Thus SIPN is a transient effect unless there are additional sinks of defects not considered in the model used for void swelling [99]. In fact a significant transient in the elongation ϵ is seen when radiation is first applied to a sample within the temperature band where void swelling happens, as illustrated in Fig. 11.20. In general, the material in the relevant temperature range follows a displacement law of form [73]:

$$\epsilon = A\sigma \left[1 - \exp\left(-\frac{\Phi t}{B} \right) \right] + C\sigma \Phi t. \qquad (11.31)$$

Stress-induced preferential absorption (SIPA) describes the bias caused by tensile stress not on the nucleation of interstitial loops but on the growth of the loops with the preferentially favored orientation by the absorption of interstitials. In this model, vacancies and interstitials are drawn into (or pushed away from) dislocations

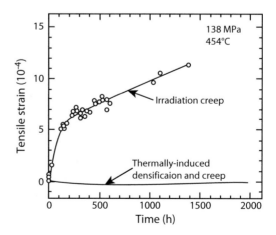

Fig. 11.20 Irradiation creep in 20% cold-worked 316 stainless steel irradiated in EBR-II. Data from [34]. From [94]. Used by permission, Springer

Fig. 11.21 Creep rate as a
function of neutron flux in
annealed 09Kh16NM3B
irradiated in BR-10. From
[94]. Used by permission,
Springer

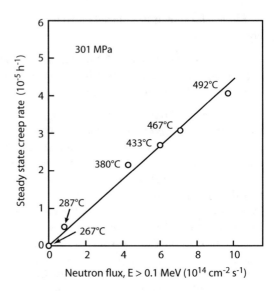

with fluxes depending on the concentrations of the vacancies and interstitials
and their diffusion coefficients. Since the vacancies and interstitials obey a rate
equation where production ($\propto \Phi$) is balanced by recombination ($\propto C_i C_v$), vacancy
and interstitial concentrations increase with neutron flux levels when irradiation
dominates over thermal production. The flux of the interstitials and vacancies into
and away from the dislocations is balanced by a climb velocity for the dislocations
which is proportional to the vacancy and interstitial concentrations and this is in
turn greater at higher flux levels. Then this climb velocity for the dislocations can be
related to the overall strain rate $\dot{\epsilon}$ and thus one arrives at a law where $\dot{\epsilon} \propto \Phi \sigma_{xy}$, i.e.
proportional to the neutron flux and to the applied stress, as found in the empirical
model shown as Eq. (11.31). Figure 11.20 shows a comparison of elongation vs.
fluence for 316 stainless steel in a fast neutron environment (the fast reactor EBR-
II) with a SIPA model for the creep growth. One can see from the data that the
actual creep rates were closer to linear growth in time than the quadratic-type growth
predicted by the SIPA model, but the overall measured elongations were fairly
consistent with the model. Irradiation creep has been found to vary with the method
of manufacture, being fairly sensitive to the amount of cold work (the thinning of
parts by mechanical rolling) and to changes in impurity makeup of the material. See
Fig. 11.21 for an example of the performance of cold-worked 316 stainless steel
to irradiation at creep-relevant temperatures. Therefore applying these results to
different alloy systems in a different neutron environment is somewhat risky.

Preferential Absorption Glide (PAG) is similar to SIPA in that it involves the
migration of vacancies and interstitials towards loops, but for the loops which are not
aligned to the applied stress but rather in the plane of the applied stress. Preferential
absorption of interstitials into the defects causing the pinning of the dislocation
enable the dislocations to climb from one pinning location and then glide to the

next with little expenditure of energy. Mansur [62] gave a simple model for the process, concentrating only on the dominant channel of interstitials being absorbed into the dislocations and thus arrived at a simple formula for the creep rate:

$$\dot{\epsilon}_{PAG} = \frac{4\Omega}{9b}(\pi L)^{1/2}\Delta Z^D D_j C_j \frac{\sigma^2}{E} \tag{11.32}$$

Here L is the dislocation line density, Ω is the atomic volume, D_j and C_j are the diffusion coefficient and concentration of interstitials, and ΔZ^D is the effective climb distance. A hallmark of this treatment is the appearance of the creep rate proportional to the square of the stress σ, unlike the linear dependence in SIPA theory. For this reason, SIPA creep is expected to dominate at low stresses and PAG creep is expected to dominate at higher stress levels.

11.1.10   Advanced Materials Concepts

Composite Materials

Composites made from dissimilar materials such as carbon fiber reinforced epoxy and fiberglass are well-known examples of composite materials in everyday life. For high-temperature applications, composite materials consisting of two different phases of the same material have emerged as candidates for both fission and fusion systems, where the heterogeneity can result in large increases in fracture toughness. An example of such a self-composite material is a SiC/SiC material produced at General Atomics and studied at UC Berkeley [31] A microstructural view of this material is shown in Fig. 11.22. In this material, one phase appears as a fiber and the other phase as the matrix. Although the present formulations do not have the fracture toughness of a metal, they are superior to the single phase ceramic material, which is probably too brittle to use as a fusion reactor first wall.

Another composite material that is a candidate for plasma-facing components is a composite of tungsten filaments in a tungsten matrix [71]. This is formed by making a braided mat of tungsten filaments, a well-developed material from the lighting industry, and building a matrix surrounding the filaments with chemical vapor deposited (CVD) tungsten, called W_f/W. This material, like the SiC/SiC material just described, has been found to have superior fracture toughness to the elemental W. Another possible use of the W filaments is to combine the braided mat of these in a copper matrix formed around the W filaments by melt infiltration. The high thermal conductivity is of great benefit in reducing thermal stress.

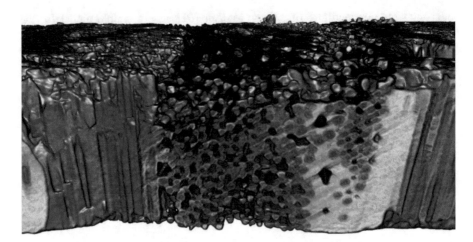

Fig. 11.22 Microstructure of SiC/SiC composite material from 3D reconstruction of a microto-mography analysis. From [31]. Used by permission, Elsevier

11.1.11 Advanced Steels for Fusion Applications

Steel is one of the oldest and best-known metal alloys. It has been produced by various methods for the last 4000 years, second only to bronze in its length of use. Yet newer steel concoctions are emerging, some driven especially by the need for fusion reactor structural materials. Two of the newest lines of inquiry involve reduced activation ferritic/ martensitic steels (RAFM), using conventional tempering or using oxide dispersion strengthening (ODS). A survey of the current state of the art is given in [104].

One feature of RAFM steels lies in their lack of nickel as in the 316 and 304 austenitic stainless steels. Ni has an inelastic (n, p) reaction at high neutron energies, which produces ^{60}Co, resulting in structures which are highly radioactive $(t_{1/2} = 5.27$ y). The ferritic/martensitic steels for fusion application generally do not contain Ni. Another feature in ferritic/martensitic steels is their greatly reduced propensity for void selling [55]. This is especially important for any steel in a fusion reactor application, since normal operation in a reactor would require operation in the range of $0.3 T_m \rightarrow 0.5 T_m$, which is the temperature range where void swelling is usually active (for steels with $T_m \approx 1370\,°C$, this corresponds to $220\,°C \rightarrow 550\,°C$).

The designs of these newer high-strength formulations are done using computational thermodynamics modeling [104] (see also http://matcalc.tuwien.ac.at/). Newer thermomechanical treatments (TMTs) (using traditional methods such as temperature cycling and hot rolling/cold rolling) have resulted in low-activation steels with yield strengths above 500 MPa at temperatures up to 500 °C. Some questions remain about weldability and production in large-scale "heats," but the results look very promising.

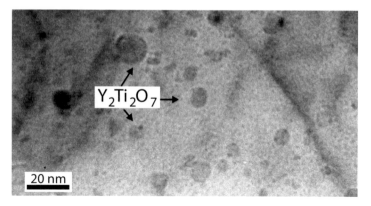

Fig. 11.23 Transmission electron micrograph of nanometer-scale oxide particles embedded in the ferritic/martensitic steel matrix. From [66]. Used with permission, Elsevier

ODS steels use nanoparticles of some metal oxide that are embedded into the metal matrix of the steel. An example is Fe-9Cr-0.1C-1.5W-0.2V-0.5Ti-0.35Y_2O_3 (wt %) made using powder metallurgy techniques [66]. This material is made by combining powdered Cr alloy steel with Ti and yttria (Y_2O_3), ball milling, hot isostatic pressing, and finally forging the alloy. The current fabrication techniques are time-consuming and not ideal for large-scale production. But these materials have been produced in small samples and have shown superb mechanical characteristics in some cases. They also show the promise of high radiation resistance. Figure 11.23 shows an electron micrograph of an ODS alloy of this type, and this shows the presence of the nanometer-size oxide inclusions, which act in a manner similar to traditional precipitates, but with much smaller particle sizes. These give the material its superior mechanical properties.

Further research will be required in order to make these materials a viable option for fusion reactor structural use. Whereas the RAFM materials developed with conventional heat treatment techniques have some problems with weldability, the problem is exacerbated in the ODS materials by the formation of coalesced larger-sized clumps of oxide during welding, which in turn can lead to brittle failure and loss of toughness. However, these materials show promise, should this problem be solved. As is the case with all candidate fusion reactor materials, the lack of an available test environment with 14 MeV neutrons places some uncertainty on the lifetime of these materials in a large, power-producing reactor.

11.2 Plasma-Surface Interaction

While fusion reactor structural lifetime will be dependent on the interaction of neutrons with the structural material for blankets and other components, interaction of the plasma with walls and divertor structure is already a problem in magnetic fusion experiments due to plasma–surface interaction issues. The issues are particularly acute for the case of divertor structures. Divertors function to scrape off the plasma outside of the closed flux surfaces, in a manner similar to the gutters on a public swimming pool. As such they can have rather large fluxes of ions impinging on the surfaces, as well as high overall heat loads from electrons, neutrons, and photons. Typically these ions have energies in the range from $100 \rightarrow 500\,\mathrm{eV}$. The ions can dislodge surface atoms in the divertor, resulting in some sputtering of the material and ultimately contributing to line radiation and charge exchange. The qualities required in a divertor are thus low ion sputtering rates and high heat flux capability. The current design for the ITER divertor uses a tungsten plasma-facing component. While W has a very high melting temperature and a low sputtering coefficient, it is susceptible to damage from ion bombardment under high heat-load conditions. Because of this material choice for ITER, W has been studied carefully in regard to its plasma–surface interaction properties in as realistic an environment as possible.

The earlier designs for ITER had a combination of carbon fiber and W materials for plasma-facing components. While each of these excels individually in one area or another, it was noted that the combination of the two in the plasma chamber could lead to trouble because of the possible interaction of the two. Sputtering of these two materials could lead to the production of tungsten carbide layers on the outside of each. While W has a low tritium permeation rate and C has high tritium permeation, WC has a low tritium diffusivity. The potential for a relatively porous carbon component to become coated with a thin layer of WC could effectively lock the tritium in place in the C component. This would lead to very high tritium inventories in the machine in a short time. Consequently, carbon structures have been replaced in the ITER design with tungsten structures, which fixes the WC problem but may introduce more demands on the divertor design. The following section discusses the current knowledge concerning the use of W as a plasma-facing component.

11.2.1 Tungsten Surface Modifications Under Ion Exposure

Both H and He plasma exposure has been studied in W plasma-facing components, and He exposure has been found to be more detrimental in that lower energy fluxes have been found to lead to surface cracking and subsequent rise in the temperature differential ΔT across the outer region of the W next to the plasma [11]. While standard operations in the ITER device will generate much larger fluences of hydrogen (i.e., D and T) ions than helium, He ions represent the "ash" of the thermonuclear process in DT fusion, and despite the low burnup rate of the fuel ions in the plasma ($f_B \sim n\tau < \sigma v > /2 \approx 0.01$), the He ions may have higher energies when they arrive at the divertor, since they are born with 3.5 MeV energy. Also, initial tests in ITER may concentrate on exploring the H mode threshold (see Chap. 8), which is more readily done in He plasmas, so the study of He surface damage is very important for the near term.

Exposure of W components to He plasma has been found to exhibit some "classical" surface modifications including blistering, bubble formation, and cracking [78]. In more recent experiments, He plasma exposure in ITER-relevant conditions has been found to cause the growth of a fuzzy layer of tungsten "nanotendrils" with no particular overall crystalline orientation [5]. An example of a test of a tungsten in a specially designed plasma surface–interaction testbed known as PISCES at UC San Diego is described in [4]. Figure 11.24 shows microphotographs of the W surface after intensive plasma bombardment. One notices the open, spongy appearance of the surface after plasma exposure. A closer look of the surface using an advanced electron microscopy technique known as transmission Kikuchi diffraction (tKD) [93] has been used to glean the crystalline orientation of the tendrils [75]. Figure 11.25 shows an application of tKD mapping of a W surface which has been exposed to He plasma. At relatively low ion fluences (10^{24} He ions m^{-2}), the polished W surfaces used in these tests become faceted and the grain orientation becomes clear. But at higher fluences (10^{26} He ions m^{-2}), the tendrils are seen growing out of each crystal face with no particular discernible preference for crystal orientation in determining the growth rate.

The overall dynamics of fuzz formation is an area of active research. The fuzz growth seems to follow a time dependence characteristic of a diffusion process, with size $\propto t^{0/5}$, but with some incubation period equivalent to a fluence $\Phi_0 = 2.5 \times 10^{24}\,\mathrm{m}^{-2}$ [77], so that

$$z = k\,(\Phi - \Phi_0)^{1/2}. \qquad (11.33)$$

Some theoretical support for this experimentally observed phenomenon has come about using large scale molecular dynamics (MD) simulations performed by Wirth et al. [43, 98]. These MD studies show that the delay in fuzz formation is probably caused by the need for the He gas in the metal to diffuse deep enough into the substrate so that bubbles formed by the nucleation of voids and gas atoms are not so close to the surface so that they burst through immediately. The challenge has been to model these systems atom-by-atom on a scale length which is hundreds of nanometers (nm), which is a daunting task considering that the lattice spacing is on the order of 0.3 nm.

11.3 Tritium Management

One can estimate the inventory of tritium required for a DT-burning fusion reactor by following the life cycle of tritium through the reactor and balance of plant. A 3000 MWth reactor (generating $\approx$1000 MWe) has a tritium burning rate given by

$$\dot{m}_T = \frac{P_f}{E_f} m_T = 5.33 \times 10^{-6}\,\mathrm{kg\ s}^{-1} = 0.46\,\mathrm{kg\ d}^{-1}. \qquad (11.34)$$

Fig. 11.24 Tungsten fuzz
following exposure to He
plasma at three temperatures:
(a) 900 K, (b) 1120 K, and
(c) 1320 K. From [4]. Used
with permission, Elsevier

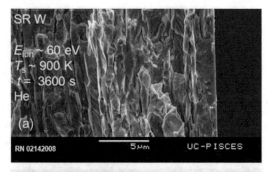

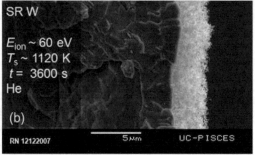

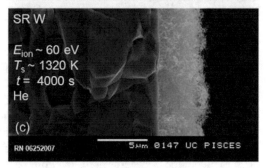

However, the burnup fraction of T in the plasma is given by

$$f_B = \frac{n_D n_T < \sigma v > \tau_p}{n_T} \approx \frac{n \tau_P}{2} < \sigma v > \approx 0.01, \tag{11.35}$$

so that the circulating inventory of tritium $\approx \dot{m}_T / f_B \approx 50 \,\mathrm{kg\,d^{-1}}$. Thus a 1-day reprocessing time represents a required inventory on the order of 50 kg. A larger inventory may be required for operational contingencies, or because additional inventory may be intentionally accumulated to use in the startup of another reactor.

It is interesting to look at this inventory in terms of radiological risk. The activity of tritium is given by

$$\mathscr{A} = -\dot{N} = \lambda N = \frac{\ln 2}{t_{1/2}} N = \frac{\ln 2}{t_{1/2}} \frac{N_A}{M} m_T = 3.59 \times 10^{17} m_T(\mathrm{kg}) \; \mathrm{Bq}, \tag{11.36}$$

or about $9700 \,\mathrm{Ci\,g^{-1}}$.

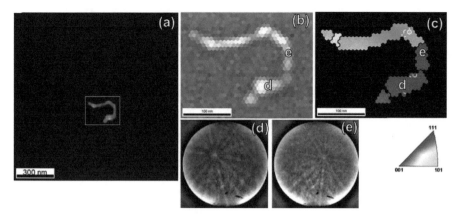

Fig. 11.25 "(**a**) SEM image of an isolated tungsten nanotendril on a continuous carbon film. The green box denotes the area of tKD mapping. (**b**) Image quality map. (**c**) Z-axis (out-of-the-page) inverse pole figure colored map. Coloration is by the inset unit triangle. Black boundaries are high angle, cyan boundaries low-angle. (**d–e**) Typical tKD patterns from the marked points. Pixel pitch: 8 nm." From [75]. Used by permission, Elsevier

Thus blanket tritium inventories in a 3000 MWth reactor will be a few $\times 10^{17}$ Bq, and onsite inventories on the order of a few $\times 10^{19}$ Bq will be required. (Note that in the older units of Curies, where 1 Ci$= 3.7 \times 10^{10}$ Bq, these are blanket and plant inventories of 10 MCi and 1 GCi, respectively.) The USEPA standard is for a maximum of 20000 pCi l^{-1} (740 Bq l^{-1}), which makes it clear that enormous isolation factors will be required to ensure that a reactor does not pose a threat to the biosphere. (For the example given here, the volume of water required for dilution of the full-plant T inventory to this standard equals 23,000 years of flow through the Tuolumne River, which supplies drinking water to the San Francisco Bay Area).

Will it be possible to limit T releases to values consistent with these standards? The answer is that probably a system can be designed that is compliant to these standards, but some attention must be paid to choices for materials and temperatures involved in the tritium related systems. We will look at the basic transport physics involved in T permeation.

11.3.1   Sievert's Law and Tritium Permeation

We can model tritium permeation through a solid wall using one of the two scenarios pictured in Fig. 11.26. For tritium in a gaseous phase intermixed with a gaseous coolant, as shown in Fig. 11.26a, it is a straightforward calculation to find the partial pressure p_{T_2} of the tritium in the gaseous phase, if the concentration of tritium is known, using the ideal gas laws. For tritium to diffuse through a solid wall, typically two rate processes are involved. One is the dissociation of the tritium dimers into

Fig. 11.26 Application of Sievert's law for tritium permeation with (**a**) a gaseous coolant and (**b**) a liquid coolant

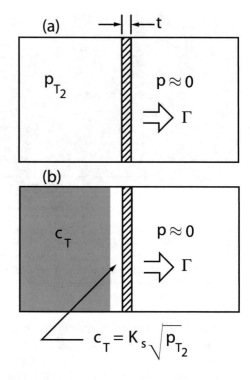

atomic tritium at the gas–wall interface. The other rate process is the diffusion of tritium through the wall. Since T atoms are much more likely to slip through the metal or ceramic wall lattice than molecules, the atomic diffusion rate is typically orders of magnitude above the diffusion rate for T molecules. The nature of the chemical reaction

$$T_2 \rightleftharpoons T + T \tag{11.37}$$

allows us to write down an equation for the concentration of tritium just inside the solid wall as

$$c_T(0) = K_S \sqrt{P_{T_2}} \tag{11.38}$$

where K_S is a solubility constant. We assume that the other side of the wall is an unconfined space with a tritium partial pressure near zero. If we assume that the flow is diffusion-limited, we have a concentration profile which is linearly decreasing from $x = 0$ to $x = t$, where t is the wall thickness. Then the flow rate is

$$\Gamma = D\nabla c_T = \frac{Dc_T}{t}, \tag{11.39}$$

and by substitution with Eq. (11.38)

$$\Gamma = \frac{DK_S\sqrt{P_{T_2}}}{t}. \tag{11.40}$$

Typically the diffusion and solubility coefficients are functions of temperature with a typical Arrhenius form:

$$D = D_0 \exp\left(-\frac{E_d}{T}\right)$$
$$K_S = K_{S0} \exp\left(-\frac{E_s}{T}\right), \tag{11.41}$$

so that we can write a temperature-dependent form for the tritium flux as

$$\Gamma = \frac{D_0 K_{S0}\sqrt{P_{T_2}}}{t} \exp\left(-\frac{(E_s + E_d)}{T}\right). \tag{11.42}$$

The situation where the tritium is dissolved in a coolant can be addressed similarly using a simple trick, as shown in Fig. 11.26b. Suppose that we have a very thin gaseous boundary between the coolant and the wall. Then we apply the same method to derive the coolant tritium partial pressure from the tritium concentration in the coolant. Then if we use a Sievert solubility constant K_S^c for the coolant, we have a repeat of Eq. (11.38), but this time in the coolant itself:

$$c_T(\text{coolant}) = K_S^c\sqrt{P_{T_2}}. \tag{11.43}$$

Then we proceed to find the tritium flux using Eq. (11.40), now obtaining:

$$\Gamma = \frac{DK_S^w c_T(\text{coolant})}{K_S^c t}. \tag{11.44}$$

Here we have used K_S^w to indicate the solubility of tritium in the wall.

11.3.2   Solubility and Diffusivity Data

FLiBe

Measurements of diffusion and solubility for hydrogen usually take the place of actual measurements with tritium because of the risk of radioactive contamination in working with tritium. While solubility properties are typically not affected by the switch in isotopes, the diffusivity changes but in a predictable way: since diffusivities are inversely proportional to the square root of mass, $D_T = D_H/\sqrt{3}$, except in some unusual cases where there is some other quantum-mechanical effect in play. In some laboratories equipped with adequate hot-cell equipment, however, T permeation constants have been measured directly. One such hot-cell study was done at Idaho National Laboratory by Calderoni et al. [13] on FLiBe, a popular fluoride salt for nuclear applications. (One should note that in addition to the tritium risk, the experiment had to be designed mindful of the beryllium toxicity as well.) The measurements of diffusion and solubility resulted in these suggested values:

$$D_{FLiBe,T} = 9.3 \times 10^{-7} \exp\left(\frac{-42 \times 10^3}{RT}\right) \, m^2 \, s^{-1}$$

with the gas constant $R = 8.3144 \, J \, mol^{-1} \, {}^\circ K$, and

$$K_{FLiBe,T} = 7.892 \times 10^{-2} \exp\left(-\frac{35.4 \times 10^3}{RT}\right) \frac{mole}{m^3 Pa}.$$

The experimental results are plotted in Figs. 11.27, 11.28, and 11.29, along with data from two earlier studies, Refs. [33] (in FLiNaK) and [2] (in FLiBe). In Fig. 11.27, one can clearly see the $\sqrt{p_{T_2}}$ dependence on the permeation rate, and the Arrhenius-like behavior of D and K_S are visible in Figs. 11.28 and 11.29 by plotting the logarithm of these quantities vs. $1/T$, which will yield straight-line plots if $\{D, K_S\} \propto \exp\left(-\{E_d, E_s\}/T\right)$.

Fig. 11.27 Permeation rate through FLiBe, showing $\sqrt{p_{T_2}}$ dependence. From [13]Used by permission, Elsevier

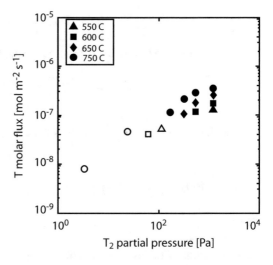

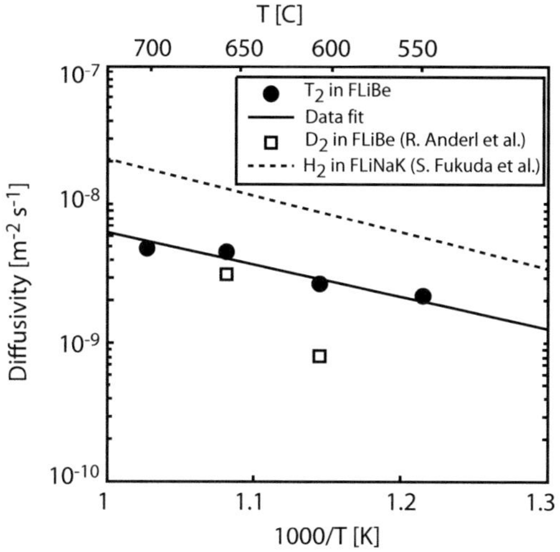

Fig. 11.28 Diffusivity of tritium in FLiBe. From [13], with data from [33] and [2]. Used by permission, Elsevier

Fig. 11.29 Solubility of tritium in FLiBe. From [13], with data from [33] and [2]. Used by permission, Elsevier

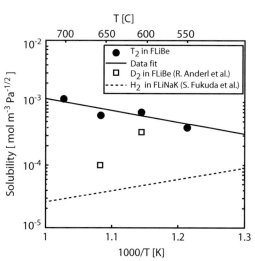

Lithium-Lead Eutectics

Lithium-lead coolants have also been suggested for fusion blankets. These have the advantage of potentially high tritium breeding ratios (TBR) due to the presence of $(n, 2n)$ and $(n, 3n)$ reactions in Pb at high energies. LiPb forms eutectic alloys, with fairly low eutectic temperatures, reducing the risk of the coolant "going solid" after an emergency shutdown. A phase diagram for $Li_x Pb$ alloy, with $x < 0.2$, is shown in Fig. 11.30.

Tritium solubility and transport in LiPb coolant has been measured and described in [45] and reviewed in [63]. However, there are major discrepancies in the reported transport values between measurements by different groups. For example, Ref. [63]

Fig. 11.30 Phase diagram
for Li$_x$Pb alloy, with $x < 0.2$.
Data from [46]. Used with
permission, Elsevier

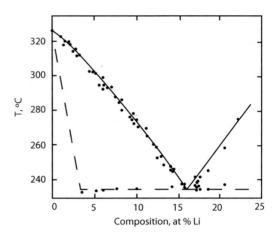

shows a scatter in the solubility parameter K_s over two orders of magnitude at
reactor-relevant temperatures, and almost a factor of five for the diffusion coefficient
D over a similar range. The measurements are difficult to make due to the
tendency for the LiPb eutectic to form an oxide layer on its surface and to pick up
contamination from materials in contact with it. This also begs the question of what
the actual conditions would be for this material in an actual fusion reactor coolant
duct. While LiPb has less chemical reactivity with air and water in the case of a
rupture, compared to Li, at the present time the uncertainties in its tritium transport
are probably too large to design a reactor with a LiPb coolant.

Tritium in Metals and Graphite

For many candidate fusion reactor wall materials, the diffusion and solubility
coefficients have been measured, but mostly with protium ($_1^1 H_2$) instead of tritium.
A summary of these measurements is given in Table 11.1. For most of these
cases, diffusion and solubility have not been directly measured with tritium, and
an estimate of the tritium diffusivity is $D_T \approx D_H/\sqrt{3}$ unless otherwise indicated.

11.4 Magnets

Fusion reactors based on a magnetic confinement scheme will almost certainly
require superconducting magnets. If we consider an acceptable steady-state power
density to be on the order of a few megawatts per cubic meter, and thus densities
in the 10^{20} m^{-3} range, then limitation of the plasma pressure to be a few percent
of the magnetic pressure thus points to the need for magnetic fields on the order
of 4.0 T or greater. Production of magnetic fields on this scale, if attempted

Table 11.1 Diffusivity and solubility for protium in various metals and classes of alloys in the absence of trapping

Alloy	Diffusivity $D = D_0 \exp(-E_D/RT)$		Solubility, Φ/D $K = K_0 \exp(-\Delta H_s/RT)$		Ref.
	D_0 $\left(\dfrac{m^2}{s}\right)$	E_D $\left(\dfrac{kJ}{mol}\right)$	K_0 $\left(\dfrac{mol\ H_2}{m^3\sqrt{MPa}}\right)$	ΔH_s $\left(\dfrac{kJ}{mol}\right)$	
Beryllium	3×10^{-11}	18.3	18.9[a]	16.8[a]	[48]
			5.9×10^{6a}	96.6[a]	[90]
Graphite	9×10^{-5}	270	19	−19.2	[3]
Aluminum	2×10^{-8}	16	46	39.7	[79, 101]
Vanadium	3×10^{-8b}	4.3[b]	138	−29	[32, 88]
RAFM steels[c]	1×10^{-7}	13.2	436	28.6	
Austenitic stainless steel	2×10^{-7}	49.3	266	6.9	[76]
Nickel	7×10^{-7}	39.5	564	15.8	[61]
Copper	1×10^{-6}	38.5	792	38.9	[6]
Zirconium	8×10^{-7}	45.3	3.4×10^7	35.8	[51, 52]
Molybdenum	4×10^{-8}	22.3	3300	37.4	[91]
Silver	9×10^{-7}	30.1	258	56.7	[50, 65]
Tungsten	6×10^{-4}	103.1	1490	100.8	[30]
Platinum	6×10^{-7}	24.7	207	46.0	[24]
Gold	5.6×10^{-8}	23.6	77,900[d]	99.4[d]	[26]

From [15]. Sandia National Laboratory
[a]The solubility of hydrogen in beryllium is very low and there is not good agreement between the few studies of the material
[b]Data for isotopes other than protium does not scale as the square root of mass
[c]Values are averaged over the data presented in Figure 12 and Figure 13 of Ref. [15]
[d]Estimated using the permeability from [14] and the quoted diffusivity

with copper or aluminum coils, will result in an energy consumption in the coils of hundreds of megawatts and thus a serious fraction, if not over 100%, of the electricity generated. In this section some basic physics for superconductivity will be presented, and radiation damage issues will be discussed. Finally, some issues related to mechanical design to account for thermal and mechanical stability of the coils will be presented. While the historical and theoretical notes here are brief and only intended to give general knowledge, there are several good textbooks on the subject, including [92].

11.4.1   Superconductivity

The earliest report of superconductivity in metals was in 1911 by Heike Kamerlingh Onnes [74], who discovered superconductivity in mercury. Shortly thereafter, superconductivity was found in many metallic elements, including lead, niobium,

aluminum, cadmium, among many others. The early set of superconducting materials exhibited what is now called "Type I" behavior, which is to say that the materials
tended to exclude magnetic fields from their interior when in a superconducting
state and returning to "normal" conditions when the magnetic field surpassed
some threshold value. This field-exclusion behavior is called the Meissner effect
and is different from the ordinary skin effect in conducting material because the
material expels magnetic flux from its interior when it becomes superconducting,
rather than just freezing it in place. Another type of superconductors were later
discovered which showed the Meissner effect at magnetic fields up to some point
(now called H_{c1}), and then started to let in quantized units of flux (called fluxons
or vortices) of magnitude $\Phi_0 \equiv h/(2e) \approx 2 \times 10^{-15}$ Wb as the field increased,
up to another value of magnetic field H_{c2} when superconductivity was destroyed.
These "newer" superconductors are now called Type II superconductors and are
the workhorse superconductors in engineering applications at present, including the
ITER tokamak as well as the Large Hadron Collider at CERN, the W7X stellarator
at IPP-Greifswald, and the K-Star tokamak in South Korea.

Fundamental to the study of superconductors are two characteristic lengths.
The first of these, the London length, pertains to the general scale length for the
variation of magnetic fields within a superconducting material [60]. The London
length concept is derived from an observation that the ground state of a collection
of superconducting electrons s must have zero net momentum, using the canonical
momentum of Hamiltonian mechanics and thus quantum theory; that is,

$$\mathbf{p} = m\mathbf{v}_s + e\mathbf{A}. \tag{11.45}$$

This leads to

$$< \mathbf{p} >= 0 \rightarrow < \mathbf{v}_s >= -\frac{e\mathbf{A}}{m}, \tag{11.46}$$

and using $\mathbf{J} = -n_s e < \mathbf{v}_s >$ and the double-curl Maxwell equation then yields

$$\nabla^2 \mathbf{A} - \frac{1}{\lambda_L^2} \mathbf{A} = 0, \tag{11.47}$$

with the London length λ_L defined by

$$\lambda_L^2 = \frac{c^2 m \epsilon_0}{n_s e^2}, \tag{11.48}$$

which has the same form as the Debye shielding equation in Chap. 3. (In fact, if
one called the quantity $\omega_{ps} = \left(n_s e^2/(m\epsilon_0)\right)^{1/2}$ the electron plasma frequency, one
would see that this is the collisionless skin depth for penetration of electromagnetic
waves in an unmagnetized plasma, but this is coincidental because (a) the medium
is magnetized and (b) there are no electric fields.)

To understand the behavior of Type I and Type II superconductors, one can observe a simple fact: single electrons are spin = 1/2 particles and thus behave as fermions. Single electrons thus obey Fermi-Dirac statistics and scatter off of lattice atoms with simple interaction characteristics which assure a transfer of energy. This leads to a simple model for the conductivity of such electrons in metallic conductors. At low temperatures, however, electrons can become paired together with a very weak binding energy $\sim 10^{-3}$ eV. The "Cooper pairs" thus generated, although not exactly quasiparticles, act in many ways as bosons with integer spin and symmetrical, rather than anti-symmetrical, wave functions and thus form a Bose-Einstein condensate, with the pairs of electrons able to share the lowest energy states in the conductor. This can result in these pairs not being able to interact with lattice atoms with enough energy to produce phonons and thus experience no drag forces, hence zero resistivity. The original paper by Leon Cooper in 1956 [19], along with contributions by John Bardeen and John Schrieffer, led to the Nobel Prize for the three in 1972. The "BCS" theory is now the basic understanding of type II superconductor behavior. A critical component in the theory is the "coherence length" ξ describing the distance between the two electrons in the Cooper pair.

Thus the two important spatial parameters, the London length λ_L and the Cooper pair coherence length ξ, form the basic criterion for Type I and Type II superconducting behavior. The relationship is as follows:

$$\xi > \sqrt{2}\lambda_L \qquad \text{Type I}$$
$$\xi < \sqrt{2}\lambda_L \qquad \text{Type II.}$$

(11.49)

The maximum fluxon density is set by the coherence length, and this gives a theoretical result for the maximum magnetic field for the material to remain superconducting:

$$B_{c2} \approx \frac{\Phi_0}{(2\xi)^2}.$$

(11.50)

The maximum current density in the superconductor is set by the limit on the sideways $\mathbf{J} \times \mathbf{B}$ force that the fluxons can experience before they can be thrust out of the preferred lattice sites, where they are considered to be "pinned" otherwise. These preferred lattice sites are typically defect sites. This can lead to a somewhat counterintuitive situation where cold work and other defect-producing mechanisms can increase the critical current density, up to a point.

Radiation Damage in Type II Superconductors

Radiation damage for Type II superconductors can be characterized by shifts in both the critical current density J_{c2} as a function of neutron fluence and by the temperature at which the material falls out of superconductivity. Figure 11.31 shows

Fig. 11.31 Radiation damage in NbTi as a function of fast neutron fluence. From [41]. Used by permission, Elsevier

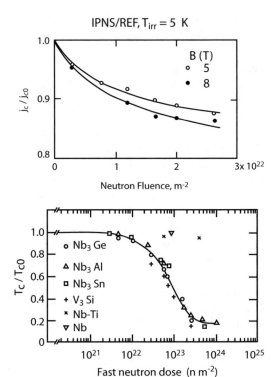

Fig. 11.32 Critical temperature T_c vs. neutron fluence for type II superconductors. From[10]. Used by permission, Elsevier

the dependence of the critical current density J_{c2} with neutron fluence for NbTi, and Fig. 11.32 shows the change in the critical temperature T_c as a function of fast neutron fluence for common Type II superconductors. One sees from this that the critical fluence before radiation damage becomes pervasive is around $10^{22}\,\mathrm{n\,m}^{-2}$, with NbTi showing somewhat more robust behavior, especially for T_c shift. Given that first wall fluxes for a 1 MW m^{-2} neutron wall loading are around $10^{18}\,\mathrm{m}^{-2}\,\mathrm{s}^{-2}$, this indicates a time to failure on order of $10^4\,A$ s, where A is the attenuation factor for the fast neutron flux. For a 30-year lifetime for the magnet system in a reactor ($t \approx 10^9$ s), this requires that $A \approx 10^5$ for that lifetime. The Li blanket can contribute to this shielding factor, but in most designs a dedicated radiation shield between the blanket and the magnet will be required.

More Recent Candidate Superconductors

More recently another class of superconductors has emerged which apparently is not explained by the conventional BCS theory. These materials have exhibited much higher critical temperatures, allowing cooling with liquid nitrogen ($T_b = 78\,^{\circ}$K) and even Freon-14 (CF$_4$, $T_b = 145\,^{\circ}$K). Figure 11.33 shows the timeline of progress towards higher critical temperatures in a variety of superconducting materials. A

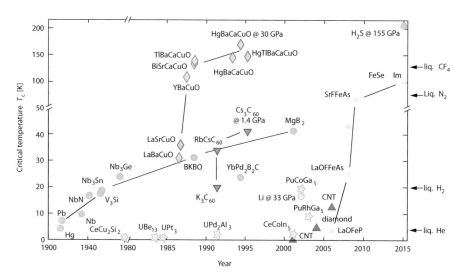

Fig. 11.33 Critical temperature T_c vs. year of discovery for some superconducting materials. From [80]

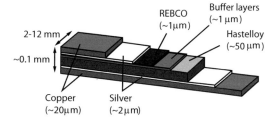

Fig. 11.34 REBCO tape superconductor for the reactor design described in [87]. Used with permission, Elsevier

general class of the materials of the barium-cupric oxide type, known as REBCO (for "rare earth barium copper oxide") have been developed by several different companies and these have shown promise as engineered materials for high-field magnets. An example of a fusion reactor design study using a REBCO toroidal field coil, along with other REBCO coils, is described in [87]. The conductor for this design uses a flat, tape-wound type of coil, which has shown promise for production of very high magnetic fields (some up to 100 T in small samples!), albeit mostly in tests at 4.2 °K. A schematic cross section for the REBCO conductor used in [87] is shown as Fig. 11.34.

11.4.2 Cryogenic Stability

Most superconducting materials have rather high resistance when they are not in the superconducting state. (In general alloys have higher electrical resistance than ele-

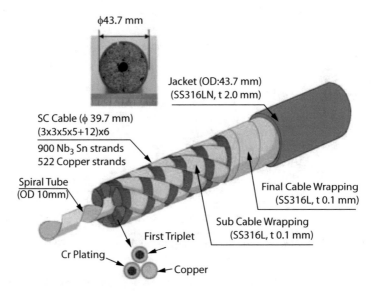

Fig. 11.35 Superconducting magnet conductor design for the toroidal field (TF) coil on ITER. From [42]. Used by permission, Elsevier

mental metals, and none of the elemental superconductors exhibit Type II behavior.) It is also important to note that the specific heat of all materials goes to zero as the temperature approaches absolute zero, and thus a very small amount of energy to be dissipated due to, for example, a small mechanical adjustment of the conductor position due to self-stress or some external impact may generate a temperature rise sufficient to drive the conductor "normal," i.e. out of superconductivity. Therefore it is especially important to stabilize the conductor to thermal transients by providing a parallel conduction path with low (but finite) resistance in the event of a going-normal type of event. To this end, the superconducting magnet conductors are typically embedded in copper, and wound with other strands of pure copper, and contained within a bath of liquid helium with good thermal transfer characteristics to the conductor. A design for the ITER toroidal field coil conductor is shown here as Fig. 11.35.

Liquid helium has a similarity to water with regard to its heat transfer characteristic in one respect. As the temperature difference ΔT between the surface and the coolant bulk rises, the heat flux q'' also increases until some critical heat flux is reached, and then the heat flux starts dropping. The point of departure from the monotonic behavior at the lower heat fluxes is caused by the change in the boiling characteristics of the fluid: at low ΔT, boiling produces small bubbles which nucleate on the surface, whereas when ΔT increases, the vapor phase creates a sideways motion under the coolant, and little of the surface is wetted by the fluid. This secondary regime, called film boiling, results in less heat transfer until some very high value of ΔT is reached. (Anyone familiar with the behavior of a drop of

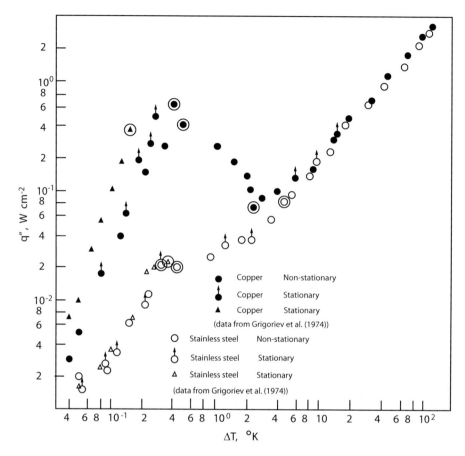

Fig. 11.36 Boiling curve for helium on copper and stainless steel. From [39], with data from [38]. Used with permission, Elsevier

water landing in a hot frying pan will recognize this condition.) Figure 11.36 shows this effect for the case of a copper surface wetted by liquid helium. (Note that this effect is apparently absent in a stainless steel-helium situation, presumably due to a different surface tension effect: in this case, there does not appear to be a nucleate boiling regime with good heat transfer.)

Because of this boiling characteristic in the helium-copper case, the engineering design of a cryogenically stable superconducting conductor generally requires that the maximum heat flux generated by a normal transient be kept below this critical heat flux q''_{crit}, which is typically taken to be $0.3 \to 0.4\,\mathrm{W\,cm^{-2}}$. Since only the Cu stabilizer is carrying the current during such a transient, the heat balance must be such that the ohmic heating per unit length $I^2\rho_s/A_s$ be less than the maximum heat removal rate which is set by the critical heat flux and the wetted perimeter p, thus:

$$\frac{I^2\rho_s}{A_s} \leq q''_{crit} p. \tag{11.51}$$

As an example of a "classically" designed superconducting magnet conductor, consider the MFTF-B magnet designed at Livermore in 1979 [44]. Here the conductor had a square cross section 12.4 mm on a side. The inner conductor, 6.5×6.5 mm, consisted of copper with 480 embedded NbTi strands which were 0.2 mm in diameter and twisted with a pitch of 1.8 m. An outer jacket of pure copper was wrapped around the inner conductor, and it had a set of perforations and internal grooves to promote coolant flow. (By taking credit for the additional grooves and perforations, the wetted perimeter was 8.17 cm rather than just 4×1.24 cm = 4.96 cm.) The overall stabilizer-to-superconductor ratio (by area) was 6.7:1. The resistance per unit length of the copper stabilizer $\rho_s/A_s = 46\,n\Omega\,cm^{-1}$ at 4.5 °K. Applying Eq. (11.51) to this design, at the operating current of 5775 A, one obtains a heat flux during a normal excursion of $0.19\,W\,cm^{-2}$, well within the boiling limit for liquid He.

The ITER TF conductor shown in Fig. 11.35 above is perhaps a bit more sophisticated in its design. In this conductor, the individual strands are not combined into a solid matrix but rather twisted in a rope-like structure with helium flowing in between the strands as well as in a central coolant duct. The cryogenic stability is determined by the effective wetted perimeter of each strand, which is taken as a fraction α of the perimeter πd of each strand [7]. For the ITER TF coil conductor design, α is taken to be 5/6. If the heat flux is taken to be $q'' = h\left(T_c - T_{op}\right)$, with T_c the conductor temperature and T_{op} the operating temperature of the coolant, then the current density in the strands is, for thermal stability:

$$J_{str} = \left(\frac{4h\alpha \left(T_c - T_{op}\right)}{\rho_{Cu}d} \frac{x}{1+x} \right)^{1/2}, \tag{11.52}$$

where x is the ratio of the areas of the copper to non-copper zones in the conductor. The Newton cooling law coefficient is taken as $h = 1000\,W\,m^{-2}\,°K^{-1}$ for Nb$_3$Sn conductors and $h = 400\,W\,m^{-2}\,°K^{-1}$ for NbTi conductors. The coolant in the ITER TF coils is designed to be at 5 °K in a supercritical state, i.e. a state above the critical point in temperature and pressure, where liquid and gaseous phases are not distinguishable, so that the boiling conditions used above do not apply. More complex modeling of the situation during a thermal transient is now done using computer codes such as Gandalf [8], which can also track the evolution of "hot spots" during a full quench, and can then account for the possibility of irreversible damage during such an event.

11.4.3 Magnet Stresses

The magnetic coils in a fusion device can generate tremendous forces. The elementary law describing an incremental force on a filamentary wire is given by the Lorentz force

$$d\mathbf{F} = Id\mathbf{l} \times \mathbf{B}. \tag{11.53}$$

It is easy to see the stress induced in a simple geometry such as a long, straight solenoid. Here one can consider the magnetic pressure to be $p = B^2/(2\mu_0)$ as shown in the earlier chapter on MHD equilibrium. We also have a relatively simple formula for the stress in the wall of a pipe of radius R and thickness t filled with a fluid with pressure p inside:

$$\sigma = \frac{pR}{t}, \tag{11.54}$$

so applying this formula to the magnetized solenoid results in an expression for the hoop stress on a solenoid:

$$\sigma = \frac{B^2 R}{2\mu_0 t}. \tag{11.55}$$

This simple expression is not directly applicable to any real magnet system, but it does reveal certain qualitative features. For one, stress generally scales with the square of the magnetic field and linearly with the radius of curvature R_c and inversely with the thickness of the load-bearing structure. For this reason, toroidal field magnets tend to be of a D-shaped design, with the least curvature (largest R_c) on the outboard side, where the field is the weakest, and the greatest curvature near the inboard side, until the minimum radial point in the coil is reached, where the coils are keyed together and the forces are radial. This type of deign is sometimes called a constant-stress design and is a good approximation of the ITER TF coil design. The actual stresses are a bit more complex than this simple picture, and the real stress distribution is a function of the material properties, as well as the magnetic forces and gravity. Figure 11.37 shows a finite element calculation for the stresses on the ITER TF coil.

11.4.4 Electrical Protection

The ITER TF coil has a total megnetic energy of $W_m = 1/2 \, LI^2 \sim 40\,\text{GJ}$ [18] and management of this huge amount of energy in case of a quench (normal incident) is a formidable task. Like any inductor, the current cannot instantaneously change, and the maximum affordable dI/dt is determined by the maximum voltage that the coil insulation can withstand without breakdown. Details of the electrical power supply design are given in [57]. In the case of a forced TF shutdown, the design maximum voltage is taken to be 4.0 kV anywhere in the system, and each component is designed for a maximum voltage of 17.5 kV and will be tested to 16 kV. Taking a maximum current in the conductor for the TF coil of 68 kA, this gives an L/R time of $\tau_{L/R} = 2W_m/(V_{max}I)$ of 300 s, or 5 min. Since the ITER TF coils are attached to both slides of a radial plate, the plate is biased to the average voltage between the two current leads on each coil through a resistive voltage divider, although a provision is made to break that connection in case of a single point coil-to-plate insulation fault.

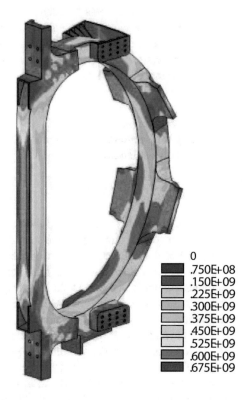

Fig. 11.37 Finite element calculation of the stresses in the ITER TF coil at the end of a plasma burn. Stresses are in Pa. From [28]. Used with permission of the author

Since the other coils on ITEER operate in a pulsed mode, these coils develop voltage even in "normal" (superconducting) conditions, and they are designed for higher breakdown voltages during transients, going as high as 28 kV in the central solenoid (CS) circuit. As these coils have less stored energy, they have been designed for faster shutdown times, and they are designed to be shut down quickly in the event of a TF fault.

11.5   Vacuum Systems

It is of utmost importance in achieving good performance in a burning fusion plasma that vacuum conditions are pristine. A typical specification for the base pressure in a tokamak plasma chamber is a maximum of 10^{-8} torr. (Note that 1 torr = 1 mm Hg = 1.3332 mBar = 133.32 Pa.) In addition, for a tritium-burning environment, all equipment must be operating with no hydrocarbon or silicone lubricants, as these become pathways for radioactive contamination. Thus all seals must be of a metallic compression type and leak checked to assure no leak larger than about 10^{-12} torr-l s^{-1}. In general, however, the main source of unwanted particles in a well-designed vacuum system will be on the surfaces of the components inside the vacuum envelope, as well as from within these materials. This may require that components be aggressively cleaned and then baked to drive water, carbon dioxide, and other gases from within the materials. For this reason, taking a large tokamak or other magnetic fusion device "up to air" is a major undertaking, and will result in weeks of time off-line. An important part of experimental design is to provide an adequate number of isolation valves, so that if one component, such as a neutral beam injector or a diagnostic device, develops a vacuum integrity problem, then that event will not compromise the entire machine to a great extent. The inventory of hydrogen isotopes within the machine must be accounted for carefully, as an up-to-air emergency incident can create explosive conditions, and tritiated water formed by such an explosion has a biological retention rate about 10^5 times higher than molecular DT or T_2.

Another problem for management of a large vacuum system actually comes about because of the large size compared to the more familiar table-top experiments. While in general the surface-to-volume ratio goes down with increased size, making the goal of achieving a low operational base pressure possible in normal operation, the time to detect even a large leak, say a one centimeter hole in the vacuum envelope, may be difficult to see immediately. This is especially true if a leak happens near a cryogenic surface, where the influent gases will be sublimated on the cold surfaces with no immediate signs of distress. (A water leak on the TFTR neutral beam injector neutralizer cell on the test stand at Berkeley over a weekend created several tons of ice.)

11.5.1 Viscous and Molecular Flow

The starting point for any discussion of vacuum technology is the determination of a key characteristic parameter. That parameter, called the Knudsen number Kn, is the ratio of the mean free path for collisions of like particles, $\lambda = (\sqrt{2}n\sigma)^{-1}$, where σ is a typical collision cross section and n is the number density of molecules in the gas, to a characteristic length (or diameter) D:

$$\text{Kn} = \frac{\lambda}{D}. \tag{11.56}$$

By assigning a characteristic molecular diameter d_m, we have a characteristic cross-sectional area πd_m^2 for two molecular centers a distance d_m apart to collide and thus we can write

$$\lambda = \frac{1}{\sqrt{2}n\sigma} = \frac{RT}{p} \cdot \frac{1}{\sqrt{2}} \cdot \frac{1}{\pi d_m^2} \equiv \frac{k_1}{p}, \tag{11.57}$$

Table 11.2 Mean free path coefficients k_1 for various gases

Gas	Chemical form	k_1(Pa m)
Hydrogen	H_2	11.5×10^{-3}
Nitrogen	N_2	5.9×10^{-3}
Oxygen	O_2	6.5×10^{-3}
Helium	He	17.5×10^{-3}
Neon	Ne	12.7×10^{-3}
Argon	Ar	6.4×10^{-3}
Air		6.7×10^{-3}
Krypton	Kr	4.9×10^{-3}
Xenon	Xe	3.6×10^{-3}
Mercury	Hg	3.1×10^{-3}
Water vapor	H_2O	6.8×10^{-3}
Carbon monoxide	CO	6.0×10^{-3}
Carbon dioxide	CO_2	4.0×10^{-3}
Hydrogen chloride	HCl	3.3×10^{-3}
Ammonia	NH_3	3.2×10^{-3}
Chlorine	Cl_2	2.1×10^{-3}

At 273.15 °K. From [49]

and the molecular size d_m can be experimentally determined from viscosity measurements of the gas. Table 11.2 lists the resulting coefficient k_1 for a fixed temperature of $0\,°C = 273.15\,°K$.

When Knudsen numbers $Kn \ll 1$, the molecular dynamics is set by the transport properties within the gas, i.e., it behaves like a classical fluid. At the other extreme, when $Kn \gg 1$, molecules interact mostly with the walls of the system, and the transport takes on a character similar to photon transport, but without significant absorption in the metallic walls.

Of particular interest is the expression for the throughput of a gas into a perfectly absorbing sheet with area A. (We can call this a "black hole" for gas particles.) The directed flux Γ_- through such a surface is given by

$$\Gamma_- = \frac{n\bar{v}}{4} = \frac{n}{4}\left(\frac{8}{\pi}\frac{k_B T}{m}\right)^{1/2}. \tag{11.58}$$

(The factor of four in the denominator is due to half of the particles in a maxwellian distribution traveling away from the surface, rather than towards it, and another factor of two is due to the average cosine $< \cos\theta >= 1/2$ for the forward-moving particles.)

We define pumping speed S_t such that the throughput of a pump Q is given by

$$Q = S_t\,(p - p_u) \tag{11.59}$$

The throughput Q, which is a measure of the total particles per second being pumped, can be written in various units, but the most convenient unit is to state it in a unit which represents pressure × volume per unit time, using the ideal gas law. If Eq. (11.59) is employed with the same pressure unit on both sides, then the pumping speed has units of volume per second, e.g liters per second or m³ per second. We then note, employing the black-hole model for a pump with an inlet of area A, the pumping speed is given simply by

$$S_t = \frac{<v>}{4} A. \tag{11.60}$$

No pump operating in the high-Kn regime can pump faster than this black-hole pumping speed. Notice, however, that this characteristic speed is huge, especially for hydrogen isotopes where $<v>$ is large. If we designate a fraction η to be the ratio of the real pumping speed to the black hole pumping speed, we can get a fair estimate for the pumping speed from a cryogenic pump with an inside panel temperature of 4.5 °K. The following example is typical.

Suppose that we have an $A = 1.0\,m^2$ He-cooled cryopanel with a pumping efficiency of 0.3 relative to the black hole pumping speed. Find the pumping speed for nitrogen (N₂) at 300 °K.

First some physical constants:

$$k_B = 1.38064852 \times 10^{-23}\,J\,°K^{-1}$$

$$u = 1.66054 \times 10^{-27}\,kg$$

$$M(N_2) = 14.0067 \times 2u = 4.65174 \times 10^{-26}\,kg$$

Then:

$$<v> = 476.173\,m\,s^{-1},$$

and then the pumping speed is

$$S_t = 0.3 \cdot A \cdot <v> / 4 = \boxed{35.713\,m^3\,s^{-1}},$$

or about 35,000 liters per second.

In association with the pumping speed concept is a similar formula associated with a pipe or duct. Here we define a conductance C such that

$$Q = C(p_2 - p_1), \tag{11.61}$$

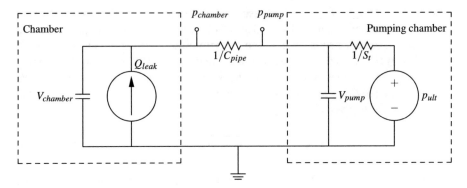

Fig. 11.38 Electrical analog of a vacuum system

where p_1 and p_2 are the pressures at the two ends of the pipe. For a vacuum system with a chamber connected to a pump through a pipe with conductance C_{pipe}, the pump-pipe combination then has an overall pressure drop $p_{chamber} - p_{ult} = Q\left(1/C_{pipe} + 1/S_t\right)$, or as an effective pumping speed

$$S_t' = \frac{S_t C_{pipe}}{S_t + C_{pipe}}, \qquad (11.62)$$

such that $p_{chamber} - p_{ult} = Q/S_t'$. Note that this formula is analogous to the formula for the resistance formed by two resistors in parallel. Similarly, the conductance of more than one conductance in hydraulic series is given by:

$$1/C_{tot} = 1/C_1 + 1/C_2 + \cdots + 1/C_n. \qquad (11.63)$$

We can also treat the volume of the evacuated regions as a sort of capacitance in the electrical analogy, since $Q = V(dp/dt)$ gives the rate of pressure change in a chamber of volume V with gas throughput Q. Thus a full circuit model, including transient conditions, is represented by Fig. 11.38.

From this model, it is easy to show the time dependence of pressure from an initial pumpout. Suppose that we ignore the volume of the pumping chamber (and also the pipe, which is not shown in this model). The "voltage" on capacitor "$V_{chamber}$" is the initial pressure p_0. The "RC" time constant is just

$$\tau = V_{chamber} \left(\frac{1}{S_t} + \frac{1}{C_{pipe}} \right), \qquad (11.64)$$

and the chamber pressure vs. time is given by

$$p_{chamber}(t) = (p_0 - p_{ult} - p_{leak}) \exp(-t/\tau) + p_{leak} + p_{ult}, \qquad (11.65)$$

with the elevation of the base pressure due to leaks p_{leak} given by

$$p_{leak} = Q_{leak} \left(\frac{1}{S_t} + \frac{1}{C_{pipe}} \right). \tag{11.66}$$

Conductances of pipes in the molecular flow regime have been available for various geometries for some time. The conductance for pipes of constant cross-sectional area can be written as

$$C = \alpha A \frac{<v>}{4}, \tag{11.67}$$

where α is a transmission probability for molecules to travel from one end of the pipe to the other without momentum exchange with the walls, so that $\alpha = 1$ for a length zero aperture and the black hole throughput is retained. For a circular pipe of diameter D and length L, the classic paper by Clausing in 1932 [17] gives the value of α:

$$\alpha = \frac{1}{1 + 3L/(4D)}, \tag{11.68}$$

and thus the conductance $\propto D^3/L$ for a long pipe. Conductances for more complex shapes such as ells, baffled pipes, and conical tapered transitions are cataloged in [40]. One important example is given here as shown as Fig. 11.39: the transmission probability of a 90-degree ell, with an arm of length $L = D$, measured from the inside of the bend, is 0.21.

Some caveats to this building-block approach should be mentioned, however. The conductance calculations were done with the assumption of a large chamber between each component in the vacuum line, so that the gas distribution becomes

Fig. 11.39 Transmission probability α for a 90-degree ell in a pipe. From [40]

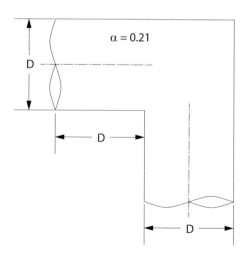

isotropic again after transmission through a single component, and connecting two components in hydraulic series does not give the exact value as if the combination of the two components were treated as a single unit. For example, if two pipes with $L/D = 10$ were connected together, the series-parallel formula given in Eq. (11.63), using Eq. (11.68) for the individual conductances, gives an overall $\alpha = 1/17$, whereas treating this as a single pipe with $L/D = 20$ gives $\alpha = 1/16$: not a big difference here, but larger inaccuracies occur with components with $L/D \sim 1$.

A second caveat involves the use of a single leak rate as is shown in Fig. 11.38. In most practical vacuum applications, the leak rate is determined by three different sources of molecules. The first are true leaks to the outside world, through small cracks in welds, imperfect seals, and so forth. These are rare in large fusion experiments after final assembly, because each component is thoroughly leak-checked, typically using helium mass spectrometry, before it is shipped and installed. (In fact it is not uncommon for suppliers to ship parts under high vacuum and with a vacuum gauge attached, with a low pressure reading required as part of the acceptance test before the part is paid for.) The second source of leaks are the so-called "virtual leaks," which happen when a small cavity exists in the vacuum wall with a tortuous path into the vacuum chamber such that the gas inventory in this cavity is pumped out slowly, e.g. days or weeks. These can be avoided by having welds only on the inside of a metallic joint (easier to accomplish these days in the advent of robotic e-beam welding), and using vented screws and other fasteners where required, and avoiding them when possible. But the third category is possibly the greatest source of particles in the long term (hours to months after initial pumpdown), and that is outgassing from surface contamination. Typical cold surfaces are covered with hundreds of monolayers of atmospheric components (one monolayer of 10 Å-diameter molecules is about 10^{18} molecules per square meter), and these are attached to the surface by intermolecular van der Waals forces. Of these, water can be the most pernicious, although baking systems to $200 \to 400\,^{\circ}$C can liberate these and other strongly attached molecules. Organic materials such as grease, fingerprints, and other detritus can be a serious source of slow outgassing. (A famous case was a large evacuated experiment at Berkeley, which could not reach high vacuum until the chamber was re-opened and a cheese sandwich left inside by a worker was removed!). A campaign to not have these contamination sources present in the first place is the best defense against these difficult sources. Vacuum vessel walls in the fusion device can benefit from helium and hydrogen discharge cleaning, where low-temperature plasmas are formed in the chamber and scrub the surface, sometimes for long periods of time (days to weeks).

Thus calculating pumpdown times from the simple model shown in Fig. 11.38 may give a short, but accurate, result for the pumpdown time to some medium-quality vacuum, say, 10^{-6} torr, but arrival at the final system ultimate base pressure of better than 10^{-8} torr may take much longer. For this reason, complete up-to-air venting is avoided as much as possible, and isolation of subsections of the device requiring repair, and then backfilling these volumes with dry nitrogen rather than ordinary air, may minimize the down-time of the device. (One might also add that in a tritium-burning fusion device, the radioactive contamination carried out as atmospheric contamination is carried in is another good reason to avoid up-to-air events, planned or otherwise.)

Table 11.3 Main parameters for the ITER vacuum system

Parameters	Value
Vacuum vessel free volume	$\sim 1350\,\text{m}^3$
Ultimate base pressure for hydrogen isotopes/ impurities	$10^{-5}/10^{-7}\,\text{Pa}$
Base pressure in the dwell phase	$10^{-4}\,\text{Pa}$
Available effective molecular pumping speed	$>30\,\text{m}^3\,\text{s}^{-1}$
Typical divertor pressure during plasma operation	$1 \rightarrow 10\,\text{Pa}$
Maximum throughput during plasma operation	$200\,\text{Pa}\,\text{m}^3\,\text{s}^{-1}$

From [20]

11.5.2   ITER Vacuum System

Table 11.3 gives the main parameters for the ITER vacuum system. The ITER vacuum system [20] uses two pumping technologies, roughly dividing the pumping between rough pumping where Kn≪ 1 and high vacuum pumps where Kn ≫ 1. For the initial pumpdown to molecular flow conditions (taken as a pressure of 10 Pa (7.5 mTorr)), a set of mechanical pumps will be used. The two pumping phases, rough pumping and high-vacuum pumping, will be discussed in turn.

Rough Pumping

While there are some technologies that meet the ITER criteria, there is no commercial product that has shown to be ideal for the mechanical pumps. However, there are some products that either can be used off-the-shelf or customized for this application. The first of these is the scroll pump. Figure 11.40 shows the basic concept. Two spiraling vanes, one fixed and one movable, create a trapped gas volume as the movable vane wobbles in a circular motion without actually rotating. Because of this lack of rotation, the mechanical motion can be imparted through a bellows capable of sideways motion, thus creating an all-metal seal. Scroll pumps are frequently used in tritium environments, but typically on a much smaller scale. Also, the close tolerances between the movable scroll and the side walls create a limitation on the compression ratio available and potentially threaten the lifetime.

Next, there are Roots blowers. Figure 11.41 shows this concept. Here two impellers work in a meshed fashion to yield an almost-positive displacement device. This relatively mature technology (popular in superchargers for internal combustion engines) is sometimes used as a first,"coarse" vacuum pumping stage in front of another mechanical pump, or in a cascade arrangement with other Roots blowers. While they have good throughput, they suffer from heat dissipation due to gas friction in the small gap between the rotors and the casing, and have very modest achievable pressure ratios. Also, the current models have lubricated seals, and some development effort would be required to create an oil-less bearing and feedthrough system, although ferrofluidic seals have been tested and shown acceptable, if not perfect, tritium leak rates.

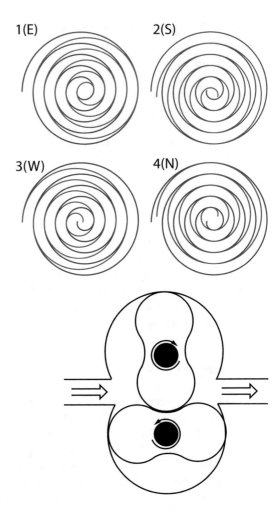

Fig. 11.40 Diagram showing one cycle of motion in a scroll pump. A stationary scroll is shown in blue. A movable scroll (red) wobbles (without rotation) clockwise (indicated north, south, east, west) and creates a crescent-shaped pocket of gas which moves clockwise and inward in a peristaltic fashion

Fig. 11.41 Roots blower mechanical design. Some designs have more than two lobes

Finally, there have been proposed designs for mercury liquid ring pumps [35]. These work by compressing pockets of gas between the vanes of a paddlewheel-type rotor eccentrically mounted in a circular chamber, which throws a uniform layer of the pump fluid towards the wall of the chamber by centrifugal force. Figure 11.42 shows a schematic of this design. While this concept is still in the prototype phase, its high performance has suggested further development. Details pertaining to the seal design and removal of mercury from the exhaust stream remain to be addressed for large-scale use on ITER.

High-Vacuum Pumping

The high-vacuum pumping system for ITER is more certain. It will definitely consist of large-area cryopanels with an outer baffle at 78 °K and an inner surface at 4.5 °K. While these panels have very high pumping speed per unit area as outlined

Fig. 11.42 Prototype mercury liquid ring roughing pump for ITER. From [35]. Used by permission, Elsevier

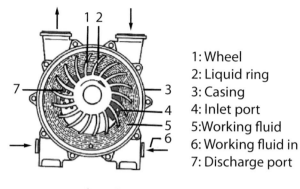

1: Wheel
2: Liquid ring
3: Casing
4: Inlet port
5: Working fluid
6: Working fluid in
7: Discharge port

Fig. 11.43 Photo of the ITER cryopump. Front 80 °K baffle/gate valve not shown. ©Fusion for Energy

4.5 °K panels (28)

80 °K outer shield   80 °K inner shield

above, they are not capable of pumping helium per se. However, by coating these inner cryogenic surfaces with activated charcoal, they become cryosorption pumps. A photo of the ITER cryopump is shown as Fig. 11.43. Significant research has gone into choosing the right cryosorption material. It has been found that activated charcoal made from a particular crop of coconuts grown on one plantation in Indonesia gives the highest absorption for helium, and the ITER organization has secured a large supply and is, as of this writing, storing them in the basement of the administration building at the facility outside Aix-en-Provence in southern France [70].

11.6   Blanket and Shield Design

11.6.1   Introduction

Fusion reactors based on the DT reaction must have a sustainable fuel cycle. This means that they must breed tritium in sufficient quantities to replenish the initial fuel load within a reasonable time frame to keep the total tritium inventory down below certain limits prescribed by general safety and regulatory amounts. Furthermore, the startup of subsequent fusion reactors will require an additional tritium investment, so that the original reactors built must breed more tritium than they consume. The tritium-breeding blanket is also the primary structure for conversion of the neutron energy into heat for the thermal energy conversion downstream. A blanket designed purely for neutron energy conversion and tritium breeding may not result in adequate attenuation of radiation levels on its outboard side, and if this is the case, then an additional nuclear shield must be provided to reduce the neutron flux to levels which the magnets can tolerate (in the case of magnetic fusion), to optical and/or other driver components (in the case of inertial fusion), and to personnel outside of the reactor (a "biological" shield) in either case.

Two primary criteria for blanket performance are thus obvious. The first is the ratio of tritium produced in the blanket to the fusion neutrons produced in the burning plasma, labeled TBR (for "tritium breeding ratio," or simply T. Clearly $T > 1$ is needed for a successful fusion power cycle, and $T > 1.2$ will satisfy the needs for expansion of the fusion reactor power capability in the future. The second parameter of interest is the energy multiplication per fusion neutron M, defined simply as

$$M = \frac{\text{Energy generated per fusion neutron}}{14.08\,\text{MeV}}. \qquad (11.69)$$

While higher values of M might seem desirable, in most reactor designs these are achieved by combining the tritium-breeding Li with other elements which have exothermic neutron capture reactions. In some cases this can lead to elevated risks from induced radioactivity in these energy-multiplying materials, and may increase the biological risk in some accident scenarios. Extreme examples are fusion–fission hybrid designs, where the fusion neutrons are used as a source for a neutron-multiplying fission subcritical assembly, multiplying the power by factors up to ten or so and also generating fresh fission nuclear fuel, such as plutonium and ^{233}U by fertile conversion. Taking this step, however, increases the risk of larger radionuclide inventories being released to the public in accidents. Some (somewhat naively) believe that these hybrid designs are safe because the fission component cannot become supercritical as it is controlled by the fusion neutron production, and the fissile blanket is merely an amplifier. However, the real issue in conventional fission power is usually control of nuclear afterheat from the fissionable core following shutdown due to the radioactivity of fission products, which runs as high as 7% of operating power immediately after shutdown and falls to half that after 1 day. Seven percent of a typical fission reactor output is thus above 200 MW, and if no heat removable is available, this will melt the core of the reactor in a few hours. (Consider Three Mile Island and Fukushima in this instance.) These "LOCA" (loss of coolant accident) events are very dangerous situations, even in a well-designed fission reactor. Loss of criticality control, causing a supercriticality event ("ATWS," for anticipated transient without scram), is even worse, as the reactor power can reach terawatts and result in expulsion of radioactive materials in cataclysmic proportions. This type of accident in this author's opinion requires a design flaw, such as was present at Chernobyl. In short, fusion breeding blankets using fissile materials may be safe to ATWS, but not LOCA.

Here we will discuss the designs for blankets and shields in two fusion reactor design studies. The first of these is from a design for a large-scale, DT-burning tokamak reactor [27, 68]. The second is for a simpler DD-based reactor design where no T breeding is required [69].

11.6.2   Design for a DT-Burning Large Tokamak Reactor: ARIES-RS

The ARIES-RS builds on a family of ARIES (for Advanced Reactor Innovation and Evaluation Study) conceptual fusion reactor design activities starting in 1988. The ARIES-RS (for Reversed Shear) design was for a tokamak with a major radius of 5.52 m which could produce 2170 MW of fusion power (2620 MW of total power). Several first wall, blanket and shield options were studied. Figure 11.44 shows the calculated performance (TBR and M) for an assortment of primary breeding blanket materials. From this set, Li was selected as the primary T breeding material, and it had the highest TBR of any of the candidate breeding materials. Vanadium alloys were chosen as the primary structural material in most of the critical areas with high heat flux and neutron flux. While ferritic steels and silicon carbide ceramic were considered as alternative structural materials, the V alloys showed the lowest

Fig. 11.44 Tritium breeding ratio (TBR) and neutron energy multiplication for various breeding blanket materials in the ARIES-RS design. From [27]. Used by permission, Elsevier

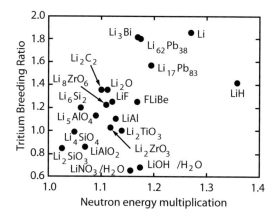

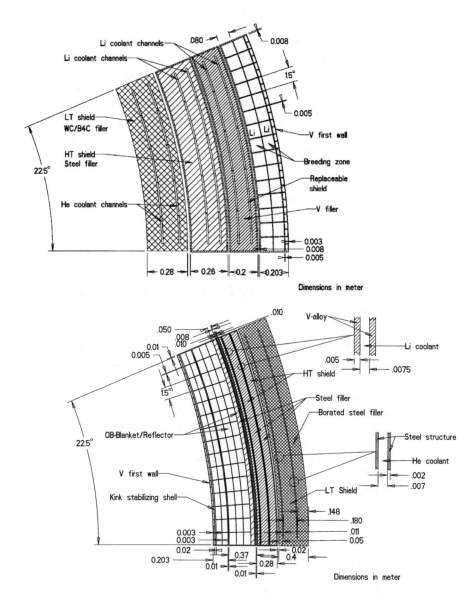

Fig. 11.45 Inboard (top) and outboard (bottom) blanket designs for the ARIES-RS blanket and radiation shield. From [27]. Used by permission, Elsevier

impact on the overall breeding ratio. Figure 11.45 shows the design of the inboard and outboard blanket. A drawback in these choices for the structural material and breeding material, V/Li, is that this combination would have very high pumping power requirements because of the high electrical conductivity of each, so it was necessary to add insulator material to the Li coolant ducts, and this opens the question of the long-term reliability of such coatings.

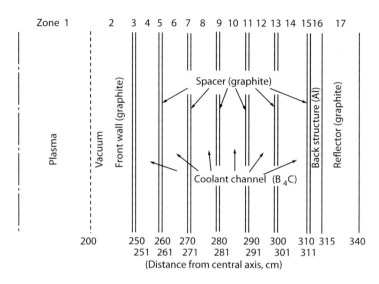

Fig. 11.46 Design for a simple non-breeding blanket. From [69]

11.6.3 A Simple Non-breeding Blanket Design

Figure 11.46 shows a simple design for a passive (non-breeding) blanket as might be used in conjunction with a catalyzed-D fusion reactor. This design was presented in Ref. [69]. The design uses a "gas-suspended" boron carbide neutron absorber and shield with a graphite reflector. In this design, the B_4C nanoparticles are suspended in an inert gas such as helium, and this gaseous slurry is circulated through a heat exchanger. Catalyzed-D, as shown in Chap. 2, relies on the production and swift removal of tritium derived for the tritium-producing branch of the D-D reaction $D + D \rightarrow T + p$, which happens for 50% of all DD fusion reactions and is quickly followed by a DT reaction with $\approx 100\times$ higher reactivity and producing a 14.08 MeV neutron. Because the other 50% of the DD reactions produce a 2.45 MeV neutron through the reaction $D + D \rightarrow{}^3 He + n$, 14.08 and 2.45 MeV neutrons are produced in equal quantities. Without the demand for T breeding in the blanket, this structure can be optimized for neutron absorption and thermal energy conversion. The goals of the blanket are thus to collect the neutron energy into the working fluid for conversion of thermal energy into electricity, and to protect the superconducting magnet which would be located just outside the blanket. (The authors of this study assumed a one-dimensional geometry, approximating a long cylindrical zone for the burning plasma.)

It is important to note that this phase of the blanket design concentrates on conversion of the neutron energy to heat and not on protection of the magnets outside the blanket, i.e. it is a blanket but not a shield. In the author's design with maximum boron concentration, they calculate that the average energy per neutron absorbed to be 7.38 MeV per neutron (a bit less than the average initial neutron

Fig. 11.47 Neutron fluence vs. total wall thickness for a stainless steel-B_4C composite wall, with data from [1]. From[10]. Used by permission, Elsevier

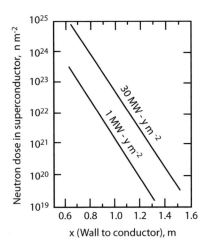

energy of $8.26\,MeV = (14.08+2.45)/2\,MeV)$ and an average neutron energy leakage of $0.06\,MeV$ per neutron.

As shown earlier, Fig. 11.32 shows the change in the critical temperature T_c as a function of fast neutron fluence for common Type II superconductors. One sees from this that the critical fluence before radiation damage becomes pervasive is around $10^{22}\,nm^{-2}$, Given that first wall fluxes for a $1\,MW\,m^{-2}$ neutron wall loading are around $10^{18}\,m^{-2}\,s^{-2}$, this gives a time to failure on order of $10^4\,A$ s, where A is the attenuation factor for the fast neutron flux. As quoted in [10], Abdou [1] shows an attenuation factor on the order of one decade per $20\,cm$ for a composite blanket structure of B_4C and stainless steel, as shown in Fig. 11.47. Thus we might estimate that the blanket design referred to above [69] might be coupled with an additional passive structure of stainless steel and boron carbide on the order of $30 \rightarrow 50\,cm$ for adequate neutron attenuation to protect the superconducting magnets.

Problems

11.1 For a superconducting magnet filament consisting of 80% copper fill on a $1.2\,cm$ square cross section, find the maximum current for thermal stability at $4.2\,°K$. Do the same for the ITER magnet filament design, accounting for the hollow cross section (cooling both inside and outside) using a critical heat flux of $0.3\,W\,cm^{-2}$.

11.2 Find the pumping speed of a turbomolecular pump in liters per second with a 10-inch diameter opening operating at 20% of the "black hole" pumping speed. Assume D_2 gas. Also find the pumping speed of the same pump with six feet of 10-inch diameter piping in front of it, with one right-angle bend.

11.3 Find the thermal stress on a 5 mm-thick plate of 304 stainless steel which is exposed to a heat flux of 200 kW per square meter. Ignore internal heating and use material properties at 300 °C.

11.4 Find the stress in the wall of a cylindrical vacuum vessel which is 5 mm thick, twenty feet in diameter, with atmospheric pressure on the outside and vacuum inside. Compare this to the yield stress of 6061-T6 aluminum alloy.

11.5 The lithium coolant in a fusion reactor has a total inventory of 5000 kg and has a 2 MCi inventory of tritium. Find the leakage rate of this tritium through $1000 \, m^2$ of 316 stainless piping surface area at 330° K with a thickness of 5 mm everywhere. Express this result in curies per day.

11.6 Find the average dpa cross section for V-20Ti alloy. Estimate the wall lifetime for 10 dpa with a wall loading of $2.0 \, MW \, m^{-2}$.

11.7 Find the pumping power required to move a $0.1 \, m \, s^{-1}$ channel of Li coolant through a 5 m section of pipe with a 30 cm diameter and 1 cm wall thickness constructed of HT-9 ferritic steel with a perpendicular magnetic field $B_\perp$ of 0.5 T.

References

1. Abdou, M.: Radiation considerations for superconducting fusion magnets. J. Nucl. Mater. **72**(1), 147–167 (1978). https://doi.org/10.1016/0022-3115(78)90398-7; http://www.sciencedirect.com/science/article/pii/0022311578903987
2. Anderl, R., Fukada, S., Smolik, G., Pawelko, R., Schuetz, S., Sharpe, J., Merrill, B., Petti, D., Nishimura, H., Terai, T., Tanaka, S.: Deuterium/tritium behavior in FLiBe and FLiBe-facing materials. J. Nucl. Mater. **329–333**(Part B), 1327–1331 (2004). Proceedings of the 11th International Conference on Fusion Reactor Materials (ICFRM-11), https://doi.org/10.1016/j.jnucmat.2004.04.220; http://www.sciencedirect.com/science/article/pii/S0022311504003952.
3. Atsumi, H., Tokura, S., Miyake, M.: Absorption and desorption of deuterium on graphite at elevated-temperatures. J. Nucl. Mater. **155**, 241–245 (1988)
4. Baldwin, M., Doerner, R.: Formation of helium induced nanostructure 'fuzz' on various tungsten grades. J. Nucl. Mater. **404**(3), 165–173 (2010). https://doi.org/10.1016/j.jnucmat.2010.06.034; http://www.sciencedirect.com/science/article/pii/S0022311510002849
5. Baldwin, M., Doerner, R., Nishijima, D., Tokunaga, K., Ueda, Y.: The effects of high fluence mixed-species (deuterium, helium, beryllium) plasma interactions with tungsten. J. Nucl. Mater. **390–391**, 886–890 (2009). Proceedings of the 18th International Conference on Plasma-Surface Interactions in Controlled Fusion Devices, https://doi.org/10.1016/j.jnucmat.2009.01.247; http://www.sciencedirect.com/science/article/pii/S0022311509002566
6. Begeal, D.: Hydrogen and deuterium permeation in copper alloys, copper-gold brazing alloys, gold, and the in situ growth of stable oxide permeation barriers. J. Vac. Sci. Technol. **15**, 1146–1154 (1978)
7. Bessette, D., Mitchell, N., Zapretilina, E., Takigami, H.: Conductors of the ITER magnets. IEEE Trans. Appl. Supercond. **11**(1), 1550–1553 (2001). https://doi.org/10.1109/77.920072
8. Bottura, L.: A numerical model for the simulation of quench in the ITER magnets. J. Comput. Phys. **125**, 26–41 (1996)

9. Brinkman, J.A.: Production of atomic displacements by high-energy particles. Am. J. Phys. **24**(4), 246–267 (1956). https://doi.org/10.1119/1.1934201

10. Brown, B.S.: Radiation effects in superconducting fusion-magnet materials. J. Nucl. Mater. **97**(1), 1–14 (1981). https://doi.org/10.1016/0022-3115(81)90411-6; http://www.sciencedirect.com/science/article/pii/0022311581904116

11. Buzi, L., Temmerman, G.D., Huisman, A., Bardin, S., Morgan, T., Rasinski, M., Pitts, R., Oost, G.V.: Response of tungsten surfaces to helium and hydrogen plasma exposure under iter relevant steady state and repetitive transient conditions. Nucl. Fusion **57**(12), 126009 (2017). http://stacks.iop.org/0029-5515/57/i=12/a=126009

12. Buzio, M., Noll, P., Raimondi, T., Riccardo, V., Sonnerup, L.: Axisymmetric and non-axisymmetric structural effects of disruption-induced electromechanical forces on the JET tokamak. In: Varandas, C., Serra, F. (eds.) Fusion Technology 1996, pp. 755–758. Elsevier, Oxford (1997). https://doi.org/10.1016/B978-0-444-82762-3.50158-0; http://www.sciencedirect.com/science/article/pii/B9780444827623501580

13. Calderoni, P., Sharpe, P., Hara, M., Oya, Y.: Measurement of tritium permeation in FLiBe (2LiF-BeF$_2$). Fusion Eng. Des. **83**(7), 1331–1334 (2008). Proceedings of the Eight International Symposium of Fusion Nuclear Technology, https://doi.org/10.1016/j.fusengdes.2008.05.016; http://www.sciencedirect.com/science/article/pii/S0920379608000926

14. Caskey, G., Derrick, R.: Trapping of deuterium during permeation through gold. Scr. Metall. **10**, 377–380 (1976)

15. Causey, R.A., Karnesky, R.A., San Marchi, C.: Tritium barriers and tritium diffusion in fusion reactors. In: Konings, R., Stoller, R. (eds.) Comprehensive Nuclear Materials, pp. 511–549. Elsevier, Amsterdam (2012). http://doi.org/10.1016/B978-0-08-056033-5.00116-6; arc.nucapt.northwestern.edu/refbase/files/Causey-2009_10704.pdf

16. Chopra, O., Shack, W.: Crack growth rates and fracture toughness of irradiated austenitic stainless steels in BWR environments. NUREG technical document, NUREG/CR-6960, pp. 1–142 (1988)

17. Clausing, P.: The flow of highly rarefied gases through tubes of arbitrary length. J. Vac. Sci. Technol. **8**(5), 636–646 (1971). https://doi.org/10.1116/1.1316379

18. Coatanea, M., Duchateau, J.L., Nicollet, S., Lacroix, B., Topin, F.: Investigations about quench detection in the ITER TF coil system. IEEE Trans. Appl. Supercond. **22**(3), 4702404–4702404 (2012). https://doi.org/10.1109/TASC.2011.2178979

19. Cooper, L.N.: Bound electron pairs in a degenerate fermi gas. Phys. Rev. **104**, 1189–1190 (1956). https://doi.org/10.1103/PhysRev.104.1189

20. Day, C., Murdoch, D.: The ITER vacuum systems. J. Phys. Conf. Ser. **114**(1), 012013 (2008). http://stacks.iop.org/1742-6596/114/i=1/a=012013

21. Di Maio, D., Roberts, S.: Measuring fracture toughness of coatings using focused-ion-beam-machined microbeams. J. Mater. Res. **20**(2), 299–302 (2005). https://doi.org/10.1557/JMR.2005.0048

22. Dolan, T.J.: Technology issues. In: Dolan, T.J. (ed.) Magnetic Fusion Technology, pp. 45–69. Springer, London (2013). https://doi.org/10.1007/978-1-4471-5556-0_2

23. Doran, D.: Neutron displacement cross sections for stainless steel and tantalum based on a Lindhard model. Nucl. Sci. Eng. **49**(2), 130–144 (1972). https://doi.org/10.13182/NSE72-A35501

24. Ebisuzaki, Y., Kass, W., O'Keeffe, M.: Solubility and diffusion of hydrogen and deuterium in platinum. J. Chem. Phys. **49**, 3329–3332 (1968)

25. Ehrlich, K.: Irradiation creep and interrelation with swelling in austenitic stainless steels. J. Nucl. Mater. **100**(1), 149–166 (1981). https://doi.org/10.1016/0022-3115(81)90531-6; http://www.sciencedirect.com/science/article/pii/0022311581905316

26. Eichenauer, W., Liebscher, D.: Measurement of the diffusion rate of hydrogen in gold. Z. Naturforsch. **17A**, 355 (1962)

27. El-Guebaly, L.: Overview of ARIES-RS neutronics and radiation shielding: key issues and main conclusions. Fusion Eng. Des. **38**(1), 139–158 (1997). https://doi.org/10.1016/S0920-3796(97)00114-2; http://www.sciencedirect.com/science/article/pii/S0920379697001142

28. Foussat, A., Lim, B., Libeyre, P., Devred, A., Mitchell, N., Oliva, A.B., Rajainmaki, H., Koizumi, N., Sgobba, S.: Engineering and materials challenges in ITER toroidal magnet system. IEEE/CSC Superconductivity News Forum (global edition) (2014)
29. Fraas, A.P., Thompson, A.S.: ORNL fusion power demonstration study: fluid flow, heat transfer, and stress analysis considerations in the design of blankets for full-scale fusion reactors. Technical report, ORNL/TM–5960, Oak Ridge National Laboratory, Oak Ridge, TN (1978). https://www.osti.gov/servlets/purl/12197857
30. Frauenfelder, R.: Solution and diffusion of hydrogen in tungsten. J. Vac. Sci. Technol. **6**, 388–397 (1969)
31. Frazer, D., Abad, M., Krumwiede, D., Back, C., Khalifa, H., Deck, C., Hosemann, P.: Localized mechanical property assessment of SiC/SiC composite materials. Compos. A: Appl. Sci. Manuf. **70**(Supplement C), 93–101 (2015). https://doi.org/10.1016/j.compositesa.2014.11.008; http://www.sciencedirect.com/science/article/pii/S1359835X14003479
32. Freudenberg, U., Vökl, J., Bressers, J., et al.: Influence of impurities on the diffusion coefficient of hydrogen and deuterium in vanadium. Scr. Metall. **12**, 165–167 (1978)
33. Fukada, S., Morisaki, A.: Hydrogen permeability through a mixed molten salt of LiF, NaF and KF (Flinak) as a heat-transfer fluid. J. Nucl. Mater. **358**(2), 235–242 (2006). https://doi.org/10.1016/j.jnucmat.2006.07.011; http://www.sciencedirect.com/science/article/pii/S0022311506004089
34. Garner, F.A.: In: Frost, B.R.T. (ed.) Materials Science and Technology, vol. 10A, p. 419. Wiley-VCH, Weinheim (1994)
35. Giegerich, T., Day, C., Jäger, M.: Mercury ring pump proof-of-principle testing in the THESEUS facility. Fusion Eng. Des. **124**, 809–813 (2017). Proceedings of the 29th Symposium on Fusion Technology (SOFT-29) Prague, 5–9 September 2016, https://doi.org/10.1016/j.fusengdes.2017.03.119; http://www.sciencedirect.com/science/article/pii/S0920379617303423
36. Gilbert, M., Dudarev, S., Zheng, S., Packer, L., Sublet, J.C.: An integrated model for materials in a fusion power plant: transmutation, gas production, and helium embrittlement under neutron irradiation. Nucl. Fusion **52**(8), 083019 (2012). http://stacks.iop.org/0029-5515/52/i=8/a=083019
37. Griffith, A.A.: VI. The phenomena of rupture and flow in solids. Philos. Trans. R. Soc. Lond. A Math. Phys. Eng. Sci. **221**(582–593), 163–198 (1921). https://doi.org/10.1098/rsta.1921.0006; http://rsta.royalsocietypublishing.org/content/221/582-593/163
38. Grigoriev, V., Pavlov, Y., Ametistov, Y.: An investigation of nucleate boiling heat teansfer of helium. In: Proceedings of 5th International Heat Transfer Conference, Tokyo, vol. 4, p. 45 (1974)
39. Grigoriev, V., Klimenko, V., Pavlo, Y., Ametistov, Y., Klimenko, A.: Characteristic curve of helium pool boiling. Cryogenics **17**(3), 155–156 (1977). https://doi.org/10.1016/0011-2275(77)90275-2; http://www.sciencedirect.com/science/article/pii/0011227577902752
40. Hablanian, M.: High-Vacuum Technology: A Practical Guide, 2nd edn. Mechanical Engineering. Taylor & Francis, Milton Park (1997). https://books.google.com/books?id=5L8uIAFm4SoC
41. Hahn, P., Hoch, H., Weber, H., Birtcher, R., Brown, B.: Simulation of fusion reactor conditions for superconducting magnet materials. J. Nucl. Mater. **141–143**, 405–409 (1986). https://doi.org/10.1016/S0022-3115(86)80074-5; http://www.sciencedirect.com/science/article/pii/S0022311586800745
42. Hamada, K., Takahashi, Y., Isono, T., Nunoya, Y., Matsui, K., Kawano, K., Oshikiri, M., Tsutsumi, F., Koizumi, N., Nakajima, H., Okuno, K., Matsuda, H., Yano, Y., Devred, A., Bessette, D.: First qualification of ITER toroidal field coil conductor jacketing. Fusion Eng. Des. **86**(6), 1506–1510 (2011). Proceedings of the 26th Symposium of Fusion Technology (SOFT-26), https://doi.org/10.1016/j.fusengdes.2010.12.054; http://www.sciencedirect.com/science/article/pii/S0920379610005995.

43. Hammond, K.D., Blondel, S., Hu, L., Maroudas, D., Wirth, B.D.: Large-scale atomistic simulations of low-energy helium implantation into tungsten single crystals. Acta Mater. **144**(Supplement C), 561–578 (2018). https://doi.org/10.1016/j.actamat.2017.09.061; http://www.sciencedirect.com/science/article/pii/S1359645417308315

44. Henning, C.D., Hodges, A.J., Sant, J.H.V., Hinkle, R.E., Horvath, J.A., Hintz, R.E., Dalder, E., Baldi, R., Tatro, R.: Mirror fusion test facility magnet. In: Proceedings of the Eighth Symposium on Engineering Problems of Fusion Research, San Francisco (1979)

45. Hoch, M.: The solubility of hydrogen, deuterium and tritium in liquid lead-lithium alloys. J. Nucl. Mater. **120**(1), 102–112 (1984). https://doi.org/10.1016/0022-3115(84)90177-6; http://www.sciencedirect.com/science/article/pii/0022311584901776

46. Hubberstey, P., Sample, T., Barker, M.G.: Is Pb-17Li really the eutectic alloy? A redetermination of the lead-rich section of the Pb-Li phase diagram (0.0 $< x_{Li}(at\%)<$ 22.1). J. Nucl. Mater. **191–194**(Part A), 283–287 (1992). Fusion Reactor Materials Part A, https://doi.org/10.1016/S0022-3115(09)80051-2; http://www.sciencedirect.com/science/article/pii/S0022311509800512

47. Irwin, G.: Fracture Dynamics. In: Fracturing of Metals. American Society for Metals, Materials Park (1948)

48. Jones, P., Gibson, R.: Hydrogen in beryllium. J. Nucl. Mater. **21**, 353–354 (1967)

49. Jousten, K.: Wutz Handbuch Vakuumtechnik. Vieweg+Teubner Verlag, Berlin (Springer Fachmedien Wiesbaden, Berlin) (2012)

50. Katsuta, H., McLellan, R.: Diffusivity of hydrogen in silver. Scr. Metall. **13**, 65–66 (1979)

51. Kearns, J.: Terminal solubility and partitioning of hydrogen in the alpha phase of zirconium, zircaloy-2 and zircaloy-4. J. Nucl. Mater. **22**, 292–303 (1967)

52. Kearns, J.: Diffusion coefficient of hydrogen in alpha zirconium, zircaloy-2 and zircaloy-4. J. Nucl. Mater. **43**, 330–338 (1972)

53. Kinchin, G., Pease, R.: The displacement of atoms in solids by radiation. Rep. Prog. Phys. **18**, 1–51 (1955). http://stacks.iop.org/0034-4885/18/i=1/a=301

54. Kitsunai, Y., Kurishita, H., Kuwabara, T., Narui, M., Hasegawa, M., Takida, T., Takebe, K.: Radiation embrittlement behavior of fine-grained molybdenum alloy with 0.2 wt% TIC addition. J. Nucl. Mater. **346**(2), 233–243 (2005). https://doi.org/10.1016/j.jnucmat.2005.06.013; http://www.sciencedirect.com/science/article/pii/S0022311505003041

55. Klueh, R., Ehrlich, K., Abe, F.: Ferritic/martensitic steels: promises and problems. J. Nucl. Mater. **191–194**, 116–124 (1992). Fusion Reactor Materials Part A, https://doi.org/10.1016/S0022-3115(09)80018-4; http://www.sciencedirect.com/science/article/pii/S0022311509800184

56. Kunieda, S., Haight, R.C., Kawano, T., Chadwick, M.B., Sterbenz, S.M., Bateman, F.B., Wasson, O.A., Grimes, S.M., Maier-Komor, P., Vonach, H., Fukahori, T., Watanabe, Y.: Measurement and model analysis of $(n, x\alpha)$ cross sections for Cr, Fe, ^{59}Co, and ^{58,60}Ni from threshold energy to 150 MeV. Phys. Rev. C **85**, 054602 (2012). https://link.aps.org/doi/10.1103/PhysRevC.85.054602

57. Libeyre, P., Bareyt, B., Benfatto, I., Bessette, D., Gribov, Y., Mitchell, N., Sborchia, C., Simon, F.: Electrical design requirements on the ITER coils. IEEE Trans. Appl. Supercond. **18**(2), 479–482 (2008). https://doi.org/10.1109/TASC.2008.920788

58. Lindhard, J., Nielsen, V., Scharff, M., Thomsen, P.: Notes on atomic collisions iii. K. Dan. Vidensk. Selsk. Mat. Fys. Medd. **33**(10), 1–42 (1963)

59. Lindhard, J., Scharff, M., Schltt, H.S.: Range concepts and heavy ion ranges. Mat. Fys. Medd. K. Dan. Vidensk. Selsk. **33**(14), 1–42 (1963)

60. London, F.: Superfluids, Volume 1. Structure of Matter Series. Wiley, Hoboken (1950). https://books.google.com/books?id=DtvvAAAAMAAJ

61. Louthan, M., Donovan, J., Caskey, G.: Hydrogen diffusion and trapping in nickel. Acta Metall. **23**, 745–749 (1975)

62. Mansur, L.K.: Irradiation creep by climb-enabled glide of dislocations resulting from preferred absorption of point defects. Philos. Mag. A **39**(4), 497–506 (1979). https://doi.org/10.1080/01418617908239286

63. Mas de les Valls, E., Sedano, L., Batet, L., Ricapito, I., Aiello, A., Gastaldi, O., Gabriel, F.: Lead-lithium eutectic material database for nuclear fusion technology. J. Nucl. Mater. **376**(3), 353–357 (2008). Heavy Liquid Metal Cooled Reactors and Related Technologies, https://doi.org/10.1016/j.jnucmat.2008.02.016; http://www.sciencedirect.com/science/article/pii/S0022311508000809

64. Matthews, J.R., Finnis, M.W.: Irradiation creep models - an overview. J. Nucl. Mater. **159**, 257–285 (1988)

65. McLellan, R.: Solid solutions of hydrogen in gold, silver and copper. J. Phys. Chem. Solids **34**, 1137–114 (1973)

66. Mo, K., Zhou, Z., Miao, Y., Yun, D., Tung, H.M., Zhang, G., Chen, W., Almer, J., Stubbins, J.F.: Synchrotron study on load partitioning between ferrite/martensite and nanoparticles of a 9Cr ODS steel. J. Nucl. Mater. **455**(1), 376–381 (2014). Proceedings of the 16th International Conference on Fusion Reactor Materials (ICFRM-16), https://doi.org/10.1016/j.jnucmat.2014.06.060; http://www.sciencedirect.com/science/article/pii/S0022311514004176

67. Morse, P., Feshbach, H.: Methods of theoretical physics, Volume 2. International Series in Pure and Applied Physics. McGraw-Hill, New York (1953). https://books.google.com/books?id=DvZQAAAAMAAJ

68. Najmabadi, F., Bathke, C.G., Billone, M.C., Blanchard, J.P., Bromberg, L., Chin, E., Cole, F.R., Crowell, J.A., Ehst, D.A., El-Guebaly, L.A., Herring, J., Hua, T.Q., Jardin, S.C., Kessel, C.E., Khater, H., Lee, V., Malang, S., Mau, T.K., Miller, R.L., Mogahed, E.A., Petrie, T.W., Reis, E.E., Schultz, J., Sidorov, M., Steiner, D., Sviatoslavsky, I.N., Sze, D.K., Thayer, R., Tillack, M.S., Titus, P., Wagner, L.M., Wang, X., Wong, C.P.: Overview of the ARIES-RS reversed-shear tokamak power plant study. Fusion Eng. Des. **38**(1), 3–25 (1997). https://doi.org/10.1016/S0920-3796(97)00110-5; http://www.sciencedirect.com/science/article/pii/S0920379697001105

69. Nakashima, H., Ohta, M.: Neutronics calculations for gas-suspended boron carbide cooling d-d fusion reactor blankets. J. Nucl. Sci. Technol. **16**(7), 513–519 (1979). https://doi.org/10.3327/jnst.16.513

70. Nathan, S.: Pump out the volume. The Engineer (17 March 2015). https://www.theengineer.co.uk/issues/march-2015-online/pump-out-the-volume/

71. Neu, R., Riesch, J., Müller, A., Balden, M., Coenen, J., Gietl, H., Höschen, T., Li, M., Wurster, S., You, J.H.: Tungsten fibre-reinforced composites for advanced plasma facing components. Nucl. Mater. Energy **12**, 1308–1313 (2017). Proceedings of the 22nd International Conference on Plasma Surface Interactions 2016, 22nd PSI, https://doi.org/10.1016/j.nme.2016.10.018; http://www.sciencedirect.com/science/article/pii/S2352179116302009

72. Norgett, M., Robinson, M., Torrens, I.: A proposed method of calculating displacement dose rates. Nucl. Eng. Des. **33**(1), 50–54 (1975). https://doi.org/10.1016/0029-5493(75)90035-7; http://www.sciencedirect.com/science/article/pii/0029549375900357

73. Olander, D.: Fundamental aspects of nuclear reactor fuel elements. Technical Information Center, U. S. Energy Research and Development Agency (1976). https://doi.org/10.2172/7343826; http://www.osti.gov/scitech/servlets/purl/7343826

74. Onnes, H.K.: Further experiments with liquid helium. D. On the change of electric resistance of pure metals at very low temperatures, etc. V. The disappearance of the resistance of mercury. Comm. Phys. Lab. Univ. Leiden **122b** (1911)

75. Parish, C.M., Wang, K., Doerner, R.P., Baldwin, M.J.: Grain orientations and grain boundaries in tungsten nonotendril fuzz grown under divertor-like conditions. Scr. Mater. **127**, 132–135 (2017). https://doi.org/10.1016/j.scriptamat.2016.09.018; http://www.sciencedirect.com/science/article/pii/S1359646216304468

76. Perng, T.P., Altstetter, C.: Effects of deformation on hydrogen permeation in austenitic stainless steels. Acta Metall. **34**, 1771–1781 (1986)

77. Petty, T., Baldwin, M., Hasan, M., Doerner, R., Bradley, J.: Tungsten 'fuzz' growth re-examined: the dependence on ion fluence in non-erosive and erosive helium plasma. Nucl. Fusion **55**(9), 093033 (2015). http://stacks.iop.org/0029-5515/55/i=9/a=093033

78. Philipps, V.: Tungsten as material for plasma-facing components in fusion devices. J. Nucl. Mater. **415**(1, Supplement), S2–S9 (2011). Proceedings of the 19th International Conference on Plasma-Surface Interactions in Controlled Fusion, https://doi.org/10.1016/j.jnucmat.2011. 01.110; http://www.sciencedirect.com/science/article/pii/S0022311511001589

79. Ransley, C., Neufeld, H.: The solubility of hydrogen in liquid and solid aluminium. J. Inst. Met. **74**, 599–620 (1948)

80. Ray, P.J.: Structural investigation of $La_{2-x}Sr_xCuO_{4+y}$-following staging as a function of temperature. Master's thesis, Niels Bohr Institute, Faculty of Science, University of Copenhagen (2015). https://doi.org/10.6084/m9.figshare.2075680.v2

81. Robinson, M.T.: The dependence of radiation effects on the primary recoil energy. In: Corbett, J.W., Ianiello, L.C. (eds.) Proceedings of Radiation-Induced Damage in Metals, CONF-710601, p. 397. USAEC Technical Information Center, Oak Ridge (1972)

82. Seeger, A.: On the theory of radiation damage and radiation hardening. In: Proceedings of the Second United Nations International Conference on the Peaceful Uses of Atomic Energy, Geneva, 1958, vol. 6, p. 250. United Nations, New York (1958)

83. Shibata, K., Ichihara, A., Kuneida, S.: Calculation of neutron nuclear data on molybdenum isotopes for JENDL-4. J. Nucl. Sci. Technol. **46**(3), 278–288 (2009). https://doi.org/10.1080/18811248.2007.9711531

84. Sinclair, G., Dolan, T.J.: Effect of stress amplitude on statistical variability in fatigue life of 75s-t6 aluminium alloy. Trans. ASME **75**, 867–872 (1953)

85. Sizyuk, T., Hassanein, A., Ulrickson, M.: Thermal analysis of new ITER FW and divertor design during VDE energy deposition. Fusion Eng. Des. **88**(3), 160–164 (2013). https://doi.org/10.1016/j.fusengdes.2013.01.088; http://www.sciencedirect.com/science/article/pii/S0920379613000987

86. Snyder, W.S., Neufeld, J.: Disordering of solids by neutron radiation. Phys. Rev. **97**, 1636–1646 (1955). https://link.aps.org/doi/10.1103/PhysRev.97.1636

87. Sorbom, B., Ball, J., Palmer, T., Mangiarotti, F., Sierchio, J., Bonoli, P., Kasten, C., Sutherland, D., Barnard, H., Haakonsen, C., Goh, J., Sung, C., Whyte, D.: ARC: A compact, high-field, fusion nuclear science facility and demonstration power plant with demountable magnets. Fusion Eng. Des. **100**, 378–405 (2015). https://doi.org/10.1016/j.fusengdes.2015. 07.008; http://www.sciencedirect.com/science/article/pii/S0920379615302337

88. Steward, S.: Review of hydrogen isotope permeability through materials. Lawrence Livermore National Laboratory report, UCRL-53441 (1983)

89. Stoneham, A.: Energy transfer between electrons and ions in collision cascades in solids. Nucl. Instrum. Methods Phys. Res. Sect. B Beam Interact. Mater. Atoms **48**(1), 389–398 (1990). https://doi.org/10.1016/0168-583X(90)90147-M; http://www.sciencedirect.com/science/article/pii/0168583X9090147M

90. Swansiger, W.: Tritium solubility in high purity beryllium. J. Vac. Sci. Technol. **A 4**, 1216–1217 (1986)

91. Tanabe, T., Yamanishi, Y., Imoto, S.: Hydrogen permeation and diffusion in molybdenum. J. Nucl. Mater. **191–194**, 439–443 (1992)

92. Tinkham, M.: Introduction to Superconductivity, 2nd edn. Dover Books on Physics. Dover Publications, New York (2004). https://books.google.com/books?id=k6AO9nRYbioC

93. Trimby, P.W.: Orientation mapping of nanostructured materials using transmission Kikuchi diffraction in the scanning electron microscope. Ultramicroscopy **120**, 16–24 (2012). https://doi.org/10.1016/j.ultramic.2012.06.004; http://www.sciencedirect.com/science/article/pii/S0304399112001258

94. Was, G.A.: Irradiation creep and growth. In: Fundamentals of Radiation Materials Science: Metals and Alloys, pp. 125–154. Springer, Berlin (2007). https://doi.org/10.1007/978-3-540-49472-0_13

95. Was, G.A.: The damage cascade. In: Fundamentals of Radiation Materials Science: Metals and Alloys, pp. 125–154. Springer, Berlin (2007). https://doi.org/10.1007/978-3-540-49472-0_3

96. Weibull, W.: A statistical distribution function of wide applicability. Trans. ASME **73**, 293 (1951)
97. Wells, H.G.: Toro-Bungay. Macmillan, Basingstoke (1909). http://www.gutenberg.org/ebooks/718
98. Wirth, B.D., Hammond, K., Krasheninnikov, S., Maroudas, D.: Challenges and opportunities of modeling plasma-surface interactions in tungsten using high-performance computing. J. Nucl. Mater. **463**(Supplement C), 30–38 (2015). Plasma-Surface Interactions 21, https://doi.org/10.1016/j.jnucmat.2014.11.072; http://www.sciencedirect.com/science/article/pii/S0022311514008757
99. Wolfer, W.: Correlation of radiation creep theory with experimental evidence. J. Nucl. Mater. **90**(1), 175–192 (1980). https://doi.org/10.1016/0022-3115(80)90255-X; http://www.sciencedirect.com/science/article/pii/002231158090255X
100. Wulff, J.: Structure and Properties of Materials: Mechanical behavior, by H.W. Hayden, L.G. Moffatt, and J. Wulff. Structure and Properties of Materials. Wiley, Hoboken (1965). https://books.google.com/books?id=pdLPBH9YbW4C
101. Young, G., Scully, J.: The diffusion and trapping of hydrogen in high purity aluminum. Acta Mater. **46**, 6337–6349 (1998)
102. Zinkle, S.J., Busby, J.T.: Structural materials for fission & fusion energy. Mater. Today **12**(11), 12–19 (2009). https://doi.org/10.1016/S1369-7021(09)70294-9; http://www.sciencedirect.com/science/article/pii/S1369702109702949
103. Zinkle, S., Farrell, K.: Void swelling and defect cluster formation in reactor-irradiated copper. J. Nucl. Mater. **168**, 262–267 (1989)
104. Zinkle, S., Boutard, J., Hoelzer, D., Kimura, A., Lindau, R., Odette, G., Rieth, M., Tan, L., Tanigawa, H.: Development of next generation tempered and ODS reduced activation ferritic/martensitic steels for fusion energy applications. Nucl. Fusion **57**(9), 092005 (2017). http://stacks.iop.org/0029-5515/57/i=9/a=092005

Chapter 12
Economics, Environmental, and Safety Issues

12.1 Introduction

This chapter is an introduction to the topics of interest outside of the physics and technology issues of nuclear fusion. There have been several review articles on these subject areas, although they have been aimed at a "moving target" as world economic conditions change, as well as the fusion concept itself, over its (now eight decades!) of development. Reference [3] provides a good overview from the 1990 perspective.

12.2 Economics and the Cost of Money

A power plant of any kind is a major works project, public or private. Here we will study the financing of such a project, using public or private funds. The first step in financing a fusion reactor, especially if it is the first of its kind, would be to set up a business entity with no other purpose than building and operating the reactor, with its only assets being the plant itself. This will prevent having the risk liability associated with the success of the construction project from extending to the parent company (if there is one) and investors.

The type of debt required for construction of a power plant is typically "non-recourse debt," meaning that loans are secured only by the assets of the construction entity. Typically, these loans are in the form of municipal bonds which are offered to financial institutions, banks, and individual investors. A notorious example was the former WWashington Public Power Supply System (now called Energy Northwest). The WPPSS (derisively nicknamed "Whoops!") put out public offerings for the financing of a project involving the construction of five nuclear reactors, and subsequently defaulted on a bond debt of USD 2.25 billion, the second largest bond default in US history [18]. Nevertheless, the amount of money in the US municipal

© Springer Nature Switzerland AG 2018
E. Morse, *Nuclear Fusion*, Graduate Texts in Physics,
https://doi.org/10.1007/978-3-319-98171-0_12

bond market is above USD 4.0 trillion, and the low overall averaged risk along with certain tax benefits attract large investors for whom a failure can be written off. Pooled diversified municipal bond funds (a type of exchange traded funds, or ETFs) are often more attractive to small investors for this reason.

Nuclear suppliers typically manufacture subsystems and are likely to be corporations that finance their research and development (R&D) efforts in addition to their manufacturing costs by the issuance of stock, corporate bonds, and reinvestment of their own funds from profits on sales of earlier products. (The first category is generally referred to as equity financing.) An example of a company with a high degree of internal R&D is Babcock and Wilcox Technologies (BWXT), a spin-off of Baccock and Wilcox, a company in existence for over 150 years and the world leader in steam boilers. A special case is Orano (formerly Areva), which manufactures nuclear components and is financed by public debt and French government participation. General Electric's nuclear division is now in a joint venture with Hitachi, and has actively pursued advanced nuclear technologies such as metal-cooled fast reactors, and one could easily see them developing components for future fusion reactors, if they decided that there was an adequate market worldwide. The combined annual revenues of the two companies is in excess of USD 200 billion. These stakeholders may become equity investors in fusion-based nuclear power plants to assure a customer for their products, in a manner similar to the relationships between the aircraft manufacturers and the airlines.

We now will explore the current models for financing of a nuclear power plant (NPP). The historical model for NPP funding is government financing. As time goes on, it has been possible for private corporations to become involved, so that the finance takes on the flavor of corporate balance sheet funding. Next, some new financial models have emerged, including the French Exeltium model, the Finnish Mankala model, vendor equity, Export credit agencies (ECA) debt and financing, and private financing with government support mechanisms [17].

12.2.1 Government Funding

The first NPPs in the UK, France, the US, and Russia used direct government financing. There were several reasons for this. Firstly, it was the government's policy to support nuclear power in each case, and the government wanted to proceed without needing to persuade private utility companies to assume the risk of the new technology. Secondly, the governments involved wanted to maintain a high level of control of these plants, including safety management, fuel supply (note that the four countries named here had already developed uranium enrichment technology because of their nuclear weapons programs), and nuclear waste issues (avoiding nuclear proliferation). As time went on, many NPPs were privatized, as the regulatory processes evolved and the governments became more confident in forming a more distant (and cost-effective) relationship with the private industry involved. However, governments are still involved in NPPs to a large extent, by remaining as the insurer of last resort, providing debt guarantees, and being heavily involved in the nuclear fuel cycle (on both ends). A rather well-developed relationship exists between the Russian energy corporation Rosatom and the Russian government, and this has created an atmosphere conducive to international sales of Russian-built reactors, although with the occasional "financial hiccups." Rosatom directly participates in the construction of ITER. The Korean nuclear supplier Kepco is in a similar position, with government backing to build NPPs in the UK.

The first commercial fusion-based NPP will probably require government financing, although it may be possible to invite other partners in. One can imagine a prestige factor in starting the first commercial fusion plant, although the board members of a traditional utility company are more likely to see the risks, financial and otherwise, that might be involved. The energy policy of a country may turn quickly. (The US election in 2016 is a prime example of that!) But the financial stability of these large governments is the "holy grail" for any power plant operator facing huge construction costs and the long intervals between the initial decision to build and the delivery of revenue electricity on the grid.

The initial supply of tritium is another reason that governments must be involved. Tritium has been produced in fairly large quantities in the countries mentioned above for their nuclear weapons programs. There are no other large sources of tritium available in these countries, although Canada has an impressive inventory from non-weapons stockpile sources and will supply ITER [8]. Tritium has been made available in the US at a fixed price and may be considered as a government-subsidized product.

From time to time the idea of a tax break (and even low-interest government loans) for "green energy" emerges. If the governments can be persuaded that a fusion reactor qualifies as green technology, then the path forward for financing a fusion NPP looks quite a bit more attractive.

12.2.2 Corporate Balance Sheet Financing

Two concepts are important in deciding on the economics of building a power plant of any kind. The first of these is the true cost of debt used to finance the construction of the plant. We assume that there is a known rate for a "risk-free" loan R_f (usually this is taken as the interest rate on US treasury bonds). To this is added a "credit spread" which reflects the market price of the bond given the amount of risk. This spread can vary widely and be dependent on many factors, including the perceived political stability of the country, the frequency of large natural disasters such as earthquakes and tidal waves, and prior history of debt service by the corporation involved. Ratings agencies such as Moody's, Standard and Poor's, and Fitch assign grades to each corporate bond issuer. In addition to secured bonds, there are typically equity pools, such as common and preferred stock. These typically have a higher rate of return but have more exposure to the investors for loss of capital due to bankruptcy and court-ordered restructuring. That being said, the overall internal cost of debt to a corporation wishing to fund a new project with external debt is given by a simple formula:

$$\text{Cost of debt} = \left(R_f + \text{Credit spread}\right)(1 - TR), \tag{12.1}$$

where TR is the tax rate being paid by the corporation. As an example, the Pacific Gas and Electric Company made a disclosure to the California Public Utilities Commission [13] that their weighted costs were 10.25% in the equity market (representing 52% of their total costs), 5.60% on preferred stock (1% of their total costs), and 4.89% on long-term debt (47% of their total costs), for an aggregate cost of debt of 7.69%. (Note especially the higher interest rate on long-term debt, reflecting a fairly high spread over Treasury rates, most likely due to a bankruptcy filing in 2001, from which they emerged in 2003.) PG&E has paid no federal income taxes for the past decade, mostly due to depreciation write-down, so we can take $TR = 0$ in Eq. (12.1), and the historical 30-year Treasury yield has been between 2.29% and 4.65%, so taking a median yield of 3.47%, so PG&E's average cost of debt works out to a credit spread of 4.22 % over the decade-average Treasury rate. (Treasury rates were higher in earlier periods, even above 10% in the 1980s.)

Another useful quantity to consider is the net present value, or NPV, of the construction project. This is a measure of the value of an investment in terms of its future rate of return r, which is the "discount rate" for future earnings reflected back to the present time (think of "I will gladly pay you Tuesday for a hamburger today," the line of the famous cartoon character "Wimpy"). The formula for NPV is

$$\text{NPV} = \sum_{t=1}^{T} \frac{c_t}{(1+r)^t} - C_0, \tag{12.2}$$

where C_0 is the capital cost of the project, c_t is the net cash inflow (revenue minus expenses, where expenses include debt service), adjusted by the tax rate TR on both, and T is the total number of periods. (For example, if the plant has a 30-year expected lifetime, and accounting periods are quarterly, then $T = 120$ and the rate $r = APR/400$, where APR is the effective annual percentage rate.)

It is important to note here treat the cost term C_0 is not just the construction cost of the plant but also the sum of the interest payments during the construction phase. If we consider a case where the loans are phased in as the money is needed to finance the construction, and assume that the debt is structured into N equal bond offerings over a period of N years, and the present value of the plant PV (actual construction costs, not counting interest) is taken at the end of construction, then the total debt cost (interest during construction, or IDC) is

$$\text{IDC} = PV \sum_{n=1}^{N} \frac{nr_{Bond}}{N}(1 - TR) = PV \frac{N+1}{2} r_{Bond}(1 - TR). \tag{12.3}$$

The total cost is then $C_0 = PV + \text{IDC}$. Interest during construction can make or break a power plant project, because delays in construction due to such factors as regulatory delays, unexpected site conditions, labor issues, and delays in securing capital (or having interest rates spiral upwards) can make a plant unprofitable before it even goes online.

A final parameter is probably the most vital in determining the decision of a governing board of a utility company to build or not. This is called the "Internal Rate of Return," or IRR. The IRR is the discount rate r at which the net present value NPV is zero. This serves as a comparison of the project's value in terms of competitive, more passive investment strategies such as buying back their stock or letting the existing fixed debt structure shrink by attrition due to the finite term length of the bonds involved. The IRR is defined by

$$\text{IRR} = r \text{ such that } NPV = 0. \tag{12.4}$$

Given these financial factors, we can now construct a metric for the value of a fusion power plant from the perspective of a utility company. Suppose that the raw construction costs for a 1 GWe plant are in the range from 2 USD to 20 USD per watt. We also assume that the sale price for the electricity generated is USD 0.15 per kilowatt-electric-hour (kWeh), and that the plant is financed over a seven-year construction period with seven bond offerings, each with an interest rate of 6.0%. We also assume that the utility company is profitable and is in a 35% tax rate bracket. This results in a markup of the plant's cost of a factor of 1.156, or 15.6% over its unburdened construction cost. (This is low compared to US experience, but representative of, say, South Korea and China.) The after-tax revenue c_t in each time period is given by the sales price of electricity generated, assuming 90% full-power availability, which on a per-quarter basis is USD 189.54 million. The after-tax debt service per quarter is $9.75C_1$ million, where C_1 is the unburdened plant cost is USD per watt. If we ignore all other cost factors such as operating personnel costs, service and maintenance, insurance, and end-of life costs, the quarter-to-quarter breakeven point would be at a value of C_1 where the difference of these two numbers is zero, or when the construction cost is USD 19.44 per watt. (While this number is higher than most fusion reactor design studies, it is important to remember that ITER, at 400 MWth and no energy conversion cycle, would have a cost per kilowatt-electric (COE) of around USD 130.00 per kWe even if it had an energy conversion system added at no cost. But it is clearly not being built with the plan to make money!)

We are now in a position to evaluate the 1 GWe reactor from the standpoint of IRR vs. the raw plant cost C_1. This is plotted in Fig. 12.1. The line where IRR=0 is marked (note that this does not have the exact intercept point as the value of C_1 given above because of the nonlinear behavior of the NPV function), and a "reasonable" value where the ICC is 10% is marked, showing that the system would have some appeal compared to other ventures, assuming that it can be shown that this option is not too risky. More realistic models, including the sensitivity to currency fluctuations, end-of life costs, service, maintenance, and personnel costs, et cetera, will drive this curve downwards, resulting in the need for less dollars per watt plant cost for a given IRR.

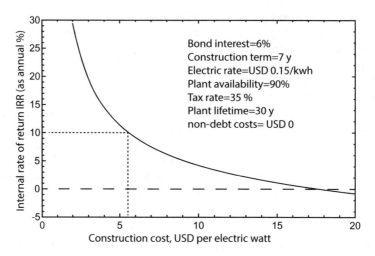

Fig. 12.1 Internal rate of return (IRR) vs. unburdened construction cost for the example given in the text

12.2.3   The Exeltium Model

Exeltium is a consortium of energy-intensive industries in France including manufacturers of steel, aluminum, chemicals, industrial gases, and paper. These industries
are very sensitive to the cost of electrical generation, and they wanted to get
stabilized pricing for electricity. Between 2005 and 2010, a syndicate of 27
industrial groups and ten banks entered into a contractual arrangement with EDF
(Électricité de France), the nuclear utility for France. This syndicate provided €4
billion up-front and then contracted for a first phase delivery of 148 TWh over 24
years for an up-front payment of €1.75 billion. The electricity price was indexed to
reflect the utility's cost for generation, but in a more stable way than the electricity
market itself. The industrial partners within the syndicate can either consume the
electricity themselves or sell it on the open market. While a second phase was
planned to start in 2011, it was not implemented because of lower-than-expected
demand for additional electricity.

12.2.4   The Mankala Model

Starting in the 1960s in Finland, a company called Oy Mankala Ab was given
approval by the government to establish a company which owns a significant
share of a power plant through equity. This legal precedent set the stage for other
"Mankala"-type syndicates. Mankala syndicates differ from the French Exeltium
model in that there is direct ownership of a portion of the power plant's assets. The

owners of the Mankala are obliged to purchase electricity from the power plant in proportion to their shareholding, but can also opt to sell the electricity on the open market. The Mankala itself operates as a limited liability corporation (LLC). The Mankala model resulted in the construction of two new NPPs in Finland in 2010, although the courts had to resolve some legal issues. It is possible that this model might be adopted in other countries in the future, and it is an intriguing option for future fusion NPPs.

12.2.5 Vendor Equity

Suppliers of nuclear technology may have the incentive to participate financially in NPPs based on new technology if the corporate balance sheet financial model looks strong and the company wants to assure their role in the manufacture of similar units in the future. However, these vendors are probably not interested in the 30-year commitment that for-profit utility companies are making, as the novelty of the product will wear off eventually and they will want to invest their capital in newer technology. Vendor equity has been used in the past in the airline industry, especially in economic downturn periods such as the market collapse in 2008, when debt and equity in the conventional markets was difficult for customers to secure. As technology vendors typically carry a fairly large debt burden themselves, using their own money to finance the purchase of their product by their customers becomes unproductive at some point. For example, the current portion of vendor financing among the aircraft manufacturer Boeing's customers is now only about 1% of the total financing base, and this is probably a reflection of the health of that industry.

12.2.6 Export Credit Agencies (ECA) Debt and Financing

Export credit agencies are non-governmental organizations that link potential exporters in a certain country with the government of that country, with the aim of stimulating export trade. The government support can consist of either direct financial support for a project or loan guarantees/credit insurance, or both. A typical scenario is that some technology with a clear benefit for the betterment of humanity (such as vaccines or safe drinking water) is made available to a third-world country, which might not be able to acquire these goods and services through an open capitalist market. Examples of ECAs include the U.S. Agency for International Development (USAID) in the US and Export Credits Guarantee Department (ECGD) in the UK. The French ECA COFACE has extended the sales of EDF's reactor technology throughout the world by providing loans from banks under French control with repayment guarantees from the French government.

Whether fusion reactor technology could flourish under this type of arrangement is less clear. As opposed to mature fission reactor technology, where getting megawatts on the grid is almost a certainty, an emerging technology such as fusion will not be without risk, and the argument for doing this in developing countries for "the betterment of mankind" is perhaps a bit thin. But for the "n-th" fusion reactor down the road, this approach might be a better fit, and might narrow the technology gap between the OECD "have" countries and the have-not countries of the world.

12.2.7   Private Financing with Government Support Mechanisms

This category of power plant financing differs from the earliest model of direct government financing in that the government only operates to stabilize the profit and loss to an energy supplier though some sort of price support for the electricity generated and for the cost of money. The government involved may stimulate the construction of a new power plant by entering into an agreement with the electrical provider to support debt coming into the project through some sort of backing of conventional bank loans through guarantees, and may also underwrite the electricity provider's profit through either a contract for difference (CfD), where the government pays in or gets out the difference between the market rate for electricity and the electricity producer's agreed-upon generating cost, or with a power purchase agreement (PPA), where the government promises to buy electricity at a fixed rate and then sells it at a negotiated price. (The latter choice has been used as a business model for large corporations such as Google to obtain electricity through solar power for its data centers, and by Anheuser Busch InBev (in Mexico) for wind power.)

This model has been very popular with "green" energy projects, but typically on a relatively modest scale by NPP standards, with most projects in the tens to hundreds of megawatt range. As fusion NPPs benefit from the larger economies of scale similar to fission power, this economic model may or may not be useful. Its success depends on the government involved and internal politics within that country, as well as that government's credit rating and capital reserves.

12.2.8   Some Concluding Remarks on Economic Competitiveness

The analysis given above views the financing of a fusion reactor from paths that have proved fruitful for financing fission reactors in the past. It should be pointed out that cost analyses of fusion conceptual designs have shown higher costs than fission reactors with similar electrical output. There could, however, be some mitigating factors making the decision to build fusion NPPs economically sound. Given here is a list of some of the possibilities that could shape the economic attractiveness of fusion NPPs.

- Compact, high power-density reactors based on alternate concepts to tokamaks (such as spheromaks, FRC configurations, heavy ion fusion, and others) might be successfully developed.
- Economy of scale might come about if many reactors were built with a common design.
- Public acceptance of fission reactors might be limited, due to public perceptions of safety or radioactive waste issues.
- Fusion-fission hybrid reactors might be deployed.
- Carbon taxes might raise the price of fossil fuel power.

12.3 Environmental Impact

12.3.1 Licensing Process

From the legal standpoint, the declaration of the environmental impact of any power plant is an integral part of the licensing process. The US version is given here, and it is similar to the process in other countries.

The US Nuclear Regulatory Commission (NRC) regulates all aspects of commercial nuclear power in the United States. It issues construction permits and operating licenses for NPPs. Since 1989, it has issued combined operating licenses (COLs) which combine the construction permit and operating license. Figure 12.2 shows the general path taken for NPP licensing and operation. Part of the license application to NRC involves the preparation of environmental impact statements (EISs), which are sent to the US Environmental Protection Agency (EPA) for review and comment. EPA must find that the plant will operate within an "ample margin of safety to protect public health" from a suite of regulations including the Clean Air Act, Clean Water Act, and the Safe Drinking Water Act.

Fig. 12.2 Licensing process for a nuclear power plant in the US. U. S. Nuclear Regulatory Commission [21]

The radioactivity issues associated with a fusion power plant can be divided into a two-by-two matrix: routine tritium releases and radiation exposure to workers to activated reactor materials (and to the public after decommissioning and burial) fill the first row, and exposure by tritium and activated reactor materials to the public in the case of an accident form the second row. The routine exposure can be calculated with some certainty, whereas the accidental releases and their consequences are a bit more difficult to estimate. We will look at the routine risks first, and then the risks posed by potential accidents.

12.3.2 Environmental Effects: Tritium

The quantitative calculations of tritium throughput through solid materials were given in Chap. 11. Here we explore the regulatory limits. For the US, regulatory limits are tied to the dose that one would receive from the release over a period of 1 year. The allowable dose is defined by three regulatory layers and two different Federal agencies. The fundamental concept in defining "what is safe" is called ALARA, for "as low as reasonably achievable." This is the point where an entity having a potential source of radiation exposure would not be required to spend more money on mitigation of the release to lower the dose. The ALARA limit is currently 3.0 millirem (mR) ($30\,\mu Sv$) per year. (Note that 1 Sievert(Sv) = $1\,J\,kg^{-1}$ of absorbed dose equivalent, with the ratio of dose equivalent to the actual dose (the RBE or Q factor) of 1.0 for beta particles such as released by tritium. $1.0\,Sv =$ 100 Rem.) This is the NRC limit as stated in the US Code of Federal Regulations (CFR) in Appendix I to 10 CFR Part 50. "If a nuclear power plant exceeds half of these radiation dose levels in a calendar quarter, the plant operator must investigate the cause(s), initiate appropriate corrective action(s), and report the action(s) to the NRC within 30 days from the end of the quarter." [22]. The EPA has a "standard" exposure limit as defined in 10 CFR 20.1301(e) of $25\,mR$ ($250\,\mu Sv$) per year, for any nuclear power plant or nuclear fuel facility. Another limit, defined by 10 CFR 20.1301(a)(1), applies to all other facilities and is $100\,mR$ ($1.0\,mSv$) per year. This allowable limit is based on the 1990 recommendations of the International Commission on Radiological Protection (ICRP). This limit is the most common one legally enforced worldwide.

How do we determine the maximum allowable leakage from a fusion power plant? The first step is to find the dose from the inhalation of one curie of tritiated water vapor, or HTO. (We choose this form rather than tritium gas, since it represents the worst-case scenario.) The dose per Becquerel is determined by the biological half-life for retention of hydrogen isotopes in the body, which is 10 days. Since this is much shorter than the radioactive half-life of 12.3 years, we can ignore the radioactive decay time. Tritium decays by beta emission, with 5.7 keV average energy going to an electron (with a range in tissue of around $6\,\mu m$) and the rest going to a neutrino with a stopping range of light years. The dose per disintegration is just $< E_{\beta^-} > /M$, where M is the mass of the human body: $M = 58$ kg average for adult women and 70 kg average for adult men. If we take $M = 63$ kg, then the dose is given by

$$D(\text{Sv per Bq}) = \frac{5.7 \cdot 1.6 \times 10^{-16}}{\lambda_{Bio} M} = 1.8 \times 10^{-11} = 66.6 \text{ Rem per Ci.} \quad (12.5)$$

Here we have used the biological decay constant $\lambda_{Bio} = \ln 2/t_{1/2, Bio}$. We note the possibility of two other pathways into the body, which are absorption through the skin (dermal transport) and uptake from home-grown food. These are typically much smaller.

Suppose that we have a person spending all of his time living near the site fence, taken to be 100 m from the source of tritium leakage from the plant. This person breathes at a normal rate of 20 liters per minute and thus consumes $20 \times 10^{-3} \times 24 \times 60 \times 365.25 = 1.052 \times 10^4 \text{ m}^3 \text{ y}^{-1}$. For this person to receive 4 mR from HTO inhalation, the maximum concentration would be $C = 4.07 \text{ nCi m}^{-3}$. If we assume that the HTO is released in a plume with the wind blowing at $v = 5 \text{ km h}^{-1}$ with a conical half-angle of $5°$, straight at the person in question, then the plume would have a footprint area of $A = \pi R^2 \theta^2 = 239.2 \text{ m}^2$ and the volumetric flow at this point would be $\dot{V} = Av$, or $332 \text{ m}^3 \text{ s}^{-1}$. Then the maximum allowable release rate Q is given by

$$Q_{max} = C_{max} \dot{V} = 1.35 \times 10^{-6} \text{ Ci s}^{-1} = 0.11 \text{ Ci d}^{-1}. \quad (12.6)$$

This is a very low tritium leakage rate in comparison with tritium facility leakage rates in the past. Notice that the allowable leakage rate scales with the square of the distance to the site fence: a one mile square site with the tritium source at its center (the situation with the RTNS-II fusion neutron source at Livermore) could be $(800/100)^2$ times higher, or about 8.8 Ci d^{-1}. Also, a full-time human inhabitant setting up camp at the site boundary would probably be escorted away by local law enforcement!

A more realistic environmental dose assessment for routine exposure must include weather data, including wind direction and speed and precipitation. A detailed study of the tritium release from the National Tritium Labeling Facility (NTLF) at Lawrence Berkeley National Laboratory is a good starting point [9]. This facility was designed with an assumption of a steady, routine release of 100 Ci y^{-1}. The study took into account the local atmospheric dispersion due to wind turbulence and diffusion as well as the historic patterns for the wind. Figure 12.3 shows the classic "wind rose" for the NTLF site. This chart shows the frequency of wind directions and speeds divided into sixteen directions and six speeds. (Note that standard meteorology usage is to define the direction that the wind is coming from). The diurnal variation of the wind direction and speed is common in a coastal area such as Berkeley, and might not be characteristic of other areas. A contour plot of the calculated dose per year at 250 m radials from the site is shown as Fig. 12.4. Note that this chart uses a "backwards" version of the wind data in Fig. 12.3 since it

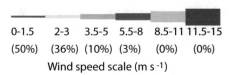

0-1.5   2-3   3.5-5   5.5-8   8.5-11 11.5-15
(50%)   (36%) (10%)   (3%)   (0%)   (0%)
Wind speed scale (m s⁻¹)
NOTE: Wind direction is the direction the wind is blowing from.

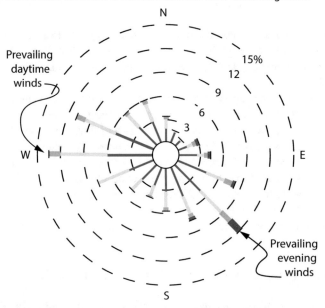

Fig. 12.3 Wind rose for calculation of atmospheric HTO transport for the Berkeley National Tritium Labeling Facility (NTLF). From [9]

is based on the wind direction "to" rather than "from." Note also that this result also includes the exposure to the average worldwide tritium concentration arising from cosmogenic tritium (about 30 MCi are in the atmosphere overall) and other sources such as nuclear testing, industrial releases, and so forth. The worldwide tritium dose is about $0.025\,\mathrm{mR\,y^{-1}}$, and this dominates the tritium dose in the LBL study at distances more than a few km from the facility.

12.3.3   Onsite Plume Model: DOE Facilities

Atmospheric dispersion modeling for radioisotope releases from Department of Energy facilities has been codified in the US by the Department of Energy [12]. The standard plume model used has the form:

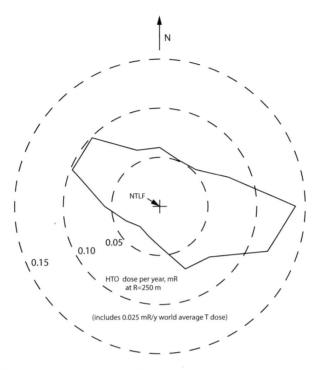

Fig. 12.4 Calculated dose from tritium at 250 m from the stack for the National Tritium Labeling Facility at Lawrence Berkeley National Laboratory, assuming a release rate of $100\,\mathrm{Ci\,y^{-1}}$. From [9]

$$\frac{\chi(x, y, z, H)}{Q} = \frac{1}{2\pi\,\sigma_y\sigma_z U}\,\exp\left(\frac{-y^2}{2\sigma_y^2}\right)\left\{\exp\left(-\frac{(z-H)^2}{2\sigma_z^2}\right) + \exp\left(-\frac{(z+H)^2}{2\sigma_z^2}\right)\right\},$$

$$(12.7)$$

where

χ	=	Concentration at location x, y, z (Bq/m^3, g/m^3, $Bq\text{-sec}/m^3$ or $g\text{-sec}/m^3$)
Q	=	Radionuclide or toxic chemical emission rate (Bq, g, Bq/sec or g/sec)
σ_y	=	Standard deviation of concentration in the horizontal direction (m)
σ_z	=	Standard deviation of concentration in the vertical direction (m)
U	=	Wind speed diluting the plume (m/s)
x	=	Downwind distance in the direction of the mean wind (m)
y	=	Distance in the horizontal plane perpendicular to the x-axis (m)
z	=	Height of the receptor (m)
H	=	Effective release height of the plume centerline (m).

A diagram depicting the model and nomenclature for this near-field plume is shown in Fig. 12.5. Here we note that there are two heights involved: the height h_0 of the emergency stack and the height H that the plume ultimately reaches. While

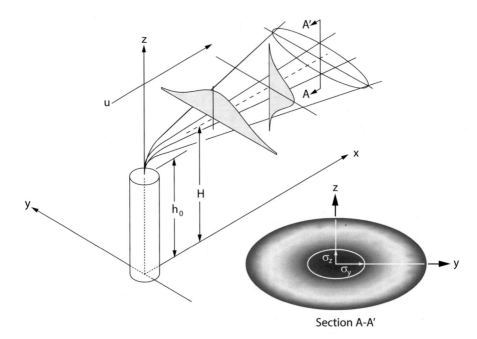

Section A-A'

Fig. 12.5 Plume model for effluent releases [12]. U. S. Environmental Protection Agency

this model is frequently used for stack gas releases from coal-fired power plants and the like where warm air is involved, this model may be uncertain for HTO release, where the temperature may or may not be above ambient. In any case, stacks are frequently accompanied by large dilution fans which can move the effluent upwards by a mechanically generated convection cell.

DOE Standard (STD) 1189–2008, "Integrating Safety into the Design Process" reads that "For the purposes of this Standard, a χ/Q value at 100 m of 3.5E-3 s/m^3 must be used for the dispersion calculation. This value is based upon NUREG-1140 (no buoyancy, F-stability, 1.0 m/s wind speed at 100 m, small building size [10 × 25 m], and 1 cm/s deposition velocity)." This has been the reference standard for near-field T releases at the US national laboratories.

12.3.4 Environmental Effects of Activated Structural Materials

At the end of life, a fusion power plant will contain a fairly large inventory of activated materials caused by neutron irradiation of the blanket components. The radioactive inventory will diminish over time due to radioactive decay. It is anticipated that at the end of life, the Li-based coolant will be removed and stored for future use, and the tritium will have also been largely removed with the coolant.

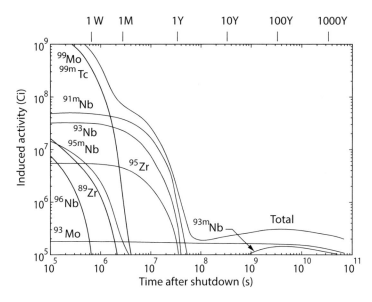

Fig. 12.6 Induced activity vs. time for a Mo first wall fusion device. Data from [6]

(Experience with JET after its tritium campaign has been valuable, and has shown that residual T in the machine was in the range of 6 g (60 kCi) after "cleanup" pure D shots [1]. While this is significant compared to the total onsite T inventory of 35 g, it is thought this has been caused by the carbon divertor in the machine, and hence the new ITER all-tungsten design. T retention will likely be much lower in ITER and future reactors.) Assuming that the tritium retention problem is solved, the other radiological concern is the inventory of activation products in the structural materials. Figure 12.6 shows a representative case, from a conceptual reactor design study called JDFR, conducted at the Japan Atomic Energy Research Institute (JAERI) in 1977 [6]. In this study, A TZM alloy blanket structure was exposed to the neutron output of a ∼1 GWe DT-burning reactor for 2 years, and the induced activity was calculated for a period to hundreds of years. Here one can see the value of leaving the reactor in place for a cooling-off period of several years. After that time, the Mo-containing blanket components (with ∼100 m^3 total volume) can be seen to have an induced activity of around 3×10^5 Ci.

One might ask how this choice of blanket material affects this estimate. In [15], a comparison study looked at 316SS, TZM, and V-20Ti alloys. Here they found that after 2 years' operation and a year or two of cooling-off, the V-20Ti and the TZM alloys were similar in their induced activity, whereas 316SS was about 100× higher induced activity. In a review paper in 1984, Kazimi [7] also noted that SiC ceramic was much lower in activation than either Mo or V alloys, but the result depends on fabrication and joining techniques used, e.g. cutting with diamond-in-metal tooling, or drag-in of impurities such as Fe during initial fabrication could increase the activation substantially. A good bit of the activation in 316SS is due to the presence of Ni, resulting in ^{60}Co production, but other competitive steels also show elevated activation through their alloying elements.

So let's return to the TZM blanket. While a good fraction of a megacurie may seem like a very large amount of radioactivity, it is important to consider that this radioactivity is spread out over a $100\,m^3$ volume and thus the specific activity is ~$3000\,Ci\,m^{-3}$; this is within the USNRC's category of waste that can be considered for direct "shallow" burial in its Class C, which is for "nuclear reactor components with between 44 and $7000\,Ci\,m^{-3}$."

So what are the risks with shallow burial? One could calculate the transport of these radionuclides by dissolution into groundwater with the potential of finding a pathway back to food, drinking water, and so on, but all of the candidate reactor materials currently being studied have very low solubilities in water, and the component characteristic dimensions are more similar to ship hulls than fine powder. (A search for a dry spot with the right soil pH would probably be required for the commissioning of the waste site and guarantee low transport.) A more reasonable question, and the one that the NRC demands an answer to, is whether an intruder into the burial site could intentionally or otherwise expose himself or others to the maximum allowable dose to members of the public, which is $0.5\,rem\,y^{-1}$. The time period for this criterion to be in effect is 100 years. The general consensus is that unearthing large reactor components is a noisy, dusty, and time-consuming process involving powered excavation tools, and would not go unnoticed for very long. So "nuclear archeology" of this kind is unlikely unless there is a total failure of historical record-keeping.

12.4   Safety Considerations

It is typical in analyzing the risks for any nuclear power plant to form a conceptual "maximum credible accident" (MCA) or "design basis accident" (DBA) and then fill in the gaps to obtain probability estimates for the MCA/DBA to occur. The useful tools in performing this type of safety assessment are fault/event tree analysis using classical probability theory, Bayesian and Markoff chain modeling, sensitivity analysis, and thermodynamic and chemical modeling of the materials involved. An example from the fission power safety world is the code MELCOR, maintained by Sandia National Laboratory and used by the USNRC to examine the safety of existing and proposed pressurized water reactors and boiling water reactors.

Caveats in using this safety analysis approach can be grouped into three general categories. One is the assumption of the DBA in the first place. The earthquake and resultant tsunami caused damage at the Fukushima Dai-Ichi NPP which was "outside the design basis" for the height of the tsunami, for example. Part of the design basis was the maximum tsunami height in the last 1000 years, but ignoring larger, earlier ones outside that window, such as the one in 869 A.D. [5]. Another category is failure of the fault/event tree modeling to properly account for hidden common-mode failure mechanisms. An example of this was the almost-failed scram of a reactor at Kahl, Germany that was traced to a set of relays in the independent power circuits for the control rod-drop electromagnets [4]. They were all provided by the same manufacturer and failed because they were all made with the wrong insulating varnish on the coils. (An ironic part of this incident was that the varnish oozed onto the electrical paddle and glued it in the "on" position because the relays were hung upside-down, so that gravity would open the electrical paddle if the relay was de-energized and the return spring was broken as a sort of "fail-safe" strategy: sometimes naive attempts at "making it safer" make things worse.) The third category is failure to completely understand the thermodynamic and chemical issues at the full complexity of the real system. An example of this would be the sodium-cooled reactor at Santa Susana, California, which suffered a partial meltdown. This was due to the contamination of the sodium coolant with an organic lubricant used in the coolant pump seals [2]. The organic lubricant coated the fuel elements and the heat exchanger and radically reduced the heat transfer characteristics between the fuel and the heat sink.

This being said, we might attempt to define the DBA for a fusion reactor. Probably the most important accident scenario to study carefully is the potential for a pipe break in a coolant channel or a rupture of a blanket component resulting in a spill of a metallic coolant such as lithium or FLiBe. These materials have exothermic reactions with air, water, and concrete. Conditions could be such that a fire results, and then portions of the blanket structural materials could become volatilized upon oxidation. Should the containment building be breached (due to, say, an aircraft impact or a hydrogen gas explosion), then these materials could become airborne. Their transport through the atmosphere and subsequent fallout are functions of wind patterns (including winds aloft), particle size and density, and their chemical affinity for water and dust.

Each phase of this accident scenario has been researched with some scrutiny. The MELCOR model mentioned above has been modified to treat Li fires in conditions relevant to air exposure of blanket-temperature Li in accident scenarios [10] and found to be in excellent agreement with laboratory benchmark tests. The potential for volatilization of fusion reactor structural materials was the subject of some MIT papers from the early 1980s [7, 14, 20] and later coupled with experiments on HT-9 and PCA steels [16] and V-15Cr-5Ti [11] at Idaho National Laboratory. It has been found that volatilization of Mo and V alloys starts at temperatures as low as 700 °C, although the V alloy is much less volatile than the Mo alloys. The HT-9 and 316SS steels are found to volatilize at temperatures between 1000 and 1300 °C, primarily because of their Mo content. The oxidation rates of these steels were also found to be about a factor of ten lower than the Mo and V alloys.

The state of the art in wide-scale atmospheric dispersion modeling has also advanced beyond the simple plume models mentioned earlier. The Department of Energy's (DOE) National Atmospheric Release Advisory Center (NARAC) has been tasked with simulation of atmospheric transport of radionuclides following a nuclear accident. Their analysis of the Fukushima disaster was published in 2012 [19] and reflects the current state of the art.

General safety guidelines for a fusion NPP

Taking into account all of the factors mentioned above, one can put together a short list of design principles which can help to mitigate the DBA given here. They are:

1. Organize tritium-containing breeder sections into small sections with isolation valves in case of pipe breaches and wall ruptures.
2. Provide metal-clad flooring (to eliminate Li-concrete reactions) and inert gas containment where potential lithium spills might occur.
3. Design with low-activation wall materials with high melting temperatures to minimize oxidation and volatilization in the event of a fire.
4. Avoid the use of water coolant where there might be the possibility of contact with tritium in a breach.
5. Provide secondary containment sufficient to withstand hydrogen gas explosions, aircraft strikes, and earthquakes.
6. Site plants in areas where the prevailing wind is away from densely populated areas.
7. Have an adequate emergency plan prepared.
8. Inform the public and news media about potential risks and evacuation strategies.
9. Work closely with law enforcement, emergency responders, and public health officials.
10. Encourage positive community relations and outreach.

Anyone reading this far in this book would probably like to see fusion power happen. For this to be so, it is vital that fusion energy is developed in a way that will meet the goal of being one of the safest, sustainable, and reliable energy sources on the planet.

Problems

12.1 Make a family of IRR vs. COE curves similar to Fig. 12.1 for the example case given in the text of a 1 GWe reactor, with interest rates ranging from 1.0% to 10.0% in one-percent increments.

12.2 Make a family of IRR vs. COE curves similar to Fig. 12.1 for the example case given in the text of a 1 GWe reactor, with tax rates ranging from zero to 35% in 5 % increments at a 6.0% interest rate.

12.3 A worker inhales air contaminated with HTO vapor which has 200 millicuries per cubic meter of HTO vapor. Assume 100% accumulation, four breaths per minute, 5 liters per breath, 10 min exposure. Calculate the whole-body dose, assuming 10-day biological half-life.

12.4 ^{48}Sc ($t_{1/2} = 1.83$ d) is produced in a V alloy blanket of a fusion device and is released in an accident with a breached containment at a rate of $1.0\,\text{kCi}\,\text{min}^{-1}$. (a) Use the DOE plume model to find the concentration at ground level 100 m from the source, with a plume height of 30 m and $1.0\,\text{m}\,\text{s}^{-1}$ wind speed. Ignore radioactive decay. Use $\sigma_z = \sigma_y$ and find a value for each consistent with the DOE "official" value of $\chi/Q = 3.5 \times 10^{-3}\,\text{s}\,\text{m}^{-3}$. (b) Now assume that σ_z and σ_y are linear functions of the distance from the accident, i.e $\sigma_z = \sigma_y = kR$. For a reactor located at Oak Ridge, TN and winds out of the west southwest (WSW), draw a "heat map" of the ^{48}Sc concentration on top of a map of the area, including radioactive decay. Locate the contour where the ^{48}Sc concentration reaches the Allowable Limit on Intake (ALI) established by the USNRC of $2 \times 10^{-9}\,\mu\text{Ci}\,\text{ml}^{-1}$.

References

1. Andrew, P., Brennan, P., Coad, J., Ehrenberg, J., Gadeberg, M., Gibson, A., Hillis, D., How, J., Jarvis, O., Jensen, H., Lässer, R., Marcus, F., Monk, R., Morgan, P., Orchard, J., Peacock, A., Pick, M., Rossi, A., Schild, P., Schunke, B., Stork, D., Pearce, R.: Tritium retention and clean-up in JET. Fusion Eng. Des. **47**(2), 233–245 (1999). https://doi.org/10.1016/S0920-3796(99)00084-8; http://www.sciencedirect.com/science/article/pii/S0920379699000848

2. Ashley, R.L., et al.: SRE fuel element damage, final report of the Atomics International Ad Hoc Committee. Technical report, NAA-SR-4488-supl. Atomics International, Canoga Park (1961). https://www.etec.energy.gov/Library/Main/Doc._No._2_SRE_Fuel_Element_Damage_Final_Report_1961_NAA-SR-4488_(suppl).pdf

3. Conn, R., Holdren, J., Sharafat, S., Steiner, D., Ehst, D., Hogan, W., Krakowski, R., Miller, R., Najmabadi, F., Schultz, K.: Economic, safety and environmental prospects of fusion reactors. Nucl. Fusion **30**(9), 1919 (1990). http://stacks.iop.org/0029-5515/30/i=9/a=015

4. Hagen, E.: Kahl relay common-mode failure. Nucl. Saf. **20**, 579–581 (1979)

5. Hollnagel, E., Fujita, Y.: The Fukushima disaster - systemic failures as the lack of resilience. Nucl. Eng. Technol. **45**(1), 13–20 (2013). https://doi.org/10.5516/NET.03.2011.078; http://www.sciencedirect.com/science/article/pii/S1738573315300024

6. Iida, H., Seki, Y., Ide, T.: Induced activity and dose rate in a fusion reactor with molybdenum blanket structure. J. Nucl. Sci. Technol. **14**(11), 836–838 (1977)

7. Kazimi, M.: Safety aspects of fusion. Nucl. Fusion **24**(11), 1461 (1984). http://stacks.iop.org/0029-5515/24/i=11/a=007

8. Kovari, M., Coleman, M., Cristescu, I., Smith, R.: Tritium resources available for fusion reactors. Nucl. Fusion **58**(2), 026010 (2018). http://stacks.iop.org/0029-5515/58/i=2/a=026010

9. McKone, T.E., Brand, K.P., Shan, C.: Environmental health risk assessment for tritium releases at the National Tritium Labeling Facility at Lawrence Berkeley National Laboratory. Technical report, LBL-37760. Lawrence Berkeley National Laboratory, Berkeley (1997). http://eta-publications.lbl.gov/sites/default/files/environmental_health-risk_assesment.pdf

10. Merrill, B.J.: A lithium-air reaction model for the melcor code for analyzing lithium fires in fusion reactors. Fusion Eng. Des. **54**(3), 485–493 (2001). https://doi.org/10.1016/S0920-3796(00)00577-9, http://www.sciencedirect.com/science/article/pii/S0920379600005779

11. Neilson, R.: Volatility of V15Cr5Ti fusion reactor alloy. J. Nucl. Mater. **141–143**, 607–610 (1986). https://doi.org/10.1016/0022-3115(86)90062-0; http://www.sciencedirect.com/science/article/pii/0022311586900620

12. Office of Nuclear Safety, Office of Environment, Health, Safety and Security: Technical report for calculations of atmospheric dispersion at onsite locations for Department of Energy nuclear facilities. Technical report, U.S. Department of Energy (2015). https://www.energy.gov/sites/prod/files/2015/04/f22/Technical-Report-for%20-Calculations-of-Atmospheric-Dispersion-at-Onsite-Locations-for-DOE-Nuclear-Facilities.pdf

13. Pacific Gas and Electric Company: Advice letter to the California Public Utilities Commision (2017). https://www.pge.com/tariffs/assets/pdf/adviceletter/GAS_3887-G.pdf

14. Piet, S.J., Kazimi, M.S., Lidsky, L.M.: Modeling of fusion activation product release and reactor damage from rapid structural oxidation. Nucl. Technol. Fusion **4**(2P3), 1115–1120 (1983). https://doi.org/10.13182/FST83-A23007

15. Piet, S.J., Kazimi, M.S., Lidsky, L.M.: Relative public health effects from accidental release of fusion structural radioactivity. Nucl. Technol. Fusion **4**(2P2), 533–538 (1983). https://doi.org/10.13182/FST83-A22918

16. Piet, S., Kraus, H., Neilson, R., Jones, J.: Oxidation/volatilization rates in air for candidate fusion reactor blanket materials, pca and ht-9. J. Nucl. Mater. **141–143**, 24–28 (1986). https://doi.org/10.1016/S0022-3115(86)80005-8; http://www.sciencedirect.com/science/article/pii/S0022311586800058

17. Reilly, F.: Innovative ways of funding nuclear power projects. World Nuclear News (18 Feb 2016). http://www.world-nuclear-news.org/V-Innovative-ways-of-funding-nuclear-power-projects-18021601.html

18. Russell, R.: Muni bonds: The most dangerous bonds to own: defaults are rare, but risks remain. U. S. News and World Report (22 Feb 2013). https://money.usnews.com/money/blogs/the-smarter-mutual-fund-investor/2013/02/22/muni-bonds-the-most-dangerous-bonds-to-own

19. Sugiyama, G., Nasstrom, J., Pobanz, B., Foster, K., Simpson, M., Vogt, P., Aluzzi, F., Homann, S.: Atmospheric dispersion modeling: challenges of the Fukushima Dai-ichi response. Health Phys. J. **102**(5), 493–508 (2012)

20. Tillack, M.S., Kazimi, M.S.: Modeling of lithium fires. Nucl. Technol. Fusion **2**(2), 233–245 (1982). https://doi.org/10.13182/FST82-A20753

21. U. S. Environmental Protection Agency: §309: reviewers guidance for new nuclear power plant environmental impact statements. EPA Publication, 315-X-08-001 (2008). https://www.epa.gov/sites/production/files/2014-08/documents/309-reviewers-guidance-for-new-nuclear-power-plant-eiss-pg.pdf

22. U. S. Nuclear Regulatory Commmission: Backgrounder on tritium, radiation protection limits, and drinking water standards (08 April 2016). https://www.nrc.gov/reading-rm/doc-collections/fact-sheets/tritium-radiation-fs.html. Accessed 30 May 2018

Glossary

adiabat constant-entropy surface, typ. in pressure-density space. 16

Alfvén wave wave in magnetized plasma described by fluid model. 136, 148

Banana regime Transport in low-collisionality regime (neoclassical transport in tokamaks). xi, 202

bang time time of maximum compression of an ICF target. 396

Bernstein wave warm plasma waves in magnetized plasma with cyclotron resonances included (RF waves). xii, 274, 287

Boltzmann collision integral expression for modification of the distribution function by collisions. 58

Boozer coordinates "natural" magnetic coordinate system (stellarator). 226

Bragg effect peaked energy deposition profile for ion beam (inertial confinement). 404

bremsstrahlung radiation from free-free electron interactions with ions. x, 75, 77, 79

catalyzed D-D D-D fusion where subsequent T and He^3 reaction products are expected to burn. ix, 35

Centurion nuclear test with small DT secondary. 345

charge exchange atomic process where electronic charge is transferred from one atom or molecule to another. 94

Child-Langmuir law current density limit in accelerators due to space charge. 249

Chirikov mapping mathematical transform mapping used to describe stochasticity. 168

coherence length distance between Cooper-paired electrons (superconductivity). 451

cold plasma waves waves in plasma where the electron and ion temperatures are assumedto be zero. 259

collisionality dimensionless ratio of collision time to bounce time (neoclassical transport in tokamaks). xi, 194

© Springer Nature Switzerland AG 2018
E. Morse, *Nuclear Fusion*, Graduate Texts in Physics,
https://doi.org/10.1007/978-3-319-98171-0

conductance ratio of gas flow to pressure drop (vacuum technology). 461

coronal equilibrium equilibrium model for radiation in optically thin plasmas. 82

cosmogenic occuring due to extraterrestrial causes. 492

creep time-dependent elongation of material due to stress/radiation damage. 434

cryosorption pump high vacuum pump with cryogenic surfaces and an activated bonding surface. 467

cutoff place where the index of refraction goes to zero (RF waves. xii, 263

Debye screening screening of individual electric charges by collective effects. ix, 42

Debye temperature temperature associayed with vibrational spectrum in a solid. 387

dielectronic recombination a free-bound transition of an electrons in an atomic system with another electronic transition within that atom simultaneously occurring. 83

displacements per atom (dpa) number of displaced atoms, divided by total number of atoms. 425

disruption large-scale MHD instability, resulting in loss of equilibrium. xi, 160

Dulong-Petit law states that molar heat capacities of all substances should be close to $3R$ for all substances and states. 389

embrittlement loss of ductility, typically due to radiation exposure. xiii, 430

energy principle method of determining stability using an integral expression giving the energy of a perturbation. x, 137, 139

ETF exchange-traded funds, a type of diverified investment. 482

exciplex excited non-dimer molecular complex. 403

fatigue failure mechanism by repetitive stress. xiii, 417

Fermi degeneracy equilibrium with Fermi spin statistics. xiii, 379

Fokker-Planck equation simplified Boltzmann transport model assuming small-angle collisions dominate. 59

fracture toughness ability of material to withstand failure by fracture. xiii, 418

Franck-Condon neutrals neutral atoms ejected from a plasma through charge exchange. 94

Frenkel pair vacancy-interstitial pair, typically caused by radiation exposure. 426

Gamow energy energy associated withquantum tunnelling (fusion reactions). 26

Gaunt factor quantum-mechanical correction to free-free radiation rate. 76

Gruneisen EOS equation of state for solid at high temperatures. 389

gyrotron high frequency RF source. 257, 327

Halbach array array of magnets producing smoothly varying fields. 254

Halite nuclear test with small DT secondary. 345

Hartmann number ratio of MHD force to viscous force. 415

helias stellarator configuration with modular coils. 218

H mode "high" confinement regime in tokamaks. 170

hohlraum "hollow room", X-ray containment structure (inertial confinement). 348

Hugill diagram Density-safety factor operational diagram describing tokamak operational space. xi, 160

Hugoniot trajectory of a shock wave through pressure vs. mass density space. 395

ideal MHD theory MHD theory with infinite conductivity. 146

inverse bremsstrahlung process by which light is absorbed in collisional plasma. xii, 352

Kelvin-Helmholtz instability instability where two fluids contact each other with different velocities. 369, 370

klystron multi-cavity RF power tube using velocity bunching (RF heating). 308

Knudsen number ratio of mean free path to system dimension (vacuum technology). 459

Langevin approximation simple collisional friction term $\propto$ velocity. 352

Langmuir wave warm unmagnetized plasma wave at high frequencies. 281

Lawson criterion parameter for basic energy balance in burning plasma. x, 97, 99, 101, 103, 105

Lindemann's law simple model for melting temperature of a solid. 390

Lindhard factor fraction of PKA energy causing displacements after accounting for electronic stopping. 423

line radiation radiation from bound-bound transitions of electrons in atomic systems. 80

London length characteristic length for fluxon distribution in superconducting materials. 450

lower hybrid frequency used for heating and current drive, typically a few GHz. 257

Lundquist number ratio of energy transport by Alfvén waves to ohmic dissipation. 149

magnetic islands closed magnetic flux surfaces over a fraction of a poloidal flux surface. xi, 155

magnetic reconnection change in magnetic topology caused by finite resistivity. xi, 146

magnetic stochasticity quasi-random magnetic line mapping following growth of magnetic perturbations. xi, 165

Mankala financing model where shareholders receive a benefit such as electricity and own the plant. 486

Mercier criterion stability criterion for modes near rational surfaces (tokamak ideal MHD stability). 150, 222

MHD fluid description of plasma. x, 109, 111

mirror machine open magnetic confinement system. 6

mix intrusion of target shell materials into fuel (inertial confinement). 346

mode conversion the transition of a wave from one branch of the dispersion relation to another (RF waves). xii, 288

mode locking lowering of tokamak plasma rotation rate caused by wall resistivity. xi, 162

Rayleigh-Taylor instability instability when heavier fluid is above a lighter fluid. xii, 368

reactivity average cross section $\times$ velocity $\sigma \times v$ averaged over velocity distribution. 26

recombination radiation radiation from free-bound transitions of electrons in atomic systems. 83

resistive tearing mode resistive MHD mode with magnetic reconnection effects. xi, 150

resonance place where the index of refraction goes to infinity (RF waves). xii, 263

resonance absorption process by which light is absorbed at the critcal density. xii, 353

Richardson number instability parameter for Kelvin-Helmholtz instability. 370

Richtmeyer-Meshkov instability instability at shock front in two fluids. 371

ring pump pump with a paddlewheel-type vane and a peripheral ring of fluid. 466

rocket equation momentum equation for an object ejecting mass. xii, 366

Roots blower rough vacuum pump with rotating dumbbell-shaped vanes. 465

Rosenbluth potentials integral-potential form of Fokker-Planck collision term. 60

rotational transform the traversal of a magnetic field in the poloidal direction as it advances in the toroidal direction. 215

Rydberg energy unit for atomic processes(=13.6 eV). 77

safety factor parameter describing toroidal-to-poloidal field strength ratio. 126

scroll pump rough vacuum pump with wobbling "scroll". 465

Sievert unit of radiation dose equivalent to 1 J kg^{-1} of X-ray dose. 490

Sievert's law permeation rate of gas through metal with $p^{1/2}$ dependence. xiii, 443

slowing down loss of particle energy by collisions. ix, 53

stellarator magnetic confinement fusion device. 5

Stix antenna antenna with with coil sections of alternating polarity. 257

Stokes diagram plots of ω vs. k for various waves, showing nonlinear coupling. xii, 361

supercritical above the critical point where liquid and gas phases are not distinguishable (thermodynamics). 456

Synchrotron radiation radiation caused by acceleration of electrons in magnetic fields. x, 96

tedrode vacuum tube with four electrodes. 299

thermal stress stress due to imbalanced expansive forces in a temperature gradient. 413

theta-pinch magnetic confinement device with azimuthal pulsed currents applied. 9

tokamak from Russian "toroidalnya kamera ee magnetnaya katushka", toroidal magnetic confinement system. 6

tunneling propagation of a wave through an evanescent region (FR heating). xii, 288

underdense place where the drive frequency is above the plasma frequency. 352

vendor equity money loaned to a customer by the vendor to pay for its own product. xiv, 487

virtual cathode region where accumulated charged particles repel others. 251

Vlasov equation collisionless Boltzmann equation. 58

void swelling increase in volume due to internal voids, usually caused by radiation exposure. 432

WKB approximation mathematical method which assumes that the wavenumber varies slowly with distance (RF waves). 272

X-mode RF wave with $\mathbf{E} \perp \mathbf{B_0}$ (RF waves). 262

Index

Printed in the United States
By Bookmasters